T0263527

HANDBOOK OF VLSI MICROLITHOGRAPHY

HANDBOOK OF
VLSI MICROLITHOGRAPHY

Principles, Technology
and Applications

Edited by

William B. Glendinning

Microlithography Consultant
Nobleboro, Maine

John N. Helbert

Advanced Technology Center
Motorola, Inc.
Mesa, Arizona

Reprint Edition

NOYES PUBLICATIONS
Westwood, New Jersey, U.S.A.

Library of Congress Catalog Card Number: 90-23646
ISBN: 0-8155-1281-3

Printed and bound in the United Kingdom
Transferred to Digital Printing, 2010

Library of Congress Cataloging-in-Publication Data

Handbook of VLSI microlithography : principles, technology, and
    applications / edited by William B. Glendinning, John N. Helbert.
            p.   cm.
    Includes bibliographical references and index.
    ISBN 0-8155-1281-3 :
    1. Integrated circuits--Very large scale integration.
2. Microlithography. I. Glendinning, William B. II. Helbert, John
N.
TK7874.H3494   1991
621.381'531--dc20   90-23646
                                          CIP

MATERIALS SCIENCE AND PROCESS TECHNOLOGY SERIES

Editors

Rointan F. Bunshah, University of California, Los Angeles *(Series Editor)*
Gary E. McGuire, Microelectronics Center of North Carolina *(Series Editor)*
Stephen M. Rossnagel, IBM Thomas J. Watson Research Center *(Consulting Editor)*

Electronic Materials and Process Technology

DEPOSITION TECHNOLOGIES FOR FILMS AND COATINGS: by Rointan F. Bunshah et al

CHEMICAL VAPOR DEPOSITION FOR MICROELECTRONICS: by Arthur Sherman

SEMICONDUCTOR MATERIALS AND PROCESS TECHNOLOGY HANDBOOK: edited by Gary E. McGuire

HYBRID MICROCIRCUIT TECHNOLOGY HANDBOOK: by James J. Licari and Leonard R. Enlow

HANDBOOK OF THIN FILM DEPOSITION PROCESSES AND TECHNIQUES: edited by Klaus K. Schuegraf

IONIZED-CLUSTER BEAM DEPOSITION AND EPITAXY: by Toshinori Takagi

DIFFUSION PHENOMENA IN THIN FILMS AND MICROELECTRONIC MATERIALS: edited by Devendra Gupta and Paul S. Ho

HANDBOOK OF CONTAMINATION CONTROL IN MICROELECTRONICS: edited by Donald L. Tolliver

HANDBOOK OF ION BEAM PROCESSING TECHNOLOGY: edited by Jerome J. Cuomo, Stephen M. Rossnagel, and Harold R. Kaufman

CHARACTERIZATION OF SEMICONDUCTOR MATERIALS—Volume 1: edited by Gary E. McGuire

HANDBOOK OF PLASMA PROCESSING TECHNOLOGY: edited by Stephen M. Rossnagel, Jerome J. Cuomo, and William D. Westwood

HANDBOOK OF SEMICONDUCTOR SILICON TECHNOLOGY: edited by William C. O'Mara, Robert B. Herring, and Lee P. Hunt

HANDBOOK OF POLYMER COATINGS FOR ELECTRONICS: by James J. Licari and Laura A. Hughes

HANDBOOK OF SPUTTER DEPOSITION TECHNOLOGY: by Kiyotaka Wasa and Shigeru Hayakawa

HANDBOOK OF VLSI MICROLITHOGRAPHY: edited by William B. Glendinning and John N. Helbert

CHEMISTRY OF SUPERCONDUCTOR MATERIALS: edited by Terrell A. Vanderah

CHEMICAL VAPOR DEPOSITION OF TUNGSTEN AND TUNGSTEN SILICIDES: by John E.J. Schmitz

(continued)

Ceramic and Other Materials—Processing and Technology

SOL-GEL TECHNOLOGY FOR THIN FILMS, FIBERS, PREFORMS, ELECTRONICS AND SPECIALTY SHAPES: edited by Lisa C. Klein

FIBER REINFORCED CERAMIC COMPOSITES: by K.S. Mazdiyasni

ADVANCED CERAMIC PROCESSING AND TECHNOLOGY—Volume 1: edited by Jon G.P. Binner

FRICTION AND WEAR TRANSITIONS OF MATERIALS: by Peter J. Blau

SHOCK WAVES FOR INDUSTRIAL APPLICATIONS: edited by Lawrence E. Murr

SPECIAL MELTING AND PROCESSING TECHNOLOGIES: edited by G.K. Bhat

CORROSION OF GLASS, CERAMICS AND CERAMIC SUPERCONDUCTORS: edited by David E. Clark and Bruce K. Zoitos

Related Titles

ADHESIVES TECHNOLOGY HANDBOOK: by Arthur H. Landrock

HANDBOOK OF THERMOSET PLASTICS: edited by Sidney H. Goodman

SURFACE PREPARATION TECHNIQUES FOR ADHESIVE BONDING: by Raymond F. Wegman

FORMULATING PLASTICS AND ELASTOMERS BY COMPUTER: by Ralph D. Hermansen

PREFACE

The chapter topics of this lithography handbook deal with the critical and enabling aspects of the intriguing task of printing very high resolution and high density integrated circuit (IC) patterns into thin resist process pattern transfer coatings. Circuit pattern density or resolution drives Dynamic Random Access Memory (DRAM) technology, which is the principal circuit density driver for the entire Very Large Scale Integrated Circuit (VLSI) industry. The book's main theme is concerned with the special printing processes created by workers striving to achieve volume high density IC chip production, with the long range goal being pattern features sizes near 0.25 μm or 256 Mbit DRAM lithography. The text is meant for a full spectrum of reader types spanning university, industrial, and government research and development scientists and production-minded engineers, technicians, and students. Specifically, we have attempted to consider the needs of the lithography-oriented student and practicing industrial engineers and technicians in developing this handbook.

The leadoff chapter focusses on the view that lithography methods (printing patterns) are pursued for the singular purpose of manufacturing IC chips in the highly competitive commercial sector, and attempts to delineate the factors determining lithographic tool selection. The reader's perspective is drawn to consider IC device electrical performance criteria versus plausible and alternative energetic, or circuit density limited, particle printing methods--visible or shorter UV optical, electron, X-ray, and ion beams. The criteria for high quality micrometer and submicrometer lithography is very simply defined by the three major patterning parameters: line/space resolution, line edge and pattern feature dimension control, which when combined with pattern to pattern alignment capability determine lithographic overlay accuracy. Patterning yield and throughput further enter in as dependent economic factors.

Resist technology has a logical, prominent, second-chapter position indicative of resist's overall importance in lithography, i.e., the end product of any IC lithography process is the patterned resist masking layer needed to delineate the VLSI circuit level. Example coverage of optical resist process optimization assures the reader a grasp of the most commonly and widely used (worldwide) lithographic process technologies. The basic resist design concepts and definitions are thoroughly covered as well as advanced lithographic processes.

Basic metrology considerations (Chapter 3) are absolutely imperative to rendering a total description of lithography methodology. The task of precisely measuring printed linewidth, or, space artifacts at submicron dimensions must be performed at present without the use of a traceable reference source--National Institute of Standards and Technology. These desirable and necessary standards must be made available in the future. However, critical and sufficient physical modelling of varied resist and IC material topological structures requires funding support and completion. Nevertheless, elucidation of optical, scanning-electron-microscope (SEM), and electrical test device linewidth measurements data present the reader with key boundary conditions essential for obtaining meaningful linewidth characterization.

The portrayal of energetic particle microlithography is totally incomplete without some detail of the actual printing tool concepts, design, construction, and performance. The printing tools are presented and described in chapters 4-7 as to their usage in the IC manufacturing world. Clearly optical lithography has been the backbone and mainstay of the world's microchip production activity and will most likely continue in this dominant role until about 1997. In the optical arena, it is found that 1X, 5X, and 10X reduction printers of the projection scanned and unscanned variety must be described in subsets according to coherent and non-coherent radiation, as well as, by wavelengths ranging from visible to deep ultraviolet. Higher resolution or more energetic sourced tools are also well described.

Next in world manufacturing usage, electron beam (e-beam) pattern printing has been vital, mostly because of its application in a pattern generation capacity for making photo masks and reticles, but also because of direct-write on-wafer device prototyping usage. The writing strategy divides e-beam printers, in general, into two groups: Gaussian beam raster scan, principally for pattern generation, and fixed or variable-shaped beam vector scan for direct-write-on-wafer applications. Subsets of the latter groups depend upon site-by-site versus write-on-the-fly substrate movements. The sophistication and complexity of e-beam printers requires diverse expertise in many technical areas such as: electrostatic and electromagnetic beam deflection, high speed beam blanking, intense electron sources, precise beam shapers, and ultra fast data flow electronics and storage. Interestingly, important special beam relationships of maximum current, density, and writing pattern path-speed require the observance of unique boundary conditions in meeting printing criteria.

On a worldwide basis, X-ray printing does not yet have high volume IC device production background examples, but high density prototype CMOS devices have been fabricated by IBM and feasibility demonstrated. The X-ray chapter presents X-ray lithography as a system approach with source, mask, aligner, and resist components. Of the competing volume manufacturing printing methods (optical and X-ray), the X-ray process is unique as a proximity and 1:1 method. As such, in order to meet the IC patterning quality criteria, extreme demands are placed on the mask fabrication process, much more so than for masks or reticles produced for the optical analogue. For economically acceptable IC production, laser/diode plasma and synchrotron ring X-ray sources must be presented as high density photon emitters. In the second part of Chapter 6, synchrotron is given special attention and presented as a unique X-ray generator with an X-ray flux collimation feature. In spite of the synchrotron's massive size and very large cost, it's multiport throughput capacity makes it viable for the very high production needs of certain industrial IC houses or possibly for multi-company or shared-company situations.

In the last of the printing tool chapters, Chapter 7, the energetic ion is depicted in a controllable, steerable, particle beam serial pattern writer performing lithography at a high mass ratio compared to an e-beam writer. The focussed ion beam not only can deposit energy to form IC pattern latent resist images, but offers as another application the direct implant of impurity ions into semiconductor wafers, obviating completely the need for any resist whatsoever and greatly simplifying the IC chip processing sequence. The versatile energetic ion plays yet another and possibly its most significant role in a "steered-beam" tool, indispensable for optical and X-ray mask repair through the precise localized oblation and/or deposition of mask absorber material.

The goal of establishing 0.35 μm IC chip production by 1995 is plagued by the constraints of yield-defect models. A small fractional-submicron mask defect population is adversely catastrophic to the mask-and-reticle-dependent energetic lithographies (optical, X-ray), and especially so for the case of 1:1 parallel reduction printing. The modernization of photo mask and reticle fabrication methods and facilities paves the way for achieving extremely accurate and defect free optical masks and reticles (<0.1/cm^2).   With defects of fractional-submicron sizes, mask and reticle repairs require fully automated "steered-beam" inspection/mapping equipment to work under full computer automation with compatible focussed ion beam repair tools.

One of the editors' purposes in assembling this book has been to accurately disseminate the results of many and varied microlithography workers. Since it is not possible in any one book to satisfy enough detail for every reader's full curiosity, we consider at least that the reader is enabled to perform his own valid analysis and make some meaningful conclusions regarding the status and trends of the vital technical thrust areas of submicron IC pattern printing technology. The editors wish to extend appreciation to various colleagues for helpful discussions and encouragement: A. Oberai, J.P. Reekstin, M. Peckerar, and many others as the lengthy lists of chapter references attest.

In addition, many individuals representing industrial, government, and university sectors have been extremely helpful in providing technical discussions, data, and figures to the chapter authors of this book. Gratitude is further extended here to those persons and their organizations. Gratitude also has been expressed via courtesy annotations in the figure captions. Finally, we commend and thank Judy Walsh for her compilation and editing skills.

Nobleboro, Maine   William B. Glendinning
Mesa, Arizona   John N. Helbert
June, 1991

CONTRIBUTORS

Phillip D. Blais
Westinghouse Electric Corp.
Advanced Technology Labs
Baltimore, MD

Franco Cerrina
University of Wisconsin
Madison, WI

William B. Glendinning
U.S. Army ETDL
Fort Monmouth, NJ

John N. Helbert
Motorola, Inc.
Advanced Technology Center
Mesa, AZ

Robert D. Larrabee
National Institute of
   Standards and Technology
Gaithersburg, MD

Loren W. Linholm
National Institute of
   Standards and Technology
Gaithersburg, MD

John Melngailis
Research Laboratory of
   Electronics
Massachusetts Institute
   of Technology
Cambridge, MA

Michael E. Michaels
Westinghouse Electric Corp.
Advanced Technology Labs
Baltimore, MD

Michael T. Postek
National Institute of
   Standards and Technology
Gaithersburg, MD

Lee H. Veneklasen
KLA Instruments, Inc.
San Jose, CA

Whitson G. Waldo
Motorola, Inc.
Chandler, AZ

NOTICE

To the best of the Publisher's knowledge the information contained in this publication is accurate; however, the Publisher assumes no responsibility nor liability for errors or any consequences arising from the use of the information contained herein. Final determination of the suitability of any information, procedure, or product for use contemplated by any user, and the manner of that use, is the sole responsibility of the user.

The book is intended for informational purposes only. The reader is warned that caution must always be exercised when dealing with VLSI microlithography chemicals, products, or procedures which might be considered hazardous. Expert advice should be obtained at all times when implementation is being considered.

Mention of trade names or commercial products does not constitute endorsement or recommendation for use by the Publisher.

CONTENTS

xviii   Contents

1

LITHOGRAPHY TOOL SELECTION STRATEGY

Phillip Blais
Michael Michaels

Westinghouse Electric Corporation
Advanced Technology Labs
Baltimore, Maryland

1.0 INTRODUCTION

Integrated Circuit (IC) fabrication requires performing a long sequence of many complex processes. Lithography, which recurs as many as 20 times, is the most important of these complex processes as it is used to define the dimensions, doping, and interconnection of each segment of each device. The domination of lithography in the total cycle time to fabricate an IC device is shown in Figure 1(1). Lithography consumes 60% of the total time required to fabricate IC devices! Labor costs are directly proportional to cycle time consumed. Selection of appropriate, and hopefully optimum, lithographic techniques and tools is shown as critical to the success of a wafer fab operation. The best choices may differ for experimental fab operations compared to high volume production fabs, but in either situation, the choice will be critical. Factors in lithographic technique and tool selection begin with a basic requirement for technical capability, continue through economic considerations, and finally end with such factors as production volume, turnaround time, product planning, process availability, etc.

2.0 STRATEGY

The definition of the method to be used for the selection of an optimum lithography system is the purpose of this chapter. There are many factors in the selection process. The primary factor is the technical requirements: are the equipment and

Wafer Processing

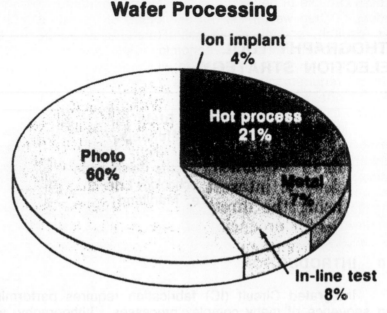

Figure 1. Lithography Dominates in Determining the Total Cycle Time for IC Processing.

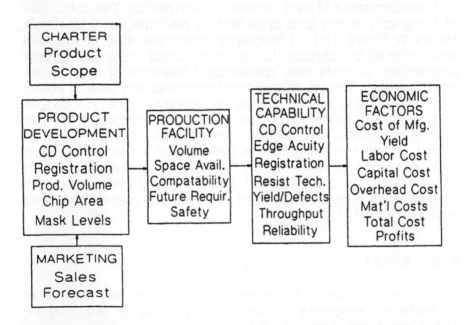

Figure 2. Block Diagram for Strategy of Tool Selection.

process capable of defining and registering the features we need to produce? Then we turn to the economics of the situation: Entry cost (capital), and operating costs (material and labor) are the major components of cost. Economics is the dominant factor once the decision has been made as to which equipment can meet the technical requirements. Within the economic constraints, we can establish the tradeoffs of capital investments, recurring cost of operation including maintenance, and acceptable yield range.

Figure 2 shows a block diagram for a prudent decision-making process. The role of charter, marketing, and production requirements in the decision process is clearly evident. These factors altogether determine requirements imposed on the lithography system to be acquired. The strategy for choosing an optimum lithography system requires these non-technical factors be considered in addition to the technical capability factor and economics.

2.1 Charter

An important input to strategy is to consider the charter. "Charter" is the overall philosophy and objective of the operation. The charter derives from the types and quantities of the IC products and the degree of difficulty of the process and product. For instance, an operation working on research and development will be at the cutting edge of the technology and therefore will be more concerned about technical capability as opposed to throughput and yield. On the other hand, an operation facility which is responsible for only one product using one IC process and expected to obtain very high volumes will use a mature process on a well defined product with emphasis on costs, yield, and throughput. One can visualize classes of applications based on volume and diversity of products and IC processes. Note: this is subjective and other ways can be used. As such, the classes are described in the following section. Note: we have chosen to describe the classes in terms of volume, which is a complex result of demand and pricing in the marketplace.

2.1.1   **Research and Development Class.** These applications are primarily driven by technical issues. They usually involve many types of IC processing with versions of each IC process such that the total number of variants could be well over ten. In addition, the number and types of products including digital devices and analog devices is usually well over ten. Such an environment will produce a very small number of any one product, a hundred or less, and will at best produce prototype devices. More often than not, such a process line will run a large number of short loop experiments as opposed to running the entire process sequence. This is the only class in which production cost is not the main consideration. Some of the primary considerations for this class are flexibility, quick turnaround time, pushing the state-of-the-art, exploratory and very small volume of end product. As such, a tradeoff of throughput versus cost of fabrication has only a secondary influence. This is not to say that cost is ignored but instead to state, that cost associated with volume is of secondary importance.

2.1.2   Very Low Volume/Many Processes and Products Class. This class of process fabrication line is often considered a development line, and not a research line. Its primary purpose is to develop a refined process sequence and to checkout some early prototyping of products which are not necessarily ready to be sold as product but could provide early product samples. This class is not dominated by economics, but is strongly influenced by it. Its primary drivers are still flexibility and technical criticality. Throughput and yield start to be more of an issue, but major issues of cost due to volume are still secondary.   Such a line usually supports several types of processes with many types of products. The product volume is still in the 100's per month at most and the versions of types of processes is still large (in the low 10's). Concerns about capital cost start to become more of an issue.

2.1.3   Low   Volume/Multiple   Processes   and Products Class. This class of fabrication line is now becoming more of a manufacturing oriented line than the lower volume classes previously discussed. Throughput vs. cost of fabrication and yields are becoming the primary drivers but still must be considered with the flexibility necessary to run a multiple number of processes and usually an even larger set of products. The processes will be mature with a moderate degree of process upgrade and incorporation of new versions of an existing process. Product volume will start to reach into the hundreds per month and could even be in the thousands. This is the first class where configuration control, certification of process and qualification of the line and products start to become very important issues. As such, the technical difficulty of doing the state-of-the-art is less a factor than in the lower volume classes.

2.1.4   Moderate   Volume/Few   Processes   and Products Class. This class is a manufacturing operation. Its process and products are mature. It has only a few different types of processes and a small number of products, usually ten or less. The volume of products becomes the primary driver to produce a cost competitive device, and is numbered in thousands of devices per month. The trade-off of throughput versus cost of fabrication dominates, and yield is the number one consideration. The cost of capital including facilities, equipment, raw materials and especially maintenance is the driving parameter for tool selection. Products need to have a long life, measured in years. The process sequence needs to become simplified to reduce the burden of fabrication.   Repeatability is more important than flexibility although the presence of both could be the most significant factor.

2.1.5 High Volume/Few Products Class. This class of fabrication line is strictly a manufacturing operation whose charter is to produce low cost product. No development and few process adjustments are made on such a fabrication line. In fact, the class has basically only one process and only a few products with product volume in the tens of thousands. The major factor is economics. The process and products are very mature and the process sequence is simplified moreso than for the lower volume classes. Performance of the products has a wide performance band and flexibility is a minor consideration. Repeatability, very short calibration and characterization time, and equipment availability (which includes reliability) are the dominant factors. Scheduling is one of continuous flow and multiple sets of equipment are the order of the day. Capital cost including facilities is a major order of business due to the large amount of equipment and the physical size of such an operation.

2.1.6 Very High Volume/One Product Class. This class is nearly totally inflexible with only one process and one product. Economics is the dominant consideration especially when considering volume. As such, throughput and yield are the two key parameters. The process and product are the most mature of any class.   Yield is paramount and throughput is a close second. Repeatability, availability (including reliability) along with the most simplified operation are the dominant issues. No development is ever done on this line which consists of the most streamlined of product flows with as few operations as can be tolerated. Efficiency and time of process are keywords. Scheduling is as straightforward as possible. Emphasis is on high yield and the shortest possible cycle time.

2.2   Marketing

Marketing is an important input to tool strategy because it provides information regarding the future technical requirements for the lithography system to be acquired. Generally, the resolution requirements for production obey an exponential relationship with time. First discovered by Gordon Moore(2), the

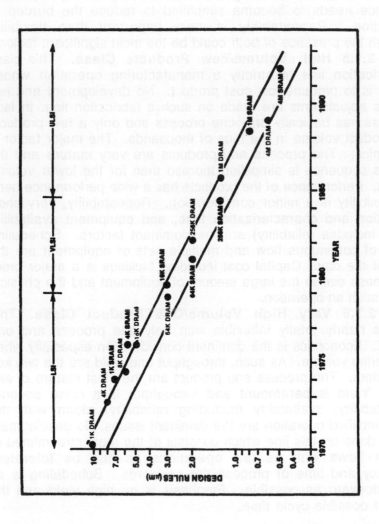

Figure 3. Minimum IC Dimensions for Future Requirements.

slope predicts halving of the critical dimension every six years. A more recent(3) extension of "Moore's Law" is shown in Figure 3. The curve shown in Figure 3 only presents an approximation to the future requirements of the industry. Chip area also increases with time to place very stringent requirements on the exposure field for the lithography equipment. The Marketing Department of the entity has the responsibility to determine the precise requirements which are necessary to reflect the charter, customer demands, and areas selected for new growth.

Unfortunately many marketing departments tend to minimize the importance of this critical function as a result of emphasis on current sales.

No IC facility can hope to become an industry leader without an annual report from Marketing which attempts to objectively predict future sales trends.   Such a report is necessary in order to formulate an action plan which is necessary to prepare for future sales. Companies which fail to provide this guidance can hardly hope to have products ready for customer evaluation in prototype equipment with assurance to the customer that deliveries will be prompt. The report should at least include sales volume as a function of minimum geometry. The distribution of chip area vs. minimum geometry should also be presented, along with sales price.   These reports in turn form the basis for evolution of the charter. A majority of the inputs to the annual report should be obtained through candid discussions with the customers throughout the year. Additional sources of information should include observations of competitor announcements/trends, government reports, cyclic patterns in book/sell ratios, etc. Supplemental data can be gathered from internal sources, e.g., process development engineers and device designers.

2.3   Product Development

Product development is where the ideas developed in Marketing are transformed into a few physical samples which can be electrically evaluated. Product development frequently works with potential customers by evaluating their needs and submitting sample IC's for evaluation.

The conversion of ideas into physical devices begins at a Computer Aided Design (CAD) system, which assists the designer in obtaining a detailed layout of the IC device which is consistent with a prescribed set of design rules. Sophisticated CAD systems are commercially available and their capability varies widely between manufacturers, depending in large part on the financial and engineering investments made. Normally, the CAD system is also capable of predicting the performance of the proposed design. Finally, the CAD system prepares a magnetic tape which contains the coordinates of all the surface features of the design. The magnetic tape is then used by the mask fabrication facility to make, inspect, and repair the masks to be used by the production facility. Alternatively, the magnetic tape can be used to pattern the wafers directly by using electron-beam lithography.

The product development group oversees the fabrication of prototype samples of the new design. Fabrication will normally be performed in either a special pilot line set up for prototype development, or in the production facility. The pilot line concept is typical for large semiconductor companies which can ill afford to disturb large production facilities. Smaller production facilities, on the other hand, normally have the flexibility to handle odd lots. Small facilities with Computer Aided Manufacturing, CAM, are especially adroit at mixing several products in a production line without processing errors. In the case where the state-of-the-art is being pressed by more demanding processing requirements, the Process Development Department will be solicited for assistance. Specifically, process development lithographers will be requested to reduce critical dimensions. The resulting efforts by the lithographer to meet smaller CDs will later provide important technical inputs for the selection of future lithography equipment.

The successful development of sample IC's with improved performance will normally result in a decision to fabricate production quantities to meet the volume of expected sales. Input information to production should include:

1.   Minimum critical dimensions, CD's
2.   Chip area
3.   Inter-level registration requirements
4.   Delivery schedule
5.   Mask levels
6.   Estimated sales price

2.4   Production Facility

Having received technical requirements for manufacturing new IC devices from Product Development, it is production's responsibility to fabricate the devices at a reasonable profit margin with the yield and volume necessary to satisfy the customers' needs.

Production is ultimately responsible for selecting the optimum lithography equipment, and begins the process of selection by determining the minimum yield necessary to make a profit. This is a relatively easy task when the new product only represents a minor variation from the present products. The present total wafer fabrication costs are first assumed to remained fixed. The fabrication cost per functional chip is estimated from the difference between the expected selling price for the IC device and the sum of the profit margin, electrical testing and packaging cost. The total necessary yield, Y_t, is then determined by:

$$[1] \quad Y_t = \frac{\text{Total Wafer Fabrication Cost}}{n\,[\text{IC Sales Price-(Profit + Test \& Packaging Costs)}]}.$$

where, n = the number of chips fabricated on a wafer, IC sales price = price as sold to the customer. The volume of production per unit time is then determined by the ratio of delivery schedule to the yield, Y_t. This is the full IC yield budget, a part of which is the lithography budget. The lithography personnel must then start translating this yield into specifications that their tools will require.

Production is fully aware of the manufacturing space available for expansion, and the cost per square foot of area required to install the lithography equipment. These facts will later be used as inputs to compare the total economic impact of acquiring each lithographic system which has been found capable of meeting all the technical requirements.

Production then sets a maximum capital appropriation which is available to purchase the lithography equipment. The appropriation is generally determined by a combination of: 1) total capital equipment funds to be made available during the year, 2) depreciation schedules, 3) available cash flow, 4) tax considerations, and 5) sales trends.

Production finally assigns an expert in lithography to determine which types of equipment and specific models are needed to technically be able to meet the functional requirements. Obviously, the lithographer will simultaneously try to stay within the capital appropriation available, installation area, and required production volume.

2.5 Technical Capability

Technical capability is an absolute requirement which describes the ability of a lithography system to produce a resist profile of given dimensions (resolution, edge acuity, CD control, etc.) properly registered to previous patterns and the effect of the tool on yield. Technical capability is a necessary but insufficient criterion for selecting a particular lithography system/tool. The technical capability of the equipment and its interactions with resist systems must be completely understood and compared with the requirements of the acquisition. This section will define how technical capability is determined from a combination of theoretical considerations, prior experience, experimental evaluation, and vendor information.

2.6   Types of Lithography

A wide variety of lithography systems exist, and a brief description of each will be discussed in this section to initiate inexperienced readers, and to provide a background for technical capability. The reader will find each system more fully described in later chapters.

This section will outline the resolution range over which each technique is technically capable of operation, and the approximate throughput in wafer-levels/hour. The types of lithography may be primarily categorized according to exposure radiation such as Optical, X-ray, E-beam and Ion Beam. Optical lithography is further subcategorized into proximity/contact printing, direct-step on a wafer, and full wafer projection. Likewise, E-Beam lithography has subcategories of direct write on a wafer, and projection printing (ELIPS). Ion-Beam lithography is being developed along three variations, namely: Focused Ion-Beam, FIB, (direct write on wafer), masked flood beam (step and repeat), and reduction stepper. The many types of lithography systems make the selection process very complex when comparing technical merit, cost of equipment, labor cost of fabrication and many other facets.

2.6.1   Optical Lithography. Optical lithography is generally the most cost effective lithography technique whenever it is capable of meeting all technical requirements. This parallel writing advantage results in high throughput with resists of moderate sensitivity. The commonly used positive type diazide-novolak resists feature high sensitivity, high contrast, excellent adhesion, good resistance to dry etching, and low cost. The primary limitations are resolution due to diffraction and substrate reflectivity. Diffraction limits resolution because the image projected on the resist becomes increasingly blurred as the dimensions decrease. Substrate reflectivity adversely affects linewidth control because of standing waves in the resist and the effect of interference at the air-resist interface on coupling of the incident exposure dose into the resist.

2.6.2 Contact/Proximity Printing. Contact printing represents an optical system with the lowest total cost since both the equipment and process are simple, and throughput is large. The primary disadvantage is low yield caused by contact between the resist and mask, and poor registration due to the extremely large field area. "Hard" contact printing yields the highest resolution but the yield is severely degraded by the repeated forceful intimate contact. Proximity printing alleviates the defect issue by not allowing the mask to directly contact the wafer. Resolution, however, is severely impaired by diffraction which increases rapidly with mask-to-resist spacing (gap). Minimum linewidth is equal to:

$$[2] \qquad LW_{min} = Q(\lambda g/2)^{1/2}$$

where Q is a constant, λ is the wavelength, and g is the mask-to-resist gap distance.

Equation 2 is only valid when $\lambda^2 \leq g\lambda << LW^2$. Adequate CD control requires a minimum value of Q equal to approximately 2.0. As an example, the minimum line/space width is approximately equal to 0.85 μm for a mask-resist gap of 1 μm using 365 nm radiation. The exposure time per wafer is of the order of 10 seconds and throughput is primarily determined by the time required for the operator to align the mask to the wafer. The throughput, with automatic cassette loading and unloading and automatic alignment, is in the vicinity of 60 wafers/hour. Contact/proximity printing is generally becoming obsolete because of the high defect density associated with physical contact of the mask with the resist and should only be considered for acquisition under unusual conditions. For example, III - V and II - VI semiconductor devices where circuit density is very low or device area is very small.

2.6.3 Full Wafer Scanning Projection Printing. Full wafer projection printing eliminates the problem of high defect density associated with contact printing, because a lens is used to image the mask onto the wafer. Full wafer projection systems normally project the image at 1X magnification using primarily reflection type lenses and a scanning slit to reduce distortion. The minimum linewidth practical for optical projection systems is equal to $k\lambda/NA$, where k is a proportionality constant, λ is the wavelength of the exposure radiation, and NA is the numerical aperture of the projection lens. The value for k has been determined by experience to be 0.8 for production worthiness using single layer resist systems. Multilayer resist systems can use a value of k=0.7. Pilot line operation with multilayer resist systems can with extra care use a k value of 0.6. The throughput is approximately 80 wafers/hour and dominated by alignment time.

2.6.4 Direct Step On a Wafer, Stepper. Direct step on a wafer extends the resolution capability of full wafer projection printing by easing the lens fabrication problem through limiting the exposure field. Reducing the exposure field decreases the focal length of the lens which makes high numerical aperture lenses easier to design, fabricate, and assemble. The lenses are normally chromatically corrected only over a narrow range of wavelengths, and interference between the incident and reflected waves causes standing waves in the resists and variable coupling between the incident radiation and the resist system. Step and scan systems have recently been developed which use a full wafer scanning type of optics to expose small areas. The advantages of broad exposure bandwidth and fewer optical surfaces are formidable. Equipment of this type is now available using deep UV radiation.

The resolution limit is equal to $k\lambda/NA$ where, k is a constant, λ is the wavelength of the exposure radiation and NA is the numerical aperture of the lens. A state-of-the-art I-line (λ=365nm) stepper with a 0.42 NA projection lens can reasonably be expected to resolve 0.69 μm with multi-layer resist systems and special care. The depth of field for this system is, however, only 1.04 μm, and this value must include resist thickness, topographical effects and the precision of the focusing mechanism. The depth of field problem poses the most serious obstacle to optical lithography in the region between 0.25 and 0.5 μm.

The throughput for an optical stepper is calculated using the following equation:

$$[3] \quad W = \frac{3600}{t_{oh} + N_{field}(t_{align} + t_{step} + t_{exp})}.$$

where W is the throughput in wafers/hour, t_{oh} is the overhead time, N_{field} is the number of exposure fields per wafer, and t_{align}, t_{step}, and t_{exp} are the times necessary to align, step the stage, and expose each field. The overhead time t_{oh} includes the time necessary to set up the lithography tool, change masks, and load wafers. The exposure time, t_{exp} (sec), is equal to the required resist dose, D_i (mJ/cm^2), divided by exposure intensity, I_i (mW/cm^2). Equations very similar to [3] can be written for every lithographic tool. Normally, the exposure time, t_{exp}, dominates the denominator, therefore, resist sensitivity and exposure brightness are very important factors in determining throughput. Throughput is typically equal to 30 to 50 150 mm wafers/hour.

2.6.5 X-Ray Lithography.

X-Ray Lithography overcomes the diffraction problems associated with optical proximity printing by using ultra short wavelengths in the region of 5 to 15 angstroms. The resolution is again limited as in contact/proximity printing by $LW \geq Q\ (\lambda g/2)^{1/2}$. In the case of X-ray lithography λ is typically equal to 10 Å, and the minimum linewidth is approximately 0.2 µm for a typical mask-wafer gap of 20 µm. The large mask-resist gap made possible with X-ray radiation eliminates the high defect density normally associated with optical contact printing. X-rays exhibit negligible reflection at resist-substrate boundaries, and excellent CD control can be maintained in simple single layer resist systems. Defects due to particulate contamination are very low since organic materials are nearly transparent to X-rays. The low absorption by resists also insures steeper profiles in thick resists. Initially, X-ray systems were developed using full wafer exposure, but fabricating large area masks proved unworkable and present systems use a step-and-repeat mode. The fabrication of 1X masks with zero defects and accurate placement of images on the mask remain the most difficult technical problems to be solved in developing production worthy systems. The future of X-ray lithography is very dependent on solving all the problems associated with mask technology. The very large costs expected for fabricating a set of defect free X-ray masks makes this lithography technique most practical when the production volume per device type is large enough to satisfactorily amortize the mask costs per IC device.

Point Source Systems. X-ray lithography was initially developed using electron impact sources of radiation, where high energy electrons bombarding the target excite the target electrons into higher orbits and X-rays are released upon relaxation. These sources are extremely inefficient ($\sim 3 \times 10^{-5}$ W/W-sr-Å for palladium) and have been replaced with more efficient ultra high temperature plasma sources. Laser or electrical discharge driven plasma sources are capable of

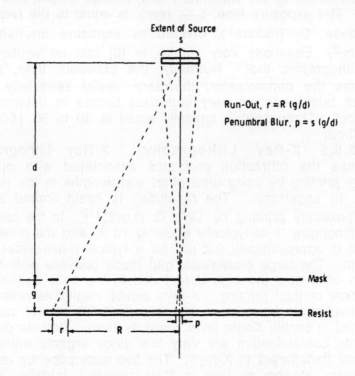

Figure 4. Penumbral Blur and Mask-Wafer Runout Which Occur When Using Point Sources of Finite Size.

delivering 20-40 mJ/cm^2 at the resist surface, and feature a source size of <200 μm in diameter. The throughput for 125 mm wafers with a resist sensitivity of 100 mJ/cm^2 is approximately 25 wafer levels per hour, using a 20x20 mm field size. The use of point sources in conjunction with finite mask-resist gaps results in penumbral blur and geometric magnification of the mask. Figure 4 depicts the phenomena of penumbral blur and geometric runout.

The runout, relative to the mask is equal to R(g/d), where R is the radius from the center of the wafer, g is the mask-resist gap, and d is the mask-source distance. The differential runout with changes in gap is equal to $\Delta r = \Delta g(R/d)$. Typically values for R and d are 4 and 30 cm, respectively. Under these conditions a registration error of 0.1 μm will occur at the edge of a 8 cm field for a 750 nm variation in mask-resist gap. An allowed variation of 750 nm places stringent demands on mask flatness and wafer substrate topography. The corresponding penumbral blur for a 200 μm source size and a mask-wafer gap of 20 μm is equal to 13 nm which is negligible compared to diffraction effects.

X-ray systems using plasma point sources will probably find niches in modest size facilities where the cost of synchrotrons is prohibitive, and the cost of masks is satisfactorily amortized.

Synchrotron Source Systems. Synchrotrons feature high intensity and excellent collimation of the radiation. High intensity results in high throughput with single layer resist systems, and high collimation minimizes the penumbral blur and run-out problems described for point sources. The output beam is in the form of a horizontal line approximately 1mm in height. Each exposure field must therefore be individually scanned by either moving the aligned mask and wafer as a unit in a vertical plane or more desirably by using a oscillating X-ray mirror to make the beam scan the exposure field. The radiant intensity for a typical storage ring is in the range of 0.25-2.5 W/mrad. The number of mrad/cm^2 of resist surface is dependent on the length of the beamline (port) and ranges between 5 to 10 mrad/cm^2 as the beamline varies from 5 m to 10 m. The intensity at the resist surface ranges between 12.5 to 25 surface ranges W/cm^2. The exposure time per exposure field is approximately 1 second for a resist sensitivity of 100 mJ/cm^2. Exposure fields of 4x4 cm are practical and the throughput for 150 mm wafers is approximately 50 wafer-levels per hour.

Mask fabrication, inspection, and repair along with a huge capital investment are the primary obstacles to widespread industrial implementation. A synchrotron costs in the vicinity of

$30M, and requires about 2-3 years to assemble and commission. A single synchrotron can easily accommodate 8 to 10 alignment machines with a combined throughput of 400 to 500 wafer-levels per hour. The high cost of mask fabrication per device type limits practical application to high volume production. Smaller volumes (100-5,000 wafers) are more economically fabricated using E-beam lithography since mask fabrication is not required. X-ray lithography, using synchrotron sources, is likely to emerge first in large corporations with large captive requirements for IC's with dimensions between 0.1 and 0.4 μm.

2.6.6 E-Beam Lithography. An electron accelerated to 25 keV has a de Broglie wavelength of only 0.074 Å! Clearly, an electron experiences negligible diffraction. A second advantage for electrons is that because of their charge they can easily be focused into a fine spot and deflected by electrostatic or magnetic fields. While the electron impact area can be precisely controlled, the electrons are easily scattered in the resist by interactions with protons and electrons in the resist. The electron beam in effect expands due to "forward scatter" in the resist. The scattering becomes more intense with atoms of high atomic weight, and substrates such as silicon and GaAs actually turn some electrons around and these "backscattered electrons" experience further scatter on their return path toward the vacuum-resist interface. The range of electron spreading is of the order of microns and two phenomena are observed. The first phenomenon is that isolated exposure areas exhibit poor edge acuity compared to the incident electron dose and a CD which is highly dependent on exposure dose and development time results. The second phenomenon is the CD of a pattern is also a function of nearby exposures. The latter phenomenon is loosely termed "proximity effect" and only becomes a serious problem in submicron lithography where adjacent patterns are very close. Partial correction for this effect can be achieved by GHOST, a technique where the whole wafer is flood exposed with a low dose which swamps out the proximity dose from adjacent patterns. Full correction is beyond the scope of this chapter, and the reader at this point only needs to know that it is computationally complex and seriously reduces the throughput of E-beam lithography.

Thermionic emission from tungsten filaments was used in early systems, but these were replaced by the higher brightness LaB_6 source, and finally by field emitter sources. The field emission type sources feature very high brightness, and low chromatic distortion (low electron energy spread) which facilitates more effective focusing. State-of-the-art beam columns typically deliver current densities, I_d, of 50-200 A/cm^2 at the resist plane at an energy of 20 keV. High resolution resists are commercially available with sensitivities, S, of 1 μC/cm^2. The exposure time per pixel is equal to S/I_d and therefore ranges between 5 to 20 ns.

Direct Write - Gaussian Beam. Direct writing with circularly shaped Gaussian beams is performed with beam diameters ranging between 0.01 μm to 0.25 μm. Two Gaussian shaped beam systems exist, namely, 1) Raster Scan, and 2) Vector Scan. Raster scan systems traverse the entire chip area on the wafer and the beam is blanked on and off as it scans a required pattern. The exposure time, T_{exp}, for a wafer is approximately equal to the number of pixels times the time to expose a pixel or:

$$[4] \qquad T_{exp} = (nA/d^2)(S/I_d)$$

where n is the # of chips/wafer, A is the total chip area, S is the resist sensitivity, d is the beam diameter, and I_d is the beam current density. The total write time increases rapidly as resolution is increased (i.e., d made smaller). In an example calculation of T_{exp} using equation 4, the conditions of a 100 mm wafer with 61 cm^2 of total write area, a resist sensitivity of 5 x 10^{-6} C/cm^2, a beam diameter of 0.1 μm with a current density of 100 A/cm^2 are employed. The exposure time per pixel is only 50 ns, but there are 6.1 x 10^{11} pixels. The total exposure time, T_{exp} is equal to 8 hrs. - 25 minutes. In addition, the total time required to process a wafer must include: 1) the time necessary to settle the deflection amplifiers prior to each exposure, 2) the stage motion time, and 3) the time to load/unload wafers. The throughput is therefore less than 0.12 wafers/hour.

Vector scan systems improve throughput by limiting the scanning area to only the required patterns. In this case, the exposure time, T_{exp}, is equal to:

$$[5] \qquad T_{exp} = (nAP/d^2)(S/I_d)$$

where P is the fraction of the combined area to be exposed - typically 20% of the total chip area. The throughput in this case is increased to 0.59 wafer-levels/hour with no loss in resolution.

Very high resolution is possible with the smallest beam diameters, but the time required to expose a pattern increases rapidly due to the increase in the number of pixels.

Direct Write - Shaped Beam. Shaped beam systems are of the vector scan type, but additional throughput is achieved by flashing the patterns by a series of relatively large rectangles. The exposure time for each rectangle (4 μm sq., maximum) is still equal to S/I_d, but tens of pixels are normally exposed simultaneously. Each rectangle is shaped by displacing a small flood beam of electrons from the center axis of the column such that they are partially intercepted by metal blades at the edges. The total electron dose per unit time is several orders of magnitude larger than Gaussian beam systems, and throughput is increased in proportion. Shaped beam systems are expected to dominate in the future, especially when modest product volume is required. European Silicon Structures (ESS) reports a throughput using the Perkin Elmer AEBLE system of 10 wafer levels per hour on 5 inch wafers at 1.2 μm geometrics for ASIC CMOS devices(4). The cost is reported as $31 per wafer level.

E-Beam Proximity Printing, EBP. Proximity systems utilize a 1X mask which is patterned with photosensitive materials (e.g., palladium.) Photo electron emission from the patterning is obtained by illuminating the patterns with ultraviolet light. The electrons are accelerated toward a resist coated wafer in a strong magnetic field to maintain collimation. These systems are not commercially ready at present and the reader need not consider this technology as practical at this time.

2.6.7 Ion Beam Lithography. Ion beams are particle in nature, and diffraction is negligible. The resolution capability of ion-beam lithography is very high. Low atomic weight ions such as H^+ are preferred for high penetration into the resist, but Ga^+ ions are more frequently used because their sources have been more fully developed with very high flux density. Ion energies of between 60 and 100 keV are typical. The ions have very high energy and each ion can create many chemical/physical events.

Focused Ion Beam, FIB. The charged nature of ions is used to focus an extremely fine source into a small spot on the wafer, and the beam is electronically deflected in a manner similar to vector scan electron-beam lithography. The relatively heavy ion is, however, more difficult to deflect and the deflection field is only of the order of 1 mm square. The throughput is adversely affected by the numerous stage motions necessary to expose a wafer.

Masked Flood Beam, MIBL. Masked ion-beam systems are essentially proximity printing and offer the prospects of high throughput by means of the parallel writing scheme. A very thin membrane mask with absorber patterns is used to shadow image a flood beam of ions. To avoid ion scatter by the membrane, "channel" membranes are used where single crystal materials are oriented to reduce the probability of ion collision with heavy atomic nuclei. Again these systems are not commercially available at present, and the reader may for the present exclude these systems from application in production.

Ion Projection Lithography - IPL. Ions, unlike photons, have an electrical charge, and they can be focused or collimated using electromagnetic or electrostatic lenses. Mask images can therefore be demagnified by lenses to project submicron images on the wafer. Mask fabrication, inspection, and repair is significantly less difficult than in 1X mask flood beam systems.

The only commercially available system of this type is a 10X reduction machine manufactured by Ion Microfabricating Systems. A resolution of 0.2 μm has been demonstrated with exposure times of 0.5 sec. in image reversal resist. The image field of the IPS 200 system is 7 x 7 mm^2, and the depth of field is ±100 μm. The throughput is approximately 490 cm^2/hr, or 3.3 six inch wafer levels/hr.

2.6.8 Lithography Support Equipment. Each lithography system requires a substantial amount of support equipment. The support equipment generally includes: 1) masks and mask making equipment, 2) mask inspection and repair equipment, 3) resist coating equipment, and 4) resist development equipment. Items 1 and 2 usually reside in mask making facilities, while items 3 and 4 reside in wafer fabrication facilities. Mask inspection equipment is usually found in both facilities.

Masks and Mask Making Equipment. The yield, therefore cost of manufacturing, is crucially dependent on the fact that the masks must be defect free and that all patterns are correctly sized and registered on the mask. The fabrication difficulty, or cost, is very dependent on the minimum dimensions required on the mask. 1X masks require high resolution and precise image placement; their cost is very high compared to 10X masks used in reduction type optical step-and-repeat systems. The cost for fabricating and inspecting a defect free 10X mask for an optical stepper is in the vicinity of $1000. Ten to fourteen masks are required to fabricate a typical IC Chip. In comparison, the cost for fabricating, inspecting, and repairing a 1X mask for x-ray lithography is estimated to be in the vicinity of $10,000 to $25,000. The total cost to fabricate IC's must take into account the substantial amounts of money invested in masks.

Mask Inspection and Repair Equipment. The yield of functional IC devices is very dependent upon the use of defect free masks. The masks must be verified that they are free of defects since no mask fabrication technique is presently able to guarantee the absence of defects. A typical mask contains some 10^8 pixels of information, each of which must be verified for existence and proper placement. The task of fabricating defect free masks is technically extremely difficult for 1X masks with submicron pixel dimensions because the probability of detecting a defect is less than unity due to limited resolution in the optical inspection systems currently being used.

Resist Coating and Development Equipment. The rightful emphasis on exposure equipment has frequently caused the lithographer to neglect the technical importance of resist coating and developing equipment. Initially coating equipment was merely a "Spinner" where individual wafers were manually placed on a vacuum chuck, the wafer was flooded with resist, and the motor was activated for several seconds. The result was poor resist thickness uniformity, both intra-wafer and inter-wafer.

In contrast, modern coating and developing track equipment is quite sophisticated and expensive, and the results are dramatically improved. The coating thickness on planar surfaces can be readily controlled to ±20 Å with defect densities of less than $0.01/cm^2$.

The improvements were largely realized by automatic wafer handling, dynamic dispense with nozzle suck-back, high acceleration digital control of motor speed, precise control of air flow, edge bead removal, etc.(5). Further, modern coaters provide improved resist adhesion through dehydration baking and HMDS (hexamethyldisilazane) treatment. The resist is also normally baked on a hot plate which provides excellent transfer of heat to the wafer, and precise temperature control to ±1°C is easily achieved. The entire mechanism is generally referred to as a "Wafer Track". Two parallel tracks are normally contained in one piece of equipment. Computer Aided Manufacturing (CAM) is currently being extended to wafer track equipment, with automatic monitoring of resist thickness, spinner speed, air flow, bubbles in resist line, hot plate temperatures, etc. The cost of a modern resist coater is in the vicinity of $200K, and with computer interfacing it can be in the vicinity of $200-600K. The cost is no longer a minor factor to the total capital budget, especially for bilevel resist systems which fully occupy a dual track machine. The throughput of such serial processing equipment is determined by the slowest step, normally the HMDS prime cycle. Typical throughput is 60 wafers/hr.

Resist development track equipment follows the serial nature of resist coating equipment. The equipment is normally automated with cassette-to-cassette operation, and with temperature control of the developer (critical to metal-ion-free MIF developers). Ultrasonic dispersion of the developer is currently being evaluated. Monitoring of the resist development rate can be performed during development to certify that the process is proceeding correctly (i.e., exposure dose, developer strength/temperature, etc. are correct). The capital cost of dual track resist development equipment is also approximately $400,000 when outfitted with CAM sensors. The throughput is approximately 70 wafers/hr., but an on vendor site throughput test should be performed using the proposed resist system to be used in production.

Resist Technology. The resolution, throughput, critical dimension control, adhesion, and post process compatibility are all very dependent on the resist system. The non-linear response of resists to exposure dose is in large part responsible for the excellent resist profiles obtained in spite of the modest imaging quality achieved by the equipment. The resolution capability of a tool and its associated resist system is directly proportional to the edge acuity of the resist profile, $\partial h/\partial x$. By simple differentials the edge acuity can be calculated:

[6]
$$\frac{\partial h}{\partial x} = \left\{\frac{\partial h}{\partial D}\right\} \left\{\frac{\partial D}{\partial x}\right\}$$

where D is the exposure dose and x is the lateral distance perpendicular to the line/space. The first term on the right side is strictly a function of the resist and its processing. The second term is equal to the gradient of the aerial image and is only a function of the exposure tool. The term $\partial h/\partial D$ is proportional to the contrast factor (gamma) of the resist system. The contrast factor for resists can be increased through the use of high exposures with weak developers, and the use of contrast enhancement materials.

High resist contrast is an important factor in achieving control of critical dimensions, as well as providing sufficient edge acuity to control the effects of resist ablation during dry etching. High resistance to thermal flow during dry etching requires resists with high glass transition temperature, T_g.

The throughput of the exposure tool is normally dominated by the resist sensitivity since exposure time is inversely proportional to resist sensitivity. The exposure time is most significant in systems which expose only a small fraction of the wafer at a time, e.g., 10X optical steppers, and direct write E-beam lithography.

2.6.9 Technical Evaluation Of Tools. The lithography engineer/scientist has primary responsibility for determining which specific types of equipment, and which models meet the technical requirements of the proposed acquisition. The necessity to meet the technical requirements is absolute. Failure to meet the technical requirements may result in financial failure of the manufacturing entity. Unlike the other decisions which involve marketing, product development, etc. the task of technical evaluation is a lonely one because of the depth of technical understanding which is required. The technical evaluation should be performed by personnel with maximum technical competence, and objectivity. The decisions must be made from a deliberate perspective of maximum emphasis on measurements and calculations, and with minimum emphasis on intuition and subjective reasoning. The following minimum parameters should be measured for the case of an optical alignment/exposure tool:

1. Adequate resist profile at minimum CD
2. CD control, normally ±10%
   a. exposure latitude
   b. focus latitude
3. Registration, normally ±0.2CD
4. Throughput, wafers/hr.
5. Maximum chip area
6. Field distortion
7. Impact on yield

The list is specific to optical steppers but very similar lists exist for all tools. All these measurements should be performed with the resist system to be used in production. The lithographer may consider new multilayer resist systems which allow less costly equipment to be purchased, but the new resist system must already be sufficiently developed to provide a meaningful evaluation, and the cost impact of more expensive resist technology must be considered.

Resist profile is especially important when dry etching or lift-off is used. The assurance of an adequate resist profile is relatively simple to perform. Scanning Electron Microscope, SEM, photographs of the resist profile as a function of exposure and focus are sufficient. Measuring the development variance in resist profile with focus is especially important since the edge acuity is a function of the gradient in the aerial image as previously given in Equation 6. The image quality must be checked in the corners of the field and compared with those in the center.

Critical Dimension, CD, control is the ability to maintain the dimensions of resist images with time over all areas of the exposure field. CD variations occur as a result of uncontrolled variations in resist thickness, exposure dose, focus, development, etc. Absolute assurance can only be made by long term statistical process control measurements of linewidth in a production environment. Partial characterization is, however, possible at vendor facilities for the first machine, or when a single machine is to be purchased. Plots of CD vs. exposure/focus can be performed at vendor facilities to simulate the probable uncontrolled variations. The incident exposure is well controlled by steppers, but the coupling of exposure dose into the resist is a strong function of resist thickness and substrate reflectivity. The linewidth can be accurately measured using the NBS crossed bridge test pattern, and electrical probing on a Prometrix LithoMap^R System. Figure 5 (a) shows the linewidth test structure which is imaged in a resist over a conductive film which has been deposited on an oxidized wafer. The Van der Pauw cross at the top measures the sheet resistance, ρ, of the conductive film, and the Kelvin bridge structure essentially measures the linewidth, L, by $R=\rho(L/A)$. The length, L, is typically 200 μm. The resist images are anisotropically transferred into the conductive film by RIE etching. Lift-off metallization is also possible, even when the resist profile is non-reentrant, if the edge acuity is high and the metallization thickness is thin (<1000 Å). Doped polysilicon and thin films of refractory metals are frequently used as conductive films for electrical measurements. Copious amounts of data are obtained for statistical analysis, and regression analysis to complex models can be accomplished. A typical exposure/focus plot obtained using a Prometrix System is shown in Figure 5 (b). Plots of this type adequately characterize the tolerance for defocus, but CD variations due to exposure are difficult to determine because of close and irregular contour spacing. A plot of the type shown in Figure 6 is more desirable in accurately characterizing this most important variable.

The percent exposure latitude, PEL, is graphically determined by finding the exposure range between +10% and -10% variance from nominal CD, and dividing by the correct nominal exposure dose. Finally, the range is multiplied by 50 to

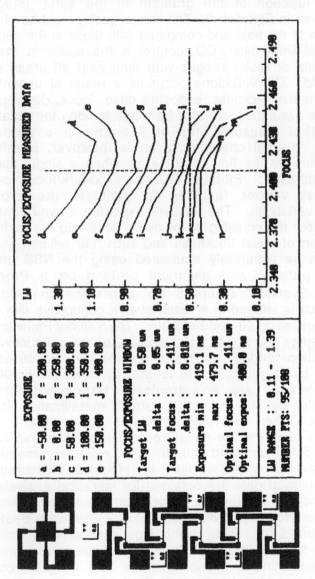

(a) Linewidth Measurement Structure   (b) Typical Focus/Exposure Plot

Figure 5.   Electrical Measurement of CD Control.

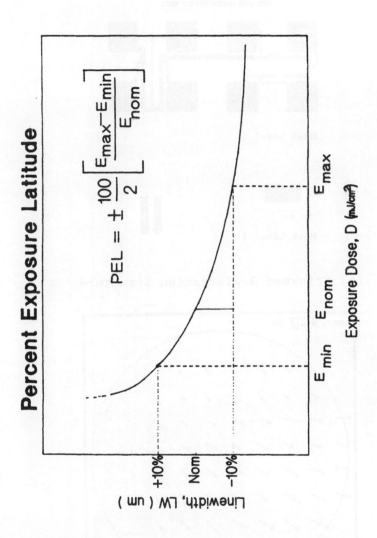

Figure 6. Characterization of CD vs. Exposure Dose.

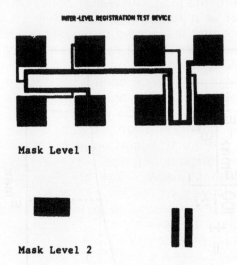

Mask Level 1

Mask Level 2

(a) Stickman Registration Structure

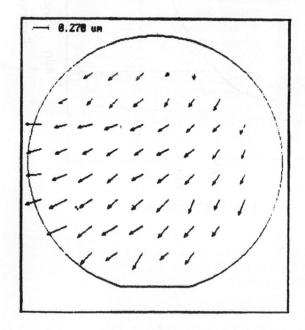

(b) Vector Map of Misregistration

Figure 7. Registration Testing Using Electrical Test Structures.

convert to ± percentage. The value considered sufficient for adequate control is a function of the resist system's ability to control reflectance and the degree of process control which exists. The exact minimum value is set by the lithographer's experience. Generally, ±12% is adequate for production using single layer resist (SLR) without reflectance control whereas ±8% is adequate for pilot line situations using rigid process control and resist systems which control the effects of reflectivity.

Inter-level registration measurements, like CD measurements, are most easily obtained using electrical test structures of the Stickman(6) type. Figure 7 (a) describes the two level Stickman structure. The structure is etched into an electrically conductive film (doped silicon) using two resist/etch cycles. The measurements are valid for intra-machine capability when both cycles are performed using the same machine. Alternatively, inter-machine (mix-and-match) lithography can be evaluated using different machines for each level. A vector plot of the misregistration at each die site is shown plotted in Figure 7 (b) for the case of a typical inter-machine test. The plot reveals an obvious stepping difference in both the X and Y directions. Subsequent computer analysis of the data will reveal the systematic sources of misregistration, e.g., alignment errors (translation and rotation), lens magnification, reticle rotation, stage orthogonality, etc. Finally, a residuals plot will be made showing that part of misregistration which is random in nature.

Registration measurements are relatively easy to perform using alignment/exposure tools at vendor facilities followed by analysis at the production facility using a Prometrix System.

The throughput of the exposure tool must not only be determined under realistic conditions (applicable resist system, etc.) but the result must be analyzed to reveal the individual time delays for each step in alignment and exposure. The throughput data can be obtained as ancillary information during CD control testing.

The maximum chip area is usually specified by the equipment vendor, but the useful area may be substantially less as a result of poor assembly and/or design. A reduction in area may occur as a result of poor exposure uniformity or radial fall-off of modulation for the aerial image. Both of these failures can be quantified through CD measurements as a function of radial distance from the center of the field.

The effect of lens distortion on registration may be revealed during the registration tests. Intra-machine testing will not reveal this factor, however, since both levels will have identical distortion. Relative distortion will be revealed when the exposure tool being evaluated is used in the second exposure/etch cycle. Absolute distortion can only be measured if the first exposure is performed in a system without any distortion. Electron-beam lithography can be considered a reasonably distortion free system, since the table's position is very precisely measured using optical interference.

The effect of defects on yield, Y, is predicted by Stapper's Model(7) as:

[7] $Y = (1 + SAD)^{1/s}$

where A is the area, D is the defect density per square cm, and S is a constant which describes the fact that the spatial distribution of the defects is not completely random. Poisson's Equation assumes that S=0, and Seed's Model assumes a value of 1.0. Westinghouse measured and found that S is normally equal to 0.44. Equation 7 is applicable for a single masking level. The total lithography yield is the product of all the Y values for all m mask levels:

[8] $Y_t = \prod_{n=1}^{m} (1 + S_n A_n D_n)^{1/s_n}$

Defect density may be measured using optical inspection equipment which identifies resist patterns not contained in the database, or by measuring the yield of metallized test patterns which contain closely spaced lines and spaces. Defects in either the lines or spaces are detected by electrically probing for opens and shorts and interpreted as missing resist patterns or excess resist, respectively, when the resist images are transferred into conductive films by etching. The defect density is calculated from the probability of non-defective test patterns by:

[9] $D = (Y^{-S}-1)/SA$

where Y is the probability of no defects (fractional yield) determined from electrical probing of the test patterns.

The impact on yield of new equipment is extremely difficult to determine using vendor facilities, since vendor facilities are seldom up to the required cleanroom standards, and the transportation of wafers generates defects. Complete assurance of yield cannot be practically made in vendor facilities. The problem is only fully remedied when a machine is purchased and installed in the manufacturing facility in anticipation of a much larger follow-on order. The high impact of yield on cost does not, however, allow the evaluator to overlook the yield factor. A partial evaluation is possible prior to purchasing a machine, when an equal number of control wafers are transported to vendor facilities and process simulated, but actually exposed and processed at the production facility. Likewise, the wafers processed at vendor facilities must undergo simulated processing in the production facility. The described pseudo evaluation technique is cumbersome and subject to unavoidable introduction of new variables. Nevertheless, it is a technique of some value which can be used when no alternative is available.

A table comparing several alignment/exposure tools will quickly reveal the strong and weak points of each tool and will provide the basic information necessary for a detailed cost analysis.

2.7 Economic Factors

2.7.1 Cost of Manufacturing. The cost of manufacturing a wafer includes all capital, direct labor, materials, and overhead costs. Direct labor normally dominates in a commercially viable manufacturing plant. While lithography is only one of many sequential processes, its repeated use on each wafer has a large weighting factor. The cost to manufacture a functional IC is:

[1 0]   Cost/Chip = $\dfrac{\text{Total fabrication cost per Chip}}{\text{Total Yield}}$

[1 1] $= \dfrac{\sum\limits_{n=1}^{m} (\text{Labor} + \text{Capital} + \text{Overhead} + \text{Material})}{\text{Total Yield}}$

where labor cost, capital cost, overhead costs, and material cost per wafer level must be summed for all major processing steps for n equal to 1 through m. The total yield is the product of the production yield, electrical yield, and packaging yield.

2.7.2 Labor Cost. Direct labor cost and indirect labor cost must both be included in labor cost. Direct labor cost is that directly applied to fabricate devices, and is primarily associated with operator cost. Direct cost is obviously very dependent on the throughput of the lithography system. Indirect labor costs includes the cost of engineering, maintenance, management, etc.

The total labor cost for lithography is implicitly dependent on the throughput of the tool and the fractional number of personnel necessary to operate the tool.

The turnaround time for processing a lot is an important factor in determining the minimum time required to completely process a lot of wafers. Turnaround time is therefore very important for quick delivery of functional ICs. The turnaround time, T_a, in hours, for a lithographic tool is equal to:

[12]
$$T_a = \sum_{n=1}^{m} \frac{\text{Lot Size (wafers)}}{\text{Throughput (wafers/hr)}}$$

where n, and m were previously defined in Equation 11.

The turnaround time for the complete IC process varies with the charter and typically ranges from 4 to 25 weeks, of which some 60% is consumed in lithography.

2.7.3 Capital Equipment Costs. Lithography equipment must be periodically replaced, primarily due to antiquation. The rapid pace of equipment development is especially true in the semiconductor industry, and lifetimes of only 5 to 10 years are normal. During this lifetime the cost of the equipment must be amortized (i.e., money must be set aside to replace it when replacement is due). The capital cost per year, C, is proportional to the total capital acquisition cost, A, multiplied by a factor which includes interest rate, i, and years of service, n:

[13]
$$C = A \ [i(1+i)^n / (1+i)^n - 1]$$

A typical example: An i-line optical stepper costs \$2M to purchase and install, the interest rate is 8% per annum, and the lifetime is 8 years. The capital cost per year is \$348,000, The capital cost per wafer level is equal to \$348,000 divided by the total wafer levels processed per year. The total wafer levels processed per year is highly dependent on the product of machine throughput and machine availability (dependent on machine reliability, stability and maintenance received).

2.7.4 Overhead Cost. Overhead cost generally includes all indirect costs necessary to fabricate the IC devices. Examples include: amortized facility cost, management, heating/air conditioning, sewer and water, quality control, purchasing, insurance, etc.

2.7.5 Material Cost. The material costs involved in making semiconductor IC devices is divided into two categories, namely: 1) Direct material costs, and 2) Indirect material costs. Direct materials includes those materials which physically become part of the IC device. Silicon wafers, metal deposition alloys, packaging materials, and diffusion/ion-implantation materials are examples of direct materials.

Indirect materials includes all materials which do not actually become part of the IC device. Examples of indirect materials are photoresist, masks, resist developers, processing gases, cleaning solvents, deionized water, and other chemicals.

The sum of direct material and indirect material costs is the total material cost. The total material costs are usually a small fraction of the total fabrication costs. Exceptions to this generality are exotic wafer materials such as GaAs, silicon-on-sapphire, and II-VI compounds.

2.7.6 Total Cost. The total cost includes the sum of the capital costs, labor costs, overhead costs, and material costs. Profits are necessary for the long term viability of any commercial manufacturing facility. The market value of an IC device is largely dependent on the competition's cost to manufacture and market elasticity. The economic health of a manufacturing facility critically depends on the ratio of its total IC cost to that of its competition.

The total cost to fabricate a wafer is dependent on minimum dimensions primarily due to increased cost of lithography. Typical costs for processing 6" (150 mm) wafers are listed in Table 1 as a function of minimum geometry(8).

These costs are for high volume facilities, running approximately 3000 to 4000 wafer starts per week, and the costs must be increased by a factor fo 2 to 3 for low production volumes and their inefficiencies(8). Electrical wafer probing will increase the wafer processing cost from $100 to $200 depending on the degree of testing performed at the wafer level. Finally, the total cost must also include the assembly, packaging and final testing costs.

DM Data Inc. continues to analyze the cost for a typical 50K gate array device using 1μm geometrics, a die area of 129 mm^2, a package cost of $25, and a probe yield of 6.5%(8). The cost per good die is determined to $71. The final cost of the assembled, packaged and tested device is projected at $240. The analysis shows that while yield dominates in determining the cost of a good die at the wafer probing stage, the total device cost is some 338% higher than that at wafer probing(8).

Table 1
Typical Total Wafer Processing Cost
VLSI Circuits - 150mm Silicon Wafers - 15 Mask Levels

Minimum Geometries (μm)	Total Fabrication Cost ($/Wafer)
1.5	300-350
1.0	400-450
0.8	500-600

The cost of lithography in this analysis appears to have minimal impact on final device cost. The assumed yield of 6.5% combined with high assembly, test and screening costs is responsible for this result. The fractional cost of lithography increases rapidly with decreasing yield. At a yield of 1%, the cost per good die jumps to $498, while the final device cost increases to $668. In this case, the lithography cost dominates since final chip processing only increases the total cost by 34%. The importance of having a strategy for selecting the optimum lithography tools increases dramatically as yield crosses below the 2% level.

2.7.7 Yield. The yield, as shown in Equation 11, is probably the most important "key" to the cost of shippable IC devices since it can range over several decades. Low yield also affects the ability to manufacture and ship IC's on time. Failure to achieve competitive yield will ultimately result in lost customers. Assuming that the IC has been properly designed, the yield is generally dependent on defects produced during processing. Defects can be classified as: 1) random point defects, and 2) non-random defects. Point defects are small (<10 μm) and randomly located. The origin of point defects is usually particulate contamination as a result of airborne dirt or more frequently as produced within the processing equipment. Non-random defects due to improper processing are usually much larger than 10 μm in diameter and they have a definite spatial relationship to the patterns on the wafer. Examples of this type include inadequate resolution, poor registration of mask levels, incomplete etching, non-uniform deposition, etc.

3.0   IMPLEMENTATION OF STRATEGY

As discussed in section 2.1, there is a large set of factors which influence the selection of lithography tools   A precise strategy must be exercised during the selection process to insure all technical requirements will be met with the lowest total cost. Also note that in general the quantification of the selection parameters is not precise but involves some subjective judgement. This section will attempt to unify all the factors into a cohesive quantative approach to selecting optimum equipment and processes.

Y. Iida(9) first attempted in 1983 to quantize a comparison of various lithographies using a figure of merit defined as follows:

$$[14] \qquad G = (T/C \cdot Y/F^2 \cdot A_v \cdot M_n)^P x (1/W^2 \cdot (1 - A/W))^R \cdot 1/T_a$$

where:

G	=	Figure of merit
T	=	Throughput, wafer-levels/hour
C	=	System cost including clean room
Y	=	Yield/unit exposure area
F	=	Normalized chip edge length
A_v	=	Equipment available, %
M_n	=	Equipment maintainability
W	=	Linewidth
A	=	Alignment accuracy
T_a	=	Turnaround time
P,R	=	Constant defining production/research environments

The first term is primarily an economic factor of great importance in a volume manufacturing entity, while the second term is a technical factor of primary importance in research and development.   In research R>1 and P<1, whereas in volume production R<1 and P>1. This approach is extremely useful in determining which type of lithography is appropriate for the application involved.

Iida compared optical, E-beam, and X-ray lithography using data available in 1982. The results are tabulated in Table 2, and plotted in Figure 8.

The curves are of course no longer accurate but their shapes and relative locations are still generally valid. The curves show that each type of lithography has a resolution region where a maximum Figure of Merit occurs.   The resolution, which is determined by where the curves cross one another, may be properly regarded as critical resolution.   The E-beam/stepper

Table 2
FIGURE OF MERIT INVOLVING TECHNICAL AND ECONOMIC FACTORS

Lithography	UV	UV Projection						EB			X-Ray	
	Contact	Refraction	Refraction					Field (mm)			S $ R	1/1
		Reduction Ratio										
		1/1	1/20	1/10	1/5			5	10	10	25	4"
Resolution (um)	2	4	6	1	2	4	0.65	1	2	0.25	0.5	1	0.25	0.5	1
Alignment (um)	1	1	1.5	0.3	0.5	0.5	0.1	0.15	0.3	0.05	0.05	0.1	0.1	0.1	0.5
Throughput (4" wafers/hr)	80	160	200	100	90	80	15	60	100	10	30	60	30	80	100
System Cost(M $)	0.1	0.1	0.1	1.0	0.35	0.2	0.75	0.75	0.75	3	3	3	1	1	0.5
Yield (%)	0.7	0.9	0.99	0.9	0.95	0.99	0.8	0.95	0.99	0.95	0.95	0.99	0.7	0.95	0.99
Figure of Merit	49	65	45	40	56	21	20	102	32	37	33	18	200	172	99

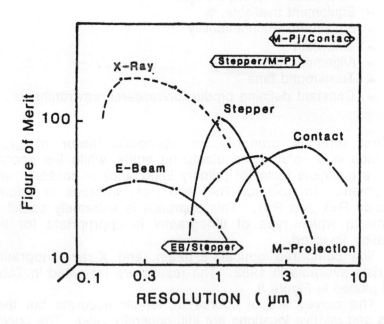

Figure 8. Figure of Merit as a Function of Resolution: Ref. Y. Iida.

critical resolution is 0.68 μm. Today, this critical resolution has shifted to the vicinity of 0.45 μm through the evolution of optical technology. The critical resolution for X-ray/stepper technology has likewise shifted from Iida's value of 0.97 μm to today's value of approximately 0.55 μm. Figure 8 incorrectly shows that E-beam lithography is not advantageous for any resolution. The primary reason for the obvious error is the fact that Iida's work does not reflect the importance of production volume. Equation 14 assumes that the lithography equipment will be 100% utilized (not including downtime). The non-recurring costs of masks, clean room space, etc. have also been ignored since these costs become less significant in very high volume production.

Figure 9 presents a picture where volume per device type is an important factor for consideration. The plot in Figure 9 shows that E-beam lithography dominates in effectiveness when production volume per device type is very low regardless of feature size, because amortized mask costs dominate total cost in small volume, but is equal to zero for E-beam lithography. The manufacture of ASIC (Application Specific Integrated Circuits) type devices is a prime example of low volume device types. E-beam's advantage rises slowly as resolution increases from 0.7 to 0.45 μm because of improved yield compared to optical lithography, and the need to use multilayer resist systems when optical lithography is pushed to the limit. The area shown advantageous to X-ray lithography assumes that mask technology will be solved in time at a reasonable cost. The increasing advantage with volume at high resolution is a result of the high fidelity of imaging in this region. Optical lithography has a commanding position in the region above 0.5 μm because of its low cost and good performance. The low volume performance decreases primarily as a result of masks costs.

D.W. Peters(10) also proposed a cost-effectiveness Figure of Merit, F, defined as follows:

[15] $$F = NQ/(C + KA)\, RP$$

where N is the real time throughput in wafer-levels/hr. including all overhead, not free-running throughput. Q is the resist process complexity factor and denotes relative yield loss (Q = 1 denotes conventional, single level, positive novolak resist processing; Q = 0.70 denotes unconventional resist materials or processing; and Q = 0.50 denotes multilayer resist processing).

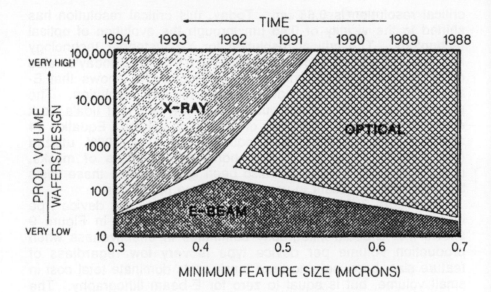

Figure 9. Optimum Lithography Technology Based on Both Minimum Feature Size and Production Volume.

Table 3
FIGURES OF MERIT FOR HIGH RESOLUTION LITHOGRAPHIES*

Parameters*	Deep-uv Steppers	E-beam direct-write	Tube-source x-ray steppers	Laser-generated plasma source X-ray steppers
R (um)	0.50	0.50	0.50	0.50
P (um)	±0.20	±0.20	±0.20	±0.10
N (wafer levels/hr)	10	2	5	48
Q	0.50	0.50	0.70	1
C ($M)	1.65	4	1.5	1.75
A (m²)	5.58	6.51	1.86	1.95
Figure of merit (/$10K)	0.28	0.02	0.22	5.36

* Reference: D. W. Peters[10]

C = Total Capital Cost
K = Clean Room Cost (~$21,500/m^2)
A = Footprint Area (m^2)
R = Working Resolution
P = Registration Accuracy

Using equation 15, Peters assembled a comparison table shown in Table 3. Peters concluded that X-ray steppers, using plasma generated X-rays, are optimum for resolving 0.5 μm geometrics. The reader is, however, warned that the values used in the Table 3 do not reflect a general concensus of lithographers, and the high cost of X-ray masks is not included. The work is nevertheless shown here as another example of models which can be used to economically compare lithographic tools. The reader may use the model with additional factors, and values obtained from personal experience.

We have discussed charter, marketing, technical requirements and economics. These factors are key to the methodology for the selection of lithography tools. First consider the charter. Once one has firmly established the charter, then the class of operation is defined. Combined with marketing, this then sets the volume of product and the price range for the product. From this, one can determine the budgeted operating cost. In parallel, marketing has defined what the product performance must be to be competitive. This in turn will lead to technical requirements such as minimum geometries, chip area, mask levels, etc. At this point, the technical requirements can define several sets of lithography tools. The remaining decision is largely determined by economics with slight adjustments from the charter. Recall that an operating budget was previously determined. Device selling price dictates what yield and throughput is required, then total cost can be computed from direct and indirect costs. One will then need to choose the tools that are affordable for the proposed life of the product or products. One last time, the charter along with marketing must be used to decide whether the selected tools are consistent with the proposed product price and operating cost. This whole strategy may take several iterations before converging. Note that what has been accomplished is a class of operations has been chosen then a series of tradeoffs accomplished within that class considering the parameters of yield, throughput, cycle time, technical requirements and indirect costs.

Let's look at two examples to apply the lithography tool selection strategy. The first example is a research and development division where responsibility is to push the state of the art in density and performance. Let's say we are trying to fabricate devices with CD's equal to 0.5 micron using chip areas of 1.0 cm on a side. First, our charter is the R&D class. We have several processes and products, but our volume is extremely low. Marketing has told us they need a certain density and transistor count to sell future products and this establishes the CD requirements. The speed of the device has also dictated the need for smaller CD's along with some other parameters which will not be discussed here. Our volume is very low, but since we have many products to support, we must have flexibility in our equipment. Cost is important, but flexibility is vital. We are, therefore, interested in reducing the various set up support functions. We will want to choose a lithography tool oriented to little or no nonrecurring cost, short cycle times, but one that is able to achieve the technical requirements. Note: amassing the final technical requirements is a long process in which each facet discussed earlier must be explored. Finally, we can select several tools which can meet the technical requirements. For our case, let's assume that an X-ray stepper, E-beam direct-write, and an i-line optical stepper are all technically qualified. However, the i-line optical stepper requires the use of multilevel resist and 0.5 micron CD's is truly pushing the state of the art. Let's say the X-ray and the optical steppers cost $2 million each while the E-beam costs $3 million. At first one might believe that the economics might eliminate the E-beam immediately. However, the E-beam needs no masks and it uses a single level resist. Both the X-ray and optical steppers need masks. The high cost of X-ray masks will be prohibitive for any operation that is continuously changing the product design. Even the optical system becomes expensive when constantly making new masks, and waiting for masks to be made causes the time from finish of design to fabrication completion to be much longer than the corresponding time for an E-beam. Here is a case where the wall clock time is as much, if not more important, than the capital cost. The cycle time is important because we are trying to establish a product to take to market and the timing of that product may be the most critical factor. The cycle time for an E-beam based product could be five weeks compared to ten weeks for an optical based system. Thus, the choice for this situation is the E-beam system.

Let's now take the other extreme. A factory is to be set up for a single process, single product operation which requires 1.25 µm CD's with chip sizes of 3 mm on a side. Note, the charter is for the Very High Volume/One Product class. Marketing has already described the product and therefore has essentially set the

the requirements for technical capability. There are obviously numerous tools which can technically do the job. This includes an optical stepper, which requires only a single level resist. The E-beam, plasma X-ray, and Ion beam are far too expensive to operate in a high volume mode and thus the optical stepper is the obvious choice. The next step is to determine what support technologies and equipment are needed to support such a high volume operation. Note, this type of operation is more concerned with recurring cost and not as much with nonrecurring cost. Thus, once the equipment and all support tools are set, there will be little or no change necessary. So the initial nonrecurring cost can be amortized over a large volume.

4.0 SUMMARY

Selection of an optimum lithography tool is a very involved process requiring knowledge and experience in several disciplines including physics, chemistry, electronics, device design, processing, marketing, manufacturing, and economics. This chapter has established a strategy by which one skilled in these disciplines can select the best tool or set of lithography tools for his situation. Although many facets make up the strategy, the technical aspect is the dominant item because the tool has to perform. Even though technical capability is the dominant facet, it is not sufficient because several types of machines will normally be technically capable, and must be combined with the economics, charter, marketing, and manifestation of economics. This is true whether the manufacturing facility is very large or simply a small laboratory in a university. Often one finds that the technical analysis has qualified several strikingly different types of lithography tools. At this point economics, whether manifested through the charter, marketing, manufacturing operations, or any combination thereof, will be the final decision maker in the selection of the best lithography for the user.

REFERENCES

1.   ICE, Lithography Dominates Wafer Processing, ICE Midterm 1988 compedium, Semiconductor International, 11:36 (1988).
2.   Moore, G. E., Remarks of Dr. Gordon E. Moore, keynote address to Kodak Interface '75, G-45 (1975).
3.   Tobey, A.C., Semiconductor Microlithography Through the Eighties, Microelectronics Manufacturing and Testing, 3:19 (1985).
4.   Burggraff, P., AEBLE - based E-beam Production Results Divulged, Semiconductor International, p.34, Nov (1989).
5.   Skidmore, K., Applying Photoresist for Optional Coatings, Semiconductor International, 2:45, Feb. (1988).
6.   Stemp, I.J., Nicholas, K.H., and Brockman, H.E., Proc. Conf. Microcircuit Engineering, '78, Cambridge England, April (1978).
7.   Stapper, C.H., Defect Density Distributions for LSI Yield Calculations, IEEE Trans. Electron Devices, ED-20: 655, (1973).
8.   Semiconductor Economics Report, Published by DM Data Inc., 6900 E. Cambelback Road, Scottsdale, 3:7, February (1989).
9.   Iida, Y., Hybrid Lithography, Nikkei Electronics, 2-1:213, (1983).
10.   Peters, D.W., Examining Competitive Submicon Lithography, Semiconductor International, 2:96, (1988).

2

RESIST TECHNOLOGY - DESIGN, PROCESSING AND APPLICATIONS

John Helbert

Motorola, Inc.
Advanced Technology Center
Mesa, Arizona

CHAPTER PREFACE

The objective of this chapter is to provide a user's view of resist technology. Other notable authors have previously provided insightful views of resist technology(1-4), but from a research or resist inventor's point of view. My intent, here, is to supplement these excellent works, not to reproduce them in another source. Some material must be rehashed for completeness, but hopefully from another complementary perspective. The emphasis of this chapter will be placed on applications of this technology to the manufacturing of prototype and production integrated circuit devices. Furthermore, a greater emphasis will be placed upon empirical resist process development to achieve reproducible and statistically controlled resist manufacturing processes.

1.0 INTRODUCTION

Organic resist technology is vital to integrated circuit(5), or more generally semiconductor device, manufacturing. Nearly every device fabrication step requires a process compatible masking layer, which is capable of providing a desired circuit level pattern. This indirect patterning is required because either the layer is not directly patternable technologically or it cannot be accomplished economically. Resist layers, as their name implies, "resist" individual layer processing steps to enable electronic devices to be fabricated vertically layer by layer on a thin silicon crystal wafer (see Figure 1)(5). These individual layers can be insulating dielectrics, semiconducting active device elements, or metallic interconnects.

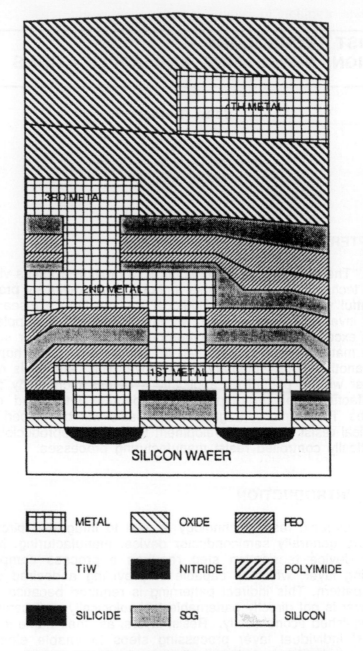

METAL	OXIDE	PEO
TiW	NITRIDE	POLYIMIDE
SILICIDE	SOG	SILICON

CROSS SECTIONAL VIEW OF VERTICAL LAYERING

Figure 1.   Cross sectional view of vertical layering in the manufacturing of silicon VLSI circuits.

In addition to providing device manufacturability from a circuit element delineation view(5), the resist process is capable of influencing device performance. The resist lithographic resolution and critical dimension (CD) control, for example, can directly influence device turn- on threshhold, speed and/or circuit density. Historically, the resist CD requirements have reduced approximately 20-30% every two years(3), thus pushing some older lithographic tools to their limits and making them obsolete. The dramatic time evolution of memory´ chip storage capability presents an adequate testimonial. Dynamic random access memory element counts have gone from 1-256 kilobits in about six years, and four megabit chips are being sampled at the time of this book preparation.

The lithographic capability of the resist is determined to a large degree by the wavelength of the electromagnetic energy source used to carry out the selective patterning process. Typically, visible light is used of wavelengths ranging from 300-420 nm. The light is imaged on the wafer through chromium metal patterned transparent glass masks using either reflective or refractive optics(3). Electromagnetic energy sources of wavelengths <310 nm can be provided by deep UV (DUV) producing systems such as high pressure mercury arc lamps or laser systems(4). Further reductions in wavelength are achieved by employing focused electron beam sources, like those found in scanning electron microscopy(i.e., 10-30 keV), or those found in soft X-ray systems (2-20 Å wavelength). Most importantly, the wavelength in most cases determines what type of resist can be employed, because the energy of the lithographic tool must be coupled to the resist to insure a conversion of electromagnetic energy to radiation chemical energy occurs. In the next section, the design of resist materials for specific radiation sources will be described. Furthermore, resist systems or processes will be described which actually extend the useful resolution or lifetime of certain lithographic aligners.

Although the lithographic properties of resists can determine circuit density and performance, the resist must first of all be device layer process compatible, or it is of academic importance only. Unfortunately, the literature abounds with resist systems of great lithographic capability, but they cannot be employed in the commercial fabrication of semiconductor devices because they are not capable of withstanding or "resisting" certain required processes. The most widely studied e-beam resist, namely poly(methyl methacrylate) (PMMA), falls into this category due to an inferior dry etch(6) capability. In the process and process applications sections of this chapter, process compatible processes will be highlighted. In this sense, the perspective is a practical user's view as opposed to a resist inventor's view, where typically little actual device fabrication experience exists and the process and manufacturing issues are not as well realized.

2.0 RESIST DESIGN

2.1 Conventional Photoresists

2.1.1 Positive Resists. These resist materials(1-4) are the workhorses of modern integrated circuit (IC) manufacturing technology. All new very large scale IC (VLSIC) fabrication lines employ high resolution positive toned material, while the older lines with more mature products still rely heavily on negative toned resists. Positive toned resists develop away to create recessed relief images in the exposed areas (see Figure 2) with safely-disposing dilute aqueous base solutions. When employed, they can be used at all device levels by simply changing the density of the Cr patterned mask or reticle by the mask shop, but the pattern information is usually digitized positive by the design group regardless of final density.

Positive photoresists are composed or formulated from several components: polymeric resins of molecular weight of the order of 1-10K, photoactive molecular organic additives (PAC), leveling agents (SLA), optional dyes to reduce substrate reflectivity effects, sensitizers and organic spinning solvents. The resin molecular weights are intentionally chosen to be low to insure solubility in the polar basic developers. The photoactive species also acts as a dissolution inhibitor, that is, it prevents development in the unirradiated regions of the film needed to resist (i.e., mask) further processes. The leveling agents prevent undulations on the resist surface by presumably plasticizing the resin or by providing a resist solution with lower surface tension to improve wetting at wafer spin.

2.1.2 PAC Influence. Photoactive compounds, or sensitizers, are usually naphthoquinone diazides (i.e., PACs) like those pictured in Figure 3(7)(8). The diazide (DAQ) moiety of this molecule absorbs in the visible region of the spectrum, but most importantly, it undergoes a photochemically-induced radiation chemical reaction, the photoelimination of the azo nitrogen, that results in a solubility change in the dissolution inhibitor photoproduct (Ref. 1-4 and references therein). It is this energy

RESIST AND ETCH PROCESSING SEQUENCE

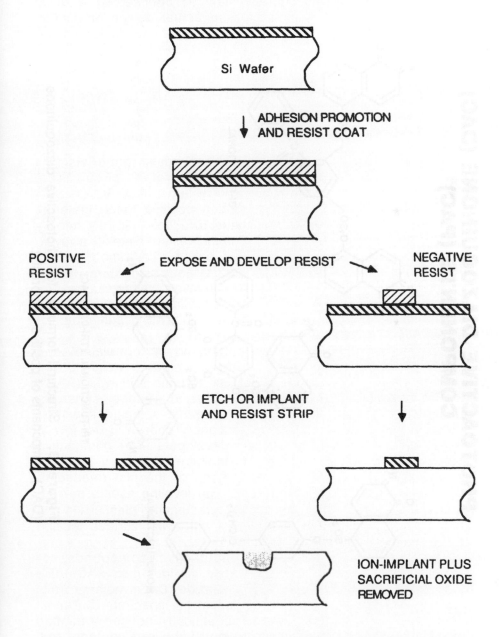

Figure 2. Resist and etching processing sequence relevant to the fabrication of VLSI circuits.

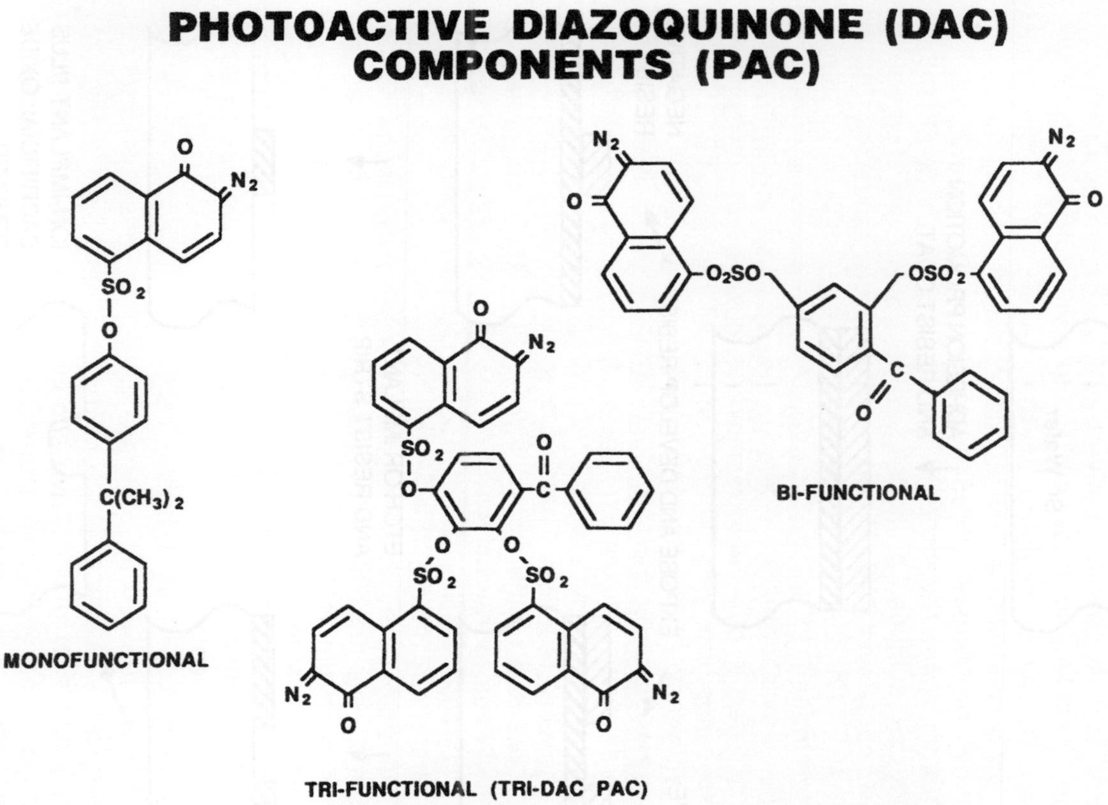
PHOTOACTIVE DIAZOQUINONE (DAC)
COMPONENTS (PAC)
MONOFUNCTIONAL
TRI-FUNCTIONAL (TRI-DAC PAC)
BI-FUNCTIONAL
SO2
O
C(CH3)2
N2
O2SO
OSO2
C

Figure 3. Structural formulae for photoactive diazoquinone (DAC) components of positive photoresist.

conversion process from electromagnetic energy to chemical reaction product which results in the observed resist behavior.

It turns out that this conversion process is fairly efficient as determined by basic quantum efficiency measurements for some PAC's. This quantity, defined as the ratio of the number of molecules reacting to photoproduct to the number of photons absorbed, $\emptyset$, can be as large as 10^6. Values larger than 1 are usually associated with a free radical chain reaction mechanism, while most photoresist photochemical reactions have values ranging from a few hundredths to a few tenths. The quantum efficiency for acetone, a model carbonyl-containing compound (i.e., C=O containing PAC) like those of Figure 3, was measured to be 0.17(9). The quantum efficiencies for the PAC's of Figure 3 were determined to all be about 0.3 at the typical optical exposure wavelengths(8). Actually, these values are quite high when compared to other energy conversion processes, thus, these photoprocesses are very energy efficient, roughly 30%. Even greater efficiency , 50%, has been observed for some resists by other researchers(10).

In the acetone example above, the light is being absorbed by the specific carbonyl chromophore group, which in turn leads to the chemical reaction. The first law of photochemistry, "only the light absorbed by the molecule (e.g., the PAC) can result in a chemical change in the molecule"(11), applies for this example and for PAC absorption in photolithography. The light, which is merely absorbed in the resin or the substrate and not at the specific chromophore, does not provide contributions to $\emptyset$. In other words, only the bleachable absorption of the resist over the exposure spectrum is important in the lithography (see Figure 4). Further, the sum of the quantum efficiencies must be 1, the second law of photochemistry(11), unless a chain reaction is involved. This definition stipulates the absorption of energy is a one-quantum process.

Absorption of light in the resist is given by the Beer-Lambert law: $I/I_o = 10^{-Ecl}$, where E is the molar extinction coefficient, c the chromophore concentration, and I the resist film thickness. Arden et al.(12) have shown high E can lead to poor resist image edge walls and larger CD variation, and should be judiciously chosen in designing the positive photoresist.

It is pretty clear the photochemical quantities of interest to photoresist design are $\emptyset$ and E. Both quantities can be measured empirically, as outlined in ref. 11. Resist sensitivity is influenced by phi, but E is merely a measure of the film absorption, and may not reflect absorption which leads to useful

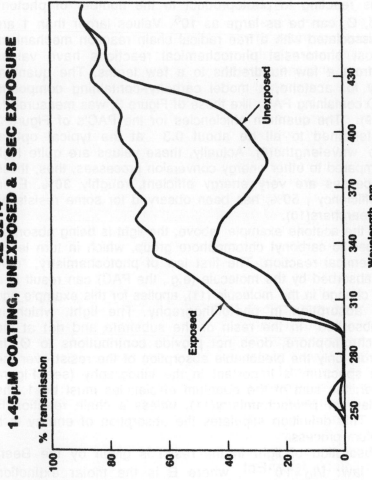

Figure 4. Bleaching curves for AZ 5214 mid -UV resists.
(Courtesy of AZ Photoresists).

radiation chemical change in the resist as a result of a photochemical reaction. For example, conventional photoresists have large E at wavelengths less than 300 nm, but are very poor resists at those wavelengths due to the high absorption of the novolac resin alone, regardless of the phi value of the PAC involved. Obviously, E must not approach 1 or the system is useless at those wavelengths, but must have some value intermediate (i.e., 0.3-0.5) so the "skin absorption effect" can be avoided. This ensures the resist image will be cleared to the substrate, and that the resist image edge wall will not be severely degraded (i.e., undercut) from normal due to the high resist absorptivity(4)(12).

The composition of the PAC can influence both the spectral response and the contrast or resolution of the resist. This is important because mask/reticle aligners operate at different wavelengths, therefore, the PAC must be designed for the wavelength characteristic of that particular aligner tool. Willson and coworkers (ref. 4 and references therein) have written key papers in the area of PAC design and have demonstrated successful PAC wavelength tuning through chemical synthesis. By adding chemical substituents to the PAC molecules at specific molecular bonding sites and by blending PAC's, they were successful at formulating a resist designed to be used with a Perkin-Elmer Micralign 500 lithographic exposure system operating at the mid UV (UV-3; 310 nm) region of the Hg lamp emission spectrum. It had bleachable absorption at the mid UV region, which was an indication the radiation chemical reaction of the diazonaphthoquinone molecule to the acid soluble product was occurring as required for image formation (see Figure 5).

Daniels and coworkers(13) have also shown the importance of PolyDAQ substitution (see Figure 3) of the PAC upon resist contrast or effective aerial image of the total resist system. In Figure 6, the theoretical polyphotolysis photoproduct modulation transfer function is compared to that provided by the phototool. The resist can be designed to provide image resolution better than the resolution limited tool performance, a result which is becoming more prevalent, that is, photolithography has gone from aligner limited with low contrast resists to resist performance or contrast limited. Furthermore, higher contrast resists and special resist processes are being developed to extend tool lifetimes in some cases.

Successful or high contrast positive resist design requires a non-linear response between exposed and unexposed resist. For any degree of polyphotolysis, q, the general dissolution rate is given by:

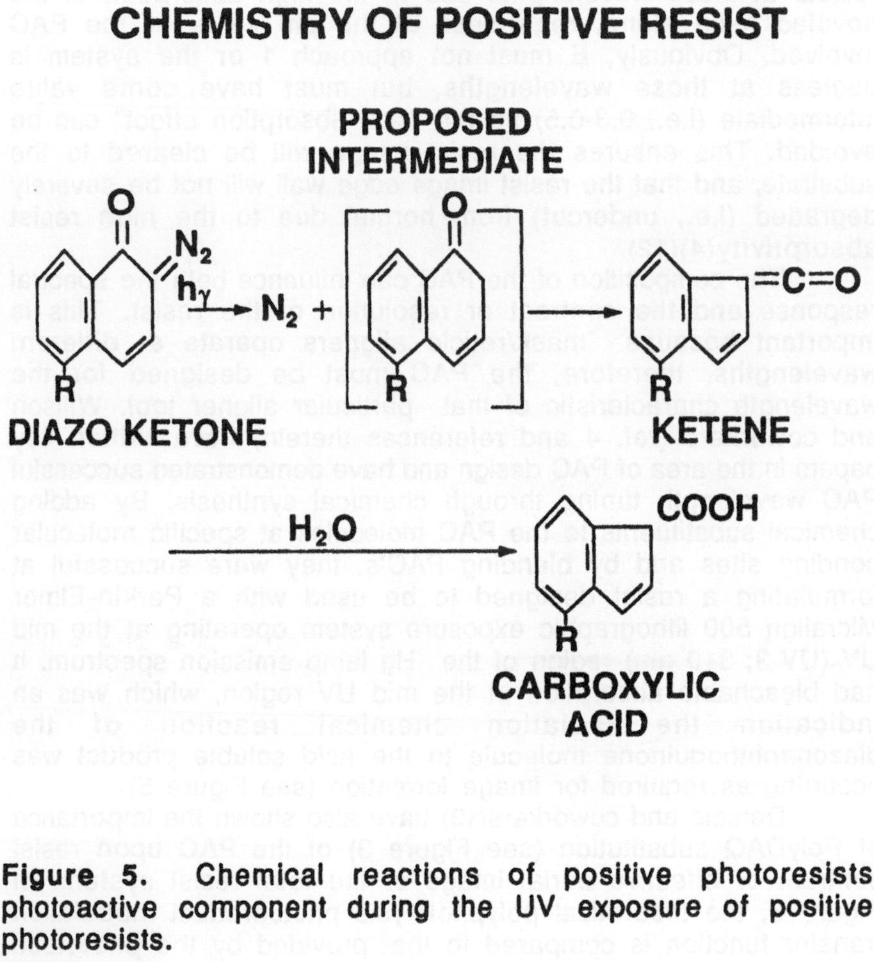

CHEMISTRY OF POSITIVE RESIST

Figure 5. Chemical reactions of positive photoresists photoactive component during the UV exposure of positive photoresists.

IMAGE MODULATION ENHANCEMENT

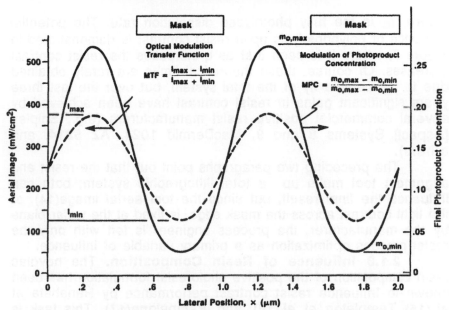

Figure 6.   Image modulation enhancement function or aerial image vs. lateral position for positive photoresist exposure. (Courtesy of Shipley Co. and SPIE ref. 13).

THEORETICAL CHARACTERISTIC CURVES

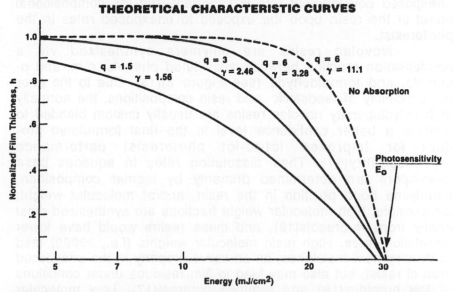

Figure 7.   Theoretical characteristic curves for positive photoresist assuming a polyphotolysis mechanism. (Courtesy of Shipley Co. and SPIE ref. 13).

[1] $R = r_0(1 - e^{-Ec})^q$

where, r_0 is the fully photolyzed dissolution rate. The potential influence of polyphotolysis upon resist contrast is demonstrated in Figure 7, where it is seen that as q increases the resist contrast increases. Of course, these theoretical limits are rarely obtained due to the complexity of the total system, but over the last three years significant gains in resist contrast have been achieved by several commercial positive resist manufacturers (e.g., Shipley [Aspect] Systems 8 and 9, MacDermid 1024, AZ 5214, and others).

The preceding two paragraphs point out that the resist and alignment tool make up a total lithographic system; both can influence the final result, but since the tool aerial image(14), or the light contrast across the mask edge, is fixed at the wafer plane by the manufacturer, the process engineer is left with only the resist process optimization as a primary variable of influence.

2.1.3 Influence of Resin Composition. The novolac resin composition of the positive photoresist formulation has been shown to influence resist contrast performance by Hanabata et al.(15) Templeton et al.(16) and Pampalone(17). This task is accomplished through resin conformational effects upon the dissolution rates at image development created by isomeric compositional effects which occur at resin synthesis. Not just the unexposed development rate is important, but the compositional effect of the resin upon the exposed to unexposed rates in the photoresist.

Novolac resins are polymers synthesized via a condensation reaction between substituted phenols, o,m, and p-cresols, and formaldehyde (see Figure 8)(15). Due to the poor reproducibility of feedstock and resin compositions, the normally high polydispersity novolac resins are usually custom blended to achieve a better confidence level in the final formulated pro-duct for improved lot-to-lot photoresist performance reproducibility(17). Their dissolution rates in aqueous base developers are determined primarily by isomer composition, methlyene bond position in the resin, and/or molecular weight; for example, high molecular weight fractions are synthesized most easily from m-cresols(18), and these resins would have lower dissolution rates. High resin molecular weights (i.e., ≥9000) lead to development resistance, an attractive property in the unexposed area of resist, but also may lead to film residues under conditions of low humidity(18) and reduced contrast(17). Low molecular

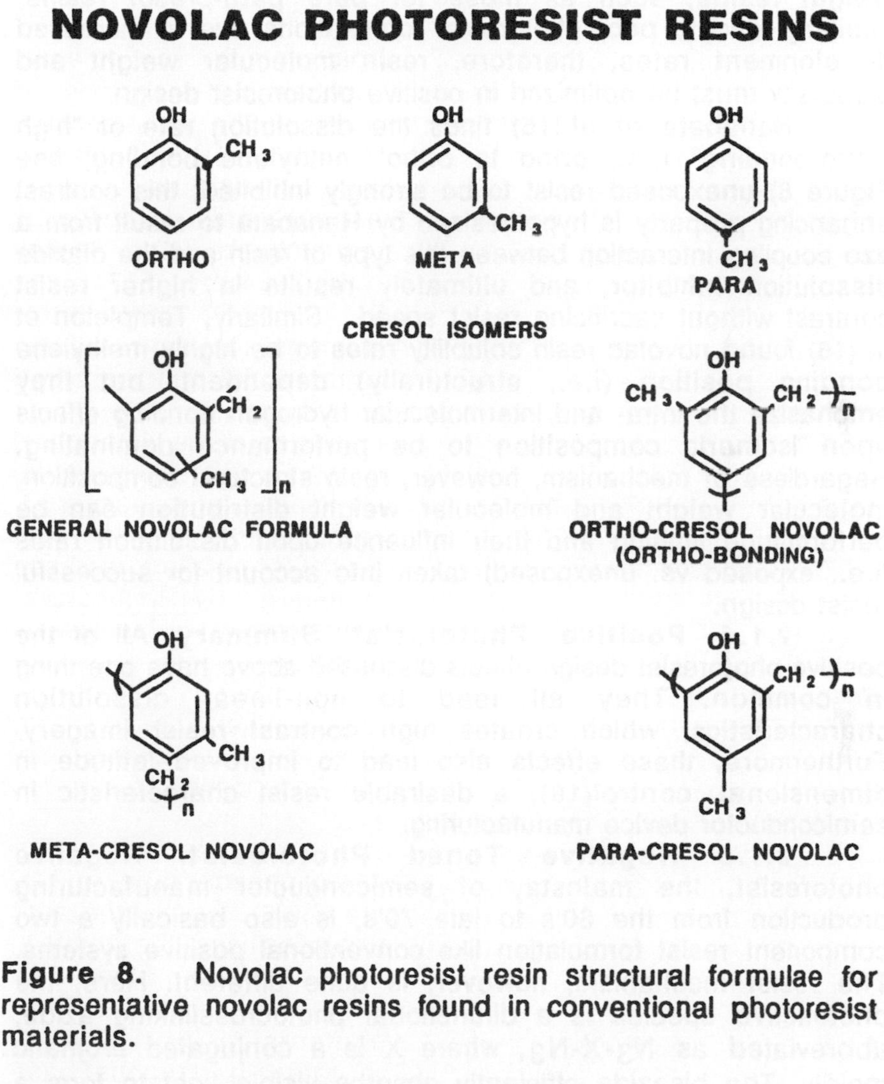

NOVOLAC PHOTORESIST RESINS

ORTHO   META   PARA

CRESOL ISOMERS

GENERAL NOVOLAC FORMULA

ORTHO-CRESOL NOVOLAC
(ORTHO-BONDING)

META-CRESOL NOVOLAC

PARA-CRESOL NOVOLAC

Figure 8. Novolac photoresist resin structural formulae for representative novolac resins found in conventional photoresist materials.

weight resins, such as those for pure para-cresol resins, similarly lead to poor photoresist formulations due to increased development rates, therefore, resin molecular weight and dispersity must be optimized in positive photoresist design.

Hanabata et al.(15) finds the dissolution rate of "high ortho-bonding" (i.e., ortho to ortho' methylene bonding; see Figure 8) unexposed resist to be strongly inhibited; this contrast enhancing property is hypothesized by Hanabata to result from a azo coupling interaction between this type of resin and the diazide dissolution inhibitor, and ultimately results in higher resist contrast without sacrificing resist speed. Similarly, Templeton et al.(16) found novolac resin solubility rates to be highly methylene bonding position (i.e., structurally) dependent, but they emphasize the intra- and intermolecular hydrogen bonding effects upon isomeric composition to be performance dominating. Regardless of mechanism, however, resin structural composition, molecular weight and molecular weight distribution can be performance limiting and their influence upon dissolution rates (i.e., exposed vs. unexposed) taken into account for successful resist design.

2.1.4 Positive Photoresist Summary. All of the positive photoresist design effects discussed above have one thing in common. They all lead to non-linear dissolution characteristics, which creates high contrast resist imagery. Furthermore, these effects also lead to improved latitude in dimensional control(19), a desirable resist characteristic in semiconductor device manufacturing.

2.1.5 Negative Toned Photoresist. Negative photoresist, the mainstay of semiconductor manufacturing production from the 60's to late 70's, is also basically a two component resist formulation like conventional positive systems. The resist mechanism, however, is quite different. Here, the photoactive species is a difunctional photocrosslinking azide, abbreviated as N_3-X-N_3, where X is a conjugated aromatic moiety. The bisazide efficiently absorbs visible light to form a very reactive nitrene, -N:, which is capable of chemically inserting into any C-H or C=C bond of the partially cyclized rubber resin to form an intermolecular crosslink between resin molecules. This crosslinking reaction creates a large increase (i.e., 2X) in the cyclized rubber binder polymer molecular weight every time two azide crosslinks occur, thus, decreasing the solubility rate of the optically exposed image substantially. The image is negative toned where the light strikes and remains following development, because of increased molecular weight due to the photocrosslinking reaction vs. the unexposed area which is completely developed away (see Figure 2).

Negative resists are designed by controlling the degree of partial cyclization of the resin and by extending the conjugation of the bis-arylazide sensitizer(3). Control of the resin cyclization reaction is thought to influence the resin molecular weight distribution which in turn influences resist contrast, while the degree of conjugation of the azide-chalcone chromophore determines the spectral wavelength absorption chacteristics of the crosslinking azide sensitizer. These photoresist formulations are generally very sensitive, because the bisarylazides have high quantum efficiency, where $\emptyset \sim 0.4$ to 0.7 and is a biphotonic average (i.e., $\emptyset_1 + \emptyset_2/2$)(20).

Unfortunately, negative resists do not withstand advanced dry etch processing (e.g., Applied Materials' reactive ion etchers) very well (see Figure 9), therefore, negative photoresists remain in use only in older production lines, where large design rules (i.e., large image sizes) are called for and wet isotropic etching is still acceptable from a process image dimension bias view. Negative photoresists suffer from low contrast generally (i.e., usually have contrast values ≤ 1.0), created at least partially by resist swelling effects which occur during development. The highest contrast negative photoresist tested at Motorola is Merck Selectilux with a contrast value of 1.7(21). Negative resists also suffer from oxygen sensitivity or reciprocity failure, which is manifested by a thinner resist image than expected due to a competing nitrene/oxygen reaction instead of the desired nitrene/resin reaction.

2.2   Deep UV Resists

Deep UV (DUV) resists, materials responsive to light with 100-248 nm wavelength, will be needed now. Lithographic alignment tools are being designed with excimer laser light sources, and beta-site machines are currently being delivered. Current photoresists are largely ineffective at these wavelengths due to the strong absorbance of the novolac resins and PAC's involved. Work on new resists has ocurred mainly in Japan(22), at AT&T Bell Labs(23) and IBM(24), but serious reservations about the device fabrication process compatibility of these new DUV model systems exists. Work with conventional positive resists has focused upon minimization of absorption through isomer resin synthesis(25). Recently, IBM workers have reported negative DUV Si-containing systems with very good properties, but

AFTER RESIST ASH REMOVAL

HUNT NEGATIVE RESIST (HNR) 100 NITRIDE ETCH TEST

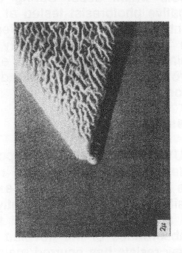

AFTER AME 8110 ETCH

Figure 9. Illustration of poor etch compatibility (reticulation or frying) of negative photoresist under AME etch conditions.

unfortunately, they require bi-layer RIE development and concerns of resist removal following etch processing remain. No commercial system has emerged to meet the projected device fabrication needs yet, but design activity for these materials at resist vendor labs is on the rise.

2.3 Radiation Resists

2.3.1 Introduction. The term radiation resist refers generically to materials that function under exposure to ionizing radiation, that is, radiation with short wavelegths such as soft X-rays, electron beams, and ion beams. Since only the relative sensitivities are changed when the radiation source is changed and not the resist process, the text in this entire chapter will be restricted to e-beam resists only. The basic resist mechanisms are unaffected by ionizing radiation source change even though the energy absorption mechanisms may change significantly. The main demand for these resists is in the area of e-beam fabrication of Cr- patterned glass or quartz photomasks and reticles(26).

For these resists, the energy is absorbed much differently than for photoresists. Here, enough energy is available to cleave any bond in the resist and initiate any possible reaction, where for photoresists only absorption at specific chromophoric sites in the resist can result in chemical reactivity. Even though the energy absorption is more uniform with depth for radiation resists and seemingly nonspecific, the radiation chemistry results are surprisingly very specific, hence, design criteria exist and will be reviewed.

2.3.2 Energy Absorption Considerations. Resist atomic composition can infuence both resolution and energy absorption. For electrons with 10-30 keV energies typical to e-beam lithography, the energy loss per unit path length is linear with resist density, the quantity Z/A (where Z and A are the average atomic and mass numbers for the resist) and the term ln E(27). The energy absorbed depends primarily upon the energy of the beam, but resist compositional effects are best for resists with the greatest H content, where Z/A is greatest. Since Z/A approaches 0.5 rapidly for elements with $Z>5$, resists with high atomic numbered compositions actually may have reduced energy transfer per unit length penetrated vertically by up to ~ 30%. More importantly, lateral scattering and backscattering effects, effects which ultimately limit resolution in e-beam lithography, increase dramatically for higher atomic numbered substituents and hence they should be avoided by resist designers(28)(29).

2.3.3 Positive resists. Positive resists have historically been designed as single component polymer type systems. The most classic examples are poly(methyl methacrylate) and poly(butene sulfone), PMMA and PBS, respectively. Their radiation chemistry was well known previous to their application as resists. Both were known to degrade, that is, produce large quantities of gaseous radiation byproducts, CO and CO_2 and SO_2, as well as exhibit reduced molecular weights under gamma-ray and e-beam exposure. In fact, both do function as positive e-beam resists[30] and[31], and the latter material is the major positive e-beam resist in use today, mostly in mask shop applications. In both resists, the design principle is the incorporation of groups that are thermodynamically favored to be split out of the molecule when irradiated; for PBS that group is in the main polymer chain while for PMMA it is in the ester side chain of the molecule. Later positive resist molecular designs involved derivatives of PMMA, namely substituted acrylates and methacrylates. These systems are represented by the general vinyl polymer structural formula:

[2] $-[CH_2-CX(CO_2Y)]-$

where, X could be CH_3 as in PMMA or any electron-withdrawing group such as a halogen or CN group, and Y could be any alkyl or halogenated alkyl group.

When polymers with this general structural formula are employed, their G_S values, polymer main chain bond scission yields/100 eV exposure dose, for degradation are large and their G_X values, intermolecular bonds formed (i.e., crosslinking) between polymer molecules/100 eV dose, are nearly zero in many cases[32][33]. This term, G_S, is analogous to $\emptyset$ for photoresists. It determines to first order the sensitivity of the resist and is approximately inversely proportional to the e-beam sensitivity[28]. Actually, the most important parameter for determining resist tone is G_S/G_X. When G_S/G_X is ≥ 4, the resist will predominantly degrade and hence be a positive resist.

The ultimate in substituted acrylates was designed/synthesized, and is currently being manufactured, by Toray in Japan as EBR-9, where X is Cl and Y is CH_2CF_3. This resist rivals PBS as a sensitive positive e-beam resist for mask making purposes, because both the X and Y substituents increase the radiation susceptibility of the resist[28].

The methacrylate polymer resist materials are characterized by fair to good sensitivity, and poor to good contrast or resolution. Unfortunately, all of these resists suffer from poor dry etch compatibility and their use in lithography is restricted to mask-making only (see Table I). This market is, however, reasonably large since most masks require a master reduction reticle for stepper repeater fabrication or are MEBES master plates themselves .

Direct-write e-beam positive resists, resists used to fabricate devices directly on silicon wafers, must be heartier or more dry process compatible(6) than the mask making resists above. As such, they have been restricted to two basic resist types: conventional positive photoresists and novolac/sulfone copolymer blends. Table II contains a representative list of these systems and their resist characteristics. Generally, these resists are less sensitive but are dry process compatible, and semiconductor devices can actually be made directly with them on silicon wafers.

Fahrenholz et al. of AT&T(34) and Hunt Chemical researchers(35) have been actively involved in designing novolac-based positive e-beam resists for direct-write circuit fabrication. The AT&T activity has focused upon two component polymer blend systems where the dissolution inhibitor is a poly (alkene sulfone) like PBS(36) and the novolac binder resin is designed to actually degrade with the inhibitor and have minimal concomitant crosslinking(34). The idea here is to minimize the competing crosslinking reaction from the resin which tends to counteract the positive action occurring in the degrading sulfone. Novolacs with bulky sustituents on the phenyl ring, n-propyl, sec-butyl, and phenyl, all produce positive acting novolac resins without any dissolution inhibitor at all. Although images of these uninhibited resins do not clear to the substrate without extensive resist loss in the unexposed areas, they provide a substantial advantage over more conventional resins that crosslink extensively over the entire dose range, which tends to conteract the positive behavior in the two component resist. Other conditions which promote positive behavior are post exposure pre-develop exposure to air (oxygen), higher resin molecular weights (i.e., limited due to solubility), and stronger base developers. As with the AT&T resists, the Hunt system also contains a proprietary resin which participates significantly to the positive e-beam behavior.

2.3.4 Negative E-beam Resists. Like positive resists, these resists can be conveniently classified between mask-making and direct-write (see Tables III and IV). Basically, these resists are direct-write compatible when they are aromatic in nature, that is, when they are polystyrenes or naphthalenes or their derivatives, and mask-making compatible when they are

TABLE I: METHACRYLATE-BASED POSITIVE E-BEAM RESISTS

RESIST	SENSITIVITY $80\ \text{MicroC/cm}^2$	CONTRAST	DRY-PROCESS COMPATIBILITY
PMMA (IBM)	80	4	NO (see Table IX)
EBR-9 (JAPAN)	12	1.4	NO
FBM-120 (JAPAN)	5	2	NO
HP POS CROSSLINKED PMMA	40	3	NO
PBS (AT&T)	1	2	NO
PMCN (ARMY)	12	1	OK
PMCA (ARMY/HONEYWELL)	16	NOT MEASURED	NO

TABLE II: NOVOLAK-BASED POSITIVE E-BEAM RESISTS

RESIST	SENSITIVITY	CONTRAST	DRY-PROCESS COMPATIBILITY
HUNT-204	50 MicroC/cm^2	2.9	YES(0.5)
PC-129	200	4	YES(0.2)
HITACHI NPR	14	0.7-0.9	YES
AZ-2400(IBM)	10	-	YES IF DUV
	25	2.5	
	50	1.6	
HUNT-1182	25	1.3-1.6	UNKNOWN
ALLIED-6010	25	1.2-4	YES
HUNT WX-214	15	2.5-4	YES IF DUV

TABLE III: MASK-MAKING NEGATIVE E-BEAM RESISTS

RESIST	SENSITIVITY	CONTRAST	DRY-PROCESS COMPATIBILITY
AT&T COP	0.8 MicroC/cm^2	1.0	NO AND SWELLING
SEL-N (JAPAN)	2	1.2	NO AND SWELLING
KODAKS	0.5	0.8-1.3	NO PLUS SOME SWELLING PLUS EDGE-SCALING
CONVENTIONAL PHOTORESISTS	1-3	0.7-1.2	NO AND SWELLING

TABLE IV: DRY-PROCESS COMPATIBLE NEGATIVE E-BEAM RESISTS

RESIST	SENSITIVITY	CONTRAST	DRY-PROCESS COMPATIBILITY
POLYSTYRENE	80 MicroC/cm^2	2.0	YES(0.05)
CMS-EX-S (JAPAN)	1.7	1.6	YES(<0.1)
MES-E (JAPAN)	1.5	1.5	YES(<0.1)
CMS-EX-R (JAPAN)	17	2.2	YES
ALPHA-M-CMS-S (JAPAN)	10	2.0	YES
AZ-1450	100	1.4	YES(<0.5)
-1450(+300)	160	2.4	
-1350(+300)	60	3.2	
GMC (AT&T)	4	1.6	YES

vinyl polymers without any unsaturated bonding except at the polymer crosslinking site.

Mask-making resists generally possess very good sensitivity, but low contrast or resolution (see Table III for examples). Their applications are restricted to making 5X or 10X reduction reticles where larger features (>4-5) microns are required. These resists withstand wet etching of thin chromium films, but also require descum processing prior to etch. These resists are basically useless for device fabrication applications using direct-write e-beam due to their poor plasma etch resistance(6). This is due to a general lack of selectivity to harsh RIE treatments and swelling behavior exhibited by these materials at feature sizes below 1 micron, the size domain where direct-write e-beam techniques are of need to improve circuit packing density.

The dry-process compatible resists shown in Table IV generally possess reduced sensitivity, but with higher contrast and resolution. The sensitivity trade-off, however, is not completely prohibitive (i.e., reasonable exposure levels <10X10^{-6} Coul/cm^2 can be employed). Most importantly, these materials are less susceptible to pattern swelling during development, and submicron images are easily obtained. As for the mask-making resists, these materials must also be descummed after development for best resolution performance. Figure 10 illustrates the effect of the oxygen RIE descum on the negative resist image foot at the base of the example image. The alpha-M-CMS, CMS, polystyrene, and AZ tone-reversed positive photoresist systems have all been used in direct-write applications to fabricate high performance MOS and Bipolar circuits.

Negative e-beam resists are designed by incorporating radiation crosslinking groups into usually single component vinyl polymer resists(1)(2). These appendage groups range typically from alpha hydrogen or halogen, to side chain epoxy(37) and allyl(38), to halogenated alkyl groups attached to styrene(39) or acrylate esters(40) (see Figure 11). These groups are all very radiation susceptible and design incorporation into the resist leads to easily crosslinkable polymers with high G_x values, and hence, good e-beam sensitivities($\leq$5X10^{-6} C/cm^2).

Recent advances in negative E-beam resist design have been in the areas of polymer blending(41) and chemical amplification(42). The blending technique, similar to that for two-component positive e-beam resists, allows the preparation of a sensitive high contrast resist from a polymer with poor

AZ 1350/1450 TONE-REVERSED NEGATIVE IMAGES

WITH DESCUM   WITHOUT DESCUM

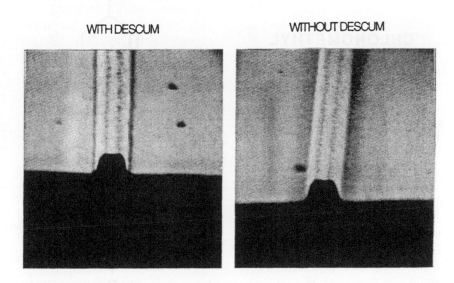

1 MICRON

Figure 10. Illustration of line edge improvement via reactive-ion-etch oxygen descum processing on negative e-beam resist images.

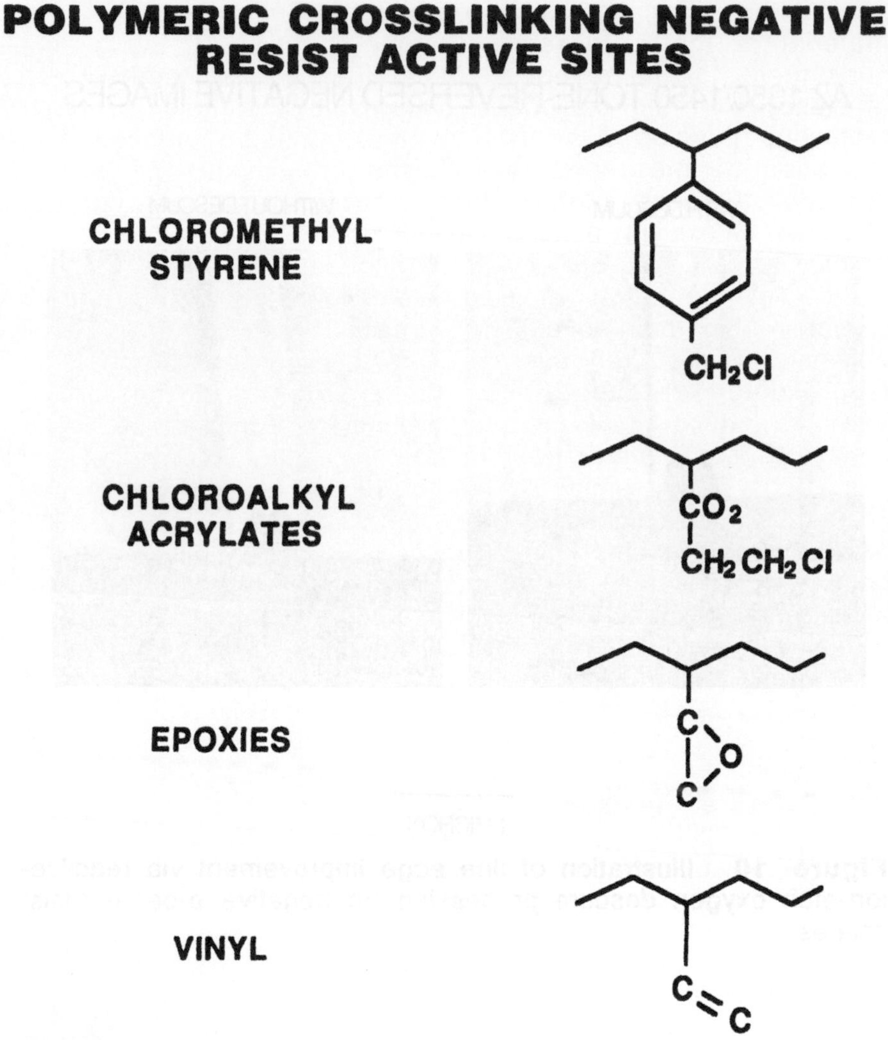
POLYMERIC CROSSLINKING NEGATIVE RESIST ACTIVE SITES
CHLOROMETHYL STYRENE
CH2Cl
CHLOROALKYL ACRYLATES
CO2
CH2 CH2 Cl
EPOXIES
C
O
C
VINYL
C
C

Figure 11. Polymeric crosslinking active sites for negative e-beam (or radiation) resists.

sensitivity.   The sensitivity requirement for the insensitive material is that it possess a good electron donating ring substituent for improved H-abstraction induced crosslinking efficiency with the coblended chloromethylstyrene. The latter technique developed at IBM involves radiation induced acid formation in the resist to catalyze crosslinking or induce degradation, hence, positive and negative behavior can both be designed. An example negative behaving system is now marketed by Shipley as SAL 601 EBR-7 or later improved derivatives(43).

The limitation of negative e-beam systems stems primarily from their advantage. Sensitive crosslinked or gelled polymers are also very susceptible to developer solvent swelling (see Figure 12) due to the three dimensional crosslinked networks formed in the irradiated polymer regions. Hence, these resists generally suffer from reduced resolution and are highly susceptible to proximity effects (cooperative exposures which occur due to backscattered electrons from adjacent lines). As a result, these systems will probably always be limited to high pattern area coverage layers requiring somewhat lower resolution as is typical of many metal layer interconnect requirements. The new Shipley acid-catalyzed resists, however, are less susceptible to this resolution limiting effect because they are base developed novolac systems (recall: positive novolac resists are non-swelling).

2.4   Future Resists

Conventional positive photoresist technology combined with multi-level processing   techniques (see applications and special processes section) will inevitably   allow the lithographic community to achieve 0.5 micron or below design rules without the extensive use of resists responsive to ionizing radiation(44). The future of optical lithography beyond 0.5 micron will require extensive development in the areas of Deep UV resists, where new resin and PAC or single component resists, and even new resist mechanisms will be   required. New advances, such as those demonstrated in DUV binder resin design by Turner et al.(45) and the chemical amplification resist mechanism by Willson et al.(42) or combinations of theoretical lithographic simulation(46) and statistical design of experiments(47) to achieve the ultimate resist design as proposed by Monohan et al., will evolve to provide the 4th or 5th generation photoresists for the 1990's. These new resists will probably be employed as top layer imaging systems for multi-layer resist processes, emphasized due to the sharp

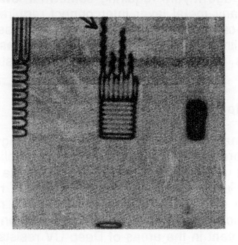

NEGATIVE E-BEAM RESIST DEVELOPMENT SWELLING

COP

PCMS

Figure 12. Optical micrographs illustrating snaking behavior, solvent developer, swelling behavior, of negative e-beam resists.

increase in numerical aperture or decreased depth of focus of proposed advanced optical lithographic systems(44).

3.0   RESIST PROCESSING

3.1   Resist Parameter Screening

Before a resist, commercial or otherwise, can be instituted into a device fabrication process flow, it must demonstrate high performance characteristics to a battery of fundamental tests. Completing these tests by no means provides an optimized resist process to be plugged into the production line. That comes later, but the results do provide the processing engineer on the IC fab line a basis for selecting one resist over another in a quantitative impartial way. Usually, the results for a new material are compared to those of an existing baseline process, whose capability may have become insufficient at one or more critical levels.

Usually, the resist vendor has spent a lot of time selecting a suitable developer for the resist to be tested, and that recommendation should be used at least initially for performance screening purposes. The investigating engineer, however, should make clear to the vendor what performance is actually being sought, that is, high contrast, speed, metal-ion free base developer or not, thermal stability, whatever. The device requirement will usually dictate the type of developer selected for testing.

3.1.1 Sensitivity and Contrast. Actual sensitivity curves are found in Figures 13 and 14 for example negative and positive systems, respectively. For these curves to have meaning, the experimental data of developer type or composition and concentration, development time and conditions, image dimension size, and the resist characteristics such as thickness must be known. The resist sensitivity for the negative system in Figure 13 is defined by that dose where 50-80% (i.e., the data may be fit to any point between 0 and 50 to 80% film retention but should be explicitly specified) of the original resist thickness is maintained; of course, the unexposed areas of resist are completely developed away and only the exposed images remain. For the positive system, sensitivity is defined as the dose where the image is cleared to the substrate (i.e., dose where $I_0-I_d/I_0=0$), a unique point from the data plot.

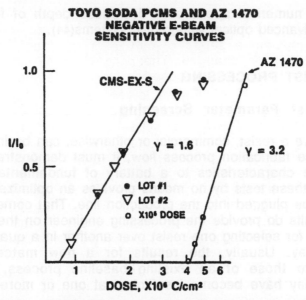

Figure 13. Characteristic sensitivity curve for two negative e-beam resists. The AZ resist is imaged density reversed with high e-beam exposure combined with a 300 mJ/cm^2 uv-4 flood exposure prior to development.

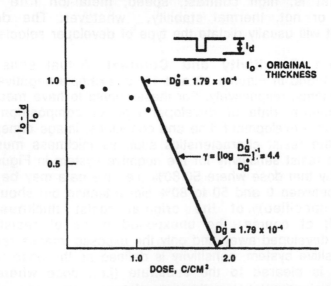

Figure 14. Characteristic sensitivity contrast curve for a positive e-beam resist example.

It should be noted here the negative resist sensitivity curve is essentially invariant with development time, where for the positive system a series of curves is obtained when the development time is changed. These curves will remain fairly parallel until appreciable unexposed resist loss begins to occur. The resulting curve never reaches I_0-I_d/I_0 values of 1.0, even at low exposure. Both negative and positive sensitivity curves shift on the exposure axis when the image CD is changed. Most lithographic tools allow for exposure to be varied across the wafer, so this data can be generated on a single wafer if desired.

Resist process contrast is given by the slope of the least squares fit of the data in Figures 13 &14. The contrast of the process is obtained by simply subtracting the log values as depicted on the figures and taking the mathematical inverse. These values are of importance because they can be used to predict edge wall angles and resist resolution(48).

To better clarify positive resist sensitivity, researchers at IBM(30) have adopted a modified sensitivity plot where I(unexposed)/Io is plotted vs. exposure (see Figure 15). Here, each data point represents an entirely different wafer and development time combination for the chosen CD to develop. Note, it would take a series of exposure response curves like Figure 14 to generate a single curve like Figures 15a and b. Since for positive resists, the full resist thickness is usually required for further process masking requirements, this method of resist sensitivity measurement is of great value. The slope of these curves, however, is not a direct measure of the image dose response and cannot be used as a measure of resist process contrast.

3.1.2 Resist Image Edge Wall. When reactive-ion etching (RIE) dry etching techniques are employed, the resist image edge wall becomes an important evaluation characteristic and should be measured. A vertical resist image edge wall is essential to RIE to minimize etch bias even for the more anisotropic processes. Unfortunately, angle measurement requires a Scanning Electron Microscopy (SEM) picture be taken edge-on with a cleaved and mounted wafer piece, which is a tedious procedure. The edge wall can be preliminarily estimated from the bulk contast as measured above(48), but in the final analysis the edge wall must be verified. Near vertical resist image edge walls (i.e.,87-90°) are specified in nearly all modern facilities, thus necessitating this resist process characteristic be known to be employed as a criterion for resist/process selection (see Figure 16 for examples).

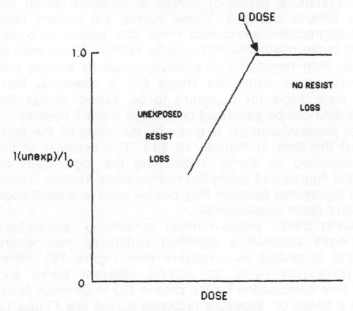

Figure 15a. Positive resist sensitivity curve for positive e-beam resists.

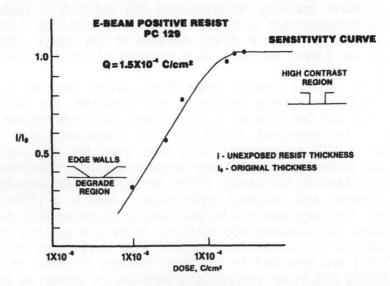

Figure 15b. Actual e-beam positive resist sensitivity curve for PC-129 positive photoresist.

RESIST IMAGE EDGE WALL PROFILES

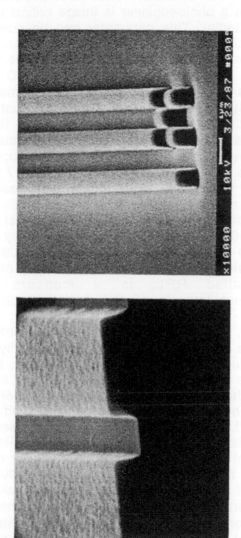

POSITIVE RESIST

NEGATIVE RESIST

Figure 16. SEM micrographs of positive and negative resist images illustrating the results of image edge wall profile angle measurement.

3.1.3 Resist CD Latitude. The most important resist process parameter to a photoengineer is image critical dimension control or CD latitude(20). This parameter can be quantitatively measured for a resist/developer combination as set forth by Walker and Helbert(49). In this method, a quantity called Δ, which is a measure of the difference of the resist image CD from that nominally exposed, is plotted vs. exposure and development time to obtain a series of characteristic curves. Figure 17 illustrates this for a positive photoresist example. Each point on this graph represents a different exposure and development time combination, which results in a cleared or fully developed image to the substrate. The solid uppermost line of Figure 17 is a least squares fit of data where the resist image just clears to the substrate; there is no data above this line because it would represent images that did not fully develop. The largest exposure on the horizontal Δ=0 line is an equivalence sensitivity condition for every plot, where sensitivities for resists can be compared on the basis of equivalent CD transfer. The |slope| of the just clearing line, called RPL, is a measure of the exposure/development latitude of the resist process being evaluated.

In Figure 18, a strong developer has been intentionally employed to illustrate the effect upon RPL and sensitivity; here an undesirably large value for RPL is obtained. Obviously, the flattest Δ, or CD, vs. dose curve is most attractive from a process stability view, and the resist process yielding that performance characteristic should be implemented onto the IC fab line.

Tables of RPL results for representative first, second and third generation resists are found in Tables V-VIII. Since these results are dependent upon testing conditions, they are not meant to be absolute as to the performance of the resists tabulated, but they are of interest for gathering some general performance trends. Second generation resists tend to be more sensitive than first generation materials, while third generation resists tend to have greater contrast as well as sensitivity without sacrificing CD latitude in the process. The effects of developer concentration and development method upon sensitivity and CD control are also evident from these tables. Developer concentration increases can lead to greater sensitivity (Table V), but usually at the expense of latitude. Development method (see Table VIII and Figure 19) can influence process contrast and should be investigated when feasible.

Figure 20 compares two real e-beam gate processes actually employed to make CMOS transistors. The process on the left had a severe exposure latitude problem, and as a result

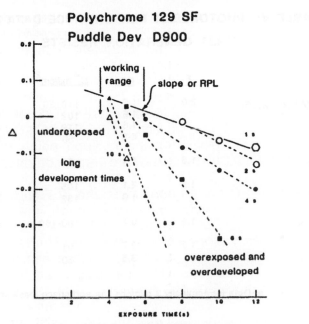

Figure 17. Characteristic sensitivity curve for positive photoresist employing CD transfer as a criterion.

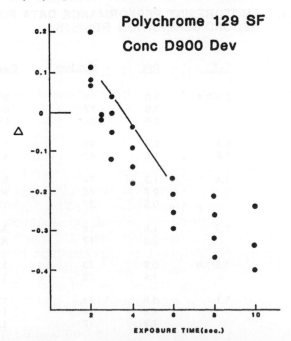

Figure 18. Delta vs. exposure characteristic curve for positive photoresist with a concentrated developer.

TABLE V: PHOTORESIST PERFORMANCE DATA FOR
FIRST GENERATION RESISTS

Resist	γ	RPL	Q_a mJ/cm^2	Developer /Method
PC-129 (Allied P2025)	2.0	1.0	110±20	D900/DIP
		1.6	102	D900/PUDDLE
		1.6	119	D910/DIP
		4-8	50	CONC 900/DIP
AZ-1350	1.5	1.4	81	1:1 MF 312/DIP
KTI-II	1.8	2.1	105	DE-3/SPIN-SPRAY
		1.0	132	DE-3/DIP
Kodak-809	1.6	0.4	100-160	809 Developer
HPR-204	1.7	2.7	91	1:1 LSI/DIP
		3.5	60	1:1 LSI/PUDDLE

a Data for nominally 3 microns line and space mask images.

TABLE VI: PHOTORESIST PERFORMANCE DATA FOR
SECOND GENERATION RESISTS

Resist	γ	RPL	Q,mJ/cm^2	Developer/Method
Allied P-5019	2.5-3.8	1.0	13	2:1 D100/DIP
		3.9	72	D150/DIP
		3.0	95	LM1500/DIP
AZ-1470	1.8	2.3	40	1:1 MF312/DIP
-4110	1.6	1.5	43	1:1 MF312/DIP
OFPR-800-1	1.4	1.3	74	NMD3-1/DIP
-800-2		2.7	75	NMD3-2/DIP
-800-1		0.5	37	1:1 MF312/DIP
KMPR-820-AA	1.2	1.6	18	AA-0980-52 /DIP
-820		2.2	17	AA-0980-52/DIP
HUNT-159	1.5-1.6	0.9	73	1:3 LSI/PUDDLE
-118		1.4	62	1:3 LSI/PUDDLE
MAC-9564	1.1	3.8	24	1:1 9562(MF)/DIP
		4.2	24	1:1 9564/DIP
-9574		1.9	12	1:1 9571/DIP

TABLE VII: 3RD GENERATION RESIST RESULTS

RESIST	$Q, mJ/cm^2$	RPL	γ	DEVELOPER
ALLIED 6010	83	1.1	1.3	LMI-600/DIP
	89	2.3	2.3	D-360/DIP
AZ 4110	45	1.0	1.5	1:1 MF 312/DIP
5214	62	2.1	2.3	1:1 MF 312/DIP
DYNACHEM				
XPR 1000	52	2.2	2.4	NMD-3/PUDDLE
1501	120	1.2	1.9	
MACDERMID 914	250	1.5	2.4	1:1 MF-62/DIP
KTI 9000	72	2.1	1.7	DE-3/SPRAY
JSR 3003A	60	0.8	1.6	JSR DEV/DIP
TOKYO OKA ONPR 830	40	0.6	1.7	NMD-3/DIP
KTI 9000	72	1.2	1.7	DE-3 Spray
Monsanto RX7	94	2.0	1.9	MFD Dip
Sumitomo PF6200	80	0.9	1.5	SOPD Dip
OFPR 800	75	1.3	1.4 (1.6)*	NMD-3/PUDDLE

*DYNACHEM DATA

TABLE VIII: DEVELOPMENT METHOD EFFECT

RESIST	METHOD	RPL $(mJ/cm^2)^{-1}$
KTI-II	SPIN/SPRAY	2.1
	DIP	1.0
HUNT-204	SPIN/SPRAY	SURFACE SPOTS
	DIP	2.7
	PUDDLE	3.5
PC-129SF	SPIN/SPRAY	SEVERE SURFACE SPOTS
	DIP	1.0
	PUDDLE	1.6

DEVELOPMENT   METHODS

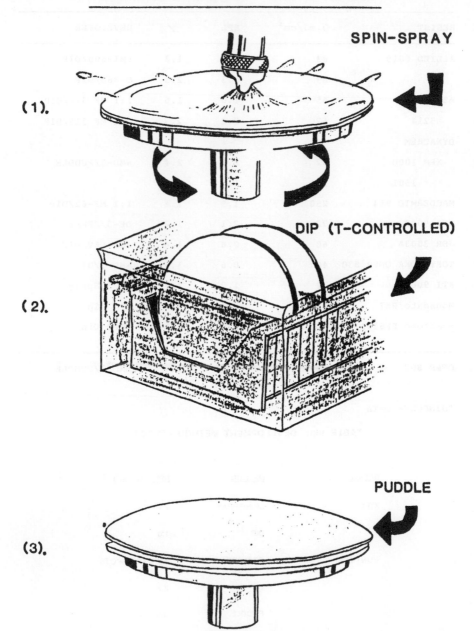

Figure   19. Development methods used to create relief images in photoresist materials.

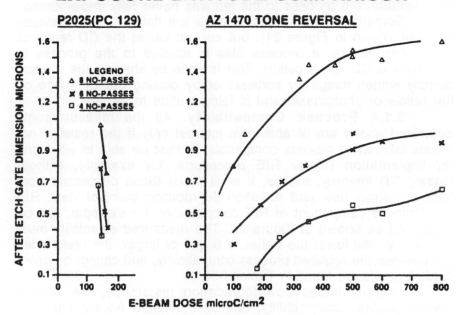

Figure 20. Exposure latitude comparison for two electron beam gate processes. The process to the left employed positive resist with unexposed pattern islands, while the process to the right is the same as that for the Figure 13 AZ process.

rework rates for this process were between 40-60%; due to this poor latitude, the process to the left was never fully implemented.
     Sometimes CD vs dose curves are flat (i.e., have latitude like that shown in Figure 21), but are not flat at the CD required. When this occurs, a process bias is applied to the process to achieve the CD specification. This is done by shaving the e-beam directly written image by software or by biasing the Cr image on the reticle or photomask, and is fairly routine to all fabs.
     3.1.4 Process Compatibility. All the measurements described above are of academic interest only if the resist is not device fabrication process compatible. It must be able to withstand Ion-Implantation (II) or RIE processes, for example, without loosing CD integrity, that is, it must resist these processes from both a thermal flow and radiation degradation point of view. RIE selectivity, a component of RIE compatibility, for example, can be measured as shown in Figure 22. The measured selectivity must be << 1, or the lower the better. If S is 1 or larger, the resist does not possess the required process compatibility, and cannot be used. Typical values are found in Tables IX & X.
     Thermal image flow/degradation resistance, the second part of process compatibility, can be estimated from polymer or resin differential scanning calorimetry (DSC) measured glass transition temperatures (tg) and thermal gravimetric analysis (TGA) measured decomposition temperatures, respectively(50). See Figures 23 and 24 for examples for pure novolac resin polymer. Resist manufacturers usually list these parameters in their technical brochures, and this additional data forms the rest of the data base required to make a resist process decision.

3.2   Resist Adhesion Requirements

     Before spinning resist onto the IC wafer in process, resist adhesion promotion is typically accomplished. This process, involving either liquid or vapor phase treatment of the wafer with hexamethyldisilazane(HMDS), has become an industrial standard. It is typically used at every lithographic step in the IC fab process, whether it accomplishes surface modification or not. HMDS processing can be carried out on automatic wafer tracks with liquid or vapor HMDS modules (e.g., GCA or SVG) or in stand alone microprocessor-controlled all stainless-steel commmercial reactor chambers (e.g., IMTEC or YES); these processes are proven for high volume production.
     Resist image adhesion is of paramount importance, because if IC circuit patterns are missing or are not the right dimension, the device will simply not function as designed. The resist systems

RESIST IMAGE CD versus DOSE

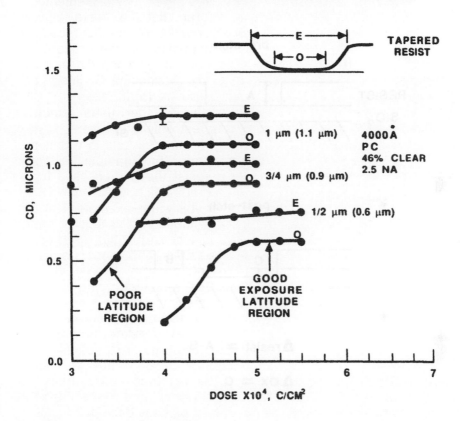

Figure 21. Resist image CD vs. dose for an example e-beam positive photoresist process.

ETCH RATE RATIO MEASUREMENT

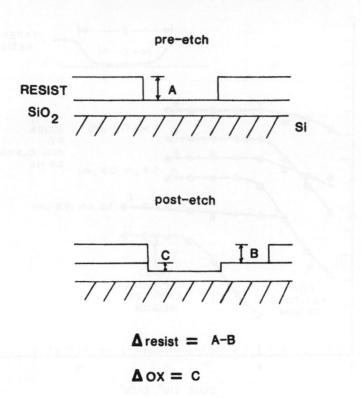

$$\Delta \text{resist} = A-B$$

$$\Delta \text{ox} = C$$

Figure 22. Etch rate ratio measurement methodology and calculations method.

TABLE IX: PLASMA ETCH RATIOS VS SIO$_2$ REFERENCE AND THERMAL PARAMETERS FOR VINYL POLYMERS

Polymer or Copolymer	Mole % ratio	t g °C	TGA °C	Plasma etch Rate Ratio
PMCN	100	120	335	0.3
PVDCN	100	165	260/500	insoluble
PVDCN-CO-MMA	52/48	155	330/500	0.4
	38/62	135	270/400	1.0
PMMA	100	107	275	0.9-1.2
PACAN-CO-MCN	50/50	131	210/500	1.0
	39/61	128	223/500	0.5
	11/89	123	216/500	0.4
PACAN	100	-	-	1.3
PCMMA	100	115	250	0.4-1.1
PCEMA	100	105	250/400	2.1
PCEMA-CO-MMA	50/50	102	293	1.4
PTFMAN-CO-MCN	12/88	123	357	0.3
PTFMAN-CO-MMA	32/68	98	295	0.3
PMCA	100	130;151	-	1.8
PTCEMA	100	123	305	2.3
PTCEMA-CO-MCA	66/34	137	286	3.5
PTCEMA-CO-MCN	50/50	130	310	0.4
PTCEMA-CO-TCECA	90/10	123	290	2.0
PTCECA	100	147	290	2.0
PTCECA-CO-MMA	29/71	130	325	1.6
PACTFEMA (EBR-9)	100	133	277	2.1
PMFA	100	131	404	0.4
PMBA	100	130	170	2.0
PBCEE	100	100	293	1.7
PBCEE-CO-MMA	25/75	78	275	1.0
	50/50	92	292	1.1
PBCME	100	72	320	2
PBCME-CO-MMA	50/50	100	310	0.9
PEMA	100	65	-	1.7
PAMBL	100	83	363	0.3
PCMS (Toyo Soda)	100	105	350	0.06
PS	100	103	322	0.1

Abbreviations and Structural Formulae:

CH$_2$=C(CH$_3$)CO$_2$CH$_2$CH$_2$CN
CEMA

CH$_2$=C(CH$_3$)CO$_2$CH$_2$CN
CMMA

CH$_2$=C(CN)$_2$
VDCN

CH$_2$=C(CH$_3$)CN
MCN

CH$_2$=C(Cl)CN
ACAN

CH$_2$=C(CF$_3$)CN
TFMAN

CH$_2$=C(Cl)CO$_2$CH$_2$CF$_3$
ACTFEMA

CH$_2$=C(Cl)CO$_2$CH$_2$CCl$_3$
TCECA

CH$_2$=C(CH$_3$)CO$_2$CCl$_3$
TCEMA

CH$_2$=C(Br)CO$_2$CH$_3$
MBA

CH$_2$=C(CO$_2$CH$_2$CH$_3$)$_2$
BCEE

CH$_2$=C(CO$_2$CH$_3$)$_2$
BCME

CH$_2$=C(H)C$_6$H$_5$CH$_2$Cl
CMS

CH$_2$=C⟨CO-O⟩
AMBL

TABLE X: POSITIVE PHOTORESIST THERMAL FLOW AND PLASMA ETCH RATE RATIO CHARACTERISTICS

Resist	TGA, ^{o}C	Image Flow Temperature, ^{o}C	PE
PC-129SF	125	140	0.2
KTI-II	133	160-170	0.4
OFPR-800	150	160-170	0.3
AZ-1350	140	140-160	0.5
AZ-2400*	145	140	0.5
KODAK-809*	-	120-130	0.3
HUNT-204	-	180	0.5

*AZ-2400, and K-809 "fry" (see Figure 9).

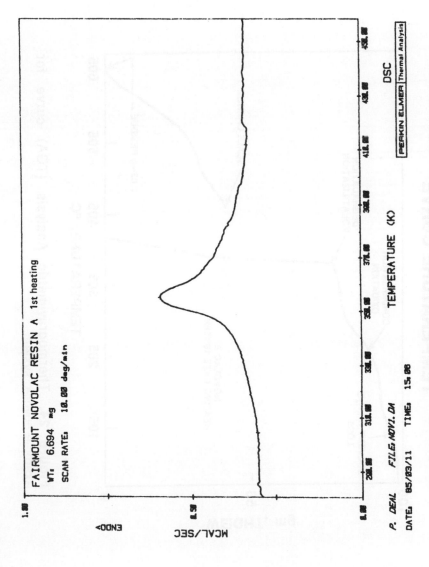

Figure 23. Differential Scanning Calorimetric (DSC) thermo-analysis curve for an example novolac resin.

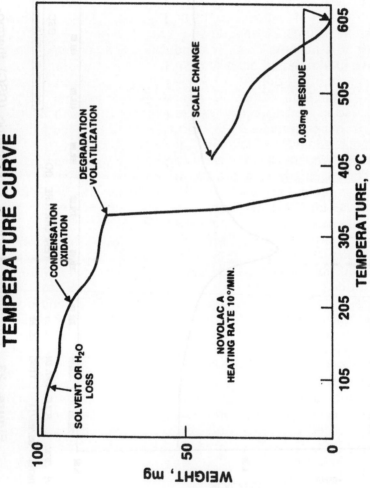

Figure 24. Thermogravimetric Analysis (TGA) curve for example novolac system.

most susceptible to resist image lifting at the development stage of processing are positive e-beam and conventional photo-   resists (51). Historically, negative photoresists also have required adhesion promotion, but in that case the goal was reduced undercut from wet etch processing and not the prevention of simple image lifting at develop(52)(51). Undercutting is less of a consideration for positive photoresists, because they are used primarily in conjunction with anisotropic dry etching processes. Examples of negative resist wet etch undercut and positive lifting at develop are shown in Figures 25 and 26.

What is accomplished by adhesion promotion treatments in IC manufacturing should actually be referred to as wafer substrate preparation, and not adhesion. Adhesion in the structural sense, as found in airplane composite material part attachment, is not accomplished in HMDS processing treatments. The term adhesion, as it is used here, refers to a more practical definition, that is, resist image adhesion. Figure 27 demonstates what is actually accomplished when a Si wafer is treated with HMDS. The ESCA(53) spectra shown clearly illustrate the removal of carbon containing adsorption species at lower binding energies in favor of the monolayer of trimethylsilyl surface reaction product(see Figures 27a and b)(54). In addition the surface is dehydrated in-situ as verified by increased water contact angle(55) and a lower O/Si ratio as measured by ESCA. Furthermore, this converted surface is stabilized for days against recontamination, therefore, the HMDS process provides a very stable surface for resist adherence. The Si 2p ESCA spectrum of Figures 27c and d verify the appearance of the trimethylsilyl silanol reaction species.

Substrate nonwetting, another adhesion problem, has been observed most frequently with mistakenly overpromoted wafers. It occurs after repeated treatments, when the wrong liquid treatment has been applied, or when vapor times exceed the optimum time for that respective substrate. It also can occur in selected circuit pattern areas and not for the whole layer, and can be characterized by higher water droplet contact angle. Although it is not generally well understood, it can be prevented by reducing prime times for vapor treatments, corrected by ion treatment of the wafer(56), and be prevented by using resist containing a spinning solvent of lower surface tension or by double resist application under dynamic spin conditions. The silicon based substrate layers, nearly all the layers encountered in IC device production except metal layers, can usually be successfully promoted against lifting by treatment with liquid silanes or silane vapor treatments at reduced pressures(52)(54)(56) (see Figure 28).

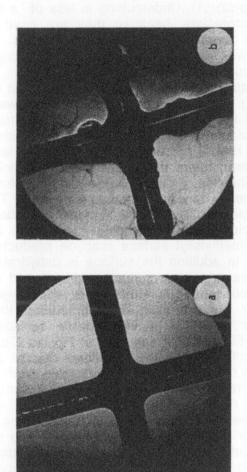

WET ETCHING ADHESION TEST ON Si(240X)

VTS   NO PROMOTER

Figure 25. Wet etching adhesion test of silicon–illustrating wet etch undercutting at the silicon resist interface during the etch process.

EFFECT OF ADHESION PROMOTER ON "LIFTING"

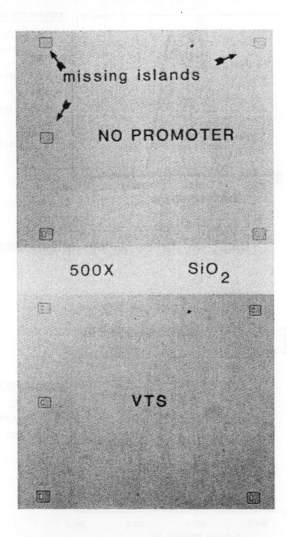

Figure 26. "Image Lifting" phenomena observed with and without adhesion promoter on SiO_2 test surfaces. This type of lifting occurs at development prior to etch processing.

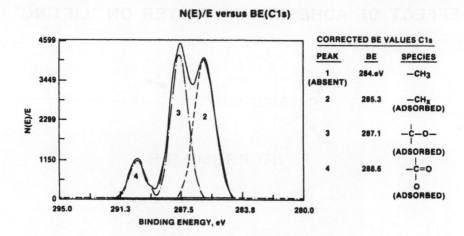

Figure 27a.   N(E)/E vs. BE for carbon 1s bare silicon wafer.

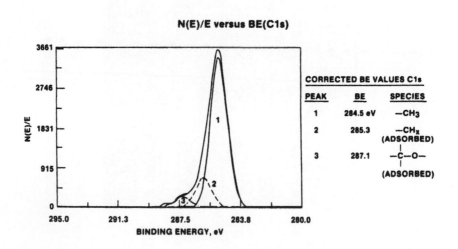

Figure 27b.   N(E)/E vs. BE carbon 1s for HMDS vapor (Star 2000) treated silicon wafer.

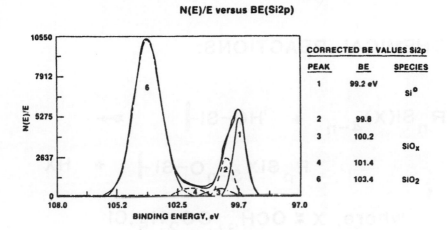

Figure 27c. N(E)/E vs. BE silicon 2p for untreated silicon wafer.

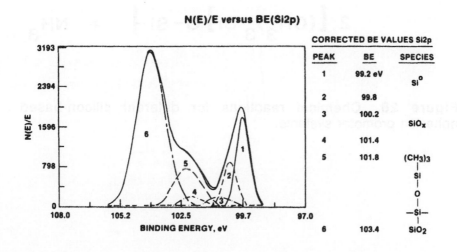

Figure 27d. N(E) vs. BE vs. for silicon 2p of HMDS treated silicon wafer.

CHEMICAL REACTIONS:

$$R_n Si(X)_{4-n} \ + \ HO-Si{-} \ \longrightarrow$$

$$R_n SiX_{3-n}O-Si{-} \ + \ HX$$

where, $X = OCH_3, OC_2H_5, Cl$

or $[(CH_3)_3 Si-]_2 NH \ + \ 2HO-Si{-} \ \longrightarrow$

$$2 \, [(CH_3)_3 Si]O-Si{-} \ + \ NH_3$$

Figure 28. Chemical reactions for different silicon-based adhesion promoter systems.

The carbon 1s   ESCA spectra are shown for a Si(100) substrate with native oxide (<50 A) in Figure 29 to illustrate the surface chemical changes between liquid and vapor promoter processes. The LPIII process, a vapor HMDS process, efficiently removes the carboxylic, etheral, and hydrocarbon impurities from the surface and replaces them with a blanket of trimethylsilyl groups comprising a monolayer (also see Figures 27 c and d).

The Mallinckrodt system, a model liquid primer with both amine and alkoxy silane molecular moieties, replaces the carbon surface species with CH_x species resulting from the condensation polymerization reaction on the surface, which produced a 20-50 Å thick adhesion promoting layer. When the Si 2p spectrum is observed, a new signal appears at 101.8 eV (see Figures 27c and d) from the $Si(CH)_3$ groups for the vapor treated HMDS substrates, while no such signal appears for the Mallinckrodt system. Hence, the two comparison systems differ in the basic method of adhesion lifting prevention even though they are both successful "lift preventing" processes. In Table XI, the ESCA results and water droplet contact angle (CA) measurements are listed for a range of representative processes. Total surface C/Si ratios from ESCA are also listed because lifting has been shown to occur when that parameter is found to be from 30-100% larger than that for primed wafers(54).

The CA measurements of Table XI indicate these treaments are also very successful at removing wafer surface water contamination, as has been verified by others (see ref. 55 and references therein). It is notable, however, that there is a correlation between CA, resist image lifting results and ESCA surface condition. If CA is less than 60 degrees, the ESCA C 1s carboxylic peak at 290 eV present, and there is a high relative total C/Si ratio, lifting or poor resist image adhesion is very likely to occur, either intermittantly or quite frequently. For semiconductor manufacturing or any manufacturing process this kind of processing uncertainty is unacceptable, and the vapor processes and some liquid treatments have created better process reproducibility. HMDS SVG, a wafer track applied liquid HMDS process, is an example of a process that worked most of the time, but provided only marginal resist image lift prevention reproducibility. Vapor temperature is seen in the Table to make little difference to the measured parameters, while time of prime does provide more attractive, i.e., higher CA values. Multiple priming and long prime times can also lead to overpromoted or

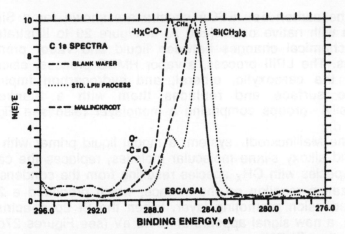

Figure 29. N(E)/E carbon 1s spectra for blank wafer and two comparison photoresist adhesion promoter processes, standard LP 3 vapor and vs. Mallinckrodt (liquid silane).

TABLE XI: ADHESION PRIME METHODS COMPARISON

METHOD	TYPE	CARBOXYLIC	TRIMETHYLSILYL	C/SI	CA
STAR 2002(5 MIN)	VAPOR(100 C)	NO	YES	0.64/NA	75
STAR 2002(90 SEC)	VAPOR(100 C)	NO	YES	0.58/NA	69
STAR 2000	VAPOR(150 C)	NO	YES	NA/0.8	77
YES LP-3	VAPOR(120 C)	NO	YES	0.54/NA	76
SVG	VAPOR(RT;760 MM)	YES; CA<70 NO; CA>70	YES	0.46/NA	75
HMDS SVG	LIQUID	YES	SMALLER	NA/1.1	58
MALLINCKRODT	LIQUID	SMALL	NO	NA/1.5	60
VIRGIN CONTROL	NA	YES	NO	1.0/1.1	24

The header "ESCA" spans the columns CARBOXYLIC and TRIMETHYLSILYL.

nonwetting (i.e., at coat) wafers, therefore, the prime time should be optimized for each substrate.

After silicon wafers have been fabricated to a certain level, they become too valuable to terminate fabrication when misprocessed. Sometimes either the wrong mask is used, the resist exposure is adjusted improperly, or the resist is improperly developed. At lithography, since the pattern is just a thin polymer layer and not an integral part of the circuit, by simply removing the misprocessed layer, the wafer can be saved and completed. A good rework process, sometimes referred to as recycle or redo, is required to accomplish this task. For device implantation, etch, or deposited layers, this flexibility is lost. As a result, rework or redo is quite common in fabrication frontends. Unfortunately, Deckert(57) has found silicon oxide wafer surfaces exhibit random photoresist adhesion variation, which is very much affected by previous chemical treatments, such as those for rework. Of course, these effects can cause defects, and the additional wafer processing required is known to statistically increase defect levels, which in turn decreases circuit electrical yield. Obviously, original and rework processes which prevent adhesion failure and are clean from a surface point of view, are very important economically. The best case situation occurs when the rework process is also the resist removal process, the one which removes the resist following layer patterning completion, thus only one process would be required.

When considering adhesion effects with rework wafers, we must first look at the effect upon substrate surface chemistry of a representative group of resist strip or removal processes. This is done in Table XII. The same wafer parameters are used as for Table XI. Oxygen plasma and sulphuric acid/peroxide are both oxidizing carbon removal techniques, Carbitol is a commercial mildly alkaline organic solvent stripper, and acetone is simply a representative organic resist solvent stripper. All but the acetone treatment restore the wafer to a state close to the original before prime and coat, but the simple acetone dissolution strip leaves the substrate in the primed, ready for recoat, state, thus saving a reprime processing step. Importantly, the other processes tend to leave the substrate less clean and with larger CA values than the virgin wafers. These observations are consistent with those reported by Deckert(57), where greater lifting or poorer adhesion was reported to occur for sulphuric/peroxide reworked wafers. Obviously, reworking wafers is not desirable, but when economically necessary, simple solvent treatments on a particle filtered wafer track like the acetone treatment are attractive and can be effective.

TABLE XII: REWORK METHOD WAFER CHARACTERIZATION

REWORK PROCESS	ESCA CARBOXYLIC	TRIMETHYLSILYL	C/SI	CA
ACETONE DISSOLUTION	-	+	NA/0.48	70
OXYGEN PLASMA	+	-	NA/0.7	39
ACETONE/PLASMA	+	-	NA/0.8	26
SULPHURIC/PEROXIDE	+	-	0.6-0.8/NA	25
CARBITOL STRIP	+	-	0.6-0.8/NA	48
VIRGIN CONTROL	+	-	0.6/1.0	25

Rework/reprimed wafer results are found in Table XIII. Here it is seen that all the rework processes return the substrate to a primed condition, but they are all a little less properly conditioned than the virgin wafer controls. Either the CA or the ESCA data are a little worse than those values for the virgin controls. Most importantly, these processes must all be concluded to provide a more particulated wafer simply due to the increased handling and processsing involved and this most likely will result in reduced circuit yield.

3.3   Resist Application

Resists are applied by wafer spinning modules either integral to a wafer track system (eg, GCA or SVG) or as individual units like those sold by Headway Research. The resist solutions with an optimally volatile spinning solvent, are applied dynamically (i.e., with the wafer spinning at 2-100 rpm) or statically to the wafer using 1-4 cc's of resist, spread across the wafer by a low frequency spin (e.g., 500 rpm), followed by a high rpm/sec ramp (e.g., 5000-40000 rpm/sec) to the final thickness determining rpm usually at ~ 5000 rpm.

The film thickness for positive photoresists can usually be approximated by:

$$[3] \qquad t = K(C)^2 / SS^{0.5}$$

where, C is the concentration of solids and SS the spin frequency in rpm. Stein(58) of Hunt Chemical has shown that K increases with the average molecular weight of the resin. The more general form of this empirical equation is:

$$[4] \qquad t = K(C)^\beta (v)^g / (SS)^a$$

where v is the solution viscosity. Log-log plots of t vs C, v, and SS provide the constants ß, g, and a from the slopes of the least squares data fits, empirically. More typically, technicians in fab lines simply run a curve of t vs. SS and fit the curve with an exponential function as shown in Figure 30 . Due to differences in coating equipment and fab ambient conditions, these curves must always be run, and vendor-provided spin curves used only as rough thickness guides. Resist vendors also adjust resist solutions to achieve the approximate thickness desired to be obtained at roughly 4-5K rpm, because thickness variation is often minimized at mid-range rpm values.

TABLE XIII: REWORK REPRIME CHARACTERIZATION

REWORK PROCESS	CARBOXYLIC	ESCA TRIMETHYLSILYL	C/SI	CA
ACETONE DISSOLUTION ONLY	NO	YES	0.48	70
SULPHURIC/PEROXIDE/SVG	NO	YES	0.48	68
SULPHURIC/PEROXIDE/2000	VERY SMALL	YES	0.59	77
CARBITOL/SVG	VERY SMALL	YES	0.56	77
CARBITOL/2000	NO	YES	0.60	70
VIRGIN/SVG	NO	YES	0.43	73
VIRGIN/2000	NO	YES	0.46	78

Data from "AZ 5206(125)"

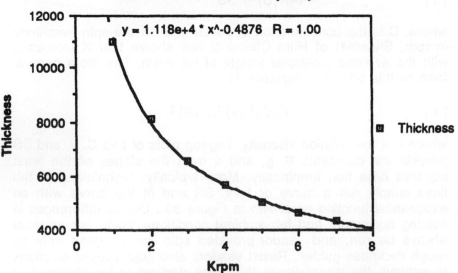

$y = 1.118e+4 * x^{-0.4876}$ $R = 1.00$

Figure 30. Thickness vs. RPM for AZ 5206 prebaked at 125°. The solid curve is an exponential fit of the experimental data.

Meyerhofer(59) has theoretically treated wafer spinning application of resist and accounted for thickness and drying times considering only centrifugal force, linear shear forces, and uniform evaporation as variables. Middleman(60a) has further shown that air flow induced by disc or wafer rotation provides the required shear stress at the liquid/air interface to enhance the rate of resist thinning on the wafer.

Of more importance than average film thickness across the wafer is the film thickness variation across the wafer measured radially from the center of the wafer. This is important, because resist image critical dimensions can vary beyond specification limits due to the resist film thickness variation. Film thickness variation depends upon the spin frequency, and an optimum frequency is usually measurable at a given solution viscosity. Typical thickness variation within wafer and wafer to wafer for 1986 vintage wafer track spinners is of the order of 20-100Å and 50-200Å, respectively which is more than adequate for CD uniformity requirements of most lithography areas. Control charts are typically used (see Figure 31) to monitor coat processes and if values for monitor wafers fall out of specification, the wafer tracks are shut down for maintenance or engineering adjustment.

Film nonuniformity, visually observed as radial stripes in the resist called striations, is prevalent when either the resist solvent is too volatile, dynamic dispense is employed, wetting additives are omitted, or at high resist concentrations. Striations are easily measured using interferometric or profilometric techniques. Orange peel, another spin problem, occurs due to rapid evaporation of volatile spinning solvents, and cloudy resist films sometimes occur when hygroscopic solvents are employed(60b).

Machine variables which have been observed to influence resist coating uniformity and average thickness are exhaust flow thru the coat module, motor frequency control and acceleration precision, dispenser type, and of course possible interactions which may occur between these and the resist variables already mentioned.

Optimizing coating processes is a complex time-consuming empirical task. It involves screening the many potentially material dependent variables through statistically designed experiments(61) to reduce or minimize process variability to achieve manufacturing success.

3.4   Prebake/Exposure/Development   Processing

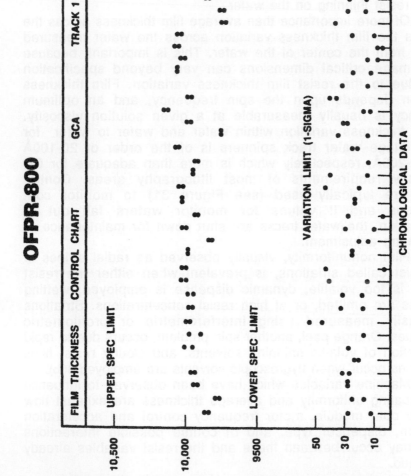

Figure 31. Thickness control chart for OFPR-800.

For positive resists, the prebake conditions, exposure and development conditions are inseparable, and together can critically determine the performance of the overall resist process. For example, sometimes interactions occur between prebake and development variables; a 2-factor interaction between prebake temperature and development method has been observed for Shipley 1400 resist. Due to this interaction, both of these variables would have to be changed in parallel to optimize this resist process. Prebake is almost always a primary variable for positive resists, because their development rates are influenced strongly by residual solvent content and the thermal history of the film. For negative resists, this is not the case-- prebake and developer compositions have less influence over resist contrast, but they can impact resist swelling behavior and hence resolution performance. Since positive resists are the evolutional choice of most new fab lines and they are most affected by these variable combinations, this section primarily addresses positive resist effects.

Variables which can influence resist image edge wall, critical dimension control, and resist sensitivity are prebake temperature and time, exposure level, developer composition and conditions, rinse composition, and fab ambient. The effect of developer composition upon CD RPL, sensitivity and contrast were demonstrated earlier in this section. All of these parameters can potentially interact, therefore, statistical engineering methods and experimental designs are invaluable in optimizing the overall resist process.

3.4.1 **Statistical Process Optimization Characterization Example.** The best way to illustrate a statistically oriented resist process optimization and its efficiency is to provide an example. In this section, a simple OFPR-800 positive photoresist process will be charaterized by generating a CD reponse surface space for the process. From that response, an optimum operating point for the resist process is obtained. Most importantly, after completing the statistically designed experiment(s), we know how the CD response varies over a much larger set of operating conditions than that unique set established as the baseline photoprocess.

Since a resist process has many different substeps, it is impractical to evaluate each variable individually. It is also unwise since the variables may interact. A statistical experimental approach, which investigates many variables simultaneously and can assign quantitative values to variable effects and interactions, is necessary.

The objective of this study was to develop and optimize a process for a photoresist (OFPR-800 manufactured by Dynachem) from both a critical dimension transfer and contrast points of view. Two of the statistical experiments--a 7-factor variable screen(62) and a 3-factor Box-Bhenken(61) response surface investigation--are described in this example. In addition, the track development process employed is characterized for completeness.

3.4.2 Background. The purpose of a variable screen experiment is to determine which of the many variables (independent variables) involved in any process step are significant, that is, which variables the engineer needs to control to optimize the overall process performance as determined by the process parameters (dependent variables). A variable screen design is a small part of a full 2^k factorial(61)(62). It is designed only to determine the significant independent variables in relatively few experimental trials or wafers, and is unable to identify any variable interactions. A full 2^k factorial can determine variable effects and interactions, but requires too many experimental runs to be practical for investigating more than 4 or 5 variables. Designs for variable screens are available in several references(61-63). It must be cautioned, to watch for main variable and two-factor interaction confounding when employing these fractional screening designs.

After identifying the significant variables for a process from a screen design, the final step is process optimization. This involves determining any independent variable interactions (non-additive responses) or non-linearity in the dependent variable response curves. Several statistical designs are available for this type of experiment, but unlike the variable screens are usually based on a full factorial experiment(61) or a higher resolution fractional factorial design. Since fewer variables are investigated, the process optimization experiment yields much more information than a variable screen design in approximately the same number of runs, because now the two factor and higher order interactions are no longer confounded in favor of screening a larger number of primary variables.

3.4.3 Experimental Designs. The resolution III screen design used in this example is given in Figure 32. The procedure for running a variable screen design is to choose two levels of interest for each variable (designated + and -, although they need not be quantitative values), and then run each sample, in random sequence, through the process determined by the screen design. Seven independent variables were examined: softbake time

SEVEN FACTOR SCREEN DESIGN

	PREBAKE TEMP. 70°C 90°C	PREBAKE TIME 60S 120S	POSTBAKE TEMP. 120° 140°	POSTBAKE TIME 60S 120S	U-1000 EXPOS. 0 +20%	SPIN CONV./SPEC	DEVELOPER METHOD MAN/TRK
1	+	+		+			+
2	+	+		+	+	+	+
3	+		+				
4			+	+	+	+	+
5	+	+			+		+
6			+		+	+	+
7		+		+	+	+	+
8							

Figure 32. Seven factor screen design for searching experimental variables of a photoresist process for significance.

and temperature, postbake time and temperature, exposure, spin method, and development method. The results were quantitatively evaluated based on one dependent variable: resist image linewidth transfer to the wafer. The main variables screened are confounded with three two-factor interactions, therefore, this design will almost always be followed by a response surface design of higher resolution or by a higher order screening design.

The purpose of a process optimization experiment is to study in more detail a small number of variables that are known to be significant in their useful range without loss of precision. Based on the results of the variable screen design described earlier, softbake temperature and exposure were chosen for further investigation. Since this experiment was run with a Perkin-Elmer 544 projection aligner, exposure tool aperture was also chosen as a related process variable to make three total variables. The dependent variables measured were resist critical dimension transfer (sidewall angles: 70-80°) and resist contrast.

The 3-variable Box-Behnken cube design employed is shown in Figure 33. The cube is defined by 3 levels each of 3 variables; the experimental points are determined by the midpoints of the 12 edges of the cube to check for response surface curvature. The center point of the cube is replicated 3 times to provide an estimate of process variability (i.e., the precision of the experiment). The effect of this design is to run a complete 2 x 2 factorial, while holding the third variable at its center point. Since the Perkin Elmer 544 has the capability of exposing a single wafer with a number of different exposure levels, this design actually examined five different exposure levels for each run instead of only three. Other process information is given in Table XIV.

Critical dimension results were evaluated by line and pitch measurements made on the 2.0µ line of a 4.0µ pitch structure. Measurements were made on a Leitz MPV-CD system. The precision capability of this tool was determined to be +/- 0.09 micron (3-sigma), a value less than required for the process tolerance. Leitz measurements of the 1.5 and 2 micron resist dimensions showed a variation of only 5% or less across the 4 " wafer. Individual line measurements were calibrated vs. a sample measured by both the Leitz and a Cambridge SEM.

3.4.4 Developer Process Characteristics. The developer process employed was a NMD-3 metal-ion free spray/puddle process at the fab temperature (70°F) on a model GCA 1006 Wafertrac. The wafer was sprayed for 2 seconds at 100 rpm, followed by a 1 sec. static spray to ensure good puddle

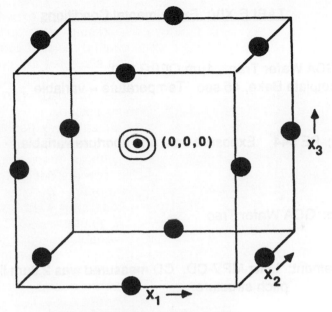

BOX BEHNKEN DESIGN CUBE

X_1	X_2	X_3
+ 1	+ 1	0
+ 1	- 1	0
- 1	+ 1	0
- 1	- 1	0
+ 1	0	+ 1
+ 1	0	- 1
- 1	0	+ 1
- 1	0	- 1
0	+ 1	+ 1
0	+ 1	- 1
0	- 1	+ 1
0	- 1	- 1
0	0	0 ⎫
0	0	0 ⎬ CENTER POINT
0	0	0 ⎭

THREE-VARIABLE BOX-BEHNKEN DESIGN EXPERIMENTS

Figure 33. Three variable Box-Behnken experimental design for probing an experimental response surface.

TABLE XIV: Experimental Conditions

Coat: GCA Wafer Trac   1μm OFPR 800
       Hotplate Bake, 45 sec. Temperature = variable

Expose: PE 544   Exposure time and aperture variable.

Develop: GCA Wafer Trac

Measurement: Leitz MPV-CD. CD measured was 2.0μm line on 4.0
              pitch structure.

formation, followed by a static 45 sec. development and 30 sec. wafer rinse.

3.4.5 Variable Screen T-test. The data analysis of the experimental design of Figure 32 is performed for each independent variable by subtracting the average linewidth from those runs for which the variable was at its low level (-) from the average linewidth from those runs for which the variable was at its high level (+). This result is designated $Y_+ - Y_-$. Linewidth measurements were made at the center and edge of each wafer.

Theoretically, the result $Y_+ - Y_-$ measures the effect on linewidth of changing the independent variable from its lower level to its higher level. In the real world, however, each process has a certain amount of variability no matter how carefully the independent variables are controlled. One way to insure the results demonstrate a real effect or are significant beyond normal intrinsic process variability is to apply a t-test to the data.

The t-test is used to compare independent sample averages from two populations (in this case, (+) and (-) levels of the independent variable) to determine if the difference between them is statistically significant(64). It works by comparing the experimental result ($Y_+ - Y_-$) to that which should have occurred based upon variability results from either both populations of data or control wafers assuming a t-distribution, which is a small sample approximation of the normal (Gaussian) distribution. The greater the result is from the control wafer population variability, the more likely the result was caused by the independent variable instead of random process variation due to lack of experimental precision. If we choose a given "confidence level" (probability that our conclusion is correct) needed, we can calculate the t-statistic from our results and compare it to the t-table entry for the appropriate number of degrees of freedom and the level of confidence. If the test statistic is greater than the corresponding table entry, then we can conclude that the independent variable is significant (that is, it must be carefully monitored to assure process control) with a corresponding level of confidence.

The test statistic for the t-test is given by:

[5] $$\text{T.S.} = Y_+ - Y_- \, / Sp \, [\frac{1}{n_+} + \frac{1}{n_-}]^{\frac{1}{2}}$$

where, Y_+ and Y_- are as defined before, n_+ and n_- are the number of samples at each level, and Sp is the pooled variation calculated from:

$$[6] \qquad S_p = \left[\frac{\sum_i (n_i-1) \ S_i^2}{\sum_i n_i - 1} \right]^{1/2} \qquad \text{where } n_i \text{ is the}$$

number of replicates at each experimental condition and S_i the cell variation between replicate observed dependent values.

   3.4.6 Results and Analysis. The results from the variable screen are given in Table XV. Prebake temperature, develop method, postbake temperature and exposure are all clearly significant in the range studied. Spin method (conventional or special) is also significant. "Special" spin was a method developed to enable easier target detection by the Ultratech stepper alignment system by removing the spread cycle at low rpm and by shortening the final spin dry cycle at the final rpm. Unfortunately, it also resulted in large resist thickness variation across the wafer, which led to the large CD variations illustrated by the values in the table. "Special" spin resist coating has been discontinued for this reason and is not recommended.

   Critical dimension vs. exposure results for the 3-factor Box-Behnken design at different softbake temperatures are graphed in Figure 34. Figure 34 shows two possible operating points for the process that will result in a critical dimension within specification: a 75°C softbake with approximately 75 millijoules/cm^2 exposure, or a 90°C softbake with approximately 96 millijoules /cm^2 exposure. The graph also suggests some variable interaction is occurring between 54-75 and 96-116 mJ/cm^2 since the three lines are less parallel over those regions. Aperture had a negligible effect on critical dimension within the range of interest (2.0 μm +/- 10%).

   In order to evaluate the experimental results quantitatively, a Yates[65] analysis was performed on the data. The Yates algorithm is a method of taking advantage of the "hidden replication" found in factorial experiments: the average result (critical dimension) of the points at which a certain factor was at its low value is subtracted from the average result of the points at which the factor was at its high value. The result is a more accurate estimate of the actual effect of that factor than would be

TABLE XV: 7-FACTOR SCREEN DESIGN RESULTS

FACTOR	CENTER ISLAND IMAGE CD	EFFECTS (Y$_+$-Y$_-$) EDGE ISLAND IMAGE CD
SPIN METHOD	-0.29μ	+0.11μ
PREBAKE TEMP	-0.11	-0.18
PREBAKE TIME	+0.03	+0.09
DEVELOP METHOD	-0.30	-0.30
POSTBAKE TEMP	-0.53	-0.35
POSTBAKE TIME	+0.03	-0.05
EXPOSURE	+0.15	+0.21

CONFIDENCE LEVEL 0.11 A 90% (T-STATISTICS)

CRITICAL DIMENSION VS. EXPOSURE (APERTURE 3)

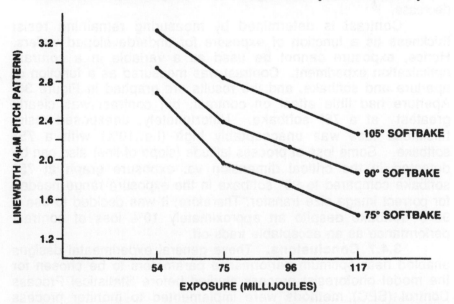

Figure 34. Critical dimension vs. exposure for OFPR-800 photoresist process.

available from only one measurement at the high and low value of each factor. The Yates algorithm is used to check for independent variable significance as well as variable interactions.

A Yates analysis can be performed for two levels of the variables only. Based on the graph shown in Figure 33, those levels closest to our expected working range were chosen. The results are shown in Table XV. The results confirm those inferred from the graph; exposure and softbake are clearly significant variables over the range studied, but aperture has a much smaller effect. All of the variable interactions are small enough to be safely ignored.

Iso-CD response surfaces have been drawn in three dimensions in Figure 35. Roughly speaking, an optimal working point would be near the center of the cube, a result that was not designed to occur intentionally. Note also, the exposure latitude for critical image control falls off rapidly at 75°C prebake, which would make operating at that prebake temperature risky; that is, if exposure unexpectedly changed a little the chance of CD failure would be great.

The second objective of this investigation was to optimize resist contrast where greatest resolution is possible. Since resist contrast also correlates with resist image sidewall angle, it is an important resist processing variable especially as linewidths decrease.

Contrast is determined by measuring remaining resist thickness as a function of exposure for underdeveloped wafers. Hence, exposure cannot be used as a variable in a contrast optimization experiment. Contrast was measured as a function of aperture and softbake, and the results are graphed in Figure 36. Aperture had little effect on contrast, but contrast was clearly greatest at a 75° softbake. Unfortunately, unexposed resist thickness loss was unacceptably high (i.e.,10%) with a 75° softbake. Some loss of process latitude (slope of line) also can be detected in the critical dimension vs. exposure graph at 75° softbake compared to 90° softbake in the exposure range needed for correct image size transfer. Therefore, it was decided to use a 90°C softbake despite an approximately 10% loss of contrast performance as an acceptable trade-off.

3.4.7 Conclusions. These general experimental designs enabled near optimum performance parameters to be chosen for the model photoresist process studied before Statistical Process Control (SPC) methods were implemented to monitor process performance. As a result of this careful process characterization, this process was successfully employed to fabricate CMOS test

CD RESPONSE SURFACE

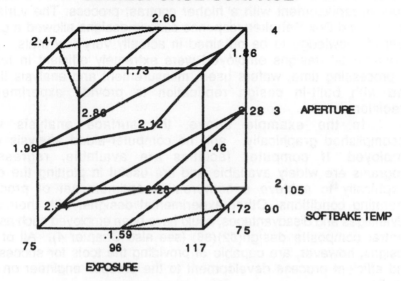

Figure 35. Iso-CD response surface for OFPR-800, generated using a Box-Behnken design.

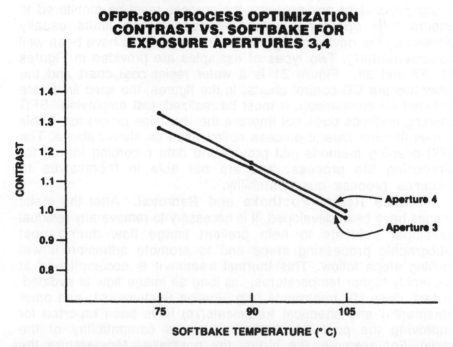

Figure 36. OFPR-800 process contrast vs. softbake temperature from Box-Behnken response surface design.

devices in SPC control over an extended period of time before process replacement with a higher contrast process. The variable screen and Box-Behnken response surface designs allowed a great deal of knowledge to be obtained in actually very few trials. The experimental designs employed were extremely efficient in terms of processing time, wafers used, measurement and analysis time, and with built-in design replication to provide experimental precision.

In the example above, the surface analysis was accomplished graphically, and no computer-aided analysis was employed. If computer facilities are available, regression programs are widely available and are useful in plotting the data graphically to achieve a compromise optimum set of process operating conditions. Other experimental designs, with their own advantages and disadvantages, could have been employed such as the central composite design(62)(66) (see also Chapter 4). All of the designs, however, are capable of providing the tools for successful and efficient process development to the process engineer on the fab line.

3.4.8 SPC Methods of Process Control. After the process optimization, whether it be a resist lithographic process, a coat process or any process, the process must be monitored to ensure it is operating within the specification limits usually dictated by the device design rules. These methods have been well documented(67). Two types of examples are provided in Figures 31, 37 and 38. Figure 31 is a wafer resist coat chart and the other two are CD control charts. In the figures, the spec limits are included for comparison. It must be realized: just employing SPC charting methods does not improve the baseline processes -- this comes through careful process optimization as shown above. The SPC charting methods just provide the data recording format for monitoring the process, and are not able in themselves to influence process quality/stability.

3.4.9 Resist Postbake and Removal. After the resist images have been developed, it is necessary to remove any residual developer solvents to help prevent image flow during post lithographic processing steps and to promote adhesion if wet etching steps follow. This thermal treatment is accomplished at preferably higher temperatures, as long as image flow is avoided. In fact, deep UV treatments (e.g., Fusion Systems)(68) and other plasma(69) and chemical treatments(70) have been reported for improving the post development process compatibility of the resist. For example, the higher the postbake temperature the more resistant the resist is to image flow and reticulation in RIE environments. Deep UV treatments and the other processes usually

PE 544 CD CONTROL CHART

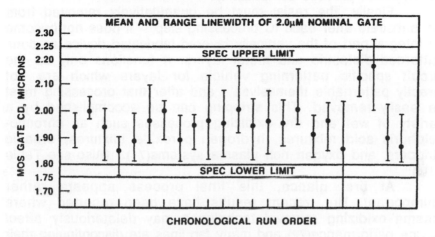

Figure 37. Two micron nominal CD control chart for OFPR-800 process using a Perkin-Elmer 544 projection printer (UV-4) as the lithographic tool.

PE 544 ISOLATION CD CONTROL CHART

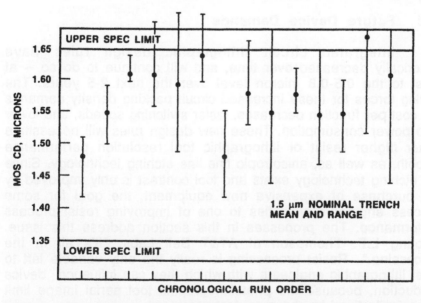

Figure 38. One and a half micron nominal CD control chart for OFPR-800 process using a Perkin-Elmer 544 projection printer (UV-4) as the lithographic tool.

allow higher postbake temperatures, hence, improved post development process compatibility.

Finally, the resist must be quantitatively removed from the substrate after each IC processing step -- it does not become an integral part of the vertically layered fabricated device as does patterned dielectric and metal layers. It functions only as the circuit specific patterning vehicle for layers which are not directly patternable themselves , and after that processing must be easily removed. This stripping can be accomplished by a variety of wet and dry oxidizing processes such as chromic-sulphuric acid mixtures, hydrogen peroxide mixtures, organic strippers, and oxygen rich plasma systems(71). (Also see Table XII.)

At first glance, this final process appears rather unimportant. But, recent results have been reported where plasma oxidizing removal techniques may deleteriously affect device performance(72) and many fab lines are discontinuing their use. One must be careful then to investigate the effects of these stripping processes upon device performance and /or surface contamination to further processing.

4.0   APPLICATIONS AND SPECIAL PROCESSES

4.1   Future Device Demands

Integrated circuit lithographic design rules have historically decreased over time, and will continue to do so -- at least to the 0.5-0.8 micron level over the next 2-5 years. The driving forces for these increased circuit packing density demands are cost per function decreases, faster switching speeds, and lower chip power consumption. These new design rules will necessitate either higher resist or lithographic tool resolution performance or both, as well as, anisotropic fine line etching technology. Since the etching technology exists and tool contrast is only improved by the purchase of expensive new equipment, the goal for some process engineers simplifies to one of improving resist process performance. The processes in this section address this issue. Quoting L.F. Thompson of AT&T Bell Labs, "it's all in the processing." Resist processing is really the only variable left to most lithographic engineers with which they can influence device production, because the photolithographic tool aerial image limit is basically fixed by the manufacturer.

Conventional single-level positive photoresist technology has recently progressed rapidly, especially in the mid-UV class, but may be incapable of providing the necessary resist imaging required for the next generation of chips. Here, resist imaging thickness is the key issue to meet RIE etch masking requirements. Therefore, new resist processing technology may be needed. Furthermore, processes which extend the performance levels of existing projection exposure tools or provide depth of focus relief are very attractive due to the increasing cost of new higher performance exposure equipment and the return on net asset demands placed upon sales of new devices to pay for these tools.

While a great deal of resist process research has been occurring to extend phototool lifetime, advances in reduction lens design and reflective optical wavelength reductions have also been occurring. As a result, optical lithography is relegating direct-write e-beam lithography to a quick turn around Application Specific IC (ASIC) role, and extensive X-ray lithography usage has been delayed years. The advanced processes in this section will address both optical and e-beam future needs only; it is assumed the e-beam processes would also be extendable to X-ray, assuming high X-ray flux.

4.2   Applications

4.2.1 Dyed and Thinned Single Layer Resist (SLR) Processes. The first and most obvious thing to do to improve resist performance on the most difficult substrates (those requiring greatest resolution or those with reflective topography as for metal or some polysilicon levels) is to reduce the resist thickness or add dyes to it, respectively. The resulting dyed and usually thick (~2 microns) material is still a single level resist process, but without the added complexity of multi-level processes. Unfortunately, resist thinning is most feasible with multi-level processes and thinning accomplishes nothing towards reflective image notching relief, the main observed problem for reflective topographical situations. Furthermore, resist thinning presents a severe problem to step coverage and metal etching because of poor selectivity, and is in fact usually prohibitive. All of these negatives aside, resist thinning has been shown by IBM researchers(73) to improve linewidth control by 15% and focus control by 35%, when and if it is feasible to do it. The latter value further reinforces the restrictive applicability of resist thinning to multi-layer processes.

The more practical solution to reflective notching problems on reflective surfaces is provided by resist dyeing. Most device areas will select this option over multi-layer Portable Conformable Masks, PCM (3), processes (see Figure 39). Dyeing

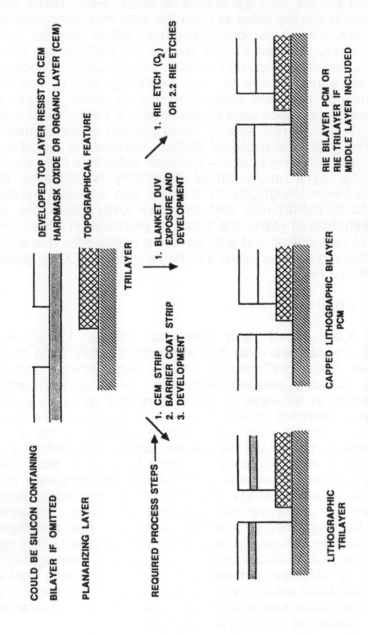

Figure 39. Multilayer resist systems illustrating different multilayer processes and PCM techiques.

resists requires a price be paid in resist contrast(74) and exposure time(75), but dyeing does provide greater process latitude (see Table XVI) and reflective notching can be effectively eliminated(76) or at least minimized. Sandia workers(77) have provided a dye system for H and G line steppers which has a very small exposure penalty, just 15 mJ/cm^2. Bolson et al.(76) have demonstrated similar results and an approximately 60% gain in exposure latitude was achieved or a reduction in K from the Raleigh resolution equation from 1.1 to 0.6 was effectively obtained. On the negative side, adding dye to the resist formulation leads to a larger standing wave foot(78) and reduced image edge walls (see Figure 40) consistent with the observed bulk contrast reductions reported by Pampalone. Most dramatically, Brown and Arnold(78) have observed a 3-fold increase in CD exposure latitude, a result which explains the wide acceptance of this technique for metal layer lithography by development and production fab lines.

4.2.2 Image Reversal (IREV). Positive photoresist image reversal (i.e., negative toned imagery from a positive toned resist) is a new processing technology which addresses the deficiencies of dyed resists. Moreover, IREV can be accomplished on dyed material to achieve the best of both worlds, namely, relief from topographical or bulk effects and reflective notching minimization. Over the past five years, many papers have been published on this subject(79-87). The salient contents of those papers will be reviewed and compared. The main focus of this section is on single level thermal and base induced reversal processes for AZ 5214 and the Genesis Star image enhancement process for Shipley 1400 positive photoresists, respectively.

Image reversal is an alternative to conventional positive photoresist technology. Briefly, a single layer of positive photoresist is exposed using a projection aligner, reversed by either doing a post exposure thermal treatment on a hotplate or by adding a base to the resist, flood exposing and developing. The result is a negative toned image with a controllable edge wall angle, something dyed resists with conventional processing cannot deliver. The IREV processes are capable of printing images previously unattainable with the given exposure tool, thus, extending the resolution and focus latitude performance of the alignment tool, and the life of it as a capital asset.

4.2.3 Thermal Image Reversal. Marriott, Garza and Spak have written the definitive paper on thermal image reversal(86). Using both theoretical PROSIM simulation and empirical methods, the AZ 5214   IREV process has been

TABLE XVI: DYED PHOTORESIST RESULTS TABLE *

RESIST	OPTICAL CONTRAST	$Q. mJ/cm^2$	RPL	DEVELOPER/MODE
OFPR 800 ON SILICON	1.3-1.6	104 (135)	1.0	NMD-3 (2.7%)/PUDDLE
OFPR 800 ON AL	-	20-88	1.6	NMD-3 (2.7%)/PUDDLE
OFPR 800 ON AL: TRILEVEL	-	123 (165)	1.1	NMD-3 (2.7%)/PUDDLE
OFPR (AR-15 DYED) ON AL	1.0	73	1.2	NMD-3 (2.7%)/PUDDLE
OFPR (MOTOROLA DYED) ON AL	1.1	166	0.9	NMD-3 (2.7%)/PUDDLE

* ULTRATECH 1000 STEPPER UTILIZED FOR TESTING

UV-3 DYED AZ-5214 PROCESS PROFILES

1.25μ Line/ Space

1.5 μ Line/ Space

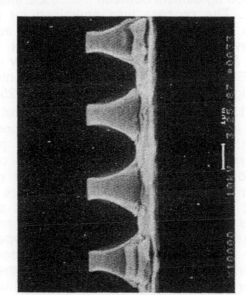

4.0 Microns

4.0 Microns

Figure 40. Edge wall profiles of heavily UV-3 dyed AZ-5214 to illustrate standing foot and poor edge wall angles consistent with low observed bulk contrast values.

optimized, and good agreement between theoretical images and real images obtained. The process was optimized for photospeed, resolution, focus and CD latitude, and vertical edge wall by adjusting the developer concentration to 0.21 N MF-312 (90 secs), setting the PEB temperature to 110 C, and by using a flood exposure after PEB. This data has been independently verified at Motorola (see Figure 41). An improvement of 1.25 micron in focus latitude, 150% improvement in CD control and a 8° improvement in image edge wall were also reported(86). AZ researchers further reported a bulk contrast performance improvement of roughly 200% for IREV AZ 5214 over that for the positive performance mode. Consistent edge wall imagery improvement can be seen when comparing Figures 41 and 42, where Figure 42 portrays images with edge wall angles more typically observed from normal positive tone performance.

   4.2.4 Base Reaction Process. Two types of base-induced positive photoresist image reversal processes have evolved, one where the basic chemical is added to the photoresist formulation prior to wafer spin(79-81,84), while the other involves a gas phase ammonia treatment of the wafer coated with the exposed resist film(82).

   The mechanistic and theoretical treatments of image reversal photoprocesses are detailed in references 80 and 83. In summary, the chemistry of both processes involves a base catalyzed thermal decomposition of the indene carboxylic acid photoproduct. Following this reaction and a flood exposure of the wafer, subsequent aqueous base development with conventional developers leaves a negative tone image in the exposed area, where indene product is rendered insoluble. Development occurs in the unexposed area because the flood exposure converts the resist in that area to the normal photochemical intermediates which are base soluble (i.e., the resist functions conventionally in these areas). The two processes differ only in the process sequencing. With the chemical additive process, the thermolysis is carried out independently before flood exposure, while the vapor process carries them out simultaneously.

   Two commercial microprocessor controlled vapor phase systems have evolved from this technology, the Yield Engineering (YES) Model 8 system and the Genesis ST-A-R 2001/2002 system. The YES 8 system delivers anhydrous ammonia gas to a heated stainless steel reaction chamber, while the 2001/2 system delivers a fresh controlled vapor fill to the temperature controlled and profiled reactor chamber every wafer load from a liquid amine decomposition subsystem.

AZ 5214 THERMAL TONE-REVERSAL PROCESS PROFILES

1.2 MICRON / UV-3/ B49B

1.8 MICRON/ UV-3/ H9P

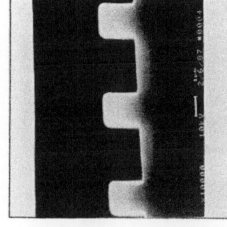
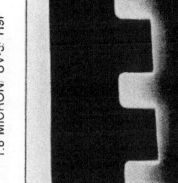

4.5 MICRONS

4.5 MICRONS

Figure 41. AZ-5214 thermal tone reversal process edge wall profiles illustrating vertical edge walls.

TYPICAL POSITIVE TONE RESIST IMAGES

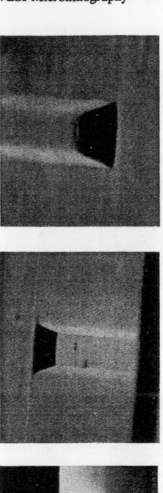

AZ 5214   OFPR-800   OFPR-800
                  NEGATIVE MASK   POSITIVE MASK

Figure 42. Edge wall profiles for conventionally developed positive photoresist images. Note, the edge walls are poorer than those for both IREV processes.

The systems vary in cost dramatically due to the base delivery technique, quality of materials and profiled temperature control systems. The ST-A-R system has sophisticated vapor delivery plumbing, combined with three zone balanced cascade temperature control and gas manifolding. YES 8, on the other hand, is a low cost system where the manufacturer claims the usage of 100% ammonia gas alleviates the control systems required for the liquid amine delivery system and the need for more exact temperature control. Although the systems differ considerably, both manufacturers claim good concentration and temperature control, low particulate contamination (YES; 5 1-micron particles/5" wafer), and good overall process performance.

4.2.5 Base Tone-Reversal Process Comparison. The solid phase base additive reversal systems reported in references 79-81 suffer from two basic problems. They generally suffer from low shelf-life(84-85) and poor resist image thickness and critical dimension uniformity across the wafer, wafer to wafer, and lot to lot. Inconsistent and sometimes incomplete reversal results were also reported in reference 81; these phenomena have been empirically verified. In addition, the base additive systems mix poorly, which probably contributes to the observed performance problems.

The vapor reversal systems are superior in these respects(82-85), and exhibit manufacturing compatibility if reduced throughput is acceptable; that is, two more processing steps are required with the commercial reversal systems vs. conventional processing. The two types of reversal processes differ mechanically only by the need to mix the additive formulation every 1-3 weeks due to shelf-life considerations, which raises concern about the overall manufacturability of that type of process, independent of irreproducibility effects(85).

4.2.6 Process Tuning. The vapor base IREV process discussed here in detail is basically the evolutionary culmination product of the efforts of references 79-82, but with further careful process and equipment optimizations carried out by Genesis researchers. The optimum time and temperature for the reversal reaction were determined to be between 30-90 minutes and 90-110°C, respectively(79-83). Of course, higher temperatures are precluded due to photoactive component thermal decomposition, and lower temperatures precluded by incomplete reversal reaction. Reference 83 clearly demonstrates the improved process contrast provided at 100°C reversal temperature. Since the remaining resist image thickness is basically invariant over these processsing ranges, other development and exposure considerations dominate the reversal behavior over these time and temperature ranges (discussed later). The final process variable level selections were made on the basis of greatest observed process contrast/resolution.

4.2.7 Resist Image Edge Wall Angle. Initially, work at Motorola on image reversal was focused upon developing a process that yielded vertical resist edge walls for resolution and RIE etching requirements. A series of wafers was run varying prebake temperature, exposures, and development time, to establish a "ball park" process. From the results of these tests and the unpublished work of others, a lower prebake temperature (70°C), shorter hotplate prebake time (30 sec.), and higher flood exposure (470 mJ/cm^2)(83) were found to yield vertical resist image edge wall profiles and 1.25 micron line-space reproduction fidelity.

With this foundation, a study was completed to determine a first iteration on first exposure and development time process parameters. An array of wafers was processed varying both first exposure and development time. The GCA Wafertrac spray/puddle method of development was employed because of its ability to deliver a uniform, reproducible, and accurately time-controlled puddle. Plots of first exposure and development time on critical dimensions show the widest exposure and development latitude at 160 mJ/cm^2 and 45 seconds, respectively. It is important to note, image CD's change less with development time increases the greater the initial imaging exposure level. A SEM cross section of a sample with these process parameters exhibits the required 90° edge wall angles (see Figures 16 and 43). As a result, these values were chosen as the optimum processing parameters for futher evaluation of processing effects upon process reproducibility. These values also represent a compromise between process latitude and throughput. The selection preference of an initial exposure level greater than 150 mJ/cm^2 was previously reported in references 81, 83, and 85. All three studies demonstrated the plateau behavior in remaining reversal image thickness at initial exposure levels greater than 150 mJ/cm^2. The dependence upon post reversal flood exposure level is discussed below.

4.2.8 Resist Contrast. A comparison of resist contrast was carried out between the standard S-1400 positive process and the image reversal process. The reversal process demonstrated a contrast of 2.2 (Figure 44), while the positive process only yielded a contrast of 1.6. These findings are very significant as resist contrast, resolution, and edge wall angle are related and

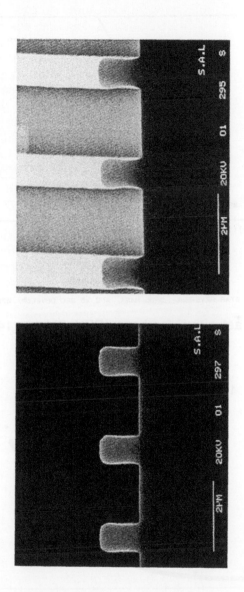

Figure 43. Base IREV process image edge wall profiles.

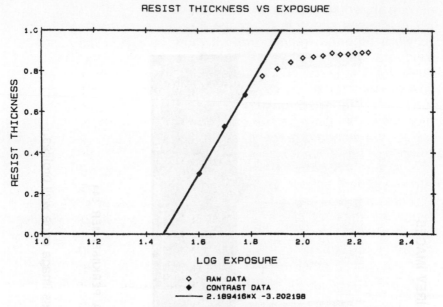

RESIST THICKNESS VS EXPOSURE

SAMPLE # 1: SHIPLEY 1400 - STAR REVERSAL, SRDL MASK, APC 45 SEC DEVELOP, 470 MJ FLOOD

Figure 44. Contrast curve for base IREV Shipley 1400 resist process.

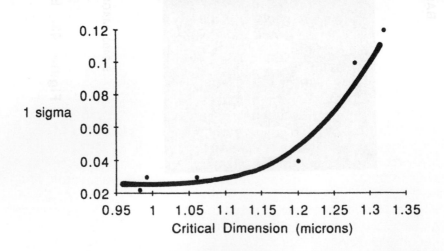

1 SIGMA VS CRITICAL DIMENSION

Figure 45. Base reversed process CD variation (1 sigma) vs. critical dimension for Shipley 1400.

well correlated(88). Contrast was previously compared for the IREV processes above in reference 82. There values of 2.38 and 1.18 were reported, respectively; it must be noted that different developers were employed. The effect, however, is still quite clear-- the image reversal process is the superior process in terms of contrast or resolution. No quantitative bulk contrast values were reported in reference 83, but the characteristic curves for the image reversed data are visually much steeper than those for the standard process, again, consistent with results cited and those of reference 83.

4.2.9 Process Latitude and CD Reproducibility. The development latitude for the image reversal process increases with increased first exposure level. At these exposure levels and greater, the reversal process becomes like that for conventional negative photoresist processes in that they can be overdeveloped with minimal impact upon CD. For normal positive resist processes this is not the case, and development times must be well-controlled for satisfactory CD performance. At lower first exposure levels, image CD's of the reversal process are more affected by development time than at greater exposure levels.

The exposure latitude (i.e., for 1st exposure) was measured for the reversal process, and the results compared to that for the normal photoprocess. Values ranging from 0.003 to 0.006 (160 mJ/cm^2) vs 0.015 micron-cm^2/mJ were observed, respectively. The resistance to CD change with exposure is 2.5-5X better for IREV. When first exposure values >160 mJ/cm^2 are employed, as in refs. 81 and 83, even better exposure latitude is observed. Thus, the base reaction reversal process possesses more than 2.5X greater exposure latitude, which translates to greater process control.

Within lot CD variation has been characterized for the base image reversal processes in references 82, 84, 85 and 89. In the earlier work(82,84), 3-sigma values of +/- 0.27 and 0.32 micron were reported for the early ST-A-R process and the Monazoline(79) additive and "Lift-off" process(84), respectively. With improved processes and equipment, came better CD control results: 0.08(85), 0.04(89), 0.08(90), and 0.06-0.09 micron for this work (see Figure 45). These latter values are far better than the typical values observed for conventional positive resist processes, which usually have values closer to those of the earlier image reversal numbers above. With the lower sub-0.1 micron 3-sigma reversal process CD variation values, the next-generation VLSI circuits should be manufacturable with reasonable yield.

A series of lots were processed according to the optimum process parameters to determine the reproducibility of generating vertical resist edge walls and critical dimension die to die, wafer to wafer, and lot to lot.   Die to die uniformity for critical dimension consistently exhibited a 3-sigma value of less than 0.1 micron, and with vertical edge wall angle. Unfortunately, small to moderate deviations were sometimes found lot to lot. Figures 45 and 46 illustrate the correlation between critical dimension and both critical dimension deviation and edge wall angle.

Due to the significant deviations of critical dimension from lot to lot, two-level 7-variable resolution III screen designs were employed to determine the variables affecting linewidth variability. Of those variables screened, development method (Wafertrack vs manual puddle), mask orientation (0° vs. 90°), first expose to ST-A-R image reversal storage time (2.5 hrs. vs. 20 hrs.), and post-flood exposure storage temperature (20° vs. 30°) did not significantly affect critical dimension control. The two remaining variables, mask type (i.e., chrome reflectivity) and HMDS vapor treatment system did reproducibly cause significant linewidth differences.

4.2.10 Mask Effects. Differences in resist image CD transferred to the wafer, as a result of mask type, were quite substantial.   Lines of 1.25 micron exposed using the bright chrome mask (BC mask) transferred to an average of 1.00 micron after image reversal. Those same lines exposed using an anti-reflective chrome mask (ARC mask) transferred to less than 0.6 micron after image reversal. This effect can be attributed to the additional exposure received by the resist due to reflections at the chrome space interface on the mask. This effect reinforces the fact that a greater first exposure increases line size, and suggests mask pattern size biasing must take chrome type into account at mask fabrication, but this is already done routinely, and should not be an additional burden of consequence[91].

The image reversal processes of this chapter are capable of 1) extending the effective resolution of the projection lithography tool utilized, 2) providing vertical edge walls for dry-etch considerations, and 3) providing superior resist image CD control vs. that of the resist when used in the conventional positive mode. Consistant with these observations, reference 83 further provides evidence of improved tool depth of focus with this process. The edge walls observed here and in references 83-85 have been theoretically accounted for in references 83 and 86 on the basis of image edge photoactive species concentration gradients, and the mechanism of the reversal processes beautifully delineated by MacDonald et al.[80].

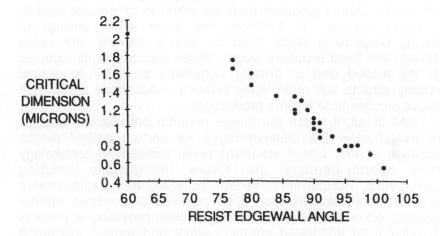

Figure 46. Critical dimension vs. edgewall angle for Shipley 1400 base IREV photoprocess.

As with any processing situation, image reversal processes have trade-off disadvantages to go along with the depth of focus, image contrast, CD control, and resist image edge wall advantages. Initial device level exposures must be equal to or greater than 2-3X normal levels.   In addition, two other wafer throughput reducing processing steps must be added, namely, the vapor treatment and flood exposure steps. These disadvantages, coupled with the added one of greater sensitivity of IREV to optical proximity effects will most likely prevent widespread application of these processes to volume production.

All in all, it is felt the image reversal process advantages may overshadow the disadvantages for certain critical device fabrication levels, where standard resist processing technology simply cannot satisfy the future lithographic imaging requirements. Furthermore, recent work(90) has demonstrated image reversal behavior for dyed positive photoresist without degrading edge wall slope advantages, thus providing a process with relief from integrated standing wave and pattern scattering effects in addition to the relief from the bulk effect provided by the undyed reversal process.

4.3   Multi-layer   Applications

4.3.1 Summary of Need.   Multi-layer processing techniques, where layers of radiation sensitive (top), non-photosensitive organic, and/or inorganic materials sandwiched together to become the total patterning layer, have become common in semiconductor and computer manufacturing R & D labs. (See Figure 39)(3).   Due to their complexity and problems that have appeared at bi-layer interfaces, these techniques have not been widely accepted in high volume production. Here, multi-layer resist processing techniques will be reviewed with emphasis upon problems associated with interfacial phenomena occurring between layers, especially for bi-layer systems.

Photolithography image edge and dimension quality is limited by two basic effects, bulk and substrate reflectivity (92,93).   The bulk effect arises when lithography patterns are required at two different topographical layer levels of the vertically-fabricated monolithic circuit.   Reflectivity effects occur when patterned areas of the circuit have different reflectivity coefficients, as well as topographical levels.

Lithographic exposure tool resolution performance can be influenced by resist processing. Stover et al.(94) have shown K from the resolution equation, $R = K \, wl/2NA$, is directly influenced by multi-layer processing: wl is the monochromatic light wavelength and NA is the numerical aperture of the projection optics lens system; K is typically 0.8 in manufacturing with single layer resist processes but can be as low as 0.5 with multilayer processes(95). Linewidth control is also directly affected by resist processing, and greater resist image critical dimension control can be achieved through multi-layer processing(93) combined with anisotropic reactive-ion etching technology.

Lin(96) has done extensive research in bi-layer systems, multi-layer processes utilizing a UV sensitive material on top with a DUV absorbing or non-absorbing system underneath. This type of multi-layer system has seen limited circuit fabricaton application, because it is plagued by deleterious interfacial layers formed between layers at coat(93,97). These problems have been solved by various treatments both before and after image formation, but bi-layer technology has taken a backseat primarily to dyed thick single-layer photoresist processes(76). Some bi-layer processes are reported to be free of interfacial mixing problems(98), but the processes described later will all be free of this yield-killing potentiality.

Tri-layer systems(99) utilize an intermediate layer between the photosensitive top layer and the virtually developer-insoluble oxygen RIE patterned bottom layer. It is usually deposited by low temperature (<250°C) thin film deposition techniques, but can also be applied as a liquid spin-on-glass solution(100-2). The middle layer can also be a spun organic polymer (also water soluble) barrier layer coating for the lithographic tri-layer processes, as opposed to RIE tri-layer processes where the middle hard mask material must be RIE etched and cannot be developed by base developers or water. While the former technique is usually void of interfacial mixing layer problems, the latter technique can exhibit this problem intermittently.

All of the multi-level processes described are successful to some degree in relieving resolution and linewidth control limitations of current single-layer optical exposure equipment. Future device fabrication requirements and the application of high numerical aperture exposure equipment, however, will most likely create the need for multi-layer processes at one or two critical device levels, and the interfacial phenomena occurring in some of these systems will have to be well understood before these needs materialize in the 1990-92 timeframe.

4.3.2 Tri-layer Gate Processes (see Figure 39). Two tri-layer processes, one an RIE and the other lithographic, have seen a lot of activity in research fab areas over the last few years(99-100) and at Motorola. The lithographic tri-layer process involves contast enhancement material, CEM(103-5) manufactured by General Electric (now Huls), which is a high extinction coefficient nitrone dyed bleaching   layer, and I-line stepper patterning. The other process involves RIE image transfer to the substrate and e-beam lithography exposure. These processes will be used as illustrative examples and are described below.

The lithographic tri-layer is composed of a conventional resist bottom layer (e.g., Kodak 820, etc.), a water soluble poly(vinyl alcohol) barrier middle layer to prevent interlayer mixing as reported for bilayer systems, and the nitrone dyed CEM 388 top layer(106). The top CEM layer is applied at 0.5 micron for depth of focus latitude relief created by the poor depth of focus (DOF) dictated by the 0.42 NA I-line lens (DOF = +/-0.5 micron). The process provides vertical edge wall images in the bottom layer (see Figure 47) following CEM and barrier coat strip on a wafer track with CEM organic stripper and water, respectively, and a normal immersion development and rinse of the bottom resist layer. The only added process steps are the two added coat and strip processes, but the image edgewall gains and CD control are substantial.

The CEM process CD control at 3-sigma for a 0.6 (gate) and 0.5 micron (trench) CD's are 0.11 and 0.10, respectively. This control compares favorably to the single layer comparison numbers in the ≥ 0.25 micron region. When CD variations are small as for this process, they allow some total overlay tolerance to be given to registration, thus providing improved total overlay performance. This is visualized in Figure 48 where the overlay of vias and metal layers is shown; the total overlay of these two circuit layers depends upon the alignment accuracy between the levels as well as the metal and via CD's and their process CD variations. The CEM process also actually allows the employment of the I-line stepper at CD's about 0.15 micron below the single layer Raleigh limit of the tool; this occurs because the CEM process increases the contrast by a factor of 1.6 over the single layer contrast value for AZ 5214 and a factor of 2.9 over the Kodak 820 reference value. Stated another way, it allows tool operation at a reduced K factor.

RIE tri-level processes are used in e-beam lithography primarily to provide relief from proximity effects, which occur from cooperative exposure from nearby pattern backscattered electrons (see Figure 49), and for RIE etch selectivity advantages. A 0.5 micron tri-level gate processing sequence is shown in Figure 50. Greeneich(107) has demonstrated the dramatic reduction in backscattering coefficient and the improvement of

FINAL IMAGE PROFILES FOR CEM 388 PROCESS

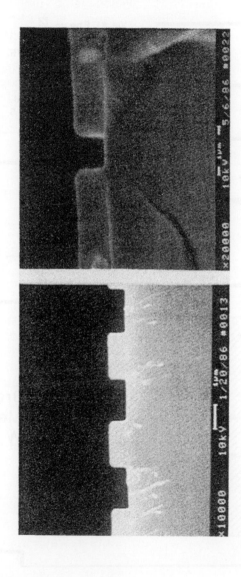

>1 MICRON SIZE

SUBMICRON SIZE

Figure 47. Edge wall profiles of the final resist image for the lithographic CEM tri-layer process. The bottom resist is Kodak 820.

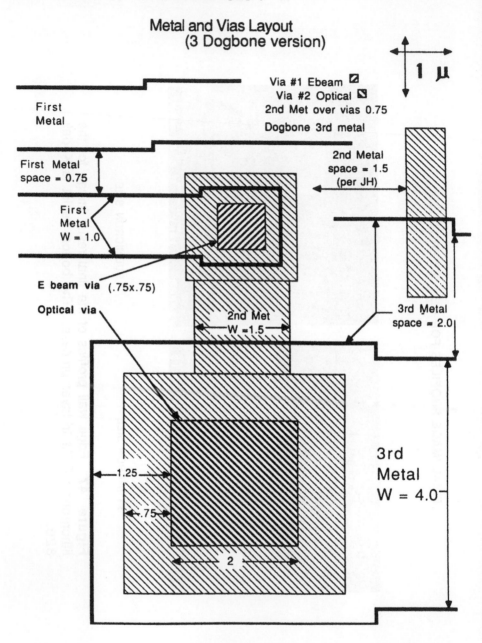

Figure 48. Metal and device level layout for hypothetical circuit. Note, the overlay tolerance would depend upon the level to level alignment and CD variation in an RMS fashion.

PROXIMITY/THICKNESS EFFECT

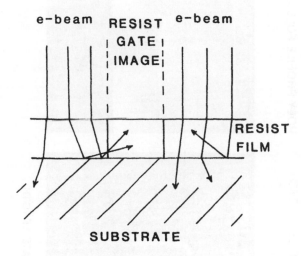

Figure 49. Proximity effect illustration for electron beam lithography cooperative exposure from adjacent written patterns.

RIE TRILEVEL PROCESSING SEQUENCE PROFILES

RESIST PROFILE   FOLLOWING TRILEVEL ETCH   POLY PROFILE FOLLOWING ETCH

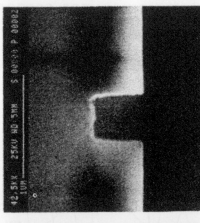

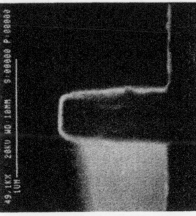

0.5 MICRON GATE LENGTH

Figure 50.   Edge wall profiles for gate lithography process.

image profile when a two micron thick planarizing layer is employed as the bottom layer. When employed as a photo process, this layer is usually a novolac or polyimide (PI) layer that can be dyed as seen for SLR processing for notching relief. For example, Bruce et al.(73) report a reduction of 30% in linewidth variation and increased depth of focus by employing a dyed PI which reduced substrate reflectivity to 4% from 33%. When the bottom layer is a conventional photoresist and is baked at high temperature, say 250°C, the bottom layer loses optical tranmission (e.g., for Hunt 204 T=0.25 at 405nm; Figure 51). Hence, even without dye the bottom layer benefits the optical lithography as well. This very process was used successully for CMOS gate and metal patterning processes employing an Ultratech 1000 1X stepper; the required 1.5 micron gate and 1.75 micron metal patterns could not be satisfactorily completed by a second generation SLR process at that time due to the reflective interference from the substrate.

When the middle layer of the tri-layer is a hard oxide layer, deposited by an ASM plasma-enhanced LPCVD unit , the process complexity is significant over SLR processing. Spin coatable intermediate layers are, however, available(100-2) thus allowing all layers to be applied by wafer track spinning. Unfortunately, some of these glasses do significantly undercut or etch during top layer development, so caution is advised with these systems. The top resists can be conventional photoresists or negative e-beam resists. When they are thin conventional resists, relief from lack of focus tolerance from high NA steppers and improved resolution is obtained.

Typical CD variation control for a tri-level gate process, using Alpha-methyl chloromethyl styrene (AMCMS-S) negative e-beam resist, is 0.05 (3-sigma) micron for the nominally written 0.5 micron gate, where the average gate dimension varies from 0.48 to 0.55 micron, lot to lot. When a positive resist, Hunt WX 214, is employed for delineating a device isolation pattern with the same RIE tri-level as above for the gate example, the CD control is typically worse at a mean = 0.57 +/-0.12 micron 3-sigma. Fortunately, the gate dimension is held tighter; this is essential to CMOS device performance because the gate length actually controls device performance.

4.4   Future Processes

With the advent of the extension of optical lithography to shorter wavelengths and higher numerical apertures to resolution values approaching 0.3 micron, the future processing needs will surely be in the multi-layer or in the surface sensitive

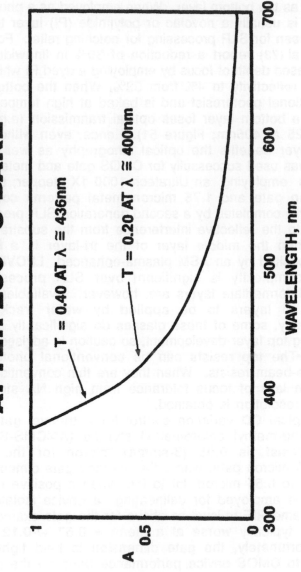

Figure 51.   Optical transmission curve for bottom layer of RIE tri-level resist process.

processing area, termed Desire processes(108-9). This processing has considerable potential, because it would allow for relief of e-beam proximity effects and photolithographic bulk and focus tolerance effects.

The latter processing allows thick SLR systems to be employed as bi-layers for RIE selectivity and bulk effect relief, and with no interlayer mixing problems. Desire processing is accomplished by post HMDS treatment of the exposed photoresist to yield an in-situ silylated bi-layer, and renders the system as a RIE bi-layer system as opposed to the more typical lithographic bi-layer systems. The system requires oxygen RIE development, like tri-level systems, of the final image.   It allows simplified processing over that for conventional bi-layer processes in that 1) the second resist spin and other top layer processing, 2) the top layer development, 3) the DUV flood exposure of the bottom layer, and 4) a subsequent separate development are eliminated and replaced by a high pressure HMDS treatment and RIE development. Obviously, Desire processes have an even greater advantage over trilayer processes due to greater overall process simplification.

Desire processing, like the other special processes of this section, possesses higher process contrast(109). This is achieved because the process images only the thin top part at the top of the thick resist layer. Furthermore, all feature sizes can be written at nearly the same exposure level, and dye can be incorporated to provide reflective notching effect relief. The process is not without problems, however, as swelling effects have been observed and reported at SPIE 1988. Although this processing is in the early stage of development, applications will probably emerge quickly, especially for pilot lines with advanced equipment.

5.0   SUMMARY AND FUTURE PREDICTIONS

Although many processing advances have been delineated in this last section, reduction of these advances to production of VLSIC's will occur with significant negative inertia. More advanced next-generation SLR systems will probably emerge first, especially at the lower wavelengths before multi-layer resist processes see extensive utilization. It should be noted that these new good SLR systems will also be needed as part of the multi-layer special processes. The high NA lithography tools may, however, accelerate the applications of these advanced processes out of forced necessity primarily for focus tolerance relief.

In the area of radiation resists, new sensitive SLR resists or surface sensitive thick resists for ASIC applications will evolve. Figure 52 illustrates clearly the need for resist speed in direct-write e-beam lithography. Furthermore, thick processes will be needed for RIE masking requirements, and multi-layer processes will be considered where SLR processes fail to provide adequate fabrication ability.

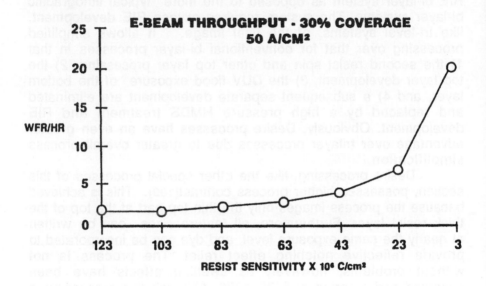

Figure 52. E-beam throughput vs. resist sensitivity for Perkin-Elmer AEBLE 150.

REFERENCES

1. Thompson, L.F. and Kerwin, R.E., Annual Review of Materials Science, 6: 267-299 (1976).
2. Bowden, M.J. and Thompson, L.F., Solid State Technology, May: 72 (1979).
3. Thompson, L.F., Willson, C.G., and Bowden, M.J., Introduction to Microlithography, Wash. D.C.: ACS Symposium Series 219 (1983).
4. Bowden, M.J., in: Materials for Microlithography, (Thompson, L.F., Willson, C.G., and Frechet, J.M.J. eds.), 266, pp 39-117, ACS Symposium Series, Wash. D.C. (1984).
5. Sze, S.M., VLSI Technology, New York: McGraw-Hill (1983).
6. Helbert, J.N., Schmidt, M.A., Malkiewicz, C., Wallace, E., and Pittman, C.U., in: Polymers in Electronics, (Davidson, T. ed.), 242, pp 91-100, ACS Symposium Series, Wash. D.C.(1984).
7. DeForest, W., Photoresist Materials and Processes, New York: McGraw-Hill (1975).
8. Kaplan, M., and Meyerhofer, D., RCA Review, 40: 167-190 (1979).
9. Barrow, G.M., Physical Chemistry, New York: McGraw-Hill (1966).
10. Watts, M.P.C., Microelectronics Manufacturing and Testing, June: 1 (1984).
11. Calvert, J.G., and Pitts, J.N., Photochemistry, New York: Wiley (1967).
12. Arden, W., Keller, H., and Mader, L., Solid State Technology, July: 143-150 (1983).
13. Trefonas, P. and Daniels, B.K., SPIE, 771: 194-210 (1987).
14. King, M.C., in: VLSI Electronics-Microstructure Science, New York: Academic (1981).
15. Hanabata, M., Furuta, A., and Uemura, Y., SPIE, 771:85-92 (1987).
16. Templeton, M.K., Szmanda, C.R., and Zampini, A., SPIE, 771:136-147 (1987).
17. Pampalone, T.R., Solid State Technology, June: 115-120 (1984).
18. Fahrenholz, S.R., Goldrick, M.R., Hellman, M.Y., Long, D.T., and Pietti, R.C., ACS Org. Coat. Plast. Chem. Preprints, 35: 306-311 (1975).

19.   Deckert, C.A. and Peters, D.A., Solid State Technology, Jan.: 76-80   (1980).
20.   Reiser, A. and Marley, R., Trans. Faraday Soc., 64: 64 (1968); Blais, P.D., Private Communication (1978).
21.   Neisius, K. and Merrem, H.J., J. Electronic Materials, 11: 761-777   (1982).
22.   Hayase, S., Onishi, Y., and Horiguchi, R., JECS, 134: 2275-2280   (1987).
23.   Reichmanis, E., Smith, B.C., Smolinsky, G., and Wilkins, C.W., JECS, 134: 653-657 (1987). Also see references therein. Wolf, T.M., Hartless, R.L., Schugard, A., and Taylor, G.N., J. Vac. Sci. Technol. B, 5:396-401 (1987).
24.   Turner, S.R., Ahn, K.D., and Willson, C.G., Chapter 17 in: Polymers for High Technology Electronics and Photonics (Bowden, M.J. and Turner, S.R., eds.), 346, pp 200-210, ACS Symposium Series, Wash. D.C. (1984).
25.   Gipstein, E., Ouano, A.C., and Thompkins, T., JECS, 129: 201-204   (1982).
26.   Fok, S. and Hong, G.H.K., in: Proceedings of Kodak Microelectronics Seminar-Interface (1983).
27.   Bethe, H.A. and Aschkin, U., in: Experimental Nuclear Physics (Segre, E., ed.), New York: Wiley (1959).
28.   Helbert, J.N., Iafrate, G.J., Pittman, C.U., and Lai, J.H., Polym. Eng. and Sci., 20:1076-1081 (1980).
29.   Iafrate, G.J., Helbert, J.N., Ballato, A.D., and McAfee, W.S., US Army R&D Tech. Report, Ft. Monmouth NJ: ECOM-4466 (1977); Iafrate, G.J., Helbert, J.N., Ballato, A.D., Cook, C.F., and McAfee, W.S., in: Proceedings of Army Science Conference, West Point NY, June   (1980).
30.   Hatzakis, M., Ting, C.H., and Viswanathan, N., in: Proceedings of Electron and Ion Beam Sci. and Techn., 6th International Conf., 542-579 (1974).
31.   Thompson, L.F. and Bowden, M.J., JECS, 120: 1304-1312 (1973).
32.   Chen, C-Y., Pittman, C.U.,and Helbert, J.N., J. Polym. Sci.: Chem. Ed., 18:169-178 (1980).
33.   Lai, J.H., Helbert, J.N., Cook, C.F., and Pittman, C.U., J. Vac. Sci. Technol., 16:1992-1995 (1979).
34.   Fahrenholz, S.R., J. Vac. Sci. Technol., 19:1111-1116 (1981).
35.   Buiguez, F., Lerme, M., Gouby, P., and Eilbeck, N., in: Proceedings of 1984 International Symposium on Electron, Ion, and Photon Beams (1984).
36.   Bowden, M.J. and Thompson, L.F., Fahrenholz, S.R., and Doerries, E.M., JECS,128: 1304-1311 (1981).

37. Thompson, L.F., Stillwagon, L.E., and Doerries, E.M., J. Vac. Sci. and Techn., 15: 938-943 (1978); Thompson, L.F., Yau, L., and Doerries, E.M., JECS, 126: 1703-1708 (1979).

38. Tan, Z.C.H., Daly, R.C., and Georgia, S.S., SPIE, 469:135-143 (1984).

39. Imamura, S., Tamamura, T., Sukegawa, K., Kogure, O., and Sugawara, S., JECS, 131:1122-1130 (1984); Saeki, H., Shigetomi, A., Watakabe, Y., and Kato, T., JECS, 133:1236-1239 (1986); Imamura, S., JECS, 126:1628-1630 (1979); JST News, 3: 56-57 (1984).

40. Taylor, G.N., Coquin, G.A., and Somekh, S., in: Technical Papers of SPE Photopolymers Principles - Processes and Materials, 130-151 (1976).

41. Tanigaki, T., Suzuki, M., and Ohnishi, Y., JECS, 133: 977-980 (1986).

42. Frechet, J.M.J., Eichler, E., Stanciulescu, M., Iizawa, T., Bouchard, F., Houlihan, F.M., and Willson, C.G., Chapter 12 in: Polymers for High Technology Electronics and Photonics (Bowden, M.J. and Turner, S.R., eds.), 346, pp 138-148, ACS Symposium Series, Wash. D.C. (1984).

43. Liu, H., deGrandpre, M., and Feely, W., Proceedings of The 31st International Symposium on Electron, Ion, and Photon Beams, Woodland Hills, CA., May (1987).

44. Brunsvold, W.R., Crockatt, D.M., Hefferson, G.J., Lyons, C.F., Optical Engineering, 26: 330-336 (1987).

45. Turner, S.R., Schleigh, W.R., Arcus, R.A., and Houle, C.G., in: Proceedings of Kodak Microelectronics Seminar-Interface (1986).

46. Oldham, W.G., Nandgaonkor, S.N., Neureuther, A.R., and O'Toole, M., IEEE Trans. Electron Devices, ED-26, 717 (1979).

47. Monohan, K., Hightower, J., Bernard, D., Cagan, M., and Dyser, D., in: Proceedings of Kodak Microelectronics Seminar-Interface (1986).

48. Blais, P.D., Solid State Technology, August, 76-79 (1977).

49. Walker, C.C., and Helbert, J.N., in: Polymers in Electronics, (Davidson, T. ed.), 242, pp 65-77, ACS Symposium Series, Wash. D.C. (1984).

50. Wong, C-P., and Bowden, M.J., in: Polymers in Electronics, (Davidson, T. ed.), 266 (see ref.4), pp 285-304; Tarascon, R., Harteney, pp. 359-388, ACS Symposium Series, Wash. D.C. (1984).

51.   Helbert, J.N., and Hughes, H.G., in: Adhesion Aspects of Polymeric Coatings, K.L. Mittal, editor, 499, Plenum Press, New York (1983).

52.   Deckert, C.A., and Peters, D.A., in Proceedings of the 1988 Kodak Microelectronics Seminar, 13 (1977); Circuits Manuf. April, (1979).

53.   Siegbahm, K., Nordling, C., Johansson, G., Hedman, J., Heden, P.F., Hamrin, K., Gelius, V., Bergmark,T., Werme, L.O., Manne, R., and Baer, Y., in: ESCA Applied to Free Molecules (North Holland, Amsterdam, 1969), Carlson, T.A., in Photoelectron and Auger Spectroscopy, Plenum Press, New York, 1975.

54.   Helbert, J.N. and Saha, N.C., Chapter 21 in: Polymers for High Technology Electronics and Photonics (Bowden, M.J. and Turner, S.R., eds.), 346, pp 250-260, ACS Symposium Series, Wash. D.C. (1984).

55.   Mittal, K.L., Solid State Technol., 89 (May 1979).

56.   Helbert, J.N., Robb, F.Y., Svechovsky, B.R., and Saha, N.C., in: Surface and Colloid Science in Computer Technology, Mittal, K.L., ed., pp 121-141, Plenum Press, New York, 1987.

57.   Deckert, C.A., and Peters, D.A., in: Adhesion Aspects of Polymeric Coatings, K.L. Mittal, editor, 469, Plenum Press, New York, 1983.

58.   Stein, A., " The Chemistry and Technology of Positive Photoresists," Technical Bulletin of Philip A. Hunt Chemical Corporation.

59.   Meyerhofer, D., in: Characteristics of Resist Films Produced by Spinning, J. Appl. Phys. 49, 3993-3997 (1978).

60a.   Middleman, S., J. Appl. Phys., 62, pp. 2530-2532, (1987).

60b.   Lai, J.H., paper at American Chemical Society National Meeting, Honolulu, April 1979.

61.   Box, G., Hunter, W., and Hunter, J., in: Statistics for Experimenters, Wiley, New York (1978).

62.   Bryce, G.R., and Collette, D.R., Semiconductor International, 71-77, (1984).

63.   Diamond, W.J., Practical Experiment Designs for Engineers and Scientists, Lifetime Learning Publications, Belmont, California, (1981).

64.   Alvarez, A., Welter, D., and Johnson, M., Solid State Technology, July, 1983.

65.   Yates, F., in: The Design and Analysis of Factorial Experiments, Bulletin 35, Imperial Bureau of Soil Science, Harpenden, Herts, England, Hafner (Macmillan).
66.   Johnson, M., and Lee, K., Solid State Technology, Sept.: 281 (1984).
67.   Campbell, D.M., and Ardehali, Z., Semiconductor International, 127-131 (1984).
68.   Matthews, J.C., and Willmott, J.I., Jr., SPIE, 470: 194-201, (1984).
69.   Ma, W. H-L., SPIE, 333: 19-23, (1982).
70.   Roberts, E.D., SPIE, 539: 124-130, (1985).
71.   Peters, D.A., and Deckert, C.A., JECS, 126: 883-885 (1979).
72.   Burggraaf, P., Semiconductor International, Sept.: 53 (1987).
73.   Bruce, J.A., Burn, J.L., Sundling, D.L., and Lee, T.N., in: Proceedings of Kodak Microelectronics Seminar-Interface (1986).
74.   Pampalone, T.R. and Kuyan, F.A., in: Proceedings of the 1985 Kodak Microelectronics Seminar, (1985).
75.   Mack, C.A., Solid State Technology, Jan.: 125-129 (1988).
76.   Bolsen, M., Buhr, G., Merrem, H.J., and van Werden, K., Solid State Technology, Feb.: 83-88, (1986).
77.   Renschler, C.L., Stiefeld, R.E., and Rodriquez, J.L., JECS, June: 1586-1587 (1987).
78.   Brown, V., and Arnold, W.H., SPIE, 539: 259-266, (1985).
79.   Moritz, H., and Paal, G., U.S. Patent No. 4,104,078, 1978.
80.   MacDonald, S., Miller, R., Willson, C., Feinberg, G., Gleason, R., Halverson, R., MacIntyre, M., and Motsiff, W., "The Production of a Negative Image in A Positive Photoresist," Proceedings of Kodak Interface, 1982.
81.   Long, M., and Newman, J., "Image Reversal Techniques with Standard Positive Photoresist," SPIE Proceedings, Vol. 469, Advances in Resist Technology, p. 189, 1984.
82.   Alling, E., and Stauffer, C., SPIE 539: 194, (1985).
83.   Klose, H., Sigush, R., and Arden, W., IEEE Transactions on Electron Devices, Vol. ED-32: 1654 (1985).
84.   Moritz, H., IEEE Transactions on Electron Devices, ED-32: 672 (1985).
85.   C. Hartglass, in: Proceedings of Kodak Interface, (1985).
86.   Marriott, V., Garza, M., and Spak, M., SPIE, 771: 221-230 (1987).

87.   Spak, M., Mammato, D., Jain, S., Durham, D., "Mechanism and Lithographic Evaluation of Image Reversal in AZ 5214 Photoresist," VII Int. Tec. Photopolymer Conf. (1985).
88.   Blais, P., Solid State Technology, August (1977).
89.   Stauffer, C., and Alling, E., ST-A-R Process Technical Bulletin #11, September (1986).
90.   Gijsen, R., Kroon, H., Vollenbroek, F., and Vervoordeldonk, R., SPIE, 631: 108, (1986).
91.   Nordstrom, T., Semiconductor International, Sept.: 158-161 (1985).
92.   Helbert, J., in: Interfacial Phenomena in the New and Emerging Technologies, (Krantz, W., and Wason, D., eds.) 2-41 - 2-44, Proceedings of the Workshop held at Department of Chemical Engineering, University of Colorado, May, (1986).
93.   Arden, W., Keller, H., and Mader, L., Solid State Technology, 143, July,(1983).
94.   Stover, H., Nagler, M., Bol, I., and Miller, V., SPIE, 470: 22 (1984).
95.   Bruning, J., ECS Proceedings of the Tutorial Symposium on Semiconductor Technology, Vol. 82-5, p. 119-137 (1982).
96.   Lin, B., Solid State Technology, May: 105-112 (1983).
97.   Lin, B., SPIE, 174: 114 (1979).
98.   Vidusek, D., and Legenza, W., SPIE, 539: 103-114, (1985).
99.   Ting, C., and Liauw, K., SPIE, 469: 24 (1984).
100.   Gellrich, N., Beneking, H., and Arden, W., J. Vac. Sci. Technol. 3: 335-338 (1985).
101.   Ray, G., Shiesen, P. Burriesci, D., O'Toole, M., and Liu, E., JECS, 129: 2152-3 (1982).
102.   Jones, S., Chapman, R., Ho, Y., and Bobbio, S., in: Proceedings of the 1986 Kodak Microelectronics Seminar, (1986).
103.   Griffing, B. and West, P., Solid State Technology, May: 152-157, (1985).
104.   Strom, D., Semiconductor International, May: 162-167, (1986).
105.   Blanco, M., Hightower, J., Cagan, M., and Monahan, K., JECS, 134: 2882-2888 (1987).
106.   Austin, J., Keller, G., Witting, G., in: Proceedings of the 1986 Kodak Microelectronics Seminar, (1986).
107.   Greeneich, J., JECS, Extended Abstract, Vol. 80-1, p. 261 (1980).

108. Roland, B., Lombaerts, R., Jakus, C., and Coopmans, F., SPIE, 771, 69 (1987).
109. Coopmans, F., and Roland, B., Solid State Technology, 93, June (1987).

3

MICROLITHOGRAPHY METROLOGY

Robert Larrabee
Loren Linholm
Michael T. Postek

National Institute of Standards and Technology
Gaithersburg, Maryland

1.0 SUBMICROMETER CRITICAL DIMENSION METROLOGY*

1.1 Introduction

The never-ending push of the semiconductor industry toward submicrometer feature sizes on integrated circuits and in discrete devices has led to the situation where the development of the techniques of fabrication of submicrometer features has exceeded the development of the metrological techniques required to accurately measure and characterize these features. At the present time there are no submicrometer critical-dimension standards available from the National Institute of Standards and Technology (NIST) for feature-size measurements on integrated circuits and thus, no way to achieve traceability to NIST for such measurements. NIST has active programs to develop the techniques for measuring such features and for certifying standards using optical microscopy and scanning electron microscopy. In addition, NIST has an active program to develop electrical dimensional metrology using specially designed test patterns. This chapter will summarize these programs and highlight some of the results obtained to date.

In the optical arena, one basic problem of submicrometer metrology arises from the diffraction of light because the feature sizes of interest are of the order of the wavelength of light used in making the measurement. A second basic problem arises from the conceptual definition of "size" when the edges of the feature are not

---

vertical and thus produce different "sizes" at the top and bottom of the feature. Ways to cope with these problems have been developed for thin-film (less than 1/4 wavelength thick) photomasks viewed in transmitted light, and standard reference materials (SRMs) for antireflecting-chrome photomasks covering the linewidth range from 0.9 to 10.8 µm have been available for about 9 years (SRM 474 and SRM 475). These SRMs are about to be replaced with a new photomask standard (SRM 473) covering the linewidth range from 0.5 to 30 µm.   The corresponding standards for integrated circuit features are currently under development and not presently available. The optical portion of this chapter describes some of the problems in making optical measurements on opaque structures (such as integrated circuits) in reflected light and what might be done to circumvent these problems until dimensional standards for integrated circuit features can be developed.

In the scanning electron microscope (SEM), the wavelength of the electrons in the "illuminating" electron beam is much smaller than the feature sizes of interest in electronic devices and circuits (e.g., the electron wavelength is about 0.4 Å at a beam energy of 1 kV and even smaller at higher beam energies). Therefore, electron diffraction effects at the specimen are usually negligible for features with dimensions and structures currently used in integrated circuits (e.g., 0.1 µm and larger). However, there is another more abiding problem in the SEM that is as serious as diffraction in the optical microscope. That problem arises because the beam electrons penetrate into, and interact with, the specimen and can emerge from points on the specimen surface that are different from the point of initial electron-beam impact and thereby produce undesired signals at those emergent points. These high energy emergent electrons can also leave the surface and can then interact with other features on the specimen or with the component parts of the microscope (i.e., anything they can happen to hit) and produce additional signals. The net effect is to produce output signals arising from points remote from the point of initial impact of the scanning beam. Consequently, the observed image-forming signal may contain (and usually does contain) information from many points on the specimen and from internal parts of the microscope and chamber. The SEM portion of this chapter describes and illustrates these and other basic problems in the SEM, points out some potential pitfalls in submicrometer SEM dimensional metrology, and outlines what NIST is currently doing to solve these problems and to develop useful SEM standards.   In addition, suggestions are made on how to make meaningful SEM measurements and achieve some measure of precision without NIST standards.

Dimensional measurements can also be made by direct electrical methods on specially designed and patterned structures of conductive thin films. These measurements are made using geometrically simple test structures, which are fabricated for this purpose with the same materials and processes used in making complex semiconductor devices. The measurements are usually performed on these test structures with a computer controlled dc parametric test system. This electrical test structure approach can result in relatively fast linewidth measurements with a precision and accuracy comparable to optical measurements for selected types of test samples. The electrical portion of this chapter discusses this technique and some recent innovative modifications of the basic test structure for specific applications.

This chapter does not discuss some of the potential or proposed techniques of dimensional metrology that might be used in the more distant future (e.g., scanning tunneling microscopy) and does not evaluate any of the proposed nonconventional modes of operation of the optical or scanning-electron microscopes (e.g., near-field optical imaging). The purpose of the chapter is to describe the general state of the art in submicrometer dimensional metrology as presently practiced by the semiconductor industry, to highlight some of the problems with such practice, and to give suggestions on how to improve the precision and accuracy of such measurements.

2.0 OPTICAL CRITICAL DIMENSION METROLOGY

2.1 Introduction

Optical microscopy has a long history of application to the measurement of small dimensions. Even today, with dimensions of interest extending into the submicrometer range, optical systems cannot be ruled out. They still have the significant advantages of high throughput, of not being restricted to conductive materials, and of not changing or destroying the specimen. However, as the dimensions of interest have evolved into the submicrometer region, optical measurements have encountered difficulties arising from the fact that the feature dimensions are becoming comparable to (or smaller than) the wavelength of the light used for their measurement, thus enhancing diffraction and scattering effects, and making the image difficult to interpret. However, the need for highly precise and accurate measurements of features with dimensions in the 0.1- to 1.0- μm range exists today, and the upper part of this range is still within the domain of optics. (The Airy disk radius is 0.45 μm for f/1 optics at the near ultraviolet wavelength of 0.366 μm). This portion of the chapter explores the general field of optical metrology in this upper range of submicrometer dimensions. It will be assumed that the emphasis is on high quality measurements with precision and accuracies in the low nanometer range. It will be seen that the attainment of this degree of precision and accuracy will depend on the understanding and control of many factors that can affect the measurement as well as on a thorough understanding of exactly what is being measured and what can go wrong with the measurement.

2.2 Precision and Accuracy

Instrumental precision (or repeatability of the measurement) is defined as the variability observed in repeated measurements on a given sample with as many as possible of the factors that affect the measurement held constant(1). The effects of unknown (or uncontrolled) sources of imprecision are often assumed to be random so the average of the series of measurements can be taken to be a good estimation of the "best" value, and the standard deviation (or 2 to 3 times the standard deviation) can be assumed to be a good estimation of the precision. However, there is no guarantee that a highly precise (reproducible) measurement is accurate (i.e., yields the correct result). Accuracy implies that there is a universal agreement upon the nature and definition of the quantity being measured and that there also exists some basic standard of comparison. Unfortunately, for the case of optical submicrometer dimensional metrology, such definitions and standards may not presently exist. However, the desired degree of precision in submicrometer optical measurements may be achievable. That is important because precision is a necessary, but not sufficient, condition for accuracy. The importance of precision is often overlooked in optical metrology as evidenced by the fact that there is often too little attention paid to the control of the factors affecting precision and to the monitoring of the instrumental precision actually achieved in practice.

There are at least four main causes of imprecision when making dimensional measurements of production specimens in an optical system(2): 1) variations in the conditions of measurement (e.g., focus), 2) perturbing environmental conditions (e.g., vibration), 3) variations in human judgement (e.g., deciding where the feature edges are located for a dimension measurement) and 4) variations in the important characteristics of the specimens being measured (e.g., index of refraction of the specimen materials). Some of these factors can be eliminated (e.g., automation can eliminate some sources of human-induced imprecision) and some can be minimized (e.g., sources of vibration can be identified and remedial measures taken). Some of the remaining (perhaps unknown) sources of imprecision are random (e.g., noise) and thus reducible to acceptable levels by averaging repeated measurements. However, if any of the remaining sources of imprecision are not random (e.g., variations in the image profile of a feature resulting from contamination deposited on the specimen), no amount of averaging will reduce them. Therefore, a well-thought-out procedure of measurement based on sound metrological principles can significantly improve precision (e.g, specifying that measurements be taken in the center of the field of view to minimize off-axis aberrations in the optical system). It is not the purpose of this chapter to list or discuss all possible sources of imprecision or to recommend a universal procedure for obtaining the best possible precision. However, one purpose is to emphasize that one very important step toward accuracy is to recognize and control all known or suspected sources of imprecision.

One does not need an accurate standard to measure instrumental precision. A typical specimen of the type to be measured (e.g., a sample of the product in question) that is known to be stable in time will often suffice. Long-term instrumental precision can be determined by repeated (e.g., hourly, daily, weekly, or monthly) measurements on this type of control specimen followed by the application of well known quality control charting (or equivalent) procedures(1-3). The spacing of the control limits can serve as a measure of the attainable instrumental precision and can provide a means of comparison to determine if the measurement is under control This type of quality control is often done to monitor specific semiconductor processing steps (that is one reason for interest in dimensional measurements), but this type of quality control is less often done for the measurement process itself. The attainment of high precision in optical submicrometer metrology is not routine and its attainment requires continued attention to the many things that can go wrong(4-7). Indeed, when something is observed to go wrong, the first question that needs to be answered is whether the problem is instrumental or not (i.e., whether the problem is in the measurement itself, or in the product being measured).

Control charting the measurement process is one way to answer this question.

The attainment of the required degree of instrumental precision does not guarantee accuracy. Given precision, there are two main sources of inaccuracy in optical submicrometer metrology(2): 1) lack of a generally accepted standard of comparison, and 2) improper use of the standards that do exist. When suitable standards are not available, there are probably good technical reasons for their unavailability (e.g., their production or certification may be beyond the current state of the art), and that reason will probably determine what can, or can not, be done to improve accuracy. The temptation is to use the best in-house control specimen as an interim standard. This may be acceptable as a temporary expedient if done correctly.

The desired degree of accuracy may be achieved if the instrument is sufficiently precise and the specimens of interest exactly match the standard of comparison in all important ways. However, in optical submicrometer dimensional metrology, one cannot always guarantee that the specimens to be measured will match the standard in all ways (e.g., in feature height, in substrate properties, in edge geometry, in the complex index of refraction of all materials in the specimen which interact with the illuminating light, etc.).   Recent efforts to model the optical interactions with integrated-circuit specimens indicate that all of these factors could be of prime importance(8). Clearly, it is inappropriate to use a thin-layer metal-on-glass photomask standard such as those of the NIST 473 and 476 series(9-10) to "calibrate" a system which subsequently will be used on other types of specimens (e.g., photoresist lines on integrated circuit structures). However, as mentioned above, it may be appropriate to use in-house control specimens as temporary calibration standards but, if and only if, they closely match the specimens to be measured and are known to be stable in time. This necessity to match the properties of the standard to the specimen of interest presents one of the more serious problems to NIST in developing optical submicrometer dimensional standards for the integrated-circuit industry. There are so many different types of specimens of interest (e.g., photoresist features with different thickness, edge geometries, and optical properties of features on a variety of substrates) that it is impossible for NIST to produce a sufficient variety of standards so that everyone will be able to find one that matches any given specimen of interest. Indeed, except for special cases where the industry has adopted standardized structures, providing matching standards for optical submicrometer dimensional metrology is probably not a practical solution to the problem of attaining accuracy! However, a deeper study of the problem leads to a suggested solution. The considerations leading up to that suggested solution are the main topics to be discussed below.

2.3 Optical Diffraction

Professor Francesco Grimaldi (1618-1663) was the first to note that there are bands of light within the shadow of a rod illuminated by a small source and his observations were the beginning of the wave theory of light(11). Today, this same effect is usually observed as the apparent ability of light to "bend" around corners of a thin sharp edge and this phenomenon is called diffraction. Diffraction effects in optical metrology become very important when the desired precision of the measurement approaches, or becomes smaller than, the wavelength of the light used to illuminate the specimen being measured. Since visible light covers the wavelength range from about 0.4 to about 0.8 μm, diffraction effects can be very large in visible-light submicrometer optical metrology. For example, diffraction effects are significant in the measurement of the width of a submicrometer-wide clear line in an otherwise opaque thin film observed in a microscope with coherent transmitted light. The light "bends" around the edges of the clear line and, by so doing, produces an image in the microscope that does not have sharp well-defined edges. The Fraunhofer diffraction from a single transmissive slit in an otherwise opaque thin film when observed as a shadow (i.e., without a microscope) has an intensity distribution of the form:

[1]   Intensity $= [(\sin x)/x]^2$

where, x is a normalized coordinate measured in the plane of the diffraction pattern in a direction perpendicular to the axis of the slit(11). Notice that the intensity distribution is without recognizable edges. The corresponding intensity distribution in the microscopic image of an edge depends on additional factors (e.g., numerical aperture of the microscope) and may not have the form of Eq. 1. However, the image of the "shadow" never has sharp well-defined edges (even in incoherent light). How then can one accurately determine edge location in a measurement of the width of lines, spaces, or more complicated patterned features? One way was suggested by the work of a number of people in the 1960s(12-21) and was pioneered at NIST as a method of calibrating a photomask standard(9)(17)(18). This method starts with a mathematical calculation of the observed intensity distribution of the light in the microscopic image of the feature of interest as a function of position across the width dimension, and then fits this computed intensity distribution to the measured intensity distribution with the desired dimension as an adjustable parameter. If agreement between the mathematical and experimental image profiles is obtained in this way, the position of the edges can be determined from the mathematical result and the desired dimension found as the distance between these edges. The photomask standards NIST produces by this technique contain opaque lines and clear spaces of varying size with their width certified by NIST. The user can calibrate his/her optical measuring system using these standards with an arbitrary (i.e., not necessarily correct) method of determining edge location (e.g., the 50% points on the image profile), but then must subsequently use this same type of photomask. Unfortunately, the intensity distribution in diffraction patterns is dependent upon the index of refraction and physical dimensions of all materials of the specimen that interact with the light, so a calibration of a measuring system with a standard of one type may not be valid for use with a specimen of another type. In addition, features with a thickness exceeding about 1/4 of the wavelength of the illuminating light cannot be considered thin in the mathematical analysis and the image profile becomes strongly dependent on the thickness of the feature (with a considerable increase in the difficulty of computing the image profiles)(8). Therefore, thin-film standards (like photomasks) cannot be used to calibrate systems used to measure features patterned in thick layers. The diffraction pattern also depends on the properties of the illumination and imaging optics of the measuring system (e.g, coherence of illumination(18) and numerical aperture of observing objective(4), so each system needs independent calibration with a standard matching the specimens of interest. However, the standards necessary for accuracy in any given case may not be available and, in these cases, precision is the best that can be obtained.

Figure 1 shows the measured intensity distribution (solid line) as a function of position in a direction perpendicular to the edge of a thin (0.15 μm thick) opaque chromium line (with an antireflecting coating) on a borosilicate glass substrate as observed in transmission with the NIST photomask calibration system(17). The light intensity has been normalized to 100% for the intensity of the light passing through the clear areas on the glass substrate. The horizontal scale is marked off by tick marks

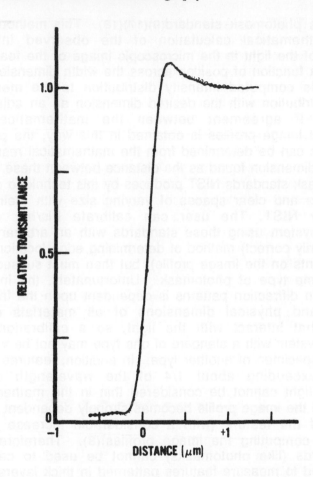

Figure 1. Comparison of experimental (line) and theoretical (open circles) profiles of an edge of an antireflecting chrome photomask line on a borosilicate glass substrate observed in transmission. The photomask was Kohler illuminated from below at a wavelength of 0.530 μm using a 0.6 N.A. objective lens as a condenser lens. The observing microscope was above the photomask and had an objective lens of 0.9 N.A. and had sufficient resolution in its image to measure this diffraction pattern. The chromium layer was approximately 0.15 μm thick, and in the theoretical model, the chrome line was assumed to have a vertical edge geometry. All other edges of lines and spaces on the photomask are sufficiently far removed from this edge so that the overlap of their diffraction patterns is negligible. Note the high degree of agreement between the theoretical predictions and the experimental profile (Figure from Reference 12 but included in a number of papers by D. Nyyssonen).

0.5 μm apart. Now consider the question: Where, on this profile, is the actual location of the edge of the line?

The open circles in Figure 1 show the results of a theoretical calculation by D. Nyyssonen of the diffraction pattern expected for this photomask specimen illuminated with the partially coherent light of the NIST photomask calibration system and observed in its imaging system(17). The theoretical curve has been similarly normalized to 100% and has been shifted horizontally to best fit the experimental data. Notice that the agreement is quite good and, since the position of the edge in the theoretical model is known, the edge on the theoretical curve can be taken to coincide with the edge on the observed profile. In this case, it is at the 25% point (after correction for extraneous light) because of the fact that the illumination light in the NIST system is effectively coherent over the effective area seen by the microscope and, when half of this area is covered by the opaque line, the electric field of the light is cut in half and the intensity (electric field squared) is reduced by a factor of 4. If the illuminating light were effectively incoherent over this area, intensities (rather than electric fields) would add, and the edge would be at the 50% point. If the coherence of the light were unknown, the actual edge location could be at an unknown position somewhere between the 25% and 50% points. For the case of the optics in the NIST photomask calibration system, the edge-location uncertainty due to a 25% to 50% uncertainty in the proper location in the image profile (see Figure 1) would amount to about 0.08 μm (i.e., 0.08 per edge or twice that for the linewidth uncertainty). This illustrates the fundamental fact that the illumination optics may be as important a factor as the imaging optics for attaining precision and accuracy(18).

Figure 2 shows calculated image intensity profiles of thick (i.e., larger than 1/4 wavelength) polysilicon lines on a silicon-dioxide-covered silicon wafer observed in reflection with small cone-angle illumination in an ideal high-resolution micro-scope(19). Notice that the width of these profiles is much larger than in the photomask example above and that their shape is much more complex with a larger number of maxima and minima. Features with heights larger than approximately 1/4 wavelength of the illuminating light introduce significant phase effects into the mathematical boundary-value problem at the feature interfaces and such features usually have wider and more complex image profiles than thin-layer features. The term "diffraction" is usually reserved for thin-layer features and the term "scattering" is used for the analogous phenomena in thick-layer features.   The distinction being that scattering involves

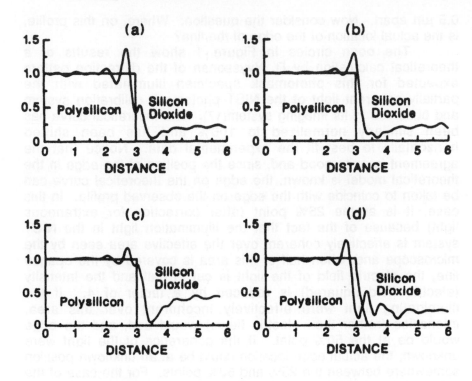

Figure 2. Theoretical profiles of one edge of a 6 μm wide and 0.6 μm thick polysilicon line on a 0.105 μm thick unpatterned layer of silicon dioxide-on-silicon wafer. The polysilicon line is assumed to be symmetric about its center and only the right half is shown in this figure. The cross-sectional shape of the line is shown superimposed on the corresponding calculated optical image profile. The dependence of the image profile on specimen geometry is illustrated for the cases of (a) vertical edges on the polysilicon line, (b) changing the thickness of the polysilicon layer to 0.65 μm, (c) restoring the original polysilicon layer thickness and changing the silicon dioxide layer thickness to 0.125 μm, and (d) restoring the original silicon dioxide thickness and changing the edge geometry of the polysilicon line to the shape shown in the figure. These profiles were computed for the case of imaging in reflected light using narrow cone-angle illumination (0.17 N.A.) at a wavelength of 0.53 μm and imaging with an objective of 0.85 N.A. (Computations by D. Nyyssonen and figure taken from Reference 2).

consideration of one more dimension than diffraction. For example, the mathematical boundary-value problem involves only one dimension (width) for diffraction from a uniform infinitely long slit in a thin layer, but two dimensions (width and height) for scattering from that same slit in a thick layer. The mathematical model(8) used to predict the expected scattering profiles from thick structures such as shown in Figure 2 is currently being tested at NIST by comparison with experiment. However, even if the model results were known to agree with experiment, the model could only be used to determine the edge locations if all the other parameters required by that model were independently known or simultaneously measured. For example, for the case shown in Figure 2 one would need to know: the complex index of refraction of the polysilicon line, the silicon dioxide layer, and the silicon substrate; the thickness of the line and the silicon dioxide layer; and the geometrical shape of the line edges. Clearly, this is a greater degree of complexity than anyone wants for a "simple" submicrometer dimensional measurement, and points out the fundamental fact that scattering effects make it difficult to locate edges in thick-layer features and thus make it difficult to measure their dimensions. It is mainly because of these difficulties that calibration methods for submicrometer dimensional standards involving thick features (e.g., photoresist on silicon wafers) are still in the research state.

Scattering effects impose a basic requirement for NIST in developing methods for certifying accurate standards for optical submicrometer metrology of thick-layer features. That requirement is the necessity for a detailed understanding, and for realistic modeling, of the scattering effects that occur in the specific specimens of interest and of the imaging of this scattered light in the actual instrument used for certification (regardless of the type of instrument or mode of operation). If this requirement is met, there will be a sound metrological basis for the criteria and thus a sound logical basis for the dimensional results given in the certification of the NIST standard. At the present time, it is not clear to what extent the user of future thick-layer dimensional standards will be burdened with the necessity to become involved in similar types of modeling.

The required modeling is not simple(8)(17) and the numerical evaluation often takes considerable time and memory on mainframe computers. Therefore, one basic principle of instrument design for practical submicrometer optical metrology of thick-layer features should be to design the measurement system for simplicity of modeling. Clearly, diffraction-limited performance is required in the optics to observe an undistorted diffraction or scattering pattern(20), and sufficient spatial resolution is required to produce a diffraction or scattering pattern that will show significant variations to changes in linewidth at the desired precision. Resolution performance beyond this may be counterproductive if it significantly increases the complexity, execution time, or computer hardware requirements of the model or increases the cost, measurement time, maintenance requirements, or complexity of operation of the instrument.

2.4   Geometry Of the Sample

In Figure 3a, a cross section of a line is shown that has an ideal structure with vertical walls, smooth edges, and a uniform height. In this case, the width of the line can be unambiguously defined as indicated by W in the figure. However, the structures of practical interest seldom, if ever, have this ideal structure. They may have ragged edges, such as shown in Figure 3b, or they may have nonvertical and possibly different right and left edge geometries as shown in Figure 3c. In cases such as these, how is linewidth to be accurately defined? Lines with ragged edges can have their width measured at a number of points along their length and the results analyzed to give an average width along with some measure of the raggedness (e.g., the standard deviation of the width measurements). If the raggedness is not severe and only an average width is required, the width can be measured in an optical system designed to simultaneously measure (and average) over a suitable length of the line. For nonvertical edge geometries, there is no way to uniquely define the "width" from first principles because, in fact, the width varies from the top to the bottom of the line. Perhaps the best approach in this case is to let the user define what is to be taken as the width based on the particular application and the reason for making the measurement. For photoresist lines, the width at the bottom of the line next to the substrate may be the "best" definition of width but, for electrically conducting lines, the average width (which, when multiplied by the height, gives the appropriate area for current flow) may be more appropriate. Alternatively, the "width of the line" might be defined as the actual width at some specified height above the substrate as shown in Figure 3c. Lines with complex edge geometries (e.g., standing-wave formations on the edges of photoresist lines) further complicate the definition and measurement of linewidth. When all these factors are considered, the ideal optical linewidth measuring system would measure the profile of the optical image of the line, but then would present the

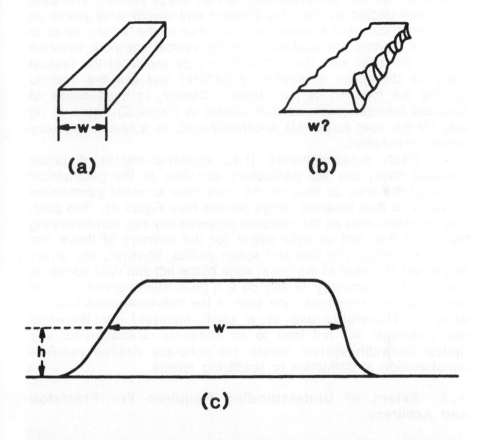

Figure 3. Schematic illustration of some of the problems encountered in defining the linewidth of nonideal structures: (a) an ideal line geometry in which the width, W, can be rigorously defined, (b) a more realistic line geometry that has nonvertical edges, dissimilar left and right edge geometries, and nonuniform cross section along its length, and (c) a typical cross section of the structure of (b) in which the linewidth at that position is defined by the geometrical width at some specified height, h, above the substrate.   The height would be chosen to give a meaningful linewidth for the intended application (e.g., small h for photoresist lines).   Note that the linewidth so defined is still a function of position along the length of the line in (b). (Figure from Reference 5).

user with a graphical representation of the geometrical cross section of the line corresponding to that image profile. The user could then decide on the edge criterion and specify what points on the geometrical cross section are to be taken as the "edges" so as to best suit his/her application. Existing optical linewidth systems present the user with the measured image profile of the feature (i.e., the diffraction or scattering pattern) and ask the user to specify the location of the edges. Clearly, in most cases of practical interest (such as that shown in Figure 2), there is no way for the user to do this accurately and, as a result, accuracy cannot be obtained.

Pitch measurements (i.e., center-to-center distance between lines) are not particularly sensitive to the geometrical shape of the lines as long as the lines have identical geometrical shapes and thus identical image profiles (see Figure 4). The pitch can be determined as the distance between any two corresponding points on their left or right edges (or the average of these two measurements). For line and space widths, however, any errors in the determination of the "true" edge of the left and right edges do not cancel by symmetry as they do in a pitch measurement, but add and produce a result with the sum of the individual edge location errors. Therefore, use of a pitch standard for linewidth measurements will not lead to an adequate calibration for any optical linewidth system where the accuracy desired requires consideration of diffraction or scattering effects.

2.5 Extent of Understanding Required For Precision and Accuracy

The diffraction or scattering effects discussed above, due to the wavelength of light being comparable to the feature sizes of interest, is a major metrological limitation to precision and accuracy in optical micrometer and submicrometer dimensional measurements. These effects can dominate the optical image profile of such features (as illustrated in Figure 2). It has been pointed out above that the detailed shape of such profiles depends on many things: the geometry and the indexes of refraction of all materials in the specimen that interact with the light; the wavelength, uniformity, coherence and direction of the illuminating light; the numerical aperture of the objective and the position of the focal plane of the microscope; the aberrations or misalignment in the imaging optics; and any vibration of the specimen relative to the component parts of the optical system(4)(5)(20). Therefore, precision in micrometer and submicrometer optical metrology can only be obtained if all such

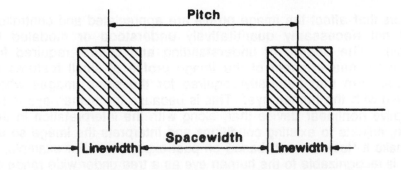

a. Ideal lines with vertical edge geometry

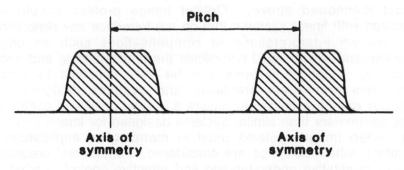

b. Lines with nonvertical edge geometry

Figure 4. Distinction between a pitch (i.e., distance between two line centers) and a linewidth and spacewidth (i.e., distance from edge-to-edge of a single feature): (a) the linewidth is only unambiguously defined for lines with vertical walls and no edge raggedness, (b) pitch is usually defined as the distance between the axes of symmetry of two lines, but it can be measured as the distance between any two similar points on the image profiles. This latter definition of pitch is appropriate for lines without an axis of symmetry, e.g., lines with dissimilar right and left edges. Note, however, that pitch is only unambiguously defined for lines with identical cross sectional shapes. (Figure from Reference 17).

factors that affect the image profile are appreciated and controlled (but not necessarily quantitatively understood or modeled in detail).   The degree of understanding and control required for precision measurement of the image profile of small features is greater than that generally required for the same images when viewed with the human eye. This is because the human eye is an adaptive nonlinear device that, along with the interpretation in the brain, adjusts to existing conditions and interprets the image so as to make it "look correct" based on past experience. For example, a tree is recognizable to the human eye as a tree under wide range of conditions of illuminating intensity and spectral composition (e.g., bright sunlight or dim moonlight).   However, the shape of the image profile of a submicrometer line may show saturation of the detector at high light levels, have an undesirable signal-to-noise level at low intensities and have a shape that depends on all the factors mentioned above.   Optical image profiles should be measured with linear systems that do not introduce any distortions (or unknown interpretations or compensations such as edge enhancement) which would complicate the understanding and need controlling.   Optical systems can be very sensitive to those parameters that affect precision, and systems designed for metrology can be expected to require a greater degree of control of those parameters than similar systems designed for image viewing (e.g., video image systems used in metrological applications). Therefore, when all things are considered, instrumental precision requires qualitative understanding and effective control of what is happening at the specimen (diffraction and scattering) and in the microscope (imaging), but not detailed modeling of these effects.

In addition to the attainment of the desired degree of instrumental precision, accuracy in optical dimensional metrology requires knowing exactly where the edges are on the measured feature profile.   There is no universal answer to the inevitable question, "Where is the edge?" The answer to that question may vary in different instruments and will depend on a quantitative understanding of many things, including all of the parameters mentioned in the above paragraph.   Indeed, it is necessary to model (i.e., quantitatively understand) the entire measurement system to answer this question.   Arbitrarily defining the edges to be at the 50% points, at the 25% points, at particular minima (or maxima) in the image profile, etc. may be correct in certain cases, but to know that it is correct requires that the particular system be modeled and its image profile calculated for each different type of sample to be measured. As mentioned above, models of this type are being developed at NIST for use in developing certification procedures for micrometer and submicrometer dimensional standards. The demonstration of a satisfactory degree of agreement between the calculated and measured image profiles under a variety of conditions (e.g., defocus, changing geometry, various materials and substrates, etc.) will provide NIST with confidence in the model and the validity of the edge criteria derived from it. The validity of the model-derived edge criteria could also be checked by measuring the features by other independent methods (assuming that independent methods known to be sufficiently accurate do, in fact, exist). Once confidence in the model is obtained, it can be used to derive edge criteria for proposed thick-layer dimensional standards and thus provide the needed sound metrological basis for their certification. However, as pointed out above, NIST cannot make a large enough variety of standards so that one of them will effectively match any given integrated-circuit feature. Therefore, any future standards certified in this way will probably have limited application. However, this approach illustrates the fundamental fact that the attainment of accuracy may require complete quantitative understanding (i.e., mathematical modeling) of the optical interactions occurring at the specimen and of the imaging of the diffracted or scattered light in the system actually used for the measurement.

Therefore, in the final analysis, instrumental precision requires qualitative identification and control of the parameters that affect the image profile. Accuracy requires considerably more in that the effect of these parameters on the image profile must be quantitatively understood and modeled in order to determine where the feature edges are located.

2.6 Equipment Requirements for Precision and Accuracy

A fundamental requirement for precision in optical submicrometer dimensional metrology is a microscope with a sensitivity capable of detecting the expected variations in the image of the features to be measured. This includes variations that are actually due to the dimensional changes of interest and, perhaps more significantly, to dimensional changes that are significantly smaller than the precision desired (i.e., 10 times smaller, because there are always contributions from other sources of precision). However, as discussed above, the image profile of a submicrometer feature observed in a microscope may depend on many things; thus, fulfilling this dimensional sensitivity requirement is not straightforward, it may be application specific, and it may require specific evaluation testing by the user. Specification of the effective numerical aperture or the conventional point-spread resolution of the optical instrument is not adequate for this purpose. Likewise, a specification of precision based on the reproducibility achieved under some set of conditions by the instrument supplier is also inadequate, because it may be inappropriate for the specific application and environmental conditions of the instrument user. Consider for a moment the fact that optical metrology tools are often used to monitor the performance of pattern generating tools and, therefore, should have an even greater precision than those pattern generating tools (but, perhaps, over a smaller field). Therefore, the same care and detail commonly used in connection with the optical pattern generation tools (e.g., step-and-repeat systems) should be applied even more strongly to the corresponding optical metrology tools. Unfortunately, because of cost considerations, this is usually not done and, as a result, the metrological results are often below expectations.

One alternative to accurate dimensional standards that match the samples of interest is to have a system that displays edge geometry information (instead of image profiles) to allow the user to define what is meant by linewidth for each particular edge-wall geometry of his/her specimens. The metrology system requirements for such a system are more stringent because it will probably be necessary to use real-time mathematical modeling to extract the desired edge information from the measured profile. If real-time mathematical modeling is required to interpret the image profile for this or any other such purpose, the instrument should be designed to accommodate this modeling and thereby reduce the computer software and hardware requirements to produce a meaningful result at a reasonable cost in a tolerable computation time. Unfortunately, most of the modeling work for existing instruments has not been done and the present efforts in this area are not large. Finally, if other parameters of the specimen must be known (e.g., feature height or index of refraction) for control purposes or to input to a model, suitable facilities for their measurement must be either included in the optical metrology instrument or provided independently.

There are a wide variety of papers on a number of optical schemes for making linewidth measurements (e.g., Fourier transform methods, confocal methods, laser-scan methods, near-field methods, phase-sensitive methods, etc.), and these schemes have not been specifically addressed in the present chapter. This is because detailed models demonstrating their capabilities and limitations are not currently available and because the present chapter was mainly intended to cover material that is generic and common to all these schemes. These papers can be found in such journals as Microlectronic Engineering, Applied Physics Letters, Ultramicroscopy, Optical Engineering, Applied Optics, Journal of the Optical Society of America, various volumes of the Proceedings of SPIE - The International Society for Optical Engineering, Proceedings of the Royal Society, etc. A few of these are given in Reference 2, but the present author is unaware of any complete bibliography as such. Perhaps the best guide for selection of the basic optical instrument for dimensional metrology is not to be initially concerned about the optical scheme used, but to ask the following very basic questions: 1) does the instrument have the precision desired for the specimens of interest in the environment in which it will be used, 2) does it have the sensitivity in that environment to detect the smallest significant change in the dimensions of the features of interest, and 3) if accuracy, in addition to precision, is required, can the feature size in question be defined geometrically for the specimens of interest and, if so, can the corresponding feature edges be located in the image profile to the accuracy required. The answers to all three of these questions should be based on sound optical and metrological arguments. If these questions cannot be satisfactorily answered, the ability of the instrument to make the required measurement is open to question.

2.7   An   Ideal   Submicrometer   Optical   Dimensional Metrology   System

The concepts necessary to operationally design an ideal submicrometer optical metrology instrument have been discussed. To be truly ideal, that instrument must have the required precision and accuracy while overcoming problems of the types discussed above. The most difficult of these problems stem from the material and edge-geometry dependence of the image profile. The problem introduced by nonvertical edge-wall geometry is basically why the ideal system will probably use an inverse approach in which the geometrical cross-sectional shape of the feature in question is computed in real time from the measured image profile. Therefore, as illustrated schematically in Figure 5, one would like an ideal system to determine and display the actual geometrical shape of the edge of the feature. It would do this by measuring the image profile (Figure 5a), and, after accepting data on everything known (or concurrently measured) about the sample (e.g., indexes of refraction), it would compute and display the edge geometry in real time (Figure 5b). The system would

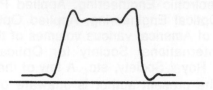

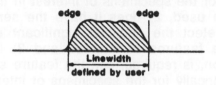

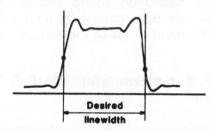

Figure 5. Schematic illustration of the ideal submicrometer dimensional metrology system: (a) measured optical profile across the image of a line. Notice that it is not obvious where the "edges" of the line are located on this diffraction or scattering pattern, (b) actual geometrical shape of the cross section of the line above as determined by solution of the inverse scattering problem. Notice that the user can now identify the points where the "edge" should be taken for the intended application, (c) the points on the original image profile corresponding to the user's definition of "edge." Notice that the desired linewidth is readily determined as the measured distance between these two points, but that these points could not be identified without solving the inverse scattering problem and presenting the actual geometrical cross-sectional shape of the line to the user. (Figure from Reference 17).

then ask the user which points on this cross section are to be considered the "edges" of the feature for the dimensional measurement (e.g., by asking the user to set video cursors on the geometrical profile as shown in figure 5b). The system would then identify the corresponding points on the measured optical profile (Figure 5c) and determine the feature dimension as the distance between these points. Notice that left-right edge symmetry is not required since the actual shape of each edge would be calculated by the system. This calculation is, in fact, the most difficult part of the process because it requires a general solution of the inverse scattering problem(21-22). This problem is not an exercise of image analysis or artificial intelligence in computer programming, but a fundamental and, for the most part, an unsolved problem in optical physics. Until this inverse problem is solved under assumptions appropriate for some real instrument and for the actual types of specimens of interest, it may be impossible to optically measure submicrometer dimensions of thick-layer specimens with the desired degree of accuracy.

This ideal system should be able to make dimensional measurements on a wide variety of specimens by solving the inverse scattering problem in real time for each particular type of specimen being measured. Given this capability, accuracy can be achieved by using a standard of one type (i.e., a standard that fully exercises the system by calibrating its basic metrology and testing its algorithms for solving the inverse scattering problem) and then using the system on a specimen of another type (i.e., the desired integrated circuit feature). This ideal system also circumvents the need to standardize on the definition of linewidth for nonvertical wall features, because that decision is made by the user in real time for his/her particular application after examination of the measured cross-sectional geometry of the specimen being measured. To be truly ideal, the system should be capable of profiling thick 3-dimensional features (such as via openings on integrated circuits) by solving the 3-dimensional inverse scattering problem in the submicrometer case where the optical interaction problem is not separable in the 2 lateral coordinates (i.e., where the diffraction or scattering effects in the 2 dimensions in the plane of the specimen overlap and interact with one another). NIST is working on these problems in two dimensions (i.e., width and height), but the magnitude of the problems and the generality of the solutions needed (e.g., applicable to a wide variety of structures, feature materials, and commercial measuring systems) will require: 1) the development of practical (i.e., nonextraneous and not overly idealized) solutions to the inverse scattering problem for some real instrument, 2) the application of this approach to the construction and certification of suitable standards, 3) the development of recommended procedures for using these standards to correctly set up and calibrate the user's systems for the intended measurements, and 4) commercially available instruments designed for real-time calculations of the inverse scattering problem.

2.8   Current Issues In the Utilization Of Optical Submicrometer Metrology

2.8.1 Accuracy Not Available. Everyone wants accurate measurements but, unfortunately, accuracy is only achieved with the expenditure of time and effort and is only possible if acceptable comparison standards are available. NIST has photomask linewidth and pitch standards available over limited ranges of dimensions, but no standards for features (such as photoresist patterns) on semiconductor wafers. Relative accuracy can be achieved with in-house standards of the desired product if the pertinent properties of that product have been shown to be stable in time (a real question with photoresist features). If absolute or relative accuracy is not available, consider the possibility that precision may suffice. In this approach to the problem, the metrologist first achieves an acceptable degree of instrumental precision and then determines experimentally that the system is making a meaningful measurement on product specimens: meaningful in the sense that the results can be shown to correlate to some measure of product performance. In this way, even though the measurement lacks accuracy, acceptable limits for process control can be empirically established. One obvious initial test of whether the measurement is meaningful is to determine if it can distinguish a potentially good product from bad. It is difficult to make general comments on methods establishing windows of acceptance in this case because they are undoubtedly application specific. Typical methods would include: correlating a measured feature size during fabrication to the electrical speed of operation of the completed electronic device, correlating the width and spacing of conducting lines (or their precursor photoresist lines) to failure due to shorts and opens, etc. Although this empirical approach is more demanding, it is likely to yield more satisfactory results than blind faith in the accuracy of the linewidth values produced by any one of the present state-of-the-art optical linewidth measuring systems. However, this approach is not a substitute for accuracy because its success depends on unknown and, perhaps, uncontrolled factors besides the measured parameter(s) that affect yield. These factors lower the correlation, diminish the value of the critical dimension measurement, and if they get out of control, can dominate the yield and destroy the previously established correlation. It is, however, something that should be done, even after accuracy is achieved, to validate the importance and justify the cost of the critical-dimension measurement in question.

2.8.2 Precision Not Attained. If the measurement does not possess the required degree of precision, then the question immediately arises as to why the measurement is being made. Indeed, if the measurement itself is not under control (or is not being monitored by acceptable quality control procedures) then there is no assurance that the required degree of instrumental precision is, in fact, being attained. Clearly, in this case, the measurement results are of limited value. Therefore, measurement precision must be monitored and, if found inadequate, must be corrected. Long-term precision is determined (and then monitored) by repeated measurements on a control specimen and then by the application of well-known quality control charting(1) or equivalent procedures to determine control limits(2). If the measuring system is found out of control, then the causes of the imprecision should be identified and corrected. In the final analysis, the significance of the results of a measurement are limited by the precision of the system making the measurement and, if that precision is unknown (or inadequate), those results cannot be trusted!

2.8.3 Improper Use of Standards. Since the standards required for accuracy in submicrometer optical metrology may not be available, there is a temptation to use whatever standards are available in whatever ways seem appropriate at the time. Unfortunately, this approach seldom leads to the desired accuracy and, in the case of NIST standards, may not lead to NIST traceability (i.e, accuracy based on an NIST standard). The methods of improperly applying standards to process control often seen by the author in integrated-circuit fabrication lines include: 1) using photomask standards for calibration of systems used to measure features on wafers, 2) measuring a single in-house specimen in a scanning electron microscope for use as an in-house all-purpose standard for calibrating optical systems, and 3) calibrating the magnification scale of an optical system by use of a pitch standard and assuming that this suffices for linewidth measurements. The best that can be said for these methods is that they are wrong! Precision and accuracy do require suitable standards but, in addition, require good metrology in their use. If the desired standard is not available, it may be because the required metrology is not available. If this is the case, accuracy may not be attainable by any means!

2.8.4 Motivation for Dimensional Measurements. As the lithographic art extends into the submicrometer region, the motivation for dimensional measurements should be re-examined in view of the increasing cost and complexity and possible impracticality of such measurements. Some very basic questions regarding critical-dimension measurements should be asked: Is the measurement being made for a sound quality control purpose? If so, how are the results being used: for feedback or feedforward purposes, for checking that the fabrication process is under control, for checking that previous sources of troubles are not becoming troublesome again, etc.? If it is not for a quality control purpose, why is the measurement being made: because specifications call for it, because it has always been made in the past, because it can be used as a diagnostic if the production line goes out of control, etc.? The answers to questions such as these can serve as a basis for determining if the measurement in question is really necessary, if the specified precision is really necessary, if accuracy is really required, etc. Quality control requires measurements and if that control is to be cost effective, the measurements should be meaningful and the results actually used for some constructive purpose. Therefore, the measurements should have a logical reason for being performed and should be based on sound metrological principles. Unfortunately, this is not always the case!

As the features decrease in size and approach tenth micrometer size, the desired control limits approach hundredth micrometer size (for 10% control limits) and that requires measurement precision (and accuracy?) at the thousandth micrometer level so as to obtain the precision necessary to compare the measurement results to the control limits. A thousandth of a micrometer is a nanometer or a distance spanned by about 10 atoms (i.e., about + or - 5 atoms on each side of the feature)! That presents a real challenge for optical metrology, even at the shorter ultraviolet wavelengths! Is the difficulty of this goal fully appreciated by the people who are demanding it, or is there a tendency in the semiconductor industry to simply extrapolate existing measurement requirements used at the larger dimensions to the smaller dimensions without consideration of the implication on the equipment cost, operator skill, environmental requirements, and throughput for such measurements? Is this extrapolation being done with the assumption that the required metrology for such atomic-scale precision and accuracy will somehow be available when it is needed? Clearly, dimensional metrology will have to evolve along with the lithographic art as the feature sizes become smaller and smaller and this involvement will undoubtedly require that a greater emphasis be placed on the need, value, cost, consideration of alternatives, and the possibility of not being able to make the desired measurements. The semiconductor industry cannot blindly assume that the required metrology will always be available when it is needed as witnessed by the fact that in the area of optical submicrometer critical-dimension metrology, it may not be available today!

2.9   Conclusion

Optical submicrometer dimensional metrology offers a number of advantages (e.g., nondestructive and fast throughput), but has a number of disadvantages (e.g., dependence of the image profile on properties of specimen materials and criticality of focus). Optical diffraction and scattering effects become dominant as the feature dimensions of interest approach and become less than the wavelength of the illuminating light.   Therefore, for accuracy in submicrometer dimensional metrology, it is necessary to model the effects of diffraction and scattering in the image and to develop a meaningful criterion of which point on the image profile corresponds to the edge of the feature being measured. Unfortunately, the necessary theoretical models do not exist today and many of the presently available instruments are not designed to facilitate such modeling. Because of this, it is not possible, at present, to attain accuracy or traceability to NIST in optical critical dimension measurements for features on integrated circuits.   Precision and a crude assessment of the accuracy may be attainable if the time and effort is taken to do the critical dimension measurement carefully and correctly.   The first step in doing this carefully and correctly is to understand the metrologically important factors in the measurements and their reduction to practice in the instrument used for the measurement. This first step is often overlooked in practice and the metrology is often not based on sound metrological principles, thereby turning critical dimension optical metrology into a poorly practiced black art rather than a science! Perhaps then, it is understandable why the industry has recently looked with disfavor at optical measurements and looked elsewhere (e.g., to scanning electron microscopy) for a more easily practiced "black art".

3.0   SCANNING ELECTRON MICROSCOPE METROLOGY

3.1   Introduction

The scanning electron microscope (SEM) has become a versatile tool for inspection and measurement for the semiconductor industry. This instrument has been considered, by some, to be the panacea for the problems described previously with the optical microscope for semiconductor inspection and measurement in the production environment. With some understanding and control of the problems associated with this instrument, it may ultimately fill this need.

The scanning electron microscope has only been in production since the early-to-mid 1960s. It has not had the 300 years of maturation of instrument design and theory characterizing the optical microscope, as described in earlier portions of this chapter. Although papers on the theory and operation of the SEM can be traced to the 1930s(23)(24), it was not until about 1968 that the first SEM became commercially available, and it was not until the 1980s that this instrument was considered by some as a serious metrology tool for semiconductor production. The attitude that the SEM could be used as a measurement tool was further fostered as device dimensions approached 1 μm where measurement in the optical microscope became difficult. Early scanning electron microscopes were used primarily as picture taking and analytical instruments and not for metrology purposes per se. These early instruments were well suited for low resolution imaging (about 30 nm resolution) and for analytical x-ray microanalysis (further information on x-ray microanalysis can be found in references 25-27). During the ensuing 25 years since its introduction, the instrument has rapidly evolved and the instrument resolution improved to where 2.0 to 4.0 nm resolution is presently routine using selected specimens(28) and higher resolution (0.8 nm) on similar samples is possible in special instruments (Table I).

The scanning electron microscope offers both advantages and disadvantages to semiconductor metrology (Table II). In the optical microscope, the effects of diffraction from the sample that are so visually apparent are not readily observed in the SEM. The majority of the images obtained are easily interpreted by even the most novice of microscopists. Therefore, a photoresist line or via opening appears like its pre-conception and not the blurry diffraction-limited image observed in the optical microscope. The clarity of the image is phenomenal under the proper conditions (Figure 6a-d). It is this main consideration that has generally led

TABLE I

Electron Source	Instrument Resolution (nm)	
	High keV	Low keV
Tungsten	3.0-5.0	>20
LaB$_6$	2.5-4.0	15-20
Field Emission (standard)	1.5-2.0	10-15
Field Emission (immersion Lens)	0.8	4.0

TABLE II

SEM vs. OPTICAL MICROSCOPE

Comparative Advantages

Potentially high resolution (0.8-200nm)

Excellent depth of field (focus)

Flexible viewing angles (0-60 degree tilt)

Readily interpreted image

Elemental analysis capability (X-ray characterization)

Minimal diffraction effects

Comparative Disadvantages

High vacuum requirement

Lower throughput than optical microscope

Electron beam/sample interactions

Potential for sample charging

Lack of a traceable linewidth standard

Possible beam-induced contamination

Possible damage due to high electron-beam energy

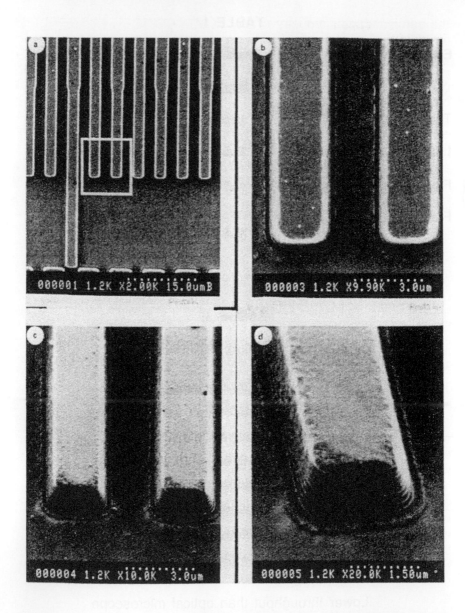

Figure 6. Scanning electron microscope inspection at low accelerating voltage (1.2 keV) of uncoated photoresist structures: (a) Magnification range typical of optical inspection instruments (Tilt=0). (b) Higher magnification of the area highlighted in the first micrograph. (c) Sample tilted to 45 degrees to view the sidewall structure. (d) High magnification showing sidewall structure.

the semiconductor industry to embrace the SEM for inspection and metrology purposes, especially where micrometer and submicrometer structures are concerned.   The ability to readily see a structure and measure it without ambiguous diffraction problems is desirable.   With this in mind, an attitude has developed that if the image can be seen better, it can be measured more precisely and, ultimately, more accurately. However, it is often forgotten that the image making up that micrograph or the linescan being measured is an electronic representation of the sample, and only that. From even before the moment that the first electron comes in proximity to the sample, biases are introduced that can seriously influence any quantitative measurements made in this instrument, biases that had little or no effect on the original uses of the SEM (i.e., imaging or elemental analysis). It will be seen in this section that the degree of precision and accuracy attainable in the SEM, just as with the optical microscope, depends upon an understanding and precise control of the instrument and those factors which can affect the measurement, as well as a complete understanding of what, in fact, is being measured.

3.2   The Basics of Scanning Electron Microscopy

The scanning electron microscope receives its name because it employs a finely focused electron beam that is moved or scanned from point to point on the specimen surface in a precise rectangular pattern called a raster pattern.   The electrons that ultimately impinge on the sample originate from an electron source or filament.   The type of SEM is defined by the type of electron source installed because each source has its own particular characteristics and requirements. Thermionic emission instruments are those where the electrons originate from a filament that is heated to a high temperature; cold field emission instruments are those that extract electrons at or near room temperature and thermally assisted field emission instruments combine the characteristics of both.   Table III compares these three types of electron sources and some of their characteristics.

The electron source (or gun) can be considered to be the "heart" of the SEM and the overall performance of the SEM directly relates to the source.   One measure of the performance characteristics of an electron gun is the measure of brightness β. Brightness is the current density of the electron beam per unit solid angle and is compared for the four types of electron emitters presently used in Table III. The design and type of electron source ultimately determine the current density of the electrons emitted

TABLE III

COMPARISON OF TRADITIONAL ELECTRON EMITTERS

USED IN SCANNING ELECTRON MICROSCOPY

	TUNGSTEN HAIR PIN	LANTHANUM HEXABORIDE	COLD FIELD EMITTER	ZR-W (100) EMITTER
TYPE OF EMISSION SOURCE	THERMIONIC	THERMIONIC	FIELD	FIELD
TEMPERATURE (K)	2650-2900	1750-2000	300	1800
BRIGHTNESS (A/CM^2 SR)	10^4-10^5	10^5-10^6	10^7-10^9	10^7-10^9
VIRTUAL SOURCE SIZE (ANGSTROMS)	1,000,000	200,000	50-100	100-200
ENERGY SPREAD(eV)	±1-2.5	±0.1-1.5	±0.1-0.2	±0.15-0.25
VACUUM (TORR)	10^{-3}-10^{-5}	10^{-5}-10^{-6}	10^{-10}-10^{-11}	<10^{-8}

from it. The larger this density for the smallest beam diameter, the better the signal-to-noise ratio (S/N) and hence the higher the limiting resolution. Some sources characteristically have a higher brightness than others, but at a price. It also can be observed from Table III that, for example, a lanthanum hexaboride filament is significantly brighter than a tungsten filament, but in order for the lanthanum hexaboride filament to operate properly, a higher vacuum is required. A similar situation exists between the comparison of lanthanum hexaboride and field emission filaments. These (and other) compromises must be considered when the particular instrument type is being evaluated.

Brightness also increases linearly with the accelerating voltage applied. This means that at high accelerating voltages (30 keV), the electron beam will be characteristically "brighter" and will have a higher signal/noise ratio than at low accelerating voltages. At higher accelerating voltages, the wavelength of the electrons is also much shorter; therefore, the effects of diffraction and other aberrations (i.e., spherical aberration) have less of an effect. The ultimate resolution of the instrument, as with the optical microscope, relates to the wavelength of the illuminating source, and thus in the SEM the resolution potentially attainable increases with accelerating voltage. An instrument can only resolve structure that is larger than the diameter of the electron beam and nominally the electron microscope can resolve about 1.5X the electron beam diameter(29). Other factors (i.e., electron beam/sample interaction effects) discussed in later portions of this chapter degrade the ultimate resolution attainable. An instrument capable of resolving 4.0 nm at 30 keV where the electron wavelength is 0.00698 nm may only be capable of resolving 15 nm at 1.0 keV accelerating voltage where the wavelength of the electrons is 0.0388 nm due to various beam broadening effects (e.g., diffraction, aberrations). Therefore, it is easy to see why the majority of scanning electron microscopy was, until a few years ago, done at high accelerating voltage (where the electron optics were optimized) on conductive or metallized samples.

Once the electron beam has been formed, it travels down the microscope column where it undergoes a multi-step demagnification with electromagnetic condenser lenses so that when it ultimately impinges on the sample, the focused spot diameter (depending upon instrument type and application) can range from about 1.0 nm to 1.0μm (at 30 keV). Proper adjustment of the accelerating voltage, beam current, and spot diameter by the operator allows optimization of the instrument for a particular sample or operation requirement.

In the electron microscope column, the electron beam is precisely deflected in the raster pattern either in an analog or digital manner depending upon the design of the particular instrument. Most newer instruments employ digital scanning so that they can use frame buffering or frame storage to improve the S/N and also incorporate auto-focus and auto-astigmatism correction(30-32). The deflection of the primary electron beam by the scanning coils in the column is synchronized with the deflection of the display cathode ray tube (CRT) so there is a point-by-point visual, albeit electronic, representation of the specimen image on the CRT screen as the electron beam scans the specimen surface. The smaller the area scanned by the electron beam, in the raster pattern, relative to the display CRT size, the higher the magnification. Therefore, this ratio must be properly calibrated to a known magnification standard in order for meaningful work to be done with this instrument, especially metrology (see SEM Standards Development). The theory of the operation of the scanning electron microscope has been covered by several authors(26)(27)(29)(33) and the reader is directed there for more in-depth coverage of this topic.

The operation of an SEM requires that at least the emission chamber and column of the instrument be under high vacuum. The level of vacuum required for the emission chamber area of each type of instrument is shown in Table III. The vacuum requirement of the sample chamber is often somewhat less stringent, being, on the average, several decades poorer in vacuum or on the order of 5.0×10^{-5} Torr (or better). High vacuum systems for the SEM include diffusion pumps, turbomolecular pumps, cryopumps and ion pumps. Depending upon the application, several different types of these pumps can be used in conjunction on the same instrument (e.g., ion pumping of the emission chamber and turbomolecular pumping of the column and specimen chamber). The high vacuum requirement in the sample chamber means that the sample must be introduced into the vacuum chamber and a pumpdown cycle must be factored into the throughput considerations. This vacuum cycle has been one of the serious drawbacks of the SEM in the semiconductor production environment since it has a direct effect upon throughput.

3.2.1 SEM Resolution. Some instruments resolve better than others (Table I) either due to engineering design or other reasons. There is, at present, no definitively accurate definition of how to quantify instrument resolution and its measurement in the SEM. The ASTM definition for resolution, as defined for optical microscopes(34), can be applied to the SEM, to obtain "the minimum distance by which two lines or points in the object must be separated before they can be revealed as separate lines or points in the image." This is essentially using a point-to-point definition (based on Raleigh's criterion) to define resolution. Put another way, this defines a pitch measurement(35-36) and not an edge-to-edge measurement (similar to a space width or a linewidth measurement) of the resolving power which is typically the manner that the "resolution of an SEM" is obtained.   Since the only practical technique is to obtain a resolution figure from edge-to-edge measurements (essentially a space width) directly from a micrograph of a suitable sample, the determination of resolution should also consider the manner by which the basic SEM signal is derived and collected relative to image statistics, method of measurement, image processing applied, noise factors, film characteristics, and exposure variables in order that precise and useful values be obtained. Once a micrograph is obtained for feature size or resolution measurement, there is the final complication which is the determination of the location of the actual edges (Figure 7a,b). While in principle this can be done on a micrograph with a ruler or microdensito-.eter, this can lead to an imprecise and undesirable situation similar to that present in the measurement of submicrometer structures with the optical microscope.   In either instance, the same edge locational problem exists. In the SEM, they are due to the electron beam/sample interaction and proximity effects, thus leading to some manner of uncertainty in the accuracy and precision of this measurement. The effects of these electron beam/sample interactions are covered later in this section.

3.2.2 Nondestructive Low Accelerating Voltage SEM Operation. Historically, scanning electron microscopy was done at relatively high accelerating voltages (typically 20 to 30 keV) in order to obtain both the best signal-to-noise ratio and instrument resolution.   As we saw earlier, the higher the accelerating voltage, the shorter the wavelength of the electron, the smaller the electron beam, and thus the better the resolution achievable.   At high accelerating voltages, nonconducting or semiconducting samples require an overcoating of gold or a similar material to provide a current path to ground and to improve the signal generation from the sample.   Otherwise, electrical charge buildup will occur on the sample (the effect of charging is discussed in a later section).   Further, early instruments were designed to accept only reasonably small samples so the large wafer samples (initially, a 50-mm diameter wafer was considered a large sample), typical of the

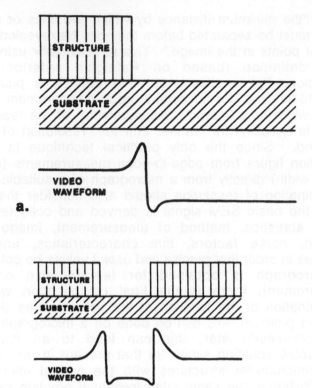

Figure 7. Edge-to-Edge resolution. (a) The scanning of the electron beam results in a video waveform derived from the collected electrons that are related to the surface structure and substrate; many of the waveforms lying close together make up the viewed image. (b) When two similar structures are viewed in proximity to each other, a similar, but not necessarily identical, mirror-image waveform is generated. Moving the structure closer together results in the eventual overlap of the waveforms making the determination of the resolution (dR) difficult. The physical location of the edge relative to the waveform is discussed in the text.

semiconductor industry, needed to be broken into smaller pieces (generally somewhat less than 25 mm) prior to inspection.   In modern semiconductor device processing, this procedure is considered a destructive technique because a broken or coated wafer cannot be processed further, and thus, all the circuits on a wafer are lost at each measurement step.   For large wafers or several inspection steps, this is a very expensive proposition. Techniques for the nondestructive "sectioning" of photoresist structures have been proposed(37) and are presently undergoing evaluation as a production procedure.   Modern on-line inspection during the production process of semiconductor devices is designed to be nondestructive which requires that the specimen be viewed in the scanning electron microscope uncoated and intact (Figure 6).   This means that modern instruments, out of necessity, be capable of handling upwards to 125 mm diameter samples (in the future, even larger wafer diameters are projected).

Another aspect of nondestructive inspection is that the sample must not be damaged or irrevocably changed by the inspection process.   High accelerating voltages interacting with the sample can damage sensitive devices(38), and low accelerating voltage inspection is thought to eliminate, or at least minimize, such damage.   Low accelerating voltage, in this case, is generally defined as being less than 2.5 keV.   A further advantage derived by operating the SEM at low accelerating voltage is that the electrons impinging on the surface of the sample have less energy, and thus have reduced specimen/electron beam interactions.   Therefore, they penetrate into the sample a shorter distance, and have a higher cross section for the production of signal-carrying electrons nearer to the surface where they can more readily escape and be collected. The techniques used in this nondestructive low accelerating voltage operation of the SEM, as used in this context, have only been in practice for about the past 3 to 5 years. Thus, this is a rapidly evolving area in scanning electron microscopy and is being applied throughout its scope of application.

SEM Metrology Instrument.   The architecture of a typical scanning electron microscope wafer-inspection instrument used in the semiconductor processing environment is similar to any modern SEM designed for low accelerating voltage operation with the exception that it is modified to accept and view large semiconductor wafers and is equipped with a clean high-vacuum system.   The instrument may also have cassette-to-cassette capabilities to facilitate wafer loading and unloading and a computer-based stage motion and video profile analysis or "linewidth" measurement system.   Two fundamentally different

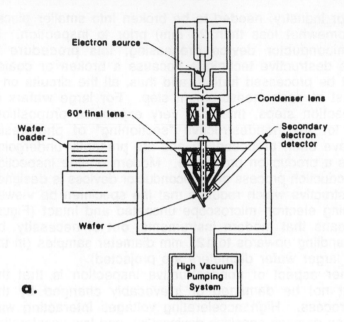

a.

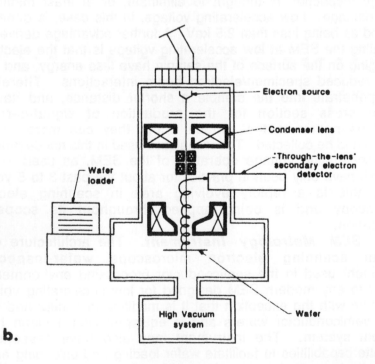

b.

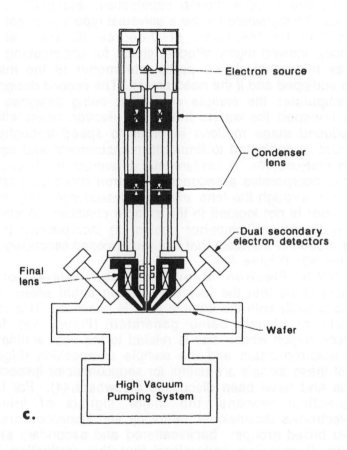

Figure 8. Diagrammatic representation of some of the basic designs of dedicated SEM instruments used for linewidth measurement: (a) Universal-type design with a conical lens where the sample can be tilted to 60 degrees (or higher) for inspection and viewed normal to the beam for linewidth measurement. (b) Flat wafer design with an in-lens detection system where the design is optimized for throughput where no sample tilt is permitted. (c) Flat wafer design with dual secondary electron detectors.

design concepts have emerged, each attempting to meet the needs of the semiconductor industry. In the first type (Figure 8a), the instrument incorporates a specimen chamber and lens design that has extensive sample motion capabilities, especially tilt. This design can be considered to be a universal type instrument where a sample can be measured at 0-degree tilt (i.e., at normal incidence), viewed highly tilted to check for undercutting or even do x-ray microanalysis on particulate matter (if the instrument was so equipped and if the need arose). The second design (Figure 8b) manipulates the sample very little, being designed only to rapidly transport the wafers below the electron beam with no tilt and reduced stage motions in order to speed throughput; this instrument is dedicated to linewidth measurement and essentially only for that use(39). A fundamental difference in this design type is that it incorporates a secondary electron detection system that allows for through-the-lens electron detection(40-41), and thus, the detector is not located in the sample chamber. A variation of this instrument and chamber design(42) incorporates the same sample motion in a system that has two opposed secondary electron detectors(43) (Figure 8c).

   3.2.3 Electron Beam/Specimen Interactions. It is well understood that the incident electron beam enters into and interacts directly with the sample as it is scanned. This results in a variety of signals being generated (Figure 9a) from an interaction region whose size is related to the accelerating voltage of the electron beam and the sample composition (Figure 9b). Many of these signals are useful for semiconductor inspection and analysis and have been discussed elsewhere(44). For historical and practical reasons, the major signals of interest to microelectronics dimensional metrology and inspection are divided into two broad groups: backscattered and secondary electrons. However, it must be understood that this distinction become extremely arbitrary, especially at low beam energies.

   Backscattered Electrons. Backscattered electrons (BSE) are those electrons that have undergone either elastic or inelastic collisions with the sample and are re-emitted with an energy that is a significant fraction (generally 50 to 80%) of the incident beam energy. This means that a 30 keV primary beam electron can produce a backscattered electron of 24 keV (or less) or a 1 keV primary electron beam can produce an 800 eV electron that can be collected and imaged or interact further with the sample, instrument or specimen chamber (see: Secondary Electrons). The measured backscattered electron yield varies with the sample, detector geometry and chemical composition of

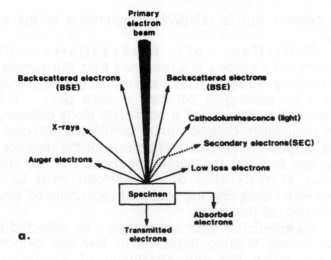

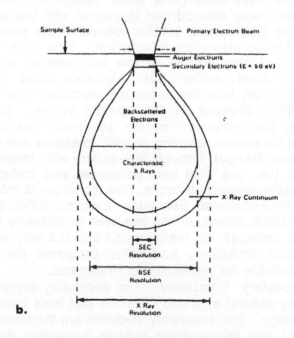

Figure 9. Types of signals generated in the scanning electron microscope. (a) Host of signals available for collection. (b) Origin of the signals in the zone of interaction.

the specimen, but is relatively independent of the accelerating voltage.

Collection   of   Backscattered   Electrons. Backscattered electrons are reemitted from the sample surface in approximately straight lines, and consequently, they must be detected by placing a detector in their path. This may be accomplished by the use of a solid state diode detector, a channel electron-multiplier detector or a scintillator detector positioned for this purpose. The size and position of the detector affects the image, and thus any measurements made from it. Therefore, the particular characteristics of the detector must be taken into account when analyzing the observed backscattered electron signal for metrological purposes.

Backscattered electrons can also be collected through the use of energy filtering detectors or low-loss detectors(45-47). Energy filtration has the advantage of discriminating those electrons that have undergone fewer sample interactions (low-loss) and thus have entered and interacted with the sample to a lesser degree (i.e., over a smaller volume of the specimen) and thus carry higher resolution information. This type of detector has been used successfully at low accelerating voltages(48), although it does suffer from signal-to-noise limitations. Discriminating BSE from the secondary electrons (see: Secondary Electrons-SEC) created at the sample is also possible by suppressing the collection of the secondary electron (SEC) generated at the sample, allowing the BSE to escape and create SEC (see: Electron Range) through interaction with internal column components (i.e., the final lens polepiece) and collecting these converted backscattered electrons. This technique is referred to as Converted Backscattered Secondary Electron (CBSE) collection (49-51). The CBSE technique has also been successfully used at low accelerating voltages for metrology(52-53), but also suffers from signal-to-noise limitations and other problems that presently make it unsuitable for production environments.

Secondary Electrons. The secondary electrons (SEC) are arbitrarily defined as those electrons that have between 1 and 50 eV of energy. The secondary electrons are the most commonly detected for low accelerating voltage inspection due to their relative ease of collection and since their signal is much stronger than any of the other types of electrons available for collection. Due to the low energy of the secondary electrons, they cannot escape from very deep in the sample, thus their information content is surface specific. Consequently, the information carried by secondary electrons potentially contains the high resolution sample information of interest for metrology.

Collection of Secondary Electrons. Secondary electrons are generally collected by the use of a scintillator-type detector of the design of Everhart and Thornley(54) or a modification of that design. Due to the low energy of the secondary electron signal, the electron paths are easily influenced by any local electric or magnetic field; therefore, this detector is equipped with a positive biased collector to attract the secondary electrons. The collection efficiency of an SEC detector relates directly to its position and potential. Detectors that have a location at some off-axis angle, as in many instruments designed to accept detectors for x-ray microanalysis, exhibit preferential detection with respect to the orientation of feature edges. In these cases, it is not possible to achieve the symmetrical video profiles necessary for precise linewidth metrology. Further, it is not easily determined if the video asymmetry demonstrated is derived from the position of the detector, from other problems introduced by the instrument's electronics, by specimen/electron beam interactions, or by a random summing of all possible problems. To compensate for an off-axis position of the secondary electron detector, the sample can be physically rotated toward it until the video profile of the line becomes symmetrical; then the structure can be aligned as designed on the display CRT by adjusting the raster pattern with digital raster rotation. Since an error can be introduced using this technique (e.g., if the sample is tilted or the raster rotation is not linear), it is much more desirable to have an on-axis detector(55) or two similar detectors on either side of the sample and the signals properly balanced and summed(43). Other detector schemes such as channelplate detectors(56)(57) and other detector schemes using quadruple detectors(58) have been proposed.

Electron Range. As shown in Figure 9, the primary electron beam can enter into the sample for some distance, even at low accelerating voltage; thus, it is important to understand and define this interaction volume. The maximum range of electrons can be approximated several ways(59). Unfortunately, due to a lack of understanding of the basic physics underlying the interaction of an electron beam in a sample, especially at low accelerating voltage, there are no equations that accurately predict electron travel in a given sample(59-60). One straightforward expression derived by Kanaya and Okayama(61) has been reported to be the most accurate presently available for approximating the range in low atomic weight elements and at low accelerating voltages. Using this equation, the calculated range of electrons in a carbon (graphite) specimen can be shown to vary from 0.03 μm at 1.0 keV to 4.4 μm at 20.0 keV. If it is considered that this calculated range approximates the boundaries of the electron trajectories as a region centered on the beam impact point (Figure 10), then it can be seen that the backscattered electrons emerging from the surface area of this region do not, in general, carry much information about the high-resolution details making up the surface topography of the specimen at the point of beam incidence. The secondary electrons, due to their inherently low energy, cannot reach the surface from deep in the specimen, so typically they escape from a region only 5 to 10 nm beneath the surface. Therefore, they do carry surface specific information in the area directly interacting with the primary electron beam. But it should be clearly understood that secondary electrons can originate from points other than the point of impact of the primary electron beam(62). Those that do originate from the point of impact are referred to as SE Type I electrons. The SE Type I electrons are the most desirable for high-resolution imaging and metrology(28)(63)(64). It would be convenient if it were possible to collect only SE Type I for metrology; unfortunately, secondary electrons are also created by re-emergent backscattered electrons at the sample surface (SE Type II) and at the polepiece of the instrument (SE Type III). Other contributions to the collected secondary electron signal include line-of-sight backscattered electrons and other sources particular to each instrument (SE Type IV). The effects of these four types of contributions to the actual image or metrology have not been fully evaluated, although recent work suggests that they contribute to a measurement broadening(52-53). Further, the smaller the diameter of the electron beam, the less pronounced is the edge broadening effect(65). Furthermore, it can be readily observed by the use of Monte Carlo calculations (Figure 11) that the interaction zone and the associated BSE generated from it provide a secondary electron signal from a much larger area than expected. Thus, the influence of the BSE on the image at magnifications where the raster pattern is greater than the interaction zone is far more significant than originally expected. Since the influence of the SE Type III electrons varies with the internal geometry of the instrument, some of the effects are instrument specific.

   Total Electron Emission. The behavior of the electrons emitted from a sample per unit beam electron (Figure 12) is extremely significant to nondestructive low accelerating voltage operation. The points where this curve crosses unity (i.e., E-1 and E-2) are the points where no net electrical charging of the sample will occur (i.e., emitted electrons equal incident electrons). During irradiation with the electron beam, an insulating sample such as photoresist or silicon dioxide can collect

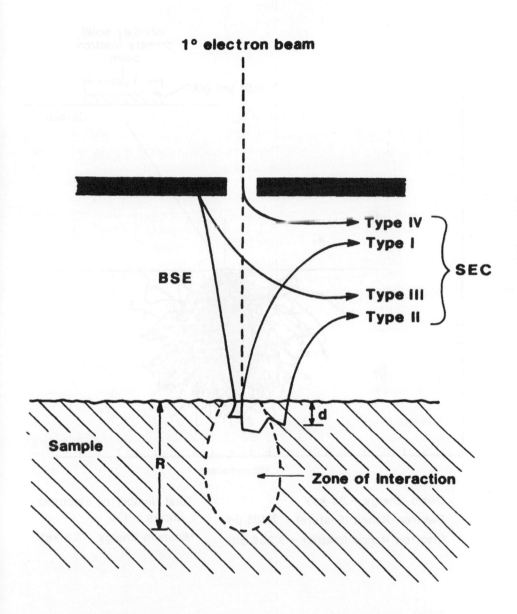

Figure 10. Origins of the various components of the secondary and backscattered electrons in the specimen chamber of the SEM.

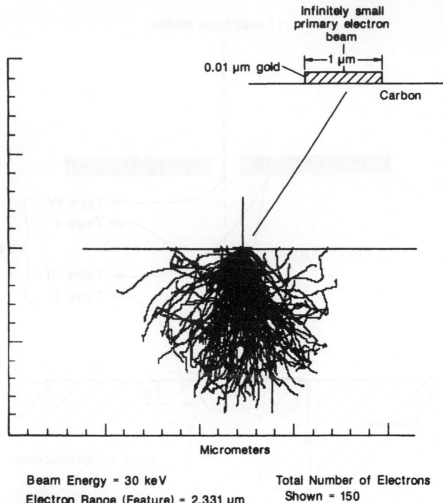

Beam Energy = 30 keV
Electron Range (Feature) = 2.331 μm
Electron Range (matrix) = 9.256 μm

Total Number of Electrons Shown = 150
Minimum Energy = 1.0 keV

Figure 11. Monte Carlo modeling of a gold island on a carbon matrix showing the extent of the signal contributions of the backscattered electrons.

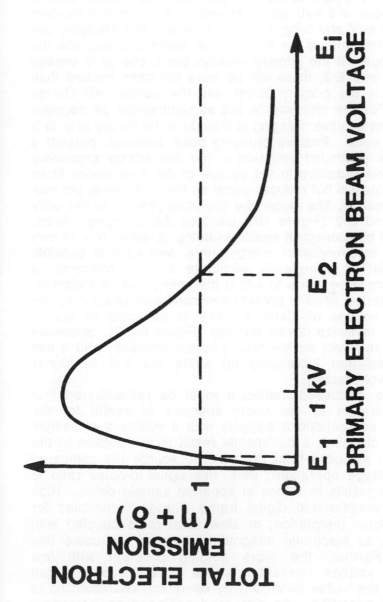

Figure 12. Variation of the total secondary plus backscattered electron yield from a specimen plotted as a function of the incident electron beam energy. The total yield is unity for two energies E-1 and E-2, called the cross-over points.

beam electrons and develop a negative charge, causing a reduction in the primary electron beam energy incident on the sample. If the primary electron beam energy is 2.5 keV and the particular sample has an E-2 point at 2.0 keV, then the sample will develop a charge to about -0.5 keV so as to reduce the effective incident energy to 2.0 keV and bring the yield to unity. This charging can have detrimental effects on the electron beam and degrade the observed image. If the primary electron beam energy is chosen between E-1 and E-2, there will be more electrons emitted than are incident in the primary beam, and the sample will charge positively. Positive charging is not as detrimental as negative charging, since positive charging is thought to be limited only to a few electron volts. Positive charging does, however, present a barrier to the continued emission of the low energy secondary electrons. This reduction in the escape of the secondaries limits the surface potential but reduces signal as these electrons are now lost to the detector. The closer the operating point is to the unity points E-1 and E-2 (Figure 13), the less the charging effects. Each material component of specimen being observed has its own total emitted electron/beam energy curve, and so it is possible that, in order to completely eliminate sample charging, a compromise must be made to adjust the voltage for all materials. For most materials used in present semiconductor processing, an accelerating voltage of about 1.0 keV is sufficient to reduce charging and minimize device damage (Figure 6a-d). Specimen tilt also has an effect on the total electron emission, and it has been reported that increasing tilt shifts the E-2 to higher accelerating voltages(66).

As we discussed earlier, it must be remembered that although operation at low beam energies is useful for the inspection of semiconductor samples with a minimum of sample damage and charging, a detrimental result is a reduction in the beam current available from the electron source (as compared with high voltage operation); thus, the signal-to-noise ratio is poorer. This results in a loss in apparent sample detail. High brightness filaments and digital frame storage techniques for multi-scan signal integration, or slow scan rates coupled with photographic or electronic integration, help to overcome this problem. Further, the more abiding problem with low accelerating voltage operation is the lower resolution (as compared to the higher beam energy operation) characteristic of this mode of operation. As discussed earlier, if an instrument equipped with a high brightness lanthanum hexaboride filament is capable of 4 nm resolution at 30 keV accelerating voltage, it may only be able to resolve about 10 to 12.5 nm of resolution (under

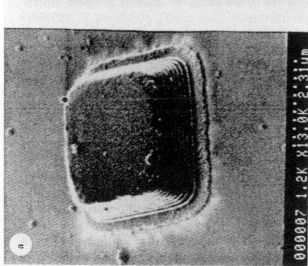

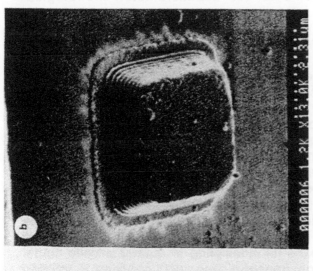

Figure 13. Scanning electron micrographs showing an illusion possible in the SEM that demonstrates that an understanding of the sample is necessary to facilitate proper image interpretation. (a) In this micrograph of uncoated photoresist, the image appears to be a positive structure. (b) In this micrograph the image appears as a contact hole. The only difference between the two micrographs is 180 degrees of raster rotation.

similar conditions) at 1 keV. This limitation is balanced by the fact that the electron beam penetrates much less into the sample at low accelerating voltage, and thus, the surface detail is enhanced, but the resolution limitation must be understood and factored into the precision requirements for submicrometer metrological applications.

3.3   Interaction Modeling

The appearance of a scanning electron micrograph is such that its interpretation seems simple. It should be understood that this interpretation may not always be correct. Care must always be taken so as not to become confused by "obvious" interpretations (Figure 13). When quantitative feature-size measurements are to be made, it is even more necessary to be able to unambiguously relate signal variations to the details of the surface morphology. Because the interaction of electrons with a solid is extremely complex (e.g., each electron may scatter several thousand times before escaping or losing its energy, and a billion or more electrons per second may hit the sample), statistical techniques are an appropriate means for attempting to mathematically model this situation. Although transport theory(67) provides a solution for simple systems, it is of little value when considering complex device geometries such as those used in the manufacture of semiconductor devices(68). The most adaptable tool, at the present time, is the "Monte Carlo" simulation technique. In this technique, the interactions are modelled and the trajectories of individual electrons are tracked through the sample and substrate (Figure 11). Because many different scattering events may occur and because there is no a priori reason to choose one over another, algorithms involving random numbers are used to select the sequence of interactions followed by any electron (hence the name, Monte Carlo). Choosing and testing the best random number generator has been a limitation to this technique, but programs are available for this purpose(69). By repeating this process for a sufficiently large number of incident electrons (usually 5000 or more), the effect of the interactions is averaged, thus giving a useful idea of the way in which electrons will behave in the solid.

The Monte Carlo technique has many benefits as well as several limitations(55). Because each electron is individually followed, everything about it (its position, energy, direction of travel, etc.) is known at all times. Therefore, it is possible to take into account the sample geometry, the position and size of detectors, and other relevant experimental parameters. The computer required for these Monte Carlo simulations is modest, so that even current high-performance desk-top personal computers can produce useful data in reasonable times. In fact, all modeling work present in this work was done on this type of computer.

In its simplest form(70), the Monte Carlo simulation allows the backscattered signal to be computed, since this only requires the program to count that fraction of the incident electrons that subsequently re-emerge from the sample for any given position of the incident beam. By further subdividing these backscattered electrons on the basis of their energy and direction of travel as they leave the sample, the effect of the detection geometry and detector efficiency on the signal profile can also be studied. However, while this information regarding the backscattered electrons is a valuable first step, under most practical conditions, it is the secondary electron signal that is most often used for metrology in the low accelerating voltage applications. Simulating this is a more difficult problem because two sets of electron trajectories, (1) those of the primary (incident) electron and (2) of the secondary electron that it generates (SE Type I), must be computed and followed. Compounding this even further is the need to also model the effects and contributions of the Type II and Type III electrons. While this is possible in the simplest cases, it is a more difficult and time-consuming approach when complex geometries are involved. For this reason, a new approach has been proposed(71-72) and is currently undergoing further development in cooperation with NIST. In this method, a simple diffusion transport model for the secondary electrons is combined with a Monte Carlo simulation for the incident electrons. This procedure allows both the secondary (SE Type I and II) and the backscattered signal profiles to be modelled simultaneously, with very little increase in computing time. Once those data are available, the effect of other signal components, such as the SE Type III signal, can also be estimated. Some experimental work comparing the modeling results with experimental data have been encouraging(68)(73).

The importance of being able to model signal profiles for some given sample geometry is that it provides a quantitative way of examining the effect of various experimental variables (such as beam energy, probe diameter, choice of signal used, etc.) on the profile produced, and gives a way of assessing how to deal with these profiles and determine a criterion of line edge detection for given edge geometries, and thus, a linewidth. However, at the present time, the Monte Carlo technique is still under development and is not yet useful for deducing the line-edge geometry from the acquired SEM video profiles. Further ramifications of the Monte Carlo modeling are discussed later.

Another approach to the modeling of the electron interaction in semiconductor samples is the use of the phenomenological approach. In this method, equation(s) are developed that fit the experimental data(55)(74)(75). This type of modeling requires that the parameters of the model be selected from physical data or from the empirical fitting of the measured profile(75), and thus, this approach can be useful in the development of edge detection algorithms. Another approach consisting of a surface integral of a probability density function describes the likelihood of the generation of a secondary electron by the primary electron beam being emitted at a given point in space(76). The advantage to this technique is a reduction in computation time. Further work in this area is also presently in progress.

3.4   Current Problems in the Use of the SEM For Metrology

There are several basic problems associated with the use of the scanning electron microscope for semiconductor metrology that a metrologist must be aware of in order to obtain significant data from the instrument. Many of these problems have been described(68) and some of them can be circumvented by understanding that they exist and can present pitfalls in the attainment of meaningful measurement data. Some of these problems are:

3.4.1   Defining the Meaning of a Linewidth.
Scanning electron microscope metrology and optical metrology have one thing in common at the present time; there is no well-defined definition of the meaning of the linewidth of most specimens(68)(73). The first consideration that must be developed and defined when describing the term "linewidth" is what is actually being measured physically. Depending upon the lithographic process level, the term "linewidth measurement" may vary relative to the structure's importance to subsequent processing steps. Many of these structures have an approximate trapezoidal cross section. Whether the critical dimension is defined as the width of the top surface edge or the base width of a line is a significant question that must be understood and designed into the measurement process (Figure 14). This also leads to problems in standardization, both from NIST's standpoint and the semiconductor industry's standpoint.

The SEM characteristically has a large depth of field (focus); therefore, the distinction of what is (or what is not) a linewidth becomes important since, if the conditions are properly

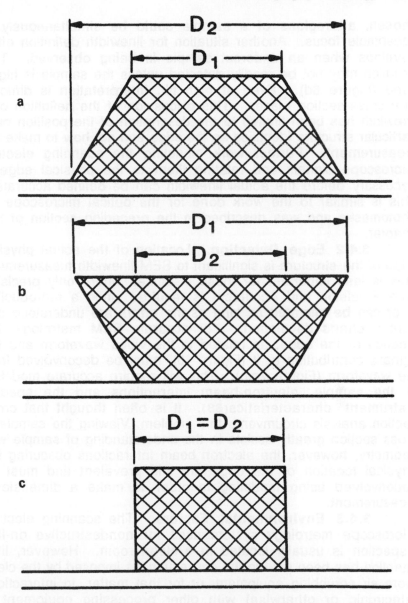

Figure 14. Drawing of a line structure in cross section demonstrating the confusion possible in determining what edge is, in fact, being measured in the SEM: (a) Trapezoidal structure where the upper width D1 is smaller than the base width D2. (b) Undercut structure where D1 is larger than D2. (c) Structure with vertical sidewalls where D1 is approximately equal to D2.

chosen, all regions of a sample could be simultaneously in acceptable focus. Another situation for linewidth definition effort develops when an undercut sample is being observed. This situation may not be readily detected unless the sample is highly tilted (Figure 6d). Even at high tilt, interpretation is difficult, with cross-section biases. This discussion of the definition of a linewidth has been limited to the description of the position on a particular structure identified as the edge and not how to make the measurement. Further work modeling the scanning electron microscope signals and relating them to the physical edge is necessary before the actual linewidth can be defined accurately. This is similar to the work done for the optical microscope for photomasks and was described in the preceding section of this chapter.

3.4.2 Edge Detection. Location of the actual physical edge of the structure is significant to SEM linewidth measurement. This is less important for those instances where only precision (see Precision and Accuracy) is required since a reproducible error can be tolerated as long as the errors are understood and control charts developed. For accurate SEM metrology, the location of the edge, as related to the video waveform and the signal's contribution to it (Figure 10), must be deconvolved from the waveform (Figure 7a,b). This will require accurate modeling of the sample, electron-beam interactions and the specific instrument characteristics(68). It is often thought that cross section analysis circumvents this problem. Viewing the sample in cross section greatly assists in the understanding of sample wall geometry, however, the electron beam interactions obscuring the physical location of the edge are still prevalent and must be deconvolved using modeling in order to make a dimensional measurement.

3.4.3 Environmental Factors. The scanning electron microscope metrology system used for nondestructive on-line inspection is usually located in a clean room. However, little attention has been paid to the consequences imposed by the clean room air scrubbing equipment, or for that matter, to interactions (electronic or otherwise) with other processing equipment in proximity to the SEM instrumentation. Vibration and stray magnetic fields associated with clean rooms impose serious problems on the electron-beam metrology instrumentation (68)(77). The SEM metrology instrument is an electronic imaging system, and as such, the problems posed by the clean room environment are readily observable in the images. It should be noted that these problems can also detrimentally affect other clean room instrumentation, but their effects may not be directly observable in real time and so the significance may be lost. In most cases surveyed, the SEM metrology instruments presently operating in the typical clean room are not performing optimally. This is usually due to two main reasons: excessive vibration and stray electromagnetic fields.   Both of these environmental problems can be eliminated given proper clean room engineering.

3.4.4 Sample Charging Effects. Sample charging and its effect on measurements made in the SEM have been studied (66)(78)(79). Negative charging results when the electron-beam voltage exceeds E-2 (Figure 12). This charging can detrimentally affect the measurement in several ways. The foremost effect is the possible deflection of the electron beam as the sample builds up an appreciable charge approaching that of the initial accelerating voltage(66)(78). This may either manifest itself as an obvious beam deflection where the image is lost or a more subtle and less obvious effect on the beam (Figure 15). The small effects are the most damaging to metrology as they may manifest themselves either as a beam deceleration or a beam deflection, neither of which is easily diagnosed. A subtle beam deflection around a line structure can move the beam a pixel point or two in the video profile, thus invalidating the critical dimension measurement. One pixel point deflection of a 1 μm line measured at 10,000X magnification with a 512 pixel point digital scan corresponds to about 30 nm linewidth error (less at higher magnification). Beam deceleration is a significant consideration since the instrument internal compensations or interaction modeling is based on the ideal accelerating voltage applied. Any unknown or uncontrolled effects applied by this factor will have significant efforts on either of these two computations. Positive charging may also have detrimental effects on the measurements, as a positively charging structure can attract information carrying secondary electrons from adjacent pixel points, thus distorting the measurement profiles. Charging may also induce permanent changes to the specimen by inducing mobile ions to translocate(71).

Proper adjustment of the accelerating voltage to the appropriate points on the total electron emission curve can minimize, if not completely eliminate, sample charging (Figures 12 and 15b). Rapid TV-rate or near TV-rate scanning is also being employed by several manufacturers to reduce charging. Under these conditions, the electron beam dwells on the sample for less time per point than in slow scan; thus, the charge has less time to develop (but signal-to-noise may be lower).   Another possible charge-reducing technique is to tilt the sample toward the secondary electron detector.   Tilting the sample permits

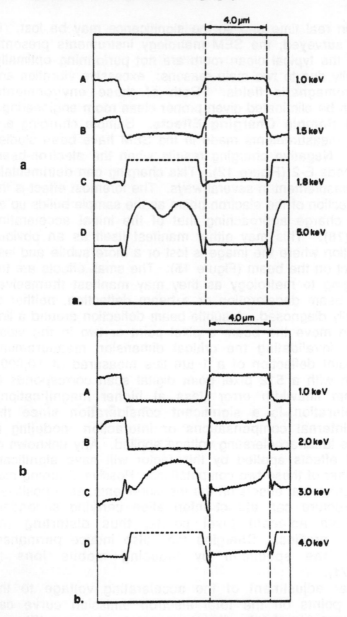

Figure 15. Sample charging and the effect on the video profile: (a) This chrome-on-glass mask sample was viewed at increasing accelerating voltages normal to the beam. No apparent charging is present in the profile at 1.0 keV; however, increasing the accelerating voltage results in deformation of the profile. (b) Tilting the same sample to 45 degrees permits the development of charging to be reduced to beyond 2.0-keV accelerating voltage.

sample inspection at higher accelerating voltages without charging effects; however, care must be taken during the critical dimension measurements to minimize possible error that tilt may introduce(66).

3.4.5 Signal Detection and Accelerating Voltage Effects.

Errors introduced to the linewidth measurement relative to the mode of signal detection and of beam acceleration voltages has been studied(50)(53). In this work, a silicon wafer sample with a silicide layer patterned with micrometer and submicrometer lines was observed and measured under controlled conditions at several accelerating voltages and electron detection conditions. A micrograph showing the effect of the choice of signal detection (secondary or backscattered electron imaging) is demonstrated in Figure 16. In that micrograph, the actual physical width of the line and the characteristics of the beam electrons were not changing as the micrograph was being taken. The only difference was the manner of signal detection in the instrument. The results of additional measurements of the secondary and backscattered signals demonstrate that, depending upon the electron detection mode used to image and measure the structure of interest, a variety of measurement results can be obtained(52). In all instances, measurements using the secondary electron signal yield a larger result (as much as 8%) than an identical measurement using the backscattered electron signal. Further, measurement-broadening effects of the beam penetration relative to the accelerating voltage are apparent. The reasons for this variability become obvious if the video waveforms are displayed (Figure 17). As can be seen, the waveforms obtained at the two different accelerating voltages are significantly different, due in part to the range of the primary beam as it penetrates into the sample between the different conditions. This clearly demonstrates that measurement criteria for each accelerating voltage must be established so that electron-beam effects can be properly accounted for and that electron beam/sample interaction effects must be understood. Changes in apparent dimension can be attributed to the uncertainties contributed by electron-beam interaction effects, solid angle of electron detection, detector sensitivity, and the criterion used to determine the edge location in the computation of linewidth. These data further suggest that if several instruments are operating on a production line, care must be exercised that all are working with the identical accelerating voltage (and other conditions), otherwise measurement variations between instruments are inevitable.

3.4.6 Sample Contamination Effects.

Semiconductor samples introduced into the SEM vary greatly in their surface

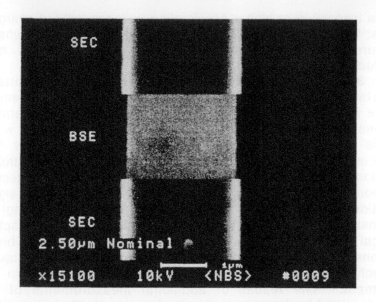

Figure 16. Effect of the mode of signal detection on the scanning electron microscope image and hence the measurements made from it. SEC-secondary electron detection; BSE-backscattered electron detection. See text for explanation.

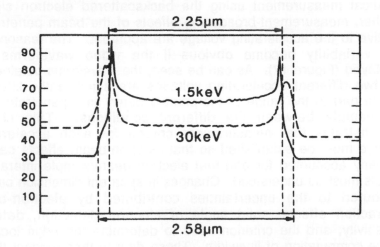

Figure 17. Overlay comparison of two digitally acquired video profiles of a nominal 2.5 μm silicide on silicon line. One profile was taken using 1.5 keV accelerating voltage and the second was done at 30 keV accelerating voltage. The automatic edge-detection threshold algorithm was arbitrarily set to 40%.

cleanliness, for SEM inspection cleanliness, in this context, is not as much a lack of particles as a chemical cleanliness. This is as much of a concern in the SEM as it is in the optical microscope. The surface contamination levels present on the sample will vary with the preceding processing steps. Residual hydrocarbons adhering to the surface or diffusing from within the structures in the vacuum can ionize as the beam scans, resulting in beam deflection or beam broadening. The electron beam can also act to decompose these hydrocarbons at the surface in the area of the raster pattern, effectively depositing a layer of carbon(68). At higher accelerating voltages, the electron beam penetrates this contamination and shows little effect. At low accelerating voltages used for nondestructive inspection, this contamination can severely alter signal generation and thus compromise data. Furthermore, this deposited layer of carbon may also affect subsequent processing of the wafer, and thus may be considered destructive at least to that inspected device(s).

3.4.7 Sample Dimensional Changes. Electron-beam irradiation can induce dimensional changes in photoresist and polymer structures(80-85). A high-resolution SEM image demonstrating a good signal-to-noise ratio can expose that sample to total electron-beam dosages higher than that typical for electron-beam lithography(80). This can have a pronounced effect on the critical dimensions by either causing the resist to swell or shrink. The authors of references 80 and 81 recently studied the dimensional stability of several commonly employed resists. This work demonstrated that even with a beam operated at 1.0 keV accelerating voltage, resist shrinkage can be induced. The rate of resist shrinkage is greatest when the electron range is approximately equal to the thickness of the resist because, under irradiation, all of the beam energy is deposited in the resist. Clearly, this is an interesting and controversial topic, and further work on this and other materials needs to be done. The possibility of dimensional changes of the sample occurring during the measurement process must be explored and care must be exercised to determine the optimum conditions where radiation damage is minimized and instrument operating conditions are optimized.

3.4.8 The SEM As a "Tool." The scanning electron microscope is a laboratory instrument. This instrument is often referred to as a "tool" by many in the semiconductor industry. Perhaps the connotation "tool" indicates the acceptance of this instrument in the production environment or perhaps the word tool refers to any of a class of instruments used by these engineers. Unfortunately, a tool can be defined in many ways; if general scientific instruments are often abused and mistreated, then tools are even more so. Scanning electron microscopes, especially those equipped as metrology instruments, are complex, expensive investments. One area that has been severely neglected by many semiconductor companies is the fact that this complex instrument must be controlled by a highly trained operator. The metrology instrument operator plays a crucial part in the success or failure of the on-line inspection program. Even the simplest of the SEM systems are far more technologically involved than their optical microscope counterparts (although in both instances highly trained individuals should be used if reliable, consistent data are expected). This is especially true where the routine instrument maintenance is concerned. Not every applicant is suited to become an SEM metrologist, and once an appropriate candidate is selected, a substantial amount of training must be invested in order for that individual to become confident and competent with the particular instrument or instruments under his supervision. Further, once an individual has proven to be an asset in that position, he must be encouraged to remain in that area and not be transferred out. Once an operator leaves the SEM metrology area, his real experience value is lost. Experience cannot be taught, only gained! This trend toward automation of the SEM inspection processes may minimize the need for a large number of trained operators at some point in the future; however, this will not be for some time.

3.5   NIST   Scanning   Electron   Microscope   SRM Development   Program

The program for the development of standard reference materials (SRMs) at the National Institute of Standards and Technology for scanning electron microscope metrology can be broken into three distinct phases: 1) development of an electron-beam instrument (or instruments) capable of certifying developed standards to an accepted referenced standard of length with a precision approaching 2.0 nm; 2) the development of appropriate electron beam/sample interaction modeling; and 3) the actual development and certification of suitable standards. Each of these three phases is undergoing simultaneous development.

For the purpose of discussing the program of standards development for the SEM, a clear distinction between the terms precision and accuracy must be made at the onset. This is necessary because in many instances these two terms have been erroneously used and treated as if they were synonymous.

3.5.1 Precision and Accuracy. In metrology(86) the term precision, often referred to as repeatability, is defined as the spread in values associated with the repeated measurements on a given sample using the same instrument under the same conditions. The assumption is that the number of measurements is large, the sample is stable over time, and that the errors introduced are random. This is essentially a measure of the repeatability of the instrumentation. Precision relates directly to at least four distinct factors: 1) Instrument; 2) Operator; 3) Environment, and 4) Sample. Many of the factors affecting SEM measurement precision have been discussed in earlier sections of this chapter. Just as in the optical microscope, to measure SEM instrument precision, it is not necessary to use an official standard. It is only necessary to use a sample that is of good quality and is stable both in the instrument and with time. This provides a measure of precision that is locally traceable, and is related only to that particular instrument and sample. Furthermore, in the SEM (due to the higher inherent resolution attainable), this precision may only relate to a given section or area of that sample because a sample may vary significantly at this level from location to location. Due to the need for stability with time, the sample materials chosen for these samples may not be identical to the typical product sample of interest (i.e., photoresist). To compare precision with more than one site or instrument would require the particular test sample to be carefully transported to the other location and then the test repeated under identical conditions. An adjunct to this would be that an organization (such as NIST) make up and test (with a single well-characterized instrument) a series of precision test samples which then could be taken to the various sites of interest and the sample precision of the instruments at those sites tested and compared with the measurement made on the original instrument.

Accuracy, on the other hand, is a far more ambiguous concept usually relating to the measurement of some agreed-upon quantity. That agreed-upon quantity (or quantities) for the SEM is one goal of the program at NIST. This goal is not necessarily identical in principle, or practice, to the goals of the present semiconductor industry, but the results are the same: that is, the production of an accurate SEM standard that can be used to determine the accuracy of semiconductor product measurements. In this way, product standardization within a company and its many sites around this country (or the world) or even between other companies can take place. Not only must the above factors affecting precision be considered as limitations of measurement accuracy but also the manner by which a given structure is being measured. Thus, a program similar to that employed for the NIST optical microscope linewidth mask standard must also be undertaken. Such a program will utilize computer modeling of the electron beam/sample interactions in order to obtain the necessary measurement accuracy. Many of those factors necessary to effectively model linewidth measurements in the SEM are not fully understood at this time(55) and approaches are being developed to quantify them(71)(87). The term accuracy should not be confused with the term precision. Scanning electron microscopes are capable of precision given that the problems outlined earlier are recognized and controlled; however, until nationally traceable SEM standards and the proper measurement procedures are available, the SEM does not have accuracy.

Just as was described earlier for the optical case, accuracy may be achieved only if the instrument making the measurement is sufficiently precise and the specimen of interest exactly matches the standard in all important ways (materials, substrate, etc.) except the dimension or dimensions being measured. One complication for linewidth metrology of thick lines (i.e., photoresist, etc.) on wafers is that, even if an acceptable standard were available composed of one set of particular materials, there is no guarantee that a given production sample will match precisely the characteristics of the standard. This is especially true because of the vast number of possible combinations of substrate and structure materials being used in semiconductor technology today (or where proprietary materials are being used). What may become feasible is the development of an accurate linewidth standard of well-established geometry and the parallel development of a computer program to handle the sample and instrumental differences between this standard and the product being measured. This problem is similar in concept to that required for the development of the computation factors for quantitative X-ray microanalysis and the programs developed at NIST (and other laboratories) to undertake this problem. A program to undertake this challenge is also being implemented.

3.5.2 Certification Instrument. The early phases of the development of a scanning electron microscope-based system for the certification of linewidth measurement standard reference materials has been reported(55). This instrument is based on a customized SEM and incorporates a piezo/interferometric stage for precise translational motion and the monitoring of distance, improved vibration isolation, microprocessor stage control system, and computer data analysis. The basic interferometer/piezo motion stage is similar in concept to the system already in use to certify the NIST SRM 484 SEM magnification standard(88), but is much more complex in its design.

3.5.3 Electron Beam Interaction Modeling. The interactions of the electron beam were discussed in earlier sections of this chapter as well as a description of the need for modeling. For accurate SEM metrology, it is important to understand the relationship of the electron beam/sample interaction to the signals detected, as this presents an analogous problem to that produced by diffraction in the optical microscope. It should also be understood that secondary electrons can originate from points other than the point of impact of the primary electron beam (Figure 10). The contribution each of these types of electrons has to the actual SEM image varies with instrument and operational parameters. The effects of these four types of contributions to the actual image or metrology have not been fully evaluated, yet they must be considered and accounted for in the model developed. Without this evaluation, the profile of the feature to be measured cannot be quantitatively analyzed to determine the accurate location of the structure's edges.

The ability to model signal profiles for some given sample geometry and composition is an important tool for linewidth metrology. Modeling provides a quantitative way of examining the effect of various experimental variables (such as beam energy, probe diameter, choice of signal used, etc.) on the profile produced, and gives a way of assessing how various algorithms deal with these profiles to determine a criterion of edge detection and, thus, a linewidth. Of course, as with any model, it must be tested by comparison of its predictions with actual experimental observations before it is used for its intended purpose.

3.5.4 SEM Standards Development. The magnification standard reference material presently available from NIST for calibrating scanning electron microscopes is SRM 484 (89). This standard provides a known pitch between gold lines in a nickel matrix and has proven useful for many SEM applications. SRM 484 has a minimum pitch spacing of about 1 μm. However, SRM 484 was developed prior to the recent interest in low accelerating voltage operation for integrated circuit wafer inspection and measurement. The present form of SRM 484 is unsuitable for use in many of the new SEM measurement instruments for two main reasons: a lack of suitable contrast in the 1.0 keV accelerating voltage range and its overall size which is not compatible with some instruments (due to the thickness which greatly exceeds that of a semiconductor wafer or mask). This sample can be chemically etched to provide contrast at low accelerating voltages, but suffers from other limitations(90). This work is the initial phase of a larger NIST program to develop suitable submicrometer standards for the SEM

based on lithographically produced samples which will eliminate the basic problems associated with SRM 484. The magnification standard under development in this program will be composed of materials that will provide suitable contrast at both high and low accelerating voltages and will have pitch spacings as large as 3000 μm and as small as 0.2 μm with several intermediate-sized pitches. This should enable the magnification calibration of SEM instruments from the lowest magnification to beyond 100,000X magnification.

The linewidth measurement standards developed for the optical microscope (SRM 473, 474, and 475) are not designed, or recommended, for calibration use in the SEM(91). The optical theory and modeling used in the certification of these optical SRMs are not directly adaptable to the SEM and, therefore, the criteria developed to determine the edge location are not applicable for anything but an optical microscope measurement. The electron-beam effects and the requirements for computer modeling in the SEM are totally different from the diffraction effects in the optical microscope. These SRMs could , however, be used to calibrate the magnification at low accelerating voltage in an SEM under conditions where the SRMs are not charging. In this mode, the magnification of the instrument could be calibrated with the pitch patterns on the SRM. However, continuing this calibration process to include linewidth measurements is not recommended, because such calibration results would be valid only for identical chrome-on-glass photomasks, and then only under those particular instrumental conditions. There is no advantage to using these SRMs for SEM pitch calibration (except for its smaller size) since they are rather expensive relative to SRM 484 and because of potential charging of its glass substrate and electrically isolated metal lines. Therefore, a need presently exists for an inexpensive low accelerating voltage magnification standard for SEM calibration and work is presently in progress at NIST to meet that need.

3.6   Conclusion

The increasing interest in nondestructive low accelerating voltage SEM operation and the application of the SEM to precise linewidth metrology, especially in the semiconductor community, has increased the need for performance samples capable of determining the state of instrument performance on a day-by-day (or perhaps even shift-by-shift) basis. Low accelerating voltage work especially associated with the on-line inspection and measurement of semiconductors is presently putting demands upon the SEM never before encountered.

It is an unfortunate fact that the majority of SEM users (in all its applications) operate the instrument under the assumption that the image that they are observing is a unique and invariant property of the sample that they are observing.   Nothing is further from the truth.   The image is, in fact, a function of all operational variables of the instrument and, as such, care must be taken when interpretation of the image or the video waveform acquired is made.   For metrology applications, such as those described here in this chapter section, the actual location of the edge for accurate linewidth measurement defines the ultimate resolution of the metrology instrument, not just the ability to discern that two edges are present.   This edge location resolution issue depends upon an entire host of factors, some of them similar to the instrument resolution issue (e.g, materials, accelerating voltage) and others characteristic of this special application of the instrument (e.g., specimen, detector location, detector characteristics).   A distinction between the conditions for good metrology and for good imaging is being developed; and these conditions are not necessarily the same for both applications. Although many of the same conditions must exist between the two applications of the SEM (especially those for high resolution in measurement situations), there are special cases where metrology applications require special conditions, such as precise and linear measurement scans, properly positioned electron detectors; and the consideration that visual appeal is not necessarily the best criterion for proper metrology.   For metrology, the relationship of the performance level of the instrument to measurement precision and accuracy must also be ascertained in order that the measurements obtained be timely and meaningful to the process. Furthermore, present linewidth measurement systems, while acquiring as few as 512 pixel units or as many as 260,000 pixel units of information, actually may use only a small fraction of these for the measurement (possibly as few as 2 pixels).   With so small a sampling, error can easily be introduced.   At the present time, product precision is the prime concern of the semiconductor industry. Until national standards for SEM linewidth measurement become available, the best that can be done is the establishment of a series of internal samples within a particular organization for determining instrumental precision.   These samples should be identical to the structures to be measured and should be used to measure process precision.   These samples should be used in conjunction with the established pitch standard(s) used to properly adjust the magnification of an instrument.   This series of well-characterized internal standards will provide an ability to determine process precision and to develop offsets to the instrument for each level to adjust for the uncompensated accuracy differences and also periodically check the instrument's measurement drift.

4.0   ELECTRICAL LINEWIDTH METROLOGY

4.1   Introduction

Integrated circuit test structures are microelectronic devices that are used to determine selected process, device, material, or circuit parameters by means of electrical tests. The structures are fabricated at the same time and by the same processes used to fabricate integrated circuits and are measured by an automatic parametric test system. Well-characterized test structures and test methods should not be sensitive to unintentional material or process variations. The results of these measurements are used in conjunction with additional nonelectrical tests to assure process control and product performance.

Test structures are powerful tools for collecting information. Coupled with a computer-controlled parametric test system, they can provide the user with information that cannot otherwise be obtained because of the nature of the measurement or the cost associated with other methods of performing the measurement.

Microelectronic test structures have been used by the integrated circuit community for many years. Traditionally, test chips containing test structures have been used to characterize the fabrication process. In some instances, they have been used to obtain parameters used in device and process simulation codes or as a tool to purchase or reject wafer lots from silicon foundries. Test chips have also been used to characterize the performance of ion implanters and lithography systems used in the fabrication of integrated circuits. A more comprehensive overview of this subject can be found elsewhere(92-94).

4.2   Electrical Linewidth Measurement

Accurate and precise determination of linewidth becomes increasingly important for process characterization and control as the design dimensions of devices decrease. Linewidth must be measured at many locations within the exposure field to fully characterize the intrafield linewidth variation associated with a specific lithography process. Measurements must also be obtained and interpreted in a relatively quick time interval in order to be of use in the manufacturing process.

The cross-bridge resistor(95) is a test structure used to determine electrical linewidth of a thin, patterned, conductive film. Electrical linewidth is defined as the effective conductive path width of a patterned line whose length is at least fifteen times longer than its width. In effect, the electrical linewidth is the path width that the electrons encounter in a conductor during the normal operation of an integrated circuit. While the electrical method provides a relatively fast and precise means of obtaining effective linewidth, it differs in concept from the more classical optical or physical definitions in that the measurement is not the distance between two well-defined points but rather the average width of the conductive strip. Ideally, an electrical measurement of linewidth would be equal to an optical or electron beam measurement for a line with perfectly vertical edges, uniform thickness, and uniform conductivity. In general, electrical linewidth measurements are not as sensitive to line edge variations as are optical or electron beam measurements. Electrical linewidth measurement also differs from optical or electron beam linewidth measurement in that the measurement itself does not directly rely on an external physical standard for calibration but on the internal calibration standards of the electrical test instruments used.

4.3 Cross-Bridge Resistor and Electrical Measurement Method

The cross-bridge resistor test structure provides very precise and relatively fast determinations of linewidth using an automated parametric test system(95-99). The structure, shown in Figure 18, is a combination of a modification to the van der Pauw resistor(100) and a single bridge resistor of design width W. In principle, by first measuring the resistance of one square of a uniformly conductive thin film and then measuring the resistance of a rectangular line of known length of the same material, one can calculate the number of square line segments comprising the line and the width of the line.

The van der Pauw resistor allows the determination of the sheet resistance of a conductive film. For a symmetrical van der Pauw cross, the sheet resistance is calculated as :

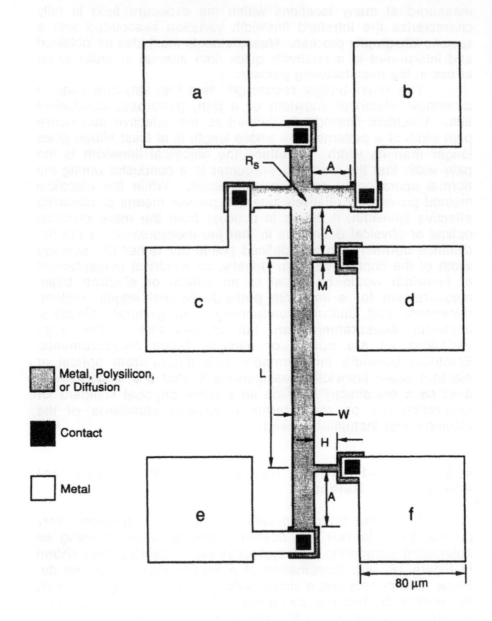

Figure 18. Cross-bridge resistor test structure.

[1] $R_S = (CF)$ V/I (note: CF is geometry dependent)

where, I is the current forced between probe pads a and b, and V is the voltage measured between c and d. Geometrical asymmetries which might affect the determination of sheet resistance have been experimentally shown to be negligible for typical devices (100).

In practice, the current is reversed and the magnitude of the two voltage measurements is averaged. By using the Kelvin four-terminal method where current is forced between two probe pads and voltage sensed at two adjacent pads, interferences caused by probe-to-probe-pad and voltage-tap resistance and by any systematic offsets in the parametric test system are eliminated. The bridge resistor is accessible through pads c through f and is used to determine the average electrical linewidth W of the conducting line based on the design center-to-center bridge length L and the sheet resistance measured at the van der Pauw cross. The linewidth W is determined as:

[2] $W = R_S L (I'/V')$

where, I' is the current forced between probe pads c and e, V' is the voltage measured between d and f, and L is the design length between the voltage taps. As with the measurement on the cross, the current is reversed and the magnitudes of the two voltages averaged. A detailed description of the test method is found in (97).

4.4   Cross-Bridge Geometrical Design Criteria

In order to minimize interferences that could affect the measurement of sheet resistance and electrical linewidth, the cross-bridge test structures, as seen in Figure 18, should be designed using the following design constraints(101).

[3] $W \geq M$

[4] $L > 15M$

[5] $H \gg M$

[6] $A > W$

where, W represents the design linewidth, L is the length between the center position of the voltage taps to the bridge, M and H are the width and length of the taps, respectively, and A is the distance between the lower voltage tap and the bottom of the bridge as shown in Figure 18.

4.5   Parametric Testing

Cross-bridge resistor test structures or chips containing these structures can be electrically tested using a wafer prober and dc voltmeters and current sources common to most electronics laboratories. Automated measurements can be made using a general-purpose commercial computer-controlled parametric test system. In order to achieve sufficient precision, a digital current source with microampere resolution and a digital voltmeter with microvolt resolution should be used. To prevent significant Joule heating while testing these structures, a measurement current is generally selected so that the voltage measured ranges between 2.0 and 20 mV. Precision of a typical test system for measuring nominal 1.0 μm linewidth test structures can be better than 0.002 μm (1 σ) as determined by repeated measurements of one cross-bridge test structure(96). The worst case measurement uncertainty of a typical system is on the order of 0.02 μm, based on the cumulative uncertainties of the test instruments at the current or voltage ranges normally used and the thermal noise encountered during wafer probing. The average measurement time of a cross-bridge resistor is generally less than 3 sec.

4.6   Electrical Measurement Results and Comparison to Optical Measurements

Linewidths derived from electrical measurement results and optical measurements agree to within the uncertainty accompanying both measurements. A plot of electrical linewidth versus design linewidth for cross-bridge resistors having design linewidths of 0.6 to 1.6 μm is shown in Figure 19. These results are from measurements on a test chip containing only cross-bridge structures, designed to evaluate the performance of a lithography process. The test chip contained a 7 by 26 array of cross-bridge resistors, with each row containing identical structures with a design linewidth range of 1.6 μm to 0.6 μm and design bridge length L of 200 μm and a 6 by 26 array of cross-bridge resistors with design linewidth of 1.0 μm and variable design bridge lengths of 20.0 μm to 200 μm arranged in rows of identical structures across the chip. The chip spanned an 8.6 mm

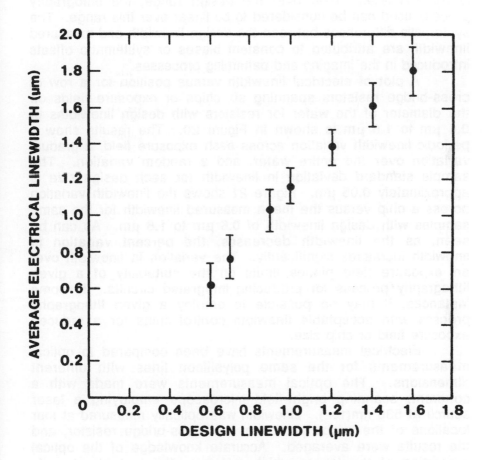

Figure 19. Average electrical linewidth versus design linewidth. Each datum and error bar (2 σ) is based on measurements made at approximately 156 measured sites.

by 9.0 mm exposure field. Details of the process used to fabricate the test chip are found in the Appendix.

Since the measured electrical linewidth seen in Figure 19 is approximately linear over the design range, the lithography process used can be considered to be linear over this range. The systematic differences between the design linewidth and measured linewidth are attributed to constant biases or systematic offsets introduced in the imaging and patterning processes.

A plot of electrical linewidth versus position for a row of cross-bridge resistors spanning six chips or exposure fields on the diameter of the wafer for resistors with design linewidths of 0.6 μm to 1.6 μm is shown in Figure 20. The results show a periodic linewidth variation across each exposure field, a gradual variation over the entire wafer, and a random variation. The sample standard deviation in linewidth for each design size is approximately 0.05 μm. Figure 21 shows the linewidth variation across a chip versus the mean measured linewidth for the same samples with design linewidth of 0.6 μm to 1.6 μm. As can be seen, as the linewidth decreases, the percent variation in linewidth increases significantly. The variation in linewidth over an exposure field places limits on the suitability of a given lithography process for producing integrated circuits. In some instances, it may be possible to employ a given lithography process with acceptable linewidth control limits for a reduced exposure field or chip size.

Electrical measurements have been compared to optical measurements for the same polysilicon lines with different dimensions. The optical measurements were made with a coherent, scanning bright-field microscope employing a laser source at 530 nm(102). Linewidth was optically measured at four locations of the bridge portion of the cross-bridge resistor, and the results were averaged. Accurate knowledge of the optical properties of the layers which make up the test structure is necessary, including thickness and complex index of refraction of the polysilicon layer and the oxide insulating layer, complex index of refraction of the silicon substrate, and edge geometry of the patterned polysilicon layer. Table IV shows a comparison between the electrical and optical linewidth measurements for cross-bridge resistors with design linewidths of 0.6 μm through 1.6 μm. Optical measurements were made along the portion of the bridge resistor between the two voltage taps and are compared to electrical measurements on the same bridge. For the samples measured, the variation in linewidth along the line was smaller than the uncertainty associated with the optical measurement. As can be seen, the electrical results are in agreement with the

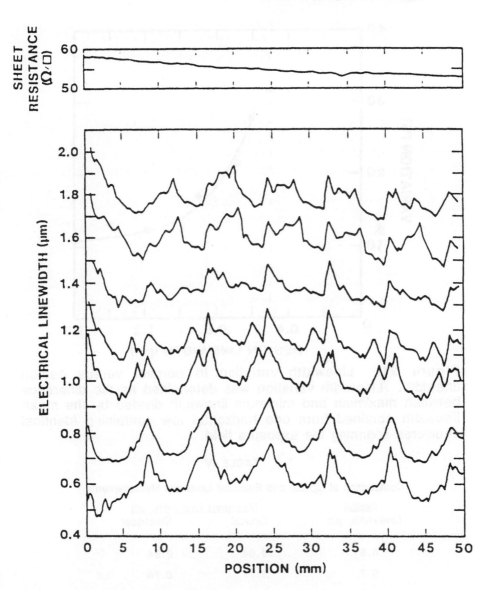

Figure 20. Sheet resistance and electrical linewidth versus wafer position for cross-bridge resistor test structures having design linewidth from 0.6 μm to 1.6 μm.

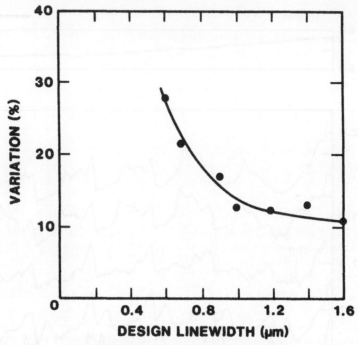

Figure 21. Linewidth variation in percent versus design linewidth. Linewidth variation was determined as the difference between maximum and minimum linewidth divided by the mean linewidth obtained from one horizontal row containing identical structures spanning the exposure field.

TABLE IV

Comparison of Optical and Electrical Linewidth Measurements

Design	Measured Linewidth, μm	
Linewidth, μm	Optical	Electrical
0.6	0.64	0.64
0.7	0.77	0.76
0.9	1.11	1.01
1.0	1.20	1.14
1.2	1.45	1.41
1.4	1.75	1.61
1.6	1.86	1.80

Electrical measurement uncertainty ±0.02 μm for samples tested.

Optical measurement uncertainty ±0.20 μm for samples tested.

optical results to the degree of uncertainty accompanying both measurements. Differences between the electrical and optical results are attributed to optical uncertainties associated with fitting the theoretical image profile to the measured profile for test samples with nonvertical edges.

For the optical measurements, the complex index of refraction of the polysilicon was measured by ellipsometry on the large area polysilicon drop-in on the same wafer. From the goodness of fit and precision of the system, the optical measurements were determined to be accurate ±0.20 μm for the samples used.

4.7   Additional Design Considerations

In general, the electrical linewidth determined from the cross-bridge measurement is not affected by the size of the van der Pauw resistor or the length of the bridge resistor. The results from two van der Pauw resistors and corresponding linewidth determinations are found in Table V. One van der Pauw resistor consisted of a simple cross formed by the intersection of the current and voltage taps with design linewidth varying from 0.7 μm to 1.6 μm. The second van der Pauw consisted of a 40 μm by 40 μm box with current and voltage tap linewidths similar to the first van der Pauw resistor. The results shown in Table V are for approximately 150 samples measured on one wafer. As can be seen, the mean sheet resistance measured from the simple cross structure is in close agreement with that of the box structure. Electrical linewidth determinations based on either van der Pauw measurement are approximately equal. These results are in agreement with previous predictions(103). The results from approximately 150 cross-bridge resistors with design widths of 1.0 μm and bridge lengths of 200 μm and 20 μm were found to have an average difference of 0.02 μm. These results indicate that the size of the van der Pauw resistor or the overall length of the bridge resistor does not significantly affect the measured electrical linewidth for the materials and geometries selected.

4.8   Cross-Bridge Modifications

Two modifications that have been made to the cross-bridge resistor include the split-cross-bridge resistor(104) and the proximity effect test structure (PETS)(105). Both are variations of the basic cross-bridge measurement technique.

The split-cross-bridge resistor, seen in Figure 22, can be used to determine the sheet resistance, electrical linewidth, and

TABLE V
Sheet Resistance and Linewidth for Cross-Bridge Resistors with Box and Cross van der Pauw Resistors

Design	Sheet Resistance, Ω		Measured Linewidth, μm	
	Box	Cross	Box	Cross
1.6	15.02 ± 0.16	14.95 ± 0.62	1.59 ± 0.02	1.59 ± 0.07
1.4	14.99 ± 0.20	14.85 ± 0.64	1.39 ± 0.03	1.38 ± 0.06
1.2	14.95 ± 0.26	15.03 ± 0.68	1.15 ± 0.02	1.16 ± 0.06
1.0	14.96 ± 0.21	14.94 ± 0.82	0.90 ± 0.02	0.91 ± 0.05
0.9	15.00 ± 0.20	15.02 ± 0.80	0.76 ± 0.03	0.77 ± 0.05
0.7	14.97 ± 0.25	14.93 ± 0.85	0.51 ± 0.03	0.50 ± 0.05

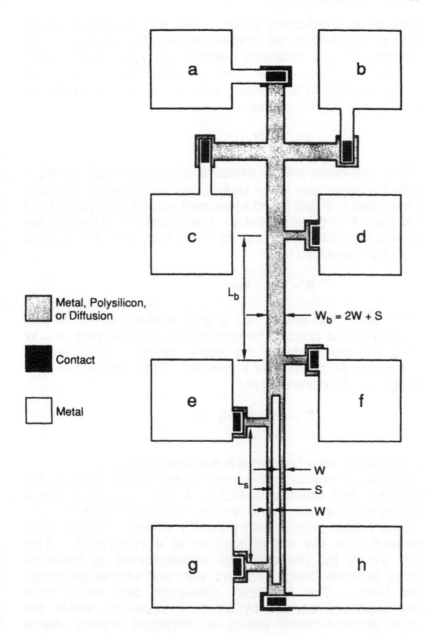

Figure 22. Split-cross-bridge resistor test structure, including bridge resistors with lengths L_b and L_s, widths W_b and W, respectively, and spacing S.

line pitch of a conducting layer. For this structure, sheet resistance is obtained in the manner previously described. The bridge portion of the structure contains an upper bridge with design linewidth W_b and a lower structure having a set of electrically parallel split bridges with equal design width W separated by spacing S such that:

[7] $W_b = 2W + S.$

The split-cross-bridge effective electrical width is $W_S = 2W$. Both the upper and lower bridges are measured by forcing a current from pad a to pad h and measuring voltage from pads d to f and from pads e to g, respectively. The respective linewidths are then W_b and W_S calculated using Eq. 7 and $W_S = 2W$. The spacing S can then be determined by:

[8] $S = W_b - W_S.$

The split-cross-bridge has a unique characteristic in that the spacing S can be checked for consistency since the pitch, P = W + S, of a line pair remains the same through the fabrication process and can be considered a constant. The pitch for the split-cross-bridge can be calculated as:

[9] $P = W_b - (W_S/2)$

and the calculated value compared to the design value.

Another extension to the cross-bridge test structure is the proximity effect test structure or PETS. A PETS, shown in Figure 23, consists of a van der Pauw resistor, an upper bridge resistor, and a lower bridge resistor with proximity inducing bars of width V separated from the bridge resistor at a spacing S. This structure allows for the electrical measurement of linewidth differences between bridge resistors with and without proximity-inducing bars. The linewidth difference can result from proximity exposure caused by electron scattering within the resist from electron beam writing or diffraction overlap effects from optical exposure.

The PETS is electrically tested in a manner similar to that for a conventional cross-bridge. First, a continuity test is made between the bottom probe pad and any other pad to assure that the proximity bars are electrically isolated from the bridge. The electrical linewidth of both the top and bottom bridge are then

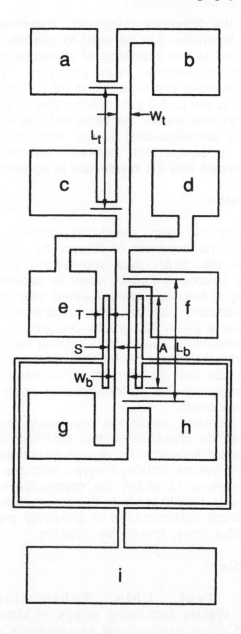

Figure 23. Proximity effect test structure (PETS), including proximity-inducing bars of length A and width T and bridge resistors, with length L_t and L_b and widths W_t and W_b, respectively, and spacing S.

determined and the difference calculated. Experimental data have shown that this technique can be used to quantify the increase in linewidth attributed to the scattering of electrons in a resist layer and substrate during electron beam exposure. Table VI shows results from a one-sided and a two-sided PETS for design spacings S ranging from 1.5 μm to 3.0 μm. Each value is an average of 25 measurements on one wafer. As expected, the magnitude of the proximity effects corresponding to the smallest spacing is the largest. Also, a comparison between optical and electrical measurements shows that the results are in agreement.

4.9   Conclusion

Microelectronic test structures provide a means for characterizing the performance of lithography processes used to image lines in the near- and sub-micron linewidth region. Arrays of these structures can be used to determine the overall spatial uniformity of the lines being defined over the full exposure field. This information can be used to determine the linewidth range over which a given lithography process can be effectively used. The importance of automated electrical testing is increasing as the size of critical features decreases. Commercial systems are available which use test structure measurements to quickly and accurately characterize the performance of complex lithography equipment and processes.

For the process conditions previously described, it has been shown that the cross-bridge test structure can be used to determine electrical linewidth for design geometries less than 1 μm and that the values obtained agree with the corresponding optical measurements to within the respective uncertainties of both measurements. Modifications to this structure have been made that allow electrical determination of proximity exposure effects, and for conductive lines, line-to-line spacing.

4.10 Appendix

4.10.1   Test   Chip   Fabrication   Process Description. Wafers containing arrays of identical test chips were fabricated using conventional low-pressure chemical vapor deposition process techniques to deposit polysilicon on a 0.15 μm layer of thermally grown silicon dioxide on 3 in. diameter wafers. Deposition was performed at 62°C for 60 minutes resulting in a 0.6 μm polysilicon film. The polysilicon layer was subsequently doped with $POCl_3$ at 950°C for 20 minutes and annealed in

TABLE VI

ΔW Measured for the Proximity Effect Test Structures

	Spacing (μm)			
	1.5	2.0	2.5	3.0
2-bar (μm)	0.58	0.35	0.21	0.11
1-bar (μm)	0.27	0.17	0.11	0.06

Control: 0.0 ± 0.04 μm

$\sigma = 0.06$ μm

nitrogen for an additional 15 minutes. Transmission Electron Microscopy analysis showed the grain size to be between 0.1 μm and 0.3 μm. The wafers were patterned using a wafer stepper exposure system containing a lens with a numerical aperture of 0.28. One large area polysilicon drop-in was included on each wafer to enable measurements to be made of the refractive index of the polysilicon film to be used in determining optical linewidth. Positive photoresist was used to delineate the patterns. Reactive ion etching was then used to delineate the polysilicon layer using a 4:1 mixture of chlorine and tribromochloromethane. The selectivity of the etch process exceeded 30:1 for pressure and power levels of 50 mT and 300 W, respectively. Cathode sheath voltages were typically 200 V to 220 V. Scanning Electron Microscopy inspection showed the polysilicon edges to be sloped at approximately 80 degrees relative to the substrate.

5.0   CONCLUSION

This chapter has discussed the issues confronting submicrometer dimensional metrology as practiced today by optical, scanning-electron and electrical techniques. Each of these techniques has its own advantages and disadvantages and there is no one technique of choice for all applications. Instrumental precision can be obtained in all three techniques with proper attention to the many factors that affect the measurement and with the practice of good quality control procedures. However, absolute dimensional accuracy for submicrometer features on semiconductor wafers is not yet possible by any of these methods! Precision is more easily attainable electrically because the method is self-calibrating on the actual specimen being measured, but absolute dimensional accuracy is elusive because the measured parameter is basically an electrically equivalent linewidth of an unknown edge geometry. In principle, optical techniques can handle all line materials and substrates, SEM techniques avoid elusive charging effects if all the specimen materials are slightly conductive or viewed at low beam energies, and electrical techniques require conductive lines on insulating substrates. Optical techniques are the first to encounter difficulties as the dimensions of interest approach submicrometer dimensions (because of diffraction with a wavelength of the illuminating light about equal to or larger than the dimensions of interest). SEM techniques have the potential of reaching further into the submicrometer range (because the edge effects due to beam-specimen interactions in the SEM generally span a smaller range about the edge than the diffraction effects in the optical microscope). Electrical techniques presently have the lead in ease and speed of measurement of submicrometer linewidths with nanometer precisions, but unlike optical and SEM techniques, require test patterns and cannot measure the actual features of interest. Optical and electrical techniques involve basic technologies that have been known and practiced for a very long time and they can be applied to the present metrological problem in relatively straightforward (but, perhaps, mathematically difficult) ways. The SEM techniques involve technologies that are relatively new and are not as well known and it is not always apparent how to even initially attack some of the present problems (e.g., solving the inverse scattering problem in the SEM). There are many variations (or modes of operation) of all three of these techniques that may have some special advantages for specific applications, and a whole chapter (or even a whole book) could be devoted to this topic alone, e.g., the aperture-scanning (or near-field) optical microscope or the holographic mode of operation of the SEM. Looking deeper into the future, the authors see a number of additional techniques that may become increasingly useful as the need for measurements moves even closer to the nanometer range or as the desired dimensional measurements become more esoteric. For example, the scanning probe techniques (e.g., scanning tunneling microscope) have the potential of atomic resolution and the acoustic microscope has the potential of seeing substructures beneath the overlying structure. In one respect, these are exciting times for submicrometer dimensional metrology because the semiconductor industry has provided the motivation and need for new measurement capabilities with precision and accuracy requirements that were not envisioned a few years ago. On the other hand, the rapid progress of the semiconductor industry to submicrometer features has led to that need faster than the basic metrological research could follow. Perhaps this is the price that must be paid by an industry that tends to take metrology for granted, and in the past, has not supported metrological research and development to the extent needed to meet its future demands.

6.0   ACKNOWLEDGEMENTS

The author wishes to thank his chapter co-authors and Diana Nyyssonen for many fruitful discussions and comments about the optical metrology issues raised in this portion of the chapter and to Jane Walters for preparation of the manuscript for publications. However, any opinions expressed are those of the current author and were gleaned from his vantage point at NIST and from discussions with people who make or use optical instruments in practical applications.

The authors would like to thank and acknowledge Dr. David Joy and Mr. William Keery for assistance in the development of the Monte Carlo program and Harry Diamond Laboratories, Adelphi, MD for the co-sponsoring of the research on the low accelerating voltage standard (MIPR no. R86-086). NIST also acknowledges the assistance of AMRAY, Inc. for their continued cooperation and assistance in the development and custom modification of the NIST Certification Instrument and Hitachi Scientific Instruments, Ltd. for the use of an S-800 Field Emission SEM for NIST research purposes.

REFERENCES

1. ASTM Manual on Presentation of Data and Control Chart Analysis, STP15D, American Society for Testing and Materials, 1916 Race Street, Philadelphia, PA 19103.
2. Nyyssonen, D., and Larrabee, R.D., Submicrometer linewidth metrology in the optical microscope. J. Res. Natl. Bur. Stand. 92: 187-203 (1987).
3. Croarkin, M.C., and Varner, R.N., Measurement assurance for dimensional measurements on integrated-circuit photomasks, NIST Tech. Note 1164 National Bureau of Standards, Gaithersburg, MD (1982)
4. Nyyssonen, D., Linewidth measurement with an optical microscope: effect of operating conditions on the image profile. Appl. Optics 16: 2223-2230 (1977).
5. Nyyssonen, D., Metrology in Microlithography, in VLSI Electronics: Microstructure Science (N.G. Einspruch, ed.), Vol. 16, pp. 265-317, New York: Academic Press (1987).
6. Standard Practice for Preparing an Optical Microscope for Dimensional Measurements, F 728-81, 1981 Annual Book of ASTM Standards.
7. Kirk, C.P., A study of the instrumental errors in linewidth and registration measurements made with an optical microscope. Proceedings SPIE 775: 51-59 (1987).
8. Nyyssonen, D. and Kirk, C.P., Modeling the optical microscope images of thick layers for the purpose of linewidth measurement. Proceedings SPIE 538: 179-187 (1985).
9. Swyt, D.A., An NIST physical standard for the calibration of photomask linewidth measuring systems. Proceedings SPIE 129: 98-105 (1978).
10. Office of Standard Reference Materials, National Institute of Standards and Technology, Room B-311, Chemistry Building, Gaithersburg, MD 20899.
11. See, for example, Hecht, E., and Zajac, A., Optics, Chap. 10 on Diffraction, pp. 329-392, Reading, Mass.: Addison-Wesley (1979), or any general textbook on optics.
12. Charman, W.N., Some experimental measurements of diffraction images in low-resolution microscopy. J. Opt. Soc. Am. 53: 410-414 (1963).
13. Charman, W.N., Diffraction images of circular objects in high-resolution microscopy. J. Opt. Soc. Am. 53: 415-419 (1963).

14.   Watrasiewicz, B.M., Image formation in microscopy at high numerical aperture. Opt. Acta 12: 167-176 (1965).

15.   Kinzly, R.E., Investigations of the influence of the degree of coherence upon images of edge objects. J. Opt. Soc. Am. 55: 1002-1007 (1965).

16.   Watrasiewicz, B.M., Theoretical calculations of straight edges in partially coherent illumination. Proc. R. Soc. London Ser. A 293: 391-400 (1966).

17.   Bullis, W.M. and Nyyssonen, D., Optical linewidth measurements on photomasks and wafers, Chapter 7 in VLSI Electronics: Microstructure Science (N.G. Einspruch, ed.), Vol. 3, pp. 301-346, New York: Academic Press (1982).

18.   Nyyssonen, D., Spatial coherence: The key to accurate optical micrometrology. Proceedings SPIE 194: 34-44 (1979).

19.   Theoretical computations of reflected-light image profiles with narrow cone-angle illumination by D. Nyyssonen published in Reference 2.

20.   Kirk, C.P., Aberration effects in an optical measuring microscope. Appl. Optics 26: 3417-3424 (1987).

21.   See, for example, feature issue on Inverse Problems in Propagation and Scattering. J. Opt. Soc. Am. 11: 1901 (1982).

22.   Larrabee, R.D., Submicrometer optical linewidth metrology. Proceedings SPIE 775: 46-50 (1987).

23.   von Ardenne, M., The scanning electron microscope: Practical construction. Z Tech. Phys. (in German) 19: 407-416 (1938).

24.   von Ardenne, M., The scanning electron microscope: Theoretical fundamentals. Z. Phys. 109: pp. 553-572 (1938).

25.   Joy, D.C., Romig, A.D., and Goldstein, J., Principles of Analytical Electron Microscopy, New York: Plenum Press (1986).

26.   Goldstein, J.I., Newbury, D.E., Echlin, P., Joy, D.C., Fiori, C., and Lifsin, E., Scanning Electron Microscopy and X-ray Microanalysis, New York: Plenum Press (1984).

27.   Postek, M.T., Howard, K.S., Johnson, A., and McMichael, K., Scanning Electron Microscopy -- A Student's Handbook, Burlington, Vermont: Ladd Research Industries (1980).

28.   Postek, M.T., Resolution and measurement in the SEM. Proceedings EMSA (G. W. Bailey, ed.), pp. 534-535 (1987).

29.   Wells, O.C., Scanning Electron Microscopy, New York:
McGraw-Hill (1974).
30.   Erasmus, S., Reduction of noise in TV rate electron
microscope images by digital filtering. J. Microsc. 127:
29-37 (1982).
31.   Erasmus, S., and Smith, K.C.A., An automatic focusing and
astigmatism correction system for the SEM and CTEM. J.
Microsc. 127: 185-199 (1982).
32.   Smith, K.C.A., On-line digital computer techniques in
electron microscopy. General introduction. J. Microsc.
127: 3-16 (1982).
33.   Newbury, D.E., Joy, D.C., Echlin, P., Fiori, C., and
Goldstein, J.I., Advanced Scanning Electron Microscopy and
X-ray Microanalysis, New York: Plenum Press (1986).
34.   Standard Definition of Terms Relating to Metallography,
1986 Annual Book of ASTM Standards, Designation E 7-
85d, pp. 12-48, American Society for Testing and
Materials, 1916 Race Street, Philadelphia, PA 19103.
35.   Jensen, S., Planar metrology in the SEM, Microbeam
Analysis 1980 (David Wittry, ed.), pp. 77-84, San
Francisco Press (1980).
36.   Jensen, S., and Swyt, D., Sub-micrometer length
metrology: problems techniques and solutions,
SEM/1980/I, SEM Inc. pp. 393-406 (1980).
37.   Schrope, D.E., Photocleave -- A method for nondestructive
sectioning of photoresist features for scanning electron
microscope inspection. Scanning Microscopy 1(3):
1055-1058 (1987).
38.   Keery, W.J., Leedy, K.O., and Galloway, K.F., Electron
beam effects on microelectronic devices, SEM/1973/IV,
IITRI Research Institute, Chicago, Illinois, pp. 507-514
(1976).
39.   Ohtaka, T., Saito, S., Furuya, T., and Yamada, O., Hitachi S-
6000 field emission CD-measurement SEM. Proceedings
SPIE 565: 205-208 (1985).
40.   Spiers, S., Look through the lens for the best surface
image. Research and Development 5: 92-95 (1987).
41.   Evins, D., and Spiers, S., Computer controlled E-beam CC-
CCD system. Proceedings SPIE 565: 217-221 (1985).
42.   Norville, R., SCANLINE -- A dedicated fab line SEM LW
measurement system. Proceedings SPIE 565: 209-216
(1985).
43.   Volbert, B., and Reimer, L., Advantages of two opposite
Everhart-Thornley detectors in SEM. SEM/1980/IV,
SEM, Inc., pp. 1-10 (1980).

44. Postek, M.T., The scanning electron microscope in the semiconductor industry. Test and Measurement World, pp. 54-75 (1983).
45. Wells, O.C., Optimizing the collector solid angle for the low loss electron image in the scanning electron microscope. Proceedings EMSA (G.W. Bailey, ed.), pp. 548-549 (1987).
46. Wells, O.C., Effects of collector take-off angle and energy filtering on the BSE image in the SEM. Scanning 2: 199-216 (1979).
47. Wells, O.C., Low loss image for surface scanning electron microscope. Appl. Phys. Lett. 19(7): 232-235 (1971).
48. Wells, O.C., Low loss electron images of uncoated photoresist in the scanning electron microscope. Appl. Phys. Lett. 49(13): 764-766 (1986).
49. Moll, S.H., Healey, F., Sullivan, B., and Johnson, W., A high efficiency nondirectional backscattered electron detection mode for SEM, SEM/1978/I, SEM Inc., pp. 310-330 (1978).
50. Moll, S.H., Healey, F., Sullivan, B., and Johnson, W., Further developments of the converted backscattered electron detector, SEM/1979/II, SEM Inc., pp. 149-154 (1979).
51. Volbert, B., and Reimer, L., Detector system for backscattered electrons by conversion to secondary electrons. Scanning 2: 238-248 (1979).
52. Postek, M.T., Keery, W.J., and Larrabee, R.D., The relationship between accelerating voltage and electron detection modes to linewidth measurements in an SEM. J. Scanning Microscopy 10: 10-18 (1988).
53. Postek, M.T., Electron detection mode and their relation to linewidth measurement in the scanning electron microscope. Proceedings EMSA/MAS (G.W. Bailey, ed.), pp. 646-649 (1986).
54. Everhart, T., and Thornley, R., Wide-band detector for microampere low energy electron currents. J. Sci. Instrum. 37: 246-248 (1960).
55. Nyyssonen, D., and Postek, M.T., SEM-based system for the calibration of linewidth SRM's for the IC industry. Proceedings SPIE 565: 180-186 (1985).
56. Russell, P.E., Namae, T., Shimada, M., and Someya, T., Development of SEM-based dedicated IC metrology system. Proceedings SPIE 480: 101-108 (1984).

57.   Russell, P.E., and Mancouso, J., Microchannel plate detector for low voltage scanning electron microscopy. J. Microsc. 140(3): 323-330 (1985).

58.   Schmid, R., and Brunner, M., Design and application of a quadrupole detector for low-voltage scanning electron microscopy. Scanning 8: 294-299 (1986).

59.   Utterback, S.G., Dimensional metrology using the scanning electron microscope. Review of NDE, pp. 1-11 (1987).

60.   Joy, D.C., Low voltage scanning electron microscopy. Institute of Physics Conference Series EMAG 87, Manchester 90:175-180 (1987).

61.   Kanaya, K., and Okayama, S., Penetration and energy-loss theory of electrons in solid targets. J. Phys. D. 5: 43-58 (1972).

62.   Drescher, H., Reimer, L., and Seidel, H., Ruckstreuko-efficient and Sekundarelektronen-Ausbeute von 10-100 keV-electronen und Beziehungen zur Raster-elektronenmikroskopie. Z. Angew. Phys. 29: 331-336 (1970).

63.   Peters, K.-R., Conditions required for high quality high magnification images in secondary electron-I scanning electron microscopy, SEM/1982/IV, SEM Inc., pp. 1359-1372 (1982).

64.   Peters, K.-R., Working at higher magnifications in scanning electron microscopy with secondary and backscattered electrons on metal coated biological specimens and imaging macromolecular cell membrane structures, SEM/1985/IV, SEM Inc., pp. 1519-1544 (1985).

65.   Matsukawa, KT., and Shimizu, R., A new type edge effect in high resolution scanning electron microscopy. Jap. J. Appl. Phys. 13 (14): 583-586 (1974)

66.   Postek, M.T., Low accelerating voltage inspection and linewidth measurement in the scanning electron microscope, SEM/1984/III, SEM Inc., pp. 1065-1074 (1985).

67.   Rez, P., A transport equation theory of beam spreading in the electron microscope. Ultramicroscopy 12: 29-38 (1983).

68.   Postek, M.T., and Joy, D.C., Submicrometer microelectronics dimensional metrology: scanning electron microscopy. J. Res. Natl. Bur. Stand. 92(3): 187-204 (1987).

69.   Kahner, D.K., Horlick, J., and Foer, D., Mathematical software in Basic: RV, generation of uniform and normal

random variables. IEEE Micro, June 1986, pp. 52-60 (1986).

70.   Hembree, KG.G., Jensen, S.W., and Marchiando, J.R., Monte Carlo simulation of submicrometer linewidth measurements in the scanning electron microscope. Microbeam Analysis 1981 (R. Geiss, ed.), San Francisco Press, pp. 123-126 (1981).

71.   Joy, D.C., Image modelling for SEM-based metrology. Proceedings EMSA/MAS (G.W. Bailey, ed.), San Francisco Press, pp. 650-651 (1986).

72.   Joy, D.C., A note on charging in low voltage SEM. Proceedings 1987 MAS Meeting, San Francisco Press, pp. 117-118 (1987).

73.   Atwood, D.K., and Joy, D.C., Improved accuracy for SEM linewidth measurements. Proceedings SPIE (see John Helbert).

74.   Swing, R.E., The theoretical basis of a new optical method for the accurate measurement of small line-widths. Proceedings SPIE 80: 65-77 (1976).

75.   Nyyssonen, D., Metrology in Microlithography, in VLSI Electronics: Microstructure Science (N.G. Einspruch, ed.), Vol. 16, pp. 265-317, New York: Academic Press (1987).

76.   Nyyssonen, D., A new approach to image modeling and edge detection in the SEM. Proceedings SPIE 921: (see John Helbert).

77.   Monahan, K., and Lim, D.S., Nanometer-resolution SEM metrology in a hostile laboratory environment. Proceedings SPIE 565: 173-175 (1985).

78.   Brunner, M., and Schmid, R., Charging effects in low-voltage SEM metrology, SEM/1986, SEM Inc., pp. 377-382 (1986).

79.   van Veld, R., and Shaffner, T., Charging effects in scanning microscopy. SEM/1971/I, IIT Research Institute, Chicago, pp. 17-24 (1971).

80.   Erasmus, S., Damage to resist structures during SEM inspection. Proceedings 30th Annual Conference on Electron, Ion and Photon Beam (in Press).

81.   Erasmus, S., Damage to resist structures during scanning electron microscopy inspection. J. Vac. Sci. Technol. B: 409-413 (1987).

82.   Kotani, H., Kawabe, M., and Namba, S., Direct writing of gratings by electron beam in poly (methyl methacrylate) wave guides. Jap. J. Appl. Phys. 18: 279-283 (1979).

83. Samoto, N., Shimizu, R., and Hashimoto, H., Changes in the volume and surface composition of polymethylmethacrylate under electron beam irradiation in lithography., Jap. J. Appl. Phys. 24: 482-486 (1985).

84. Reimer, L., and Schmidt, A., The shrinkage of bulk polymers by radiation damage in an SEM. Scanning 7: 47-53 (1985).

85. Reimer, L., Irradiation changes in organic and inorganic objects. Lab Invest. 14 (6): 344-358 (1965).

86. Manual on presentation of data and control chart analysis, STP 15D, American Society for Testing and Materials, 1916 Race Street, Philadelphia, PA 19103 (1949).

87. Joy, D.C., A model for calculating secondary and backscattered electron yields. J. Microsoc. 147(1): 51-64 (1987).

88. Hembre, G.G., A metrology electron microscope system. Proceedings EMSA/MAS (G.W. Bailey, ed.), pp. 644-645 (1987).

89. The literature accompanying NIST Standard Reference Material 484.

90. Postek, M.T., Low accelerating voltage pitch standard based on the modification of NIST SRM 484, NISTIR 87-3665, National Institute of Standards and Technology, Gaithersburg, Maryland (October 1987).

91. The handbook accompanying NIST Standard Reference Material 474 or 475.

92. Buehler, M.G., Microelectronic Test Chips for VLSI Electronics, VLSI Electronics: Microstructure Science, Vol. 6, pp. 529-576, Academic Press, Inc. (1983).

93. Buehler, M.G., and Linholm, L.W., Toward a Standard Test Chip Methodology for Reliable, Custom Integrated Circuits. Proceedings 1981 Custom Integrated Circuits Conference, pp. 142-146 (1981).

94. Carver, G.P., Linholm, L.W., and Russell, T.J., Use of Microelectronic Test Structures to Characterize IC Materials, Processes, and Processing Equipment, Solid State Technol. 23: 85-92 (1980).

95. Buehler, M.G., Grant, S.D., and Thurber, W.R., Bridge and van der Pauw Sheet Resistor for Characterizing the Line Width of Conducting Layers, J. Electrochem. Soc. 125: 650-654 (1978).

96. Linholm, L.W., Yen, D., and Cresswell, M.W., Electrical Linewidth Measurement in the Near- and Sub-Micron Linewidth Region, VLSI Science and Technology/1985

(W.M. Bullis and S. Broydo, eds.), pp. 299-308, Pennington, N.J.: Electrochemical Society (May 1985).

97.   Hasan, T.F., and Perloff, D.S., Automated Electrical Measurement Techniques to Control VLSI Linewidth, Resistivity, and Registration. Test and Measurement World 5: 78-90 (1985).

98.   Yen, D., Linholm, L.W., and Buehler, M.G., A Cross-Bridge Test Structure for Evaluating the Linewidth Uniformity of an Integrated Circuit Lithography System. J. Electrochem. Soc. 129: 2313-2318 (1982).

99.   Hasan, T.F., Perloff, D.S., and Mallory, C.L., Test Vehicles for the Measurement and Analysis of VLSI Lithographic and Etching Parameters, Semiconductor Silicon/1981 (H.R. Huff, R.J. Kriegler, and Y. Takeishi, eds.), pp. 868-881, Pennington, N.J.: Electrochemical Society (1981).

100.   Buehler, M.G., and Thurber, W.R., An Experimental Study of Various Cross Sheet Resistor Test Structures. J. Electrochem. Soc. 125: 645-650 (1978).

101.   Carver, G.P., Mattis, R.L., and Buehler, M.G., Design Considerations for the Cross-Bridge Sheet Resistor, NISTIR 82-2548, National Institute of Standards and Technology, Washington, D.C. (July 1982).

102.   Nyyssonen, D., NBSIR, to be published.

103.   David, J.M., and Buehler, M.G., A Numerical Analysis of Various Cross Sheet Resistor Test Structures. Solid-State Electron. 20: 539-543 (1977).

104.   Buehler, M.G., and Hershey, C.W., The Split-Cross-Bridge Resistor for Measuring the Sheet Resistance, Linewidth, and Line Spacing of Conducting Layers. IEEE Trans. Electron Devices ED-33: 1572-1579 (1986).

105.   Yen, D., Linholm, L.W., and Glendinning, W.B., An Electrical Test Structure for Proximity Effects Measurement and Correction, J. Electrochem. Soc. 132: 1726-1729 (1985).

4

TECHNIQUES AND TOOLS
FOR OPTICAL LITHOGRAPHY

Whit Waldo

Motorola, Inc.
PCRL
Chandler, Arizona

1.0   INTRODUCTION

The image field of a step and repeat projection aligner (a.k.a. stepper) generally is much smaller than the wafer on which the pattern is transferred. A stepper exposes a field on the wafer, then steps to the next specified site on the wafer and repeats the operation. The schematic of Figure 1 is of the most common stepper in use, where the wafer image is demagnified from the reticle pattern. The light source shown is for a mercury arc lamp, but a laser source could be substituted. Another common stepper has unity magnification and a catadioptric design (see Figure 2). This means its construction includes both reflective mirrors and refractive lens elements. The design and construction of the equipment from vendors may be different, but the advantages to a volume production fab are usually a matter of degree and the dollar cost of a wafer processed through with acceptable quality may be difficult to calculate a priori. An advanced technology involves scanning the reticle pattern and transferring the demagnified image to a stepped wafer. Figure 3 shows several different stepper configurations.

A technical discussion of optical steppers deals with image quality and image placement. Image quality is concerned with the dimensional control of features of a desired size. Metal oxide semiconductor scaling rules(1) generally maintain that final patterned linewidths should be controlled to within 10% of the nominal feature size, and the accuracy of the placement of features should be within 20-25% of the minimum feature size. For a resist image coated to thickness z and of width x, the slope of the resist edge is $\partial z/\partial x$. For exact differentials:

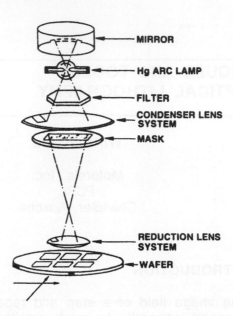

Figure 1. Schematic of a mercury arc illuminated reduction stepper. Reprinted with permission of the American Chemical Society (from L.F. Thompson and M.J. Bowden, Introduction to Microlithography, American Chemical Society, Advances in Chem. Ser., Vol. 219, 1983).

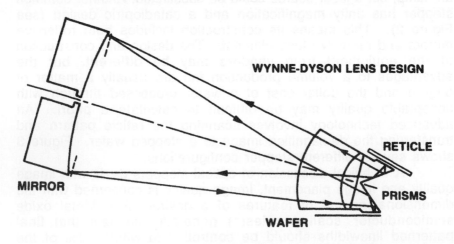

Figure 2. Schematic of a unity magnification stepper. Reprinted with permission of Ultratech Stepper, a unit of General Signal.

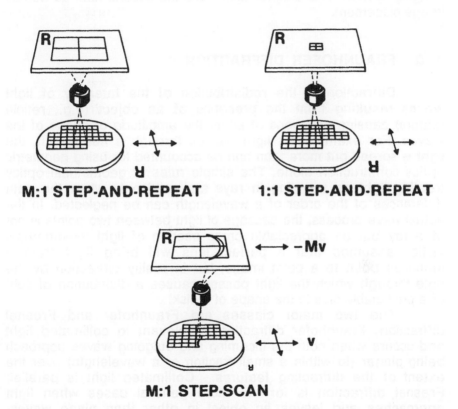

Figure 3. Schematic of different stepper configurations. Reprinted with permission of Semiconductor International (from J.H. Bruning, Semiconductor International, p. 137, April, 1981).

[1]
$$\frac{\partial z}{\partial x} = \frac{\partial z}{\partial E} \cdot \frac{\partial E}{\partial x}$$

The term $\partial z/\partial E$ reflects the processing of the resist after the latent image exists, while $\partial E/\partial x$ depends on the object and the imaging system. This chapter addresses the second term as well as image placement.

2.0   FRAUNHOFER DIFFRACTION

Diffraction is the redistribution of the intensity of light waves resulting from the presence of an object (e.g., reticle feature) causing variations of either the amplitude or phase of the waves. For example, as light passes through a narrow slit, the light is spread out more than can be accounted for using geometric optics construction alone. The simple rules of geometrical optics treating light as traveling in rays are valid only when the path differences of the order of a wavelength can be neglected. In the actual wave process, the passage of light between two points is not of a ray but an appreciable cross section of light. Geometrical optics assumes that a perfect lens will bring light from a luminous point to a point image, but in reality diffraction by the hole through which the light passes causes a distribution of light of a predictable size in the shape of a disk.

The two major classes are Fraunhofer and Fresnel diffraction. Fraunhofer diffraction is relevant to collimated light and occurs when both the incoming and outgoing waves approach being planar (to within a small fraction of a wavelength) over the extent of the diffracting features.   Collimated light is parallel. Fresnel diffraction is for the more general cases when light approaches and leaves an object in other than plane waves. Fraunhofer effects are described by the amplitude and phase of the light diffracted in a particular direction from the aperture. The diffracted light traveling in a particular direction is brought to a point in the focal plane of a converging lens. Figure 4 illustrates this for a point source.

A lens is considered diffraction limited when it has residual aberrations and errors from manufacturing which are negligible compared with the diffraction effects. These aberrations and manufacturing errors will be discussed in more detail below.

2.1   Diffraction Through a Rectangular Aperture

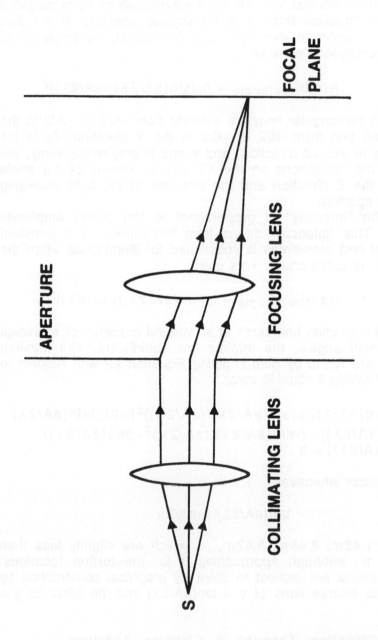

Figure 4.   System for Fraunhofer diffraction of a point source.

Although it does not describe stepper performance, Fraunhofer diffraction by a rectangular aperture is instructive(2). For collimated light having a wavelength of λ, propagated in the +Z direction through a rectangular aperture in the Z=0 plane falling upon an image plane at Z=constant, the amplitude of the electromagnetic wave is:

[2] $A(a,b)_{rectangle} = A_0[\sin(aA/2\lambda)/(aA/2\lambda)]^2$

where the rectangular aperture extends from -A/2 to +A/2 in the X direction and from -B/2 to +B/2 in the Y direction, A_0 is the amplitude in the +Z direction, and a and b are, respectively, the sines of the projections on the XZ and YZ planes of the angle between the Z direction and the direction of the light emerging from the aperture.

The "intensity" is proportional to the scalar amplitude squared. This "intensity" differs from illuminance by a constant coefficient and sometimes is substituted for illuminance when the coefficient is unimportant. This gives:

[3] $I(a,b)_{rectangle} = I_0[\sin(aA/2\lambda)/(aA/2\lambda)]^2$.

The function has zeros at all integral multiples of π except 0. For small angles, the minima are equidistant. The maxima locations are found by differentiating Equation [3] with respect to aA/λ and setting it equal to zero:

[4] $d/d(aA/\lambda)\{I_0[\sin(aA/2\lambda)/(aA/2\lambda)]^2\}=2I_0[\sin(aA/2\lambda)$
       $/(aA/2\lambda)] \cdot [-\sin(aA/2\lambda)/(aA/2\lambda)^2+\cos(aA/2\lambda)/$
       $(aA/2\lambda)] = 0$.

Maxima occur whenever:

[5] $\tan(aA/2\lambda) = aA/2\lambda$

or, at 0, 1.43π, 2.46π, 3.47π, ..., which are slightly less than $(K+ 1/2)$ π, although approaching it for the further locations. These maxima are easiest to solve by graphical construction by finding the intersections of $y = \tan(aA/2\lambda)$ and the other for $y = aA/2\lambda$.

2.2   Diffraction Through a Circular Aperture

Fraunhofer diffraction by a circular aperture describes the performance of a stepper since the lenses and stops are circular. The rectangular aperture is replaced conceptually by an approximation for a circular aperture; i.e., the aperture is divided into a series of narrow rectangular strips of equal width to calculate the effect at off-axis points. Since these strips are not of equal length, their amplitudes are unequal too, requiring the use of Bessel functions for their addition. The intensity for a circular aperture of radius r, centered at the origin, in a plane normal to the Z axis, is:

[6] $I(\rho)circle = I_0[2J_1(\rho r/\lambda)/(\rho r/\lambda)]^2$

where, J_1 is the first-order Bessel function and ρ is $\sin\theta$, the sine of the angle of deviation from the optical axis to the edge of the objective. This is illustrated in Figure 5.

2.3 Airy Disk

The image of an ideal point source through a circular aperture is illustrated schematically. This function has zeros in a single set of concentric rings with differences in the radial parameter, ρ, slightly greater than $\lambda/2r$. The bright central maximum first was noted by British mathematician and astronomer Sir George Biddell Airy, and is known as Airy's disk (see Figure 6). The bright central disk is surrounded by a number of fainter rings. Neither the disk nor the rings have intensities that are defined sharply but instead are shaded at the edges. The rings are separated by circles of zero intensity. About 85% of the energy entering the optical system is concentrated in the Airy disk, while the other 15% is spread through the rings(3). K. Strehl, a contemporary of John Rayleigh, noticed that small aberrations in the lens decrease the proportion of the energy in the central disk while that in the rings increases(4). The Strehl intensity ratio measures the relative intensity at the principal maximum of the diffraction pattern with and without aberrations (see Figure 7).

For diffraction through a circular aperture, the distance from the bright central spot to the first zero is given by $1.22\lambda f/d$, where f is the focal length and d is the diameter of the objective lens. Notice the maxima are not located symmetrically about the minima.

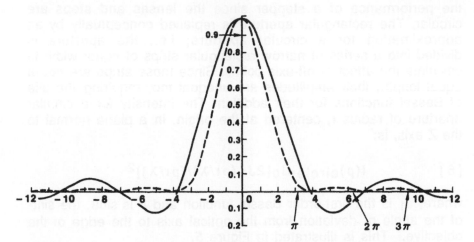

Figure 5. Amplitude (solid line) and intensity distribution (dashed line) in Fraunhofer diffraction by a circular aperture.

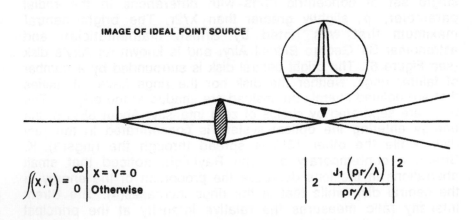

Figure 6. Image of an ideal point source. Reprinted with permission of Kodak Microelectronics Seminar (from K.A. Snow, Kodak Microelectronics Seminar, p. 83, 1975).

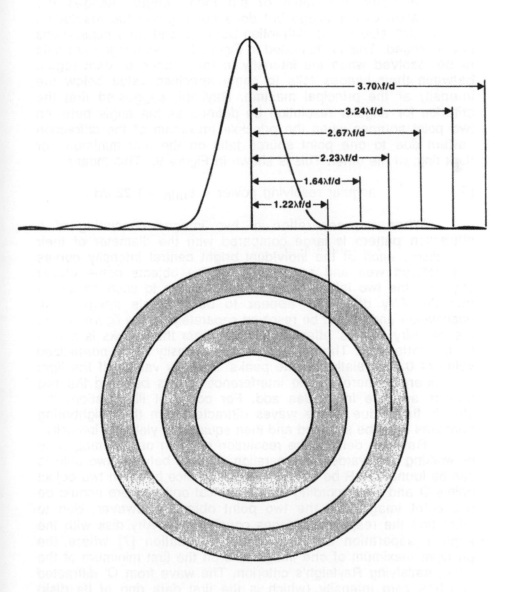

Figure 7.   Intensity distribution and its Airy disk.

3.0   THEORETICAL RESOLUTION LIMIT

Changing the value of the focal length changes the magnification of the image but does not improve the resolution. The relative size of the diffraction patterns and their separations is unchanged. This is illustrated in Figure 8. Two images are said to be resolved when the intensity in the shaded or dark region between their images falls to some specified value below the intensity at the principal maxima. Rayleigh suggested that the criterion for angular resolution be defined as the angle between two point sources when the principle maximum of the diffraction pattern due to one point source falls on the first minimum, or dark ring, of the other. This is shown in Figure 9. This means:

[7]   angular resolving power = θ_{min} = 1.22λ/d .

When the separation of two images consisting of a diffraction pattern is large compared with the diameter of their Airy disks, each of the individual bright central intensity curves are defined well and separated. As the objects come closer together, the two intensity curves will overlap to such an extent that the Airy disks will appear to be a single image upon observation and cannot be resolved separately. In the figure above, the two Airy images' effective intensity near their peaks is shown by the dotted line. The minimum of the intensity has a normalized value of 0.735 relative to the peaks' intensity values. If the light is incoherent, there are no interference effects between the two images and the intensities add. For coherent illumination, the electric fields due to the waves diffracted from the neighboring apertures must be summed and then squared to yield the intensity.

Rayleigh defined the resolution criterion of two images so by working backwards the separation distance between two objects can be found. Let R be the separation distance between two object points O and O'. According to geometrical optics, there should be two point images for the two point objects. However, due to diffraction the respective images consist of an Airy disk with the angular separation angle defined by Equation [7] where the principal maximum of one image falls on the first minimum of the other, satisfying Rayleigh's criterion. The wave from O' diffracted to I has zero intensity (which is the first dark ring of its disk) and the extreme rays O'BI and OAI differ in path length by 1.22λ. From the Figure 10, O'B is longer than OB or OA by R·sin i, and O'A is shorter by R·sin i. This means the path difference of the extreme rays from O' is 2R·sin i, and equating this to 1.22λ:

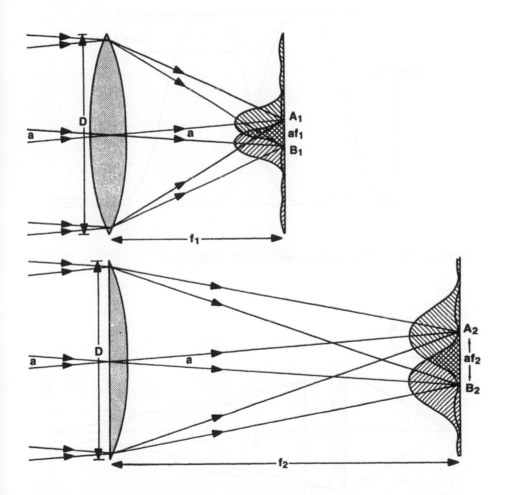

Figure 8. Magnification and resolution. The lower lens is twice the focal length of the upper ($f_2=2f_1$), so the images formed at A_2 and B_2 are twice as far apart as A_1 and B_1. The diffraction patterns caused by the equal apertures D simply scale up in linear size so there is no gain in resolution. Reprinted with permission of Reference 6.

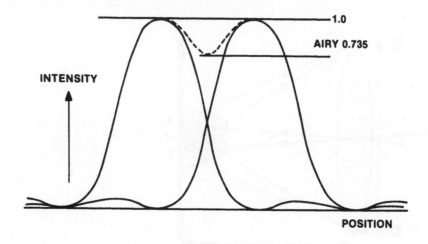

Figure 9.   Rayleigh's resolution criterion.

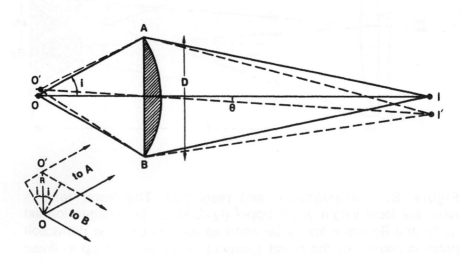

Figure 10. Resolving power of a lens. Reprinted with permission of Reference 46.

[8] $R = (1.22\lambda) / (2 \sin i)$.

Since the index of refraction, n, between the object and the objective may not be unity (i.e., of a vacuum):

[9] $R = (1.22\lambda) / (2n \cdot \sin i)$

which simplifies to:

[10] $R = (0.61\lambda) / (n \cdot \sin i)$.

The German physicist, Ernst Abbe proposed that $n \cdot \sin i$ be known as the numerical aperture (NA) of the objective. Light diffracted from the reticle is collected by the objective lens for imaging if the beams are within the acceptance angle of the objective. In practice, the largest value of the numerical aperture obtainable is about 1.6, restricted by the limited availability of immersion fluids with an index of refraction greater than 1.5. So, the theoretical resolution limit between two object points is:

[11] $R = (0.61\lambda) / NA$.

Equation [11] assumes the light scattered by two object points is independent in phase. Abbe knew this assumption was inappropriate for two points illuminated by light from a condenser (i.e., they were not self-luminous), so the resolution limit was given by:

[12] $R = (0.50\lambda) / NA$.

Actual descriptions of microlithographic resolution treat the coefficients in Equations [11] and [12] as a variable, k_1, dependent upon the object feature size and shape, the chemical process used and the condition of the image plane (e.g., substrate reflectivity, topographical flatness and planarity, defocus of image plane, etc.). A k_1 value equal to 0.61 typically is considered the resolution limit under pilot line conditions, although resolution with smaller k_1 values has been demonstrated.

[13] $R = (k_1\lambda) / NA$.

The typical microlithographic resolution limit in a production environment assumes a value of k_1 of 0.81. This is shown in Figure 11 where two points are considered resolved when the maximum of the first diffraction pattern is superimposed to the first secondary maximum of the second diffraction pattern.

4.0 DIFFRACTION GRATINGS

Diffraction gratings are useful for further understanding projection imaging systems like steppers. A diffraction grating has a number of parallel equidistant fine slits located in the same plane through which light passes. A transmissive diffraction grating is shown in Figure 12 where the zero, first, and second diffracted orders are illustrated. The angle of departure of the orders from the grating depends upon the spatial frequency of the grating.   A coarse grating having few slits per unit width will have many orders collected by the objective while a fine grating having many slits per unit width will have fewer orders collected. For high frequency gratings, the principal intensity maxima become higher and more narrow, since the intensity in the principal maxima is proportional to the square of the number of slits due to constructive interference of the light from the two slits (see Figure 13). However, the secondary maxima between principals are suppressed, since information about these orders is missing due to destructive interference. Destructive interference occurs when certain order spectra have positions corresponding to the minima of the diffraction pattern for a single slit.

Simple harmonic motion can be represented by either a sine function or a cosine function:

[14] $y = A \sin(\omega t + \phi)$
                     $y = A \cos(\omega t + \phi)$

where, A is the amplitude of the wave, ω is the angular velocity, t is the time, and ϕ is the phase (where $\omega t = \phi$). These functions can be combined into complex numbers. The two parts can be drawn along orthogonal axes with a real axis, x and an imaginary axis, y. The complex point P has rectangular coordinates (x, y). A can be represented in polar coordinates (t, ϕ), too.

[15] $\sin \phi = y/A$ and $\cos \phi = x/A$.

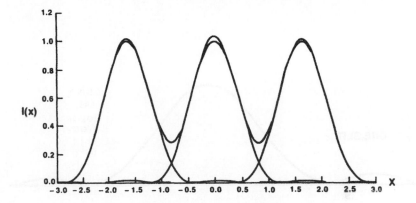

Figure   11. Volume production resolution criterion.

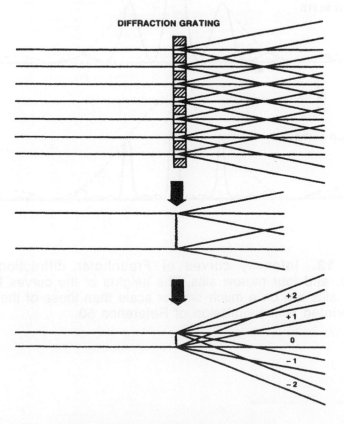

Figure   12. Transmissive diffraction gratings.

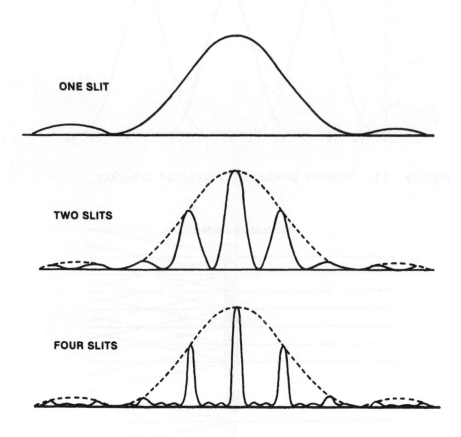

Figure 13. Intensity curves of Fraunhofer diffraction from one, two, and four narrow slits. The heights of the curves for two and four slits are on a much smaller scale than those of the single slit. Reprinted with permission of Reference 50.

Since sin ϕ = cos (90°- ϕ), the sine and cosine functions are essentially the same except for a 90° phase difference.

If A is the resultant of two orthogonal vectors a and b, where, i stands for a single counterclockwise 90° rotation, then A can be represented in polar coordinates (t, ϕ):

[16] $A = a + ib = t \cos \phi + it \sin \phi$.

Euler's formula says:

[17] $\cos \phi + i \sin \phi = e^{i\phi}$.

Combining Equations [16] and [17] gives the complex transmission function of a diffraction grating:

[18] $A = te^{i\phi}$.

where t is the amplitude transmittance of the grating and ϕ is the phase shift of the grating. According to Fourier optics theory(5), the Fraunhofer diffraction pattern of a diffraction grating is proportional to the Fourier transform of the transmission function. For a unity amplitude narrow bandwidth collimated light source incident on a reticle grating of a 1:1 projection system, the light is split into many beams of amplitudes $M_n/2$, where M_n is the nth Fourier coefficient of the reticle diffraction grating. All beams of diffraction order $m < NA \cdot p/\lambda$ pass through the lens and converge at the image plane, where NA is the numerical aperture of the lens, p is the grating period, and λ is the light's wavelength. Notice from Figure 14 that many higher diffracted orders are lost, depending upon the numerical aperture of the lens. The higher the objective numerical aperture, the more diffracted orders are collected for any grating. Although the intensity contribution of the higher diffracted orders decreases rather rapidly, these higher orders still represent information content of the object structure.

5.0 FOURIER SYNTHESIS

The French physicist and mathematician Baron Jean Baptiste Joseph Fourier developed a theory stating that any periodic function can be represented as the sum of a number of sine and cosine functions with appropriate amplitudes and frequencies:

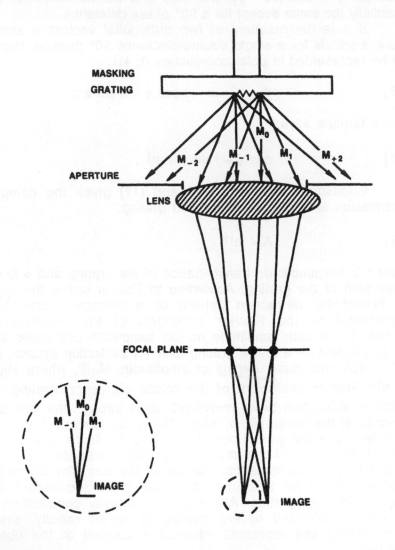

Figure 14. Mask grating with diffracted light projected through the lens and the converging grating orders.

[19] $y = A_0 + A_1\sin\omega t + A_2\sin2\omega t + A_3\sin3\omega t + \ldots$
             $+ A_1\cos\omega t + A_2\cos2\omega t + A_3\cos3\omega t + \ldots$

This is known as a Fourier series where y is the displacement of the resultant wave at a time t, A_0 is a constant (required for functions other than sine or cosine waves that have a nonzero average value), the A and B coefficients are the amplitudes of the component waves, and ω is the angular velocity. The resultant wave is represented in sound as the collection of the fundamental note and its various harmonics (integral multiples of the frequency of the original periodic function).

In general, the series:

[20] $A_0/2 + \sum\limits_{n=1}^{\infty} (A_n\cos Kx + B_n\sin Kx)$

represents the Fourier series of the function f(x), where:

[21] $A_n = 1/\pi \int\limits_{-\pi}^{\pi} f(x)\cos Kx\,dx \qquad K=1,2,\ldots$

[22] $B_n = 1/\pi \int\limits_{-\pi}^{\pi} f(x)\sin Kx\,dx \quad K=1,2,\ldots$

General problems of a periodic nature can be solved with these series as well as waves of a nonperiodic nature with substitution of more complex Fourier integrals.

By including a very large number of terms, Fourier's theorem makes it possible to synthesize any waveform, including a square wave(6). A mask of bar targets might be regarded as such a square wave spatial profile. The Fourier synthesis for this object would use only the odd harmonics. To generate a sawtooth pattern, as illustrated in Figure 15, the Fourier synthesis would use the fundamental and all of its harmonics. A very good synthesis of a sawtooth object using only a few terms is illustrated. A better approximation would be obtained by adding more terms. The period of the wave is p, so its spatial frequency, ν, is 1/p. For example, the period, or pitch, consisting of a 1.0 μm line and a 1.0 μm space is 2.0 μm. The corresponding frequency, typically

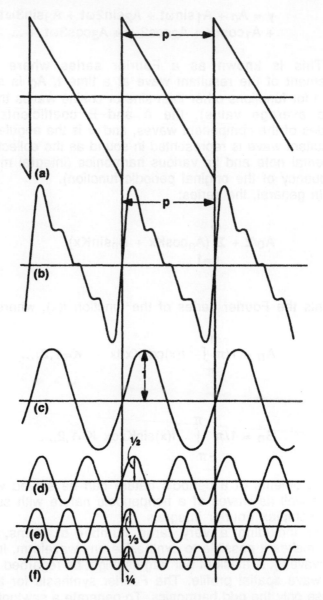

Figure 15. An example of Fourier transformation. The spectrum of the saw tooth periodic function is shown in (a). The individual sine waves are shown in (c), (d), (e), and (f). The sum of these four sine waves is (b). The addition of more sine waves of suitable harmonic frequencies would improve the approximation to the saw tooth. Reprinted with permission of Reference 6.

expressed in units of cycles per mm, is 500. The appropriate harmonic frequencies are 2/p, 3/p, 4/p, etc.

6.0   ABBE'S THEORY OF IMAGE FORMATION

Abbe's theory of image formation coupled Fraunhofer's diffraction theory and the ideas of spatial Fourier synthesis to state that any object can be regarded as the collection of superimposed sine or cosine profile diffraction gratings. Light from a point source is collimated by passing through a condenser lens. Some of the incident light on the object is passed through and forms the zero-order maximum, containing most of the luminous energy. Other incident light on the object is diffracted, forming the higher orders. The wider angles of diffraction are associated with the finer details, and it is the fine detail which is lost because these are not collected by the limited numerical aperture of the objective lens. The number of diffracted orders collected by the lens limits the information about the object which is transferred to the image. Lower frequency gratings have orders that are diffracted by small angles so many orders can be collected by the objective lens. This will result in faithful recreation in the image of the object. By contrast, higher frequency gratings diffract orders by larger angles, resulting in relatively fewer collected orders for a given numerical aperture objective. This degrades the image synthesis.

7.0   INTRODUCTION TO TRANSFER FUNCTIONS

Transfer functions are a measure of the quality of information transfer. This type of information exchange occurs in photolithography by replicating the pattern on a mask in photoresist on a wafer.   Generally, there is some loss of information with each transfer operation.

7.1   Spread Functions

Ideally, the image would be replicated as a square wave image for a square wave object. Due to Fraunhofer diffraction, we know that the image will be diffraction limited, at best. The intensity distribution in the image of an incoherently illuminated pinhole is known as the point spread function, $S(y,z)$, of the lens. The point spread function for optical lithography is not Gaussian(7), but is similar to that shape. As seen in Figure 16.

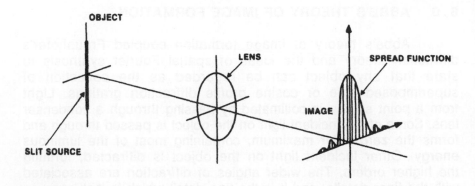

Figure 16. The image of an extended object represented as the superposition of spread functions. Reprinted with permission of Reference 8.

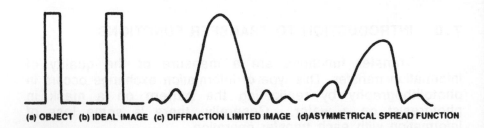

(a) OBJECT (b) IDEAL IMAGE (c) DIFFRACTION LIMITED IMAGE (d)ASYMMETRICAL SPREAD FUNCTION

Figure 17. Image of a narrow slit. Reprinted with permission of Reference 8.

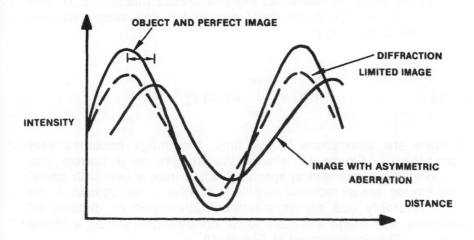

Figure 18. Intensity distribution in the image of a sine wave test object. Reprinted with permission of Reference 8.

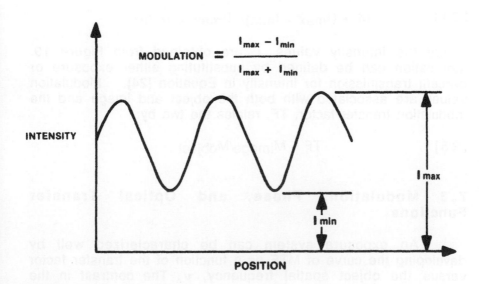

Figure 19. Modulation relationship to sinusoidal intensity distribution.

If the pinhole is replaced by a narrow slit, the distribution of light in the image is known as the line spread function, S(z). The line spread function is the point spread function integrated over the length of the slit, y:

[23]
$$S(z) = \int_{-\infty}^{+\infty} S(y,z)\, dy \ .$$

If there are aberrations in the lens, the image becomes very complex(8). Figure 17 shows the images of a narrow slit, including the asymmetrical spread function from a lens with coma. The images are of reduced amplitude relative to the objects. If the lens assembly has an asymmetrical aberration or decentered elements, the image will have lower contrast and exhibit a phase change. This is illustrated in Figure 18.

7.2 Modulation

The modulation or contrast is defined by(4):

[24]
$$M = (I_{max} - I_{min})/\ (I_{max} + I_{min})$$

where the intensity values, I, are obtained from Figure 19. Modulation can be defined by substituting either exposure or percent transmission for intensity in Equation [24].   Modulation values are associated with both the object and image and the modulation transfer factor, TF, relates the two by:

[25]
$$TF = M_{image}/M_{object}\ .$$

7.3 Modulation, Phase, and Optical Transfer Functions

An exposure system can be characterized well by developing the curve of MTF as a function of the transfer factor versus the object spatial frequency, ν. The contrast in the projected image is proportional to the MTF value for a given frequency. A lens cannot be described by a single value of TF, but by the modulation transfer function, MTF, showing that:

[26]
$$MTF = TF(\nu)\ .$$

The optical transfer function (OTF) of a lens is the Fourier transform of the spread function of that lens describing the amplitude and phase change from the object to the image(4)(9):

$$[27] \qquad OTF = \int_{-\infty}^{+\infty} S(z)e^{2\pi i \nu z} \, dz \, .$$

The phase change mentioned earlier caused by an asymmetrical aberration or decentered lens elements is dependent on the spatial frequency and contributes to the phase transfer function, ϕTF. The optical transfer function relates the amplitude change described by the MTF and the phase change described by the ϕTF:

$$[28] \qquad OTF = MTF \cdot e^{i\phi(\nu)}$$

where the ϕTF is described by the exponential term with phase transfer factor, $\phi = f(\nu)$. This transform can be expressed as a complex function consisting of a real plus an imaginary part(9).

$$[29] \qquad OTF = R(\nu) + i \, I(\nu) \, .$$

Then the MTF can be expressed in a form of:

$$[30] \qquad MTF = \sqrt{[\, R(\nu)^2 + i \, I(\nu)^2 \,]}$$

and:

$$[31] \qquad \phi(\nu) = \tan^{-1} [R(\nu)/I(\nu)] \, .$$

It is important to recognize that even though the MTF and ϕTF are presented usually as separate graphs they should not be considered independent of each other. Both are required to describe the OTF.

The optical transfer function also can be defined as the ratio of the Fourier transform of the light distribution in the image to the Fourier transform of the light distribution in the object(8):

$$[32] \quad OTF = \frac{\text{Fourier transform of image intensity distribution}}{\text{Fourier transform of object intensity distribution}} \, .$$

7.4 Cascading Linear Functions

Multiple information transfer operations occur in photolithography. The information collected by the lens from diffracted orders of light limits the number of terms in the Fourier series, and so limits the image quality. A second information transfer operation takes place in the resist system as the latent image is processed. Each of these informational transfer operations might, therefore, involve some loss of information so that the developed photoresist image might not be recreated like the mask as a square wave profile, but rather suffer some loss of contrast. One of the greatest advantages of using transfer functions is the cascading property that permits the lens transfer function to be combined with that of the photographic film(8).

$$[33] \qquad TF(\nu)_{total} = TF_{lens}(\nu) \cdot TF_{resist}(\nu) .$$

The transfer function for the lens must be for the lens system. The MTF contributions of individual lens elements do not cascade to form the stepper lens system MTF in a form as described by Equation [33]. The overall correction of the lens system depends on the aberration introduced by one element to compensate for that introduced by another element in the system. The MTF of each element can be low but the combination of elements results in a much higher MTF value for the lens system. Cascading of elements assumes a negligible contribution of a phase shift, which would make the system nonlinear, so description by its MTF would be incomplete (without the ϕTF). Also, the assumption that the transfer function of the resist is linear in response is only approximate(10), but this inaccuracy may be insignificant.

7.5 Illumination Degree of Coherence

The illumination used in optical steppers is approximately Kohler where each element of the spatially incoherent exposure source is imaged through a condenser in the entrance pupil of the projection lens with emerging plane or almost plane waves(11)(12). This is shown in Figure 20. At every point on the mask, the principal illumination direction is nominally toward the same point in the projection lens entrance pupil, so the diffraction patterns from objects at all points in the field coincide(13). Kohler illumination permits uniform illumination from a source that is not uniform. The illumination of optical imaging systems is described as either coherent or incoherent in

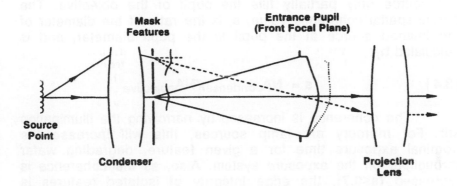

Figure 20. Kohler illumination shown image-side telecentric. Reprinted with permission of Reference 13.

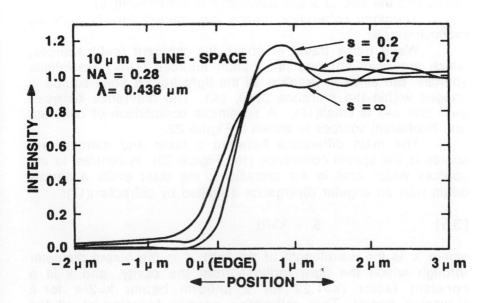

Figure 21. Edge intensity. Reprinted with permission of Reference 14.

amplitude depending upon the extent to which the source image fills the entrance pupil of the objective. Steppers are designed so the source only partially fills the pupil of the objective. The partial spatial coherence value, s, is the ratio of the diameter of the imaged source at the pupil to the pupil diameter, and is calculated by:

[34] $s = NA_{condenser}/NA_{objective}$.

The coherence is increased by narrowing the illumination slit. For mercury arc lamp sources, this will increase the nominal exposure time for a given feature, degrading wafer throughput on the exposure system. Also, as the coherence is increased ($s \leq 0.7$), the edge integrity of isolated features is degraded by a phenomenon known as ringing(14). Ringing is an undulation in the intensity at the edge, dI/dx, instead of a constant value (see Figure 21). The partial coherence value influences not only the modulation rate but also the profile itself of the aerial image reconstituted by the lens, and it is necessary to know the intensity distribution I(x) exactly in order to compute the edge profile and the size of a line developed in photoresist(15).

Temporal coherence means the emitted light is perfectly monochromatic.

Wave trains travel in phase for coherent (s=0) sources, which are point sources. Incoherent sources (s=∞) are infinite sources, but since in practice all the light from a finite source is imaged within the entrance pupil, s≤1. The difference between s=∞ and s=1 is small(16). A schematic comparison of coherent and incoherent sources is shown in Figure 22.

The main difference between a laser and mercury arc source is the spatial coherence (see Figure 23). In contrast to arc sources which emit in 4π steradians, the laser emits a narrow beam with an angular divergence θ limited by diffraction(17):

[35] $\theta = k\lambda/d$

where λ is the wavelength of the light, d is the output diameter through which the light radiates from the cavity, and k is a constant factor (k=1.22 for a uniform beam; k=2/π for a Gaussian beam). This collimation permits focusing of all the emitted energy into a small spot whose size (of approximately a wavelength) is limited only by focusing lens diffraction. In comparison to mercury arc lamps, there is no illuminance penalty for increasing the coherence. The MTF curves for an imaging

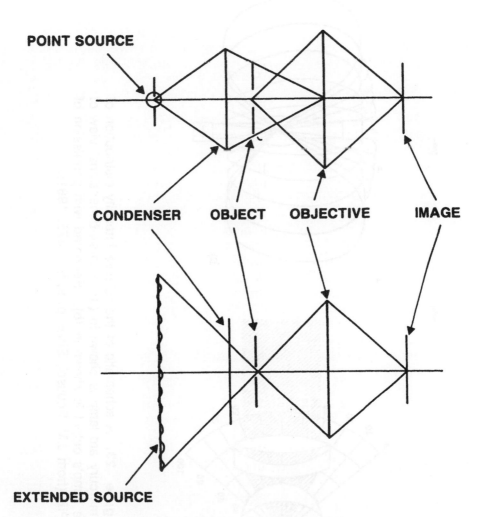

Figure 22. Schematic of coherent and incoherent projection printers. Reprinted with permission of Reference 19.

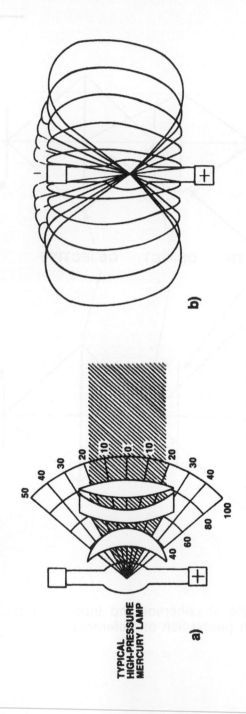

Figure 23. A schematic of the relative intensity distribution of a mercury arc lamp is shown in (a). A three-dimensional view of the lamp's output is shown in (b). Reprinted with permission of SPIE (from J.M. Roussel, SPIE, p. 9, Vol. 275, 1981).

system with different filling ratios(16)(18) are shown in Figure 24.

For coherent illumination, the MTF response is a step function of unity until the cutoff frequency is reached, when the response immediately drops to zero. Coherent systems have sharper image edge gradients and smaller point spread functions, but they also have more image degradation resulting from diffraction(19). Returning to diffraction gratings(11), Figure 25 shows a coherent plane wave normally incident on a mask which contains a grating pattern of pitch p with equal lines and spaces. The frequency, ν, equals the inverse of the pitch, p. The undiffracted component of light passing through the grating contains no information about n, the frequency of the grating. Information is contained only in the diffracted light. The direction of the first diffraction peak is given by the grating formula:

$$[36] \qquad n \cdot p \cdot \sin \theta = N\lambda \quad \text{(N=1 for first diffracted order)}$$

so that:

$$[37] \qquad \nu = n \cdot \sin \theta / \lambda .$$

If the light diffracted into direction θ is to reach the image plane, it must be collected by the lens, accepting light of all angles $\theta \leq i$, since the numerical aperture of the lens was defined previously as $n \cdot \sin i$. Therefore, the highest grating frequency which can be imaged by an optical system with coherent illumination is:

$$[38] \qquad \nu_{max} = n \cdot \sin i / \lambda \quad = NA/\lambda .$$

For incoherent illumination (s=∞), the stepper has a linear response to intensity and the cutoff frequency, ν_{max}, is:

$$[39] \qquad \nu_{max} = 2NA/\lambda .$$

Notice that the cutoff frequency for incoherent illumination occurs at twice the spatial frequency as that for coherent light.

The phase of the transfer function can be expected to vary rapidly as the spatial frequency approaches the MTF cutoff(19). The incoherent MTF for a circular pupil is given by(5):

$$[40] \quad TF(\nu) = \frac{2}{\pi} \cdot [\cos^{-1}(\nu/\nu_{max}) - \nu/\nu_{max} \cdot \sqrt{1 - (\nu/\nu_{max})^2}]$$

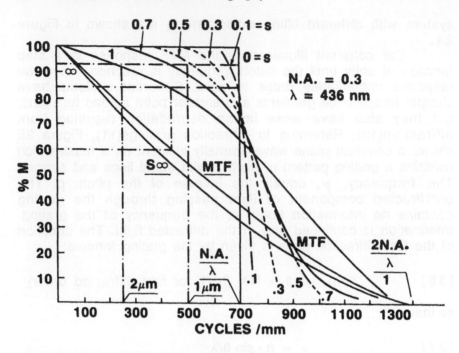

Figure 24. Effect of degree of coherence on MTF. Reprinted with permission of Reference 15.

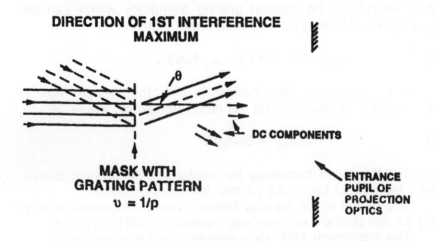

Figure 25. Diffraction of coherent and incoherent light by a grating pattern. Reprinted with permission of Reference 115.

where, ν_{max} is defined by Equation [39], and this may be approximated by the expression(16):

[41] $TF(\nu) \approx 1 - 4 \cdot \sin(\nu\lambda/2NA)/\pi$

for all but the highest spatial frequencies. The angles are expressed in radians. For minimum features, this can be further approximated by(20):

[42] $TF = 1 - 1/(\pi k_1)$

where k_1 is the Rayleigh resolution coefficient from Equation [13].

There are several technical difficulties with Figure 24 (4)(21). When the illumination is partially coherent, the system is no longer linear in either intensity or amplitude, so the imagery of real (square-wave) objects of arbitrary shape cannot be reconstructed by the MTF alone(22)(23) since phase information is unaccounted for. MTF curves are shown then only as a point of reference. However, for lithography near the resolution limit, only the fundamental frequency of the mask pattern reaches the image plane(16), and the MTF is valid for describing the modulation of the fundamental frequency of a bar target of a specified period(22). When the mask features are relatively large, the MTF becomes a meaningless metric, but the imaging begins to approach ideal so sophisticated prediction methods like the MTF are unnecessary. The MTF applies only to objects which have sinusoidal transmission characteristics(23) such as Figure 26. It is more difficult to fabricate high quality sinusoidal transmission targets than square wave bar targets(24). If the square wave, MTF_{sq}, curve is known, the sine wave, $MTF_{sine\ wave}$, curve can be calculated by(25):

[43] $TF_{sine\ wave}(\nu)=\pi/4\cdot[TF_{sq}(\nu)+TF_{sq}(3\nu)/3-TF_{sq}(5\nu)/5$

$+ TF_{sq}(7\nu)/7 - TF_{sq}(11\nu)/11 + ...]$.

7.6   Wavelength Effect on MTF

The benefits of shortening the wavelength to improve image quality at higher spatial frequencies are apparent by examining the MTF response(22) in Figure 27.

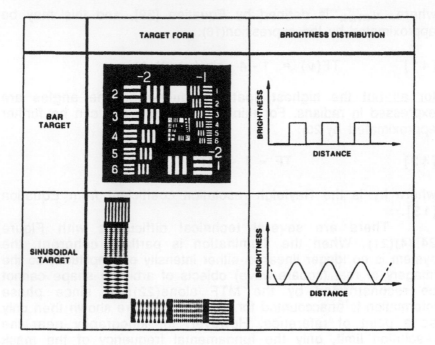

Figure 26. Comparison of bar and sinusoidal targets. Reprinted with permission of Reference 24.

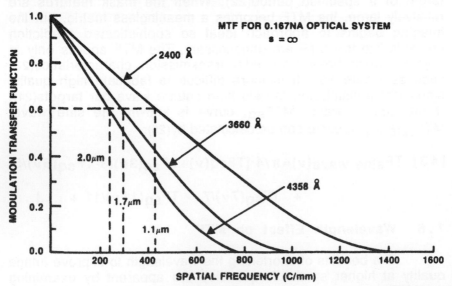

Figure 27. Incoherent MTF as a function of wavelength. Reprinted with permission of Reference 22.

7.7   Depth of Focus

The depth of focus of an optical system at the Rayleigh resolution limit is referred to as a Rayleigh unit of defocus and was defined by Strehl as:

[44] $DOF = \pm\lambda/(2NA^2)$.

In practice, this depth of focus at the resolution limit is not reached usually but the general form of the depth of focus description remains:

[45] $DOF = \pm k_2\lambda/NA^2$

where k_2 is determined empirically for the lithography process in use. Equation [45] still has serious deficiencies since it does not recognize the depth of focus latitude increases for features larger than the Rayleigh resolution size, nor the dependence on feature shape or aspect ratio even at the Rayleigh resolution size, nor the illumination conditions.

The effect of defocus on the incoherent MTF as a function of spatial frequency has been calculated(26) and is related to the MTF by the curve DOF^2. For a defocus equal to k times the Rayleigh unit of defocus, the incoherent modulation for any feature size is reduced by k^2 times the appropriate ordinate of the DOF^2 curve. Notice in Figure 28 the DOF^2 curve peaks near an operating resolution factor of $k_1=0.61$. For lower spatial frequencies, the DOF^2 is linear. Recalling the MTF curve for incoherent illumination is approximately linear in this region, the reduction in MTF due to defocus is inversely proportional to the first power rather than to the square of the numerical aperture(21). The deleterious effect of defocus for high spatial frequencies can be minimized by increasing the coherence of the light or by shortening the actinic wavelength.

7.8   Diffraction Limited Resolution

Previously, it was stated that a lens is considered diffraction limited when it has residual aberrations and manufacturing errors which are negligible compared with the diffraction effects. This will be expanded now, following Rayleigh, to designate a system as diffraction limited if the MTF lies above the curve corresponding to an optical path difference (OPD)

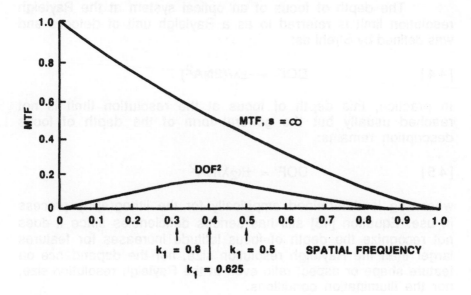

Figure 28. Effect of defocus on MTF. Reprinted with permission of Reference 21.

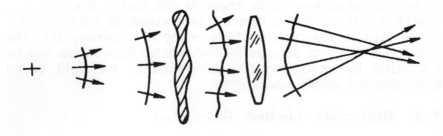

Figure 29. Effect of optical wavefront disturbances. Reprinted with permission of Reference 58.

= $\lambda/4(4)(19)$. This corresponds to a Strehl ratio of about 0.8. Using the concept of rays instead of wavefronts in Figure 29, many rays pass through an optical system from an object point to form an image point. Ideally, the optical distance will be the same along each ray. Due to imperfections, the relative paths may differ by many wavelengths. The optical path length is the product of the refractive index for each segment and the geometric length of the segment(4).

7.9 Minimum MTF Requirement

It is generally accepted that MTF values of at least 40-60% are required for a minimum working feature in positive resist(11)(22), so it can be seen that at this level of performance there is some advantage for resolving higher spatial frequencies using more coherent light. These MTF limits arise because the minimum size printable feature using any combination of photoresist and lens must satisfy(27):

[46] $$TF_{optical} \geq M_{resist}$$

where:

[47] $$M_{resist} = (E_{100} - E_0) / (E_{100} + E_0)$$

which follows Equation [24]. E_{100} is the minimum exposure energy for 100% exposure and E_0 is the maximum exposure energy for zero exposure. Exposure time or intensity can be substituted for energy in the calculation. A performance parameter of resist is its gamma, γ, value defined as:

[48] $$\gamma = [\log(E_{100}/E_0)]^{-1}$$

so Equation [46] becomes:

[49] $$M_{resist} = [10^{(1/\gamma)}-1] / [10^{(1/\gamma)}+1].$$

The theoretical basis for using γ as a figure of merit is presented elsewhere(28-31). Resist γ has been shown to be directly related to resist profile, resolution, and linewidth control (12)(33a,33b,34). However, the usefulness of γ for process development is questioned(35-38) since surface inhibition effects are not always well accounted for.

Although γ is an indirect measure of critical dimension control, it is easier to measure than image linewidths. Equally important, critical dimension experiments testing nested variances require equal cell biases to avoid an important confounding factor. This greatly increases the difficulty of linewidth experiments(39).

The resist characteristic curve is generated by making open frame (i.e., exposures with no pattern) exposures of increasing dose into a resist film coated on a bare silicon wafer. Some data are shown in Figure 30.

The value of γ of the resist also can be calculated from the slope of the least squares fitted line for normalized film thicknesses between 15-80%(39). These criteria are somewhat arbitrary, but at thicknesses >80% the shoulder effects adversely affect the calculations, since the shoulder is not part of the linear response. Also, at film thicknesses <15% the data tend to be noisy. This technique is contrary to others(37-40).

8.0   APPLICATION OF TRANSFER FUNCTIONS

MTF values can be obtained without performing laser interferometric bench tests. The object contrast is obtained easiest from measurements of the percentage transmission on a mask with a series of resolution structures used for the different spatial frequencies and application of Equation [24](41). Otherwise, the object can be scanned with a slit and photoelectric device to find I_{max} and I_{min}(42).

The image contrast can be determined experimentally by using a thin layer of photoresist as a threshold detector(19)(23). The thin resist layer makes this method independent of process parameters like absolute exposure, time, intensity, resist bake times and temperatures, and developer conditions. The image contrast in a pattern of equal lines and spaces can be calculated by noting the exposure time at which the spaces begin to clear, T_1, and the exposure time at which the lines completely disappear, T_2. The image contrast, γ, is calculated by:

[50] $$\gamma = (T_2 - T_1)/ (T_2 + T_1) .$$

Equation [50] follows from the definition of Equation [24]. Knowledge of both the object modulation and image modulation for different spatial frequencies allows the MTF curve to be generated.

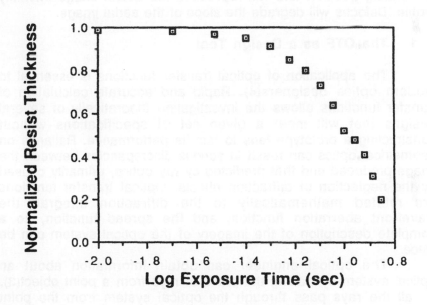

Figure 30. Resist characteristic curve.

Another method of finding the image contrast is to record the image on film and scan it using a microdensitometer to determine the maximum and minimum density values of the image(42). This method is limited to lower spatial frequencies depending upon the noise introduced to the measurements by light scattered in the film.

Alternatively, a small detector could scan the illumination profile at the image plane and determine the image contrast (43)(44). The fluorescent intensity from a scanning wafer of special construction is detected by an off-axis photometer versus the grating position and will reproduce the actual image intensity profile. Defocus will degrade the slope of the aerial image.

8.1   The OTF as a Design Tool

The application of optical transfer functions is essential to modern optics designers(4). Rapid and accurate calculation of transfer functions allows the investigation theoretically of several designs that will meet a given set of specifications without constructing a prototype lens to test its performance. Reliance on geometrical optics can result in serious discrepancies between the image produced and that predicted by ray optics, primarily caused by the neglection of diffraction effects. Optical transfer functions are related mathematically to the diffraction integral, the wavefront aberration function, and the spread function, so a complete description of the imagery of the optical system can be made.

The optical engineer can obtain information about an optical system by tracing the paths of rays from a point object(3). If all the rays pass through the optical system from the point object and converge to a point image, the system is free of aberrations. In a wave front description of imaging, periodic oscillations leave the point object traveling with equal speed in all directions. On any spherical surface whose center is at the object point, all the oscillations are in phase (see Figure 31). This surface is called a wave front. Light rays are normal to the wave front. If the system is free of aberrations, the wave front emerging from it is spherical with its center at the image point since all its normals pass through the image point. If the system has aberrations, the departure of the wave front from sphericity can be mapped by tracing rays through the system. B.R.A. Nijboer defined the wave aberration function as the difference between the wave surface and the reference sphere. Aberrations caused by surface irregularities or glass inhomogeneities are responsible for smearing the image or dislocating it from ideal.

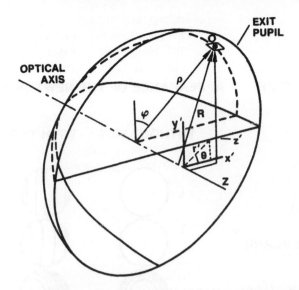

Figure 31. The spherical wavefront at the exit pupil.

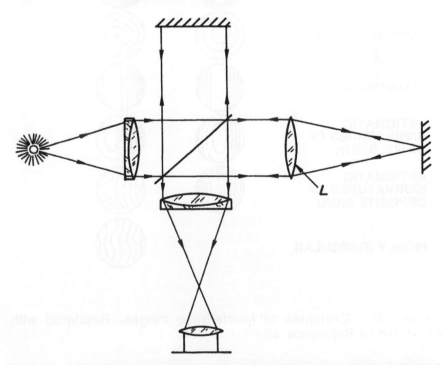

Figure 32. Twyman-Green interferometer for testing lens L. Reprinted with permission of Prentice-Hall (from J.R. Meyer-Arendt, Introduction to Classical and Modern Optics, 2nd Ed., Prentice-Hall, 1984).

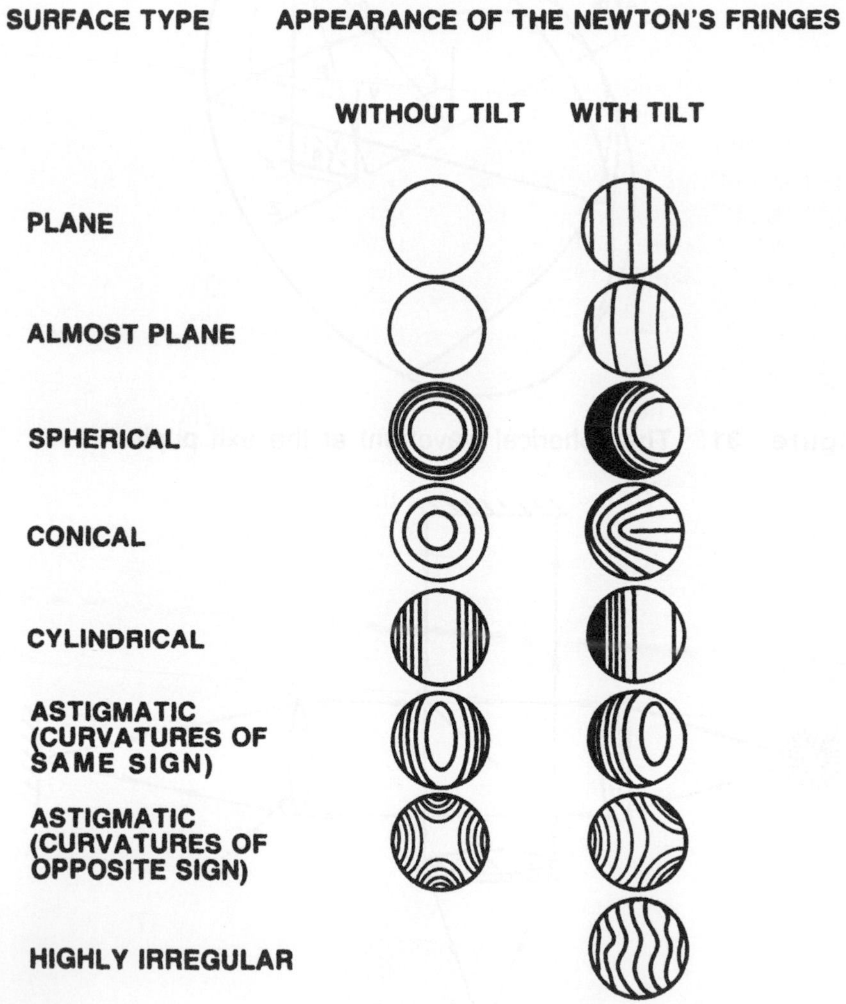

SURFACE TYPE   APPEARANCE OF THE NEWTON'S FRINGES

WITHOUT TILT   WITH TILT

PLANE

ALMOST PLANE

SPHERICAL

CONICAL

CYLINDRICAL

ASTIGMATIC
(CURVATURES OF
SAME SIGN)

ASTIGMATIC
(CURVATURES OF
OPPOSITE SIGN)

HIGHLY IRREGULAR

Figure 33. Examples of interference fringes. Reprinted with permission of Reference 45.

If the optical rays are close to the axis and almost parallel to the axis, they are called paraxial rays. The shortcomings of geometrical optics are most pronounced for rays that are relatively remote from the axis in moderate or high numerical aperture lenses and it is under these conditions that the optical transfer function is most useful(4). Information about the wave front deformations, $\Delta(x,y)$, in terms of its statistical properties can be obtained by performing laser interferometric measurements. These measurements can be made on individual lens elements to control the manufacturing process and also on the lens assembly.

8.2 Laser Interferometry

A laser interferometer splits monochromatic light into two beams. Interference, causing bright and dark fringes, takes place when the beams are reunited after traveling over different paths. Changes of position can be measured on a scale in terms of wavelength by counting fringes, allowing precise measurements. The Twyman-Green interferometer shown in Figure 32 can test the transmission properties of transparent optical components, so it is particularly useful for testing lenses, prisms, and mirrors(45)(46). The light source must be well-collimated, coherent, and monochromatic. Light exiting the collimating lens is divided in amplitude by the beam splitter. The lens, L, to be tested is placed in one of the arms and the mirror behind it is chosen so that the waves reflected by the mirror and repassing through the lens are collimated with plane wave fronts. These waves are brought to interference with the reference plane waves from the other arm of the interferometer. In practice, the mirror for the reference light will be tilted slightly so if the lens is optically perfect, the fringe pattern will be uniform and parallel. Any local variation of the optical path will produce fringes of a different pattern (other than uniform and parallel) in the corresponding part of the field. These fringes are essentially the contour lines of the distorted wave front. The entire field of the test lens can be mapped this way. Errors in the test surface can be measured in the presence of large errors in the interferometric optics by subtraction of the wave front obtained from a perfect standard of the same radius(45). Ideally, the wavelength of the interferometer matches the actinic wavelength.

Examples of fringes for different surfaces with reference to a standard flat are illustrated in Figure 33.

8.3 Wave Aberration Function Modeling

Near the end of its manufacture, the reduction lens assembly of a stepper can be tested in a similar fashion (see Figure 34). To evaluate the wave front quality, the shape of the wave fronts emanating from the lens assembly is modeled mathematically(4)(45)(47). The fringes generated by the Twyman-Green interferometer are collected by video, digitized, and analyzed using commerically available software. Mathematical expression of the wave aberration function has evolved to the use of power series expansion or Zernike circle polynomials. The polynomials are orthogonal so one advantage is both the infinite series and any of the terms can be described by a best least squares regression of the data. This means any of the individual coefficients, related to particular aberrations, can be removed from the data to illustrate residual aberrations(45). In practice, an infinite series is truncated to a manageable number of terms for calculation, but higher order aberrations must be considered in highly corrected optical systems.

The wave aberration function depends on the coordinates where the ray passes through the pupil sphere and where the object point is located in the object plane(4). The wave aberration function can be expressed as:

$$[51] \qquad W = W(r, \rho, \theta)$$

where r is the distance from the axis to the object point, and ρ and θ are the polar coordinates. For an axially symmetric wave front(46), the wave front is described by(4):

$$[52] \qquad W(r, \rho, \theta) = f(_aC_{bc}r^a\rho^b\cos^c\theta) \, .$$

The terms of the wave aberration function series are grouped according to their order(4):

$$[53] \qquad \text{order} = (\text{sum of the powers of } r \text{ and } \rho) - 1$$

Some of the lens aberrations are listed in Table 1.

Another aberration that must be considered in refractive systems is chromatic aberration. The index of refraction of a lens varies with wavelength causing images produced by different colors to come to focus at different planes. Shorter wavelength light comes to focus closer to the lens than longer wavelength light, as shown in Figure 35. This is called longitudinal chromatic aberration since the error is measured along the axis. Focal

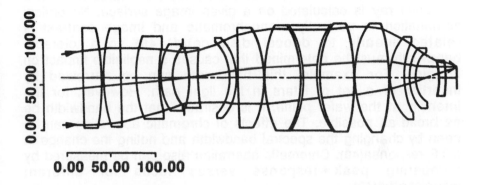

Figure 34. Schematic of a 5X, h-line, 14 element lens design. Reprinted with permission of KTI Microelectronics Seminar (from I. Friedman, A. Offner, and H. Sewell, KTI Microelectronics Seminar, Nov., 1987).

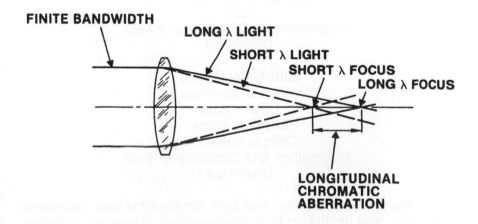

Figure 35. Longitudinal chromatic aberration. Reprinted with permission of Prentice-Hall (from J.R. Meyer-Arendt, Introduction to Classical and Modern Optics, 2nd Ed., Prentice-Hall, 1984).

lengths vary with wavelength, so the shorter wavelength light is of smaller magnification than the longer wavelength light. This is called transverse chromatic aberration since the displacement of a particular ray is calculated on a given image surface. No optical illumination is perfectly monochromatic and image contrast is related linearly to d(focal distance)/dλ(48) so a tolerable bandwidth must be determined that causes a negligible impact on imaging. For example, this bandwidth can be narrowed by insertion of a set of filters in the light path. However, for the intensity at the wafer plane, it is desirable that the bandwidth be as broad as possible. The effects of chromatic aberration can be seen by changing the spectral bandwidth and noting the change in MTF response(49). Chromatic aberration also can be measured by comparing peak response versus focus for different wavelengths(49).

Table 1. Lens Aberrations

<u>First Order Terms</u>
Focus

<u>Third Order or Seidel Aberrations</u>
Spherical aberration
Coma
Astigmatism and Curvature of field
Distortion

<u>Fifth Order Terms</u>
Spherical aberration
Linear coma
Elliptical coma
Oblique spherical
Astigmatism and Curvature of field
Distortion

Spherical aberration has both longitudinal and transverse properties and is the only monochromatic aberration affecting axial images. Rays passing through the spherical surface at different radial points come to different focus points but spherical aberration is independent of field angle so it affects all points at the same radial distance in the field similarly(50). This gives a blur circle centered around the ideal image point, where the size of the blur circle is the same everywhere in the field, but grows with the third and fifth power of the aperture sin i(51). Spherical aberration would produce an effect with the same symmetries as defocus and is partially compensated by and difficult to distinguish from defocus(52), but some distinguishing characteristics are the outer zones of the lens carry more energy per unit change in y and diffraction effects can become large with respect to the residual geometric aberration(50). For the special case of parallel incident light, Figure 36 (a) shows the paraxial focal point F' and the focal points A, B, and C for zones of increasing image height, h. Figure 36 (b) illustrates the difference between longitudinal spherical aberration and lateral spherical aberration for the object point M on axis and its paraxial image point M'. The image distance for an oblique ray traversing the lens at an image height h from the axis is s_h', and s_p' is the image distance for paraxial rays.

Whereas spherical aberration is a difference in axial location of the image for different radial zones of the lens, coma appears as a difference in magnification for different parts of the lens(46)(50). Coma causes an asymmetric aberration pattern with a size proportional to the distance from the object point to the axis, and the square of the aperture(51). Coma derives its name from the comet tail flare apparent in the well-focused image of a symmetric object point located just off the lens axis (see Figure 37). In practice, asymmetries show up more clearly in the defocused image, especially as the off-axis angle is increased. The direction of the influence changes with radial position in the field(52). Sources of the aberration can be the lens design, poor centering of the elements during assembly, asymmetrical polishing of surfaces, or nonuniform distribution of the refractive index in a lens component(45). As with astigmatism, the tangential coma is three times larger than the sagittal coma(50). Coma is described easily by monitoring the phase transfer function since it produces a lateral shift of the image, which is a measure of the asymmetry in the line spread function(49)(52). A lens system free of both spherical aberration and coma is said to be aplanatic.

The size of the aberration pattern for astigmatism and field curvature is proportional to the square of the distance from the object point to the axis and the first power of the aperture(51). Field curvature is the curvature of best focus planes along different radial positions (from on-axis to the field edge) for sagittal and tangential features. Field curvature is illustrated in Figure 38, where the best imagery will occur at the nominal focus values corresponding to the zone of least confusion. The degree of coherence of illumination changes the field curvature(48). At each radial position, sagittal and tangential features with astigmatism

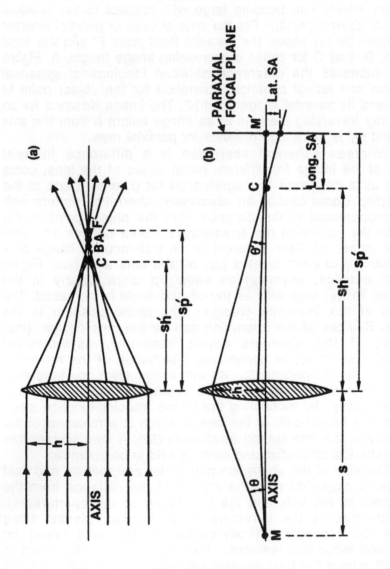

Figure 36. Lateral and longitudinal spherical aberration. Reprinted with permission of Reference 46.

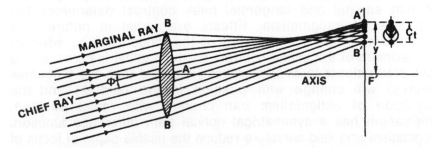

Figure 37. The tangential fan of rays for a lens with coma. The coma shown is negative since the two rays through the margin coming together at B' are of lower magnification relative to central rays. Reprinted with permission of Reference 46.

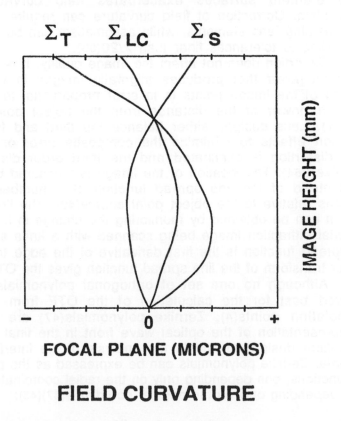

FIELD CURVATURE

Figure 38. An example of field curvature.

come to best focus at different focal planes. The defocus separation between sagittal and tangential peak contrast determines the magnitude of astigmatism. Effects are radial in nature with astigmatism, where the direction of the influence (i.e., whether the maximum of the image contrast for the sagittal features is at a focus plane closer to or farther from, the lens than the tangential features) will change with position in the field(52) and the magnitude of astigmatism can change with radial position. Astigmatism has a symmetrical optical path difference function. Astigmatism and field curvature reduce the usable depth of focus of the lens since only the intersection of the resolution-defocus space across the entire field being imaged of tangential and sagittal features is available.

The surfaces of the individual elements in a lens assembly are ground to $\leq \lambda/20$ of the desired sphericity. Cylindrical lenses have different focal lengths across each axis. Any cylindricity of these element surfaces exacerbates field curvature and astigmatism. Correction of field curvature can require additional compensating lens elements, while astigmatism can be corrected by grinding to tolerances finer than $\lambda/20(53)$.

Distortion does not affect the image quality. It is a radially symmetric error that produces stigmatic images in which the location of the image points is in error proportional to the third and fifth power of the distance from the object point to the axis(51). Lens designs either balance the third and fifth order distortion effects to minimize the composite error or the fifth order distortion is eliminated and the third order distortion is minimized(54). The location of the image is measured by finding the centroid of the line spread function at a number of field positions relative to the object point source(49). The line spread function can be obtained by monitoring the change in intensity of an aerial diffraction image being scanned with a knife edge. The line spread function is the first derivative of the edge trace. The Fourier transform of the line spread function gives the OTF.

Although no one set of orthogonal polynomials can be declared best for the calculation of the OTF from a set of interpolation points(4), Zernike polynomials(47) are preferred for representation of the optical wave front in the final phase of the optical design(55) and in the analysis of the interferometric test data. Zernike polynomials can be expressed as the product of two functions, one depending only on the radial coordinate and the other depending only on the angular coordinate(47)(45):

[54] $$Z_n{}^l = R_n{}^l(\rho)e^{il\theta}$$

where n is the radial degree of the polynomial, and l is the angular dependence parameter. The coordinate ρ is the normalized radial distance, and θ is the angle from the y axis. The radial polynomials $R_n^l(\rho)$ are functions of ρ alone.

Zernike polynomials are a set of complete orthogonal polynomials defined on a unit circle with several attractive properties(47). They are related to the classical understanding of geometrical aberrations developed by L. Seidel in terms of rays instead of wave fronts. Not only can the wave front be represented faithfully, but the polynomials are able to give the shape of the wave front relative to the reference sphere everywhere over the exit pupil. The mathematical form of the Zernike polynomial is preserved when a rotation with pivot at the center of the circle is applied to the wave front function(45). Expansion of the wave front in terms of Zernike polynomials eases the balancing of aberrations of different orders against each other in order to obtain the maximum Strehl intensity(45).

In order to calculate the OTF, knowledge is needed of the wave front aberration. However, particular properties of the optical system can be gotten directly from the wave front aberration coefficients without going to the much greater labor of computing the OTF(55)(51).

During final assembly of a reduction lens, minimization of aberrations is obtained by rotating, translating, or tilting selected elements in an iterative process with testing feedback.

The unity magnification Wynne-Dyson system has perfect imagery on-axis and for off-axis points in the sagittal plane. By using a meniscus lens coupled to a material of lower refractive index, correction can be obtained to a higher order and for greater field sizes than with a single component. Practical considerations require an air gap between the reticle and the first lens surface. This is achieved by introducing a third component of higher refractive index which introduces aberrations in the opposite sense to those of the air gap(56). The object and image focal planes are separated by means of folding prisms that reduce the available field to something less than half due to vignetting.

The introduction of aspherical optical elements in lithography equipment will bring a relatively dramatic improvement in imaging quality.

8.4   Aerial Image Intensity Distribution

Instead of evaluation by MTF for partially coherent illumination, for any specific pattern the intensity distribution of the aerial image can be computed(21), as illustrated in Figure 39. For partially coherent illumination, a quantity S can thus be computed that is comparable to the MTF in the case of incoherent illumination as a measure of the recordability of a particular pattern. The value of S is given by:

[55] $S = \Delta I/2rI_r$

where I_r is the intensity at which the width of the aerial image is equal to the scaled linewidth of the object. $\Delta I/I_r$ is the fractional intensity change corresponding to the fractional dimensional change. Here, r is the fractional recorded linewidth variation. It should be noted that although the quantity S is similar to the MTF as a measure of recordability since it is the slope of the intensity variation of the aerial image at the recording point, its value can be greater than unity since with partial coherence the intensity variation curve is generally not a sine wave. The MTF measures both slope and contrast, whereas the quantity S measures only slope. Defocus will degrade the image contrast.

Alternatively, the theoretical image contrast can be calculated with algorithms including the effects of partial coherence in the illumination and defocus aberration(57).

The experimentally measured contrast can be degraded from ideal by many factors, including light scattered within the lens assembly during transmission (flare), residual aberrations from both design and manufacturing, chromatic aberration caused by inadequate source filtering, and wafer plane defocus. Scattered light commonly encountered in optical systems is the result of small, very rapid wavefront disturbances(58). Selection of the narrow band filter involves tradeoffs between image contrast and illumination intensity (see Figure 40). As the quasimonochromatic actinic light is narrowed, exposure times increase and standing wave effects are exacerbated. The importance of image contrast is related directly to linewidth, L, control(23), as shown in Figure 41. Image contrast above 90% may be required for submicron lithography. The familiar criterion of 60% MTF for good imaging corresponds to an image contrast of 95% at a partial coherence value of 0.7. Operation at lower image contrast values will require better process control and image contrasts as low as 70% may be useful for volume manufacturing(59).

9.0   NUMERICAL AND STATISTICAL METHODS

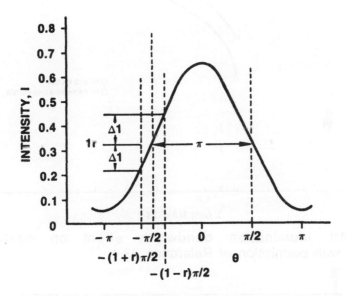

Figure 39. Aerial image intensity distribution of the image of a square wave object with partially coherent illumination. Reprinted with permission of Reference 21.

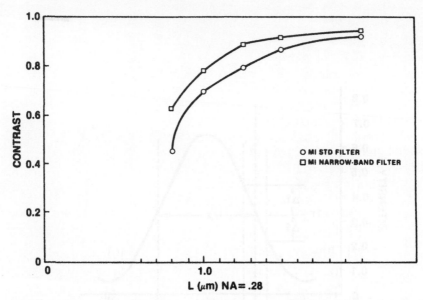

Figure 40. Illumination bandwidth effect on contrast. Reprinted with permission of Reference 23.

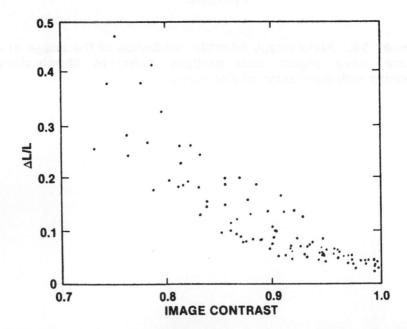

Figure 41. Effect of contrast on linewidth control. Reprinted with permission of Reference 23.

Numerical and statistical methods are invaluable tools of engineering that permit quantification of phenomena.

9.1 Data Regression

Numerical methods can be used to advantage in characterizing the performance of steppers. The method of least squares(60) allows linear and quadratic (or higher order) polynomials to be regressed from data to describe their performance. These two lower order polynomials are useful in many circumstances. The general equation for a straight line has the form:

[56] $$y = a + bx.$$

The error criterion to be minimized for the linear fit for N data points would be:

[57] $$\sum_{x=1}^{N} e^2(x) = \sum_{x=1}^{N} [y(x) - a - bx]^2.$$

Simultaneous equations are generated by taking partial derivatives of Equation [57] with respect to a and b and then setting the two resulting equations to zero. The simultaneous equation solution results in(61) a slope, b, of:

[58] $$b = \frac{N \Sigma xy - \Sigma x \Sigma y}{N \Sigma x^2 - (\Sigma x)^2}$$

where the Σ notation represents the summation of terms from x=1 to N. The intercept, a, is given by:

[59] $$a = (\Sigma y - b \Sigma x) / N.$$

The general expression for a second order (quadratic) polynomial is:

[60] $$y = a + bx + cx^2.$$

The error criterion to be minimized for the quadratic fit would be:

[61]
$$\sum_{x=1}^{N} e^2(x) = \sum_{x=1}^{N} [y(x) - a - bx - cx^2]^2 .$$

Simultaneous equations are generated by taking partial derivatives of Equation [61] with respect to a, b, and c and then setting the three resulting equations to zero. The simultaneous equation solution is a little awkward to apply but results in(62):

[62]   1. $b = \dfrac{\gamma\delta - \theta\alpha}{\gamma\beta - \alpha^2}$

where:

$$\gamma = (\Sigma\ x^2)^2 - N\ \Sigma\ x^4$$

$$\delta = \Sigma\ x\ \Sigma\ y(x) - N\ \Sigma\ xy(x)$$

$$\theta = \Sigma\ x^2\ \Sigma\ y(x) - N\ \Sigma\ x^2 y(x)$$

$$\alpha = \Sigma\ x\ \Sigma\ x^2 - N\ \Sigma\ x^3$$

$$\beta = (\Sigma\ x)^2 - N\ \Sigma\ x^2$$

and the Σ notation represents the summation of terms from x=1 to N.

2.   Once b has been determined, c is found from:

[63] $c = \dfrac{\theta - b\alpha}{\gamma}$.

3.   The intercept a can be found from:

[64] $a = (\Sigma\ y(x) - b\ \Sigma\ x - c\ \Sigma\ x^2)/N$.

The maximum or minimum for each curve is found by solving the first derivative with respect to x set equal to zero, i.e.,

[65] $\dfrac{dy}{dx} = 0 = b + 2cx$.

9.2   F-test and T-test

The most common experiments test a process under two different conditions, generally two different levels of a single process factor. These are called B versus C tests. A comparison of variances uses the F-test to determine if the difference between the variances (i.e., the square of the standard deviation values) is statistically significant at a given confidence level. A comparison of means uses the t-test to determine if the difference between the mean values is statistically significant at a given confidence level. Both the F-test and the t-test assume that independent random samples are evaluated from normally distributed populations. Even if a test is statistically significant, a subjective engineering judgement must be made about the importance of the results, i.e., whether the observed differences are important.

There are two widely used measures of variability. One is the range and the other is the standard deviation. The range is easy to compute and is the difference between the high and low observed test values. For small sample sizes, the estimate of the standard deviation, s, is used to approximate the population standard deviation, σ. The estimate of the standard deviation is more complicated to compute than the range:

$$[66] \qquad s^2 = \sum_{i=1}^{N} (x_i - \bar{x})^2 / (n-1)$$

where x_i is the i-th value of x, $\bar{x}$ is the average value of all the x_i's, and n is the sample size or number of observations. Small sample sizes have $n \leq 30$, although populations are described by sample sizes at least a couple of orders of magnitude larger.

Comparison of standard deviations in a B versus C test is done with an F-test. The shape of the F-test curve is asymmetrical (i.e., the distribution of variances is not a normal distribution) and depends upon the degrees of freedom (n_i-1, or the sample size less one) of test condition i. The F-test is:

$$[67] \qquad F_{exp} = s_1^2 / s_2^2$$

where $s_1 > s_2$. F_{exp} is compared against F, the tabulated value (available in any statistics text) dependent upon the degrees of freedom and a, which is the experiment risk (the confidence coefficient is $1-\alpha$). For two tailed tests (where there is a possibility of either a ± variation about a mean), the F factor must reference a risk of $\alpha/2$ in the tables. If more than two variances are being compared, F_{max} tables are referenced instead of F tables. If $F_{exp} \leq F$, there is insufficient evidence to indicate a difference in the population variances.

The t-distribution and the standard normal distribution follow essentially the same curve. The distribution of $t_{\alpha/2,n-1}$ in:

[68] $$\Delta \bar{x} \geq t_{\alpha/2,n-1} \cdot s_{p,n-1} \cdot \sqrt{(1/n_1 + 1/n_2)}$$

for samples drawn from a normally distributed population was discovered by W.S. Gosset and published in 1908 under the pen name of Student. He referred to the quantity under study as t and it has been known ever since as Student's t(63). A normal frequency curve is centered on the population mean, μ, and is symmetrical about this point. In a normal distribution, the area under the curve for $\mu \pm 1\sigma$ is 0.6826, for $\mu \pm 2\sigma$ is 0.9544, and for $\mu \pm 3\sigma$ is 0.9973, where σ is the population standard deviation. In Equation [65], $\Delta \bar{x}$ is the difference between the estimates of the population means, $t_{\alpha/2,n-1}$ is a value gotten from a Student t table, α is the experiment risk (the confidence coefficient is 1-α), the pooled estimate of the standard deviation of $\bar{x}$ is $s_{p,n-1} \cdot \sqrt{(1/n_1+1/n_2)}$, n is the weighted average of the sample size, and n_i is the sample size of test group i. For two tailed tests (where there is a possibility of either a ± variation about a mean), the t factor must reference a risk of $\alpha/2$ in the tables; the value of t from the tables also depends on the degrees of freedom, which is the weighted sample size less one, (n-1). If the sample sizes of the two test groups are different (i.e., $n_1 \neq n_2$), the pooled estimate of the standard deviation, s_p, must be derived from the two individual estimates of the standard deviation, s_i, by:

[69] $$s_p^2 = [(n_1-1)s_1^2 + (n_2-1) s_2^2]/[(n_1-1)+ (n_2-1)] .$$

The weighted average of the sample sizes, n, can be calculated many ways. One way is the harmonic mean:

[70] $$n = \left\{ \left[\sum_{i=1}^{L} (1/n_i) \right]/L \right\}^{-1}$$

where L is the number of different test groups (e.g., L=2 for a B versus C test).

Determining the appropriate sample size is a key question in any engineering test. It should be obvious now that a few iterations with Equations [67] or [68] will yield information about a suitable initial value for the sample size if some preliminary information is known or an educated guess can be made about the expected differences between the variances or means, respectively.

9.3 Multifactor Experiments

In practice, it is rare that a single factor dominates a process so that a simple B versus C test over two levels will adequately characterize the process. Usually, there are many factors which must be tested so that a process can be developed with adequate latitude to work under volume manufacturing conditions. Multifactor tests are used to determine whether one or more factors are acting individually or together to improve the process.

9.3.1 Blocking, Randomization, and Replicates. A block is a group of experimental runs made under similar test conditions. For example, blocking the equipment would call for the same equipment set to be used for all the runs. This eliminates a family of variation caused by equipment-to-equipment variation. Blocking also may call for use of the same batch of materials to be used or the tests to be run under the same time frame. The objective of blocking is to reduce the experimental error by making test comparisons under homogeneous conditions. Block what you can and randomize what you cannot. The objective of randomization is to distribute the effects of uncontrolled variables randomly throughout the experimental runs and to reduce or eliminate systematic errors. Random order tables are one source available for randomization of runs. Replication of runs is used to estimate the experimental error by repeating each individual cell of test conditions. Replication increases the precision of estimates and improves the chances of making meaningful statistical comparisons.

9.3.2 Experimental Designs. There are many experimental designs for multifactor tests(64). Only two will be discussed here.

Factorial Experiments. A factorial experiment is described as a 2^N experiment, for 2 levels and N factors. With only a single replication, there are 2^N cells in the factorial design. The total number of experimental runs is #Replicates·2N. The simplest full factorial test is a 2X2 matrix with two levels tested for two experimental factors, or a 2^2. The two levels are designated typically +1 and -1. The next simplest full factorial (i.e., all possible cell combinations are tested) is a 2^3, which can be visualized as a cube. The cube edges represent the +1 and -1 levels (the cube center is 0, and is untested). Each parallel plane of the cube represents one of the three factors.

Some advantages of a 2^N experiment include its greater efficiency versus testing a process by manipulating one factor at a time while holding all the other factors constant. A 2^N experiment allows interactions between factors (i.e., the combined effect on a process of a number of factors at a certain test level might be undetected except at that special combination) to be detected and estimated.

Interactions between factors are observed graphically in Figure 42 as an interception between the linear responses of two factors. Mathematically, the interaction is a difference in values of the slopes of the responses of one factor tested over two levels of a second factor.

Full factorial designs follow the pattern of Table 2.

Table 2. Factorial Design Generator

LEVELS

Test Cell	Factor 1	Factor 2		Factor 3
1	-	-	|	-
2	+	-	|	-
3	-	+	|	-
4	+	+	|	-

2^2 _____

5	-	-		+
6	+	-		+
7	-	+		+
8	+	+		+

2^3 _____

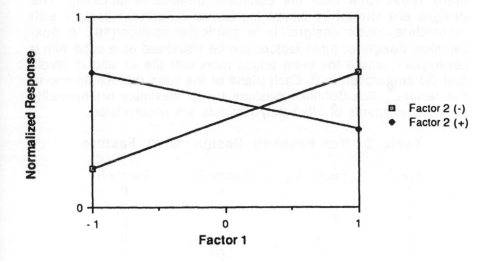

Figure   42. Interaction between two factors.

Response Surface Experiments. One type of response surface experiment tests factors over three levels. These experiments can be saturated and are described by the notation 3^N. Saturated designs are not usually run except for N=2 factors. Nonsaturated designs over three levels are called Box-Behnken tests(65). In general, this class of second-order designs requires many fewer runs than the complete three-level factorials. The designs are formed by combining two-level factorial designs with incomplete block designs in a particular manner(66). A Box-Behnken design for three factors can be visualized as a cube with a centerpoint, where the cube edges represent the +1 and -1 levels and the centerpoint is 0. Each plane of the cube represents one of the factors. Box-Behnken designs try to maximize orthogonality so the coefficients of fitted polynominals are uncorrelated.

Table 3. Box-Behnken Design for 3 Factors

Cell	Factor 1	Factor 2	Factor 3
1	-	-	0
2	+	-	0
3	-	+	0
4	+	+	0
5	-	0	-
6	+	0	-
7	-	0	+
8	+	0	+
9	0	-	-
10	0	+	-
11	0	-	+
12	0	+	+
13	0	0	0

9.4   Analysis of Experiments

Analysis of factorial(64) and surface response designs(65) are described elsewhere. Usually, commercial software packages are used for response surface analysis because of their convenience. Analysis of different factors is facilitated by coding the levels, i.e., instead of using the actual units of interest (e.g., for develop time, seconds) the levels are normalized (e.g., factorials use +1 and -1 levels, Box-Behnken tests use +1, 0, and -1 levels, etc.) so the difference in magnitudes of levels between the factors do not present computational difficulties. Also, analysis of variance (ANOVA) methods are designed to measure differences between means since the assumption is made that the distributions are normal. If the outputs of the test are variances, which have the asymmetrical distribution described by the F-test, analysis of the data using a transformation (e.g., taking the natural logarithm of the standard deviation values) may be called for. If successful, the transformed distribution will be normal.

Factorial designs with replications run in random order allow testing of hysteresis (i.e., comparison of cell variation to find if the response, after displacement of factor levels, can reproduce its original performance) and linear effects (i.e., are the values of the intercept and slope reasonable?). Response surface designs have an added advantage of testing for nonlinear effects (i.e., typically responses are described by second order polynomials) so optimization can be performed by finding the maximum or minimum of the response between levels. Finally, no analysis is complete without graphically analyzing the results (remember the cliche: a picture is worth a thousand words).

9.5   Process Control

There are many statistical tools available to control a process so the product is made with high quality. Quality is the conformance to target performance levels and statistical process control uses methods to reduce process variations so continuous improvement is possible.

9.5.1. Pareto Charts. Alfredo Pareto was an Italian sociologist who noted in a study that 20% of the citizens controlled 80% of the wealth. This observation applies as a causal generalization to a lot of manufacturing processes. Concentration on a few critical factors gives the most benefit for reducing process variability. Pareto charts can be used to manage engineering priorities. An example is shown in Figure 43.

9.5.2 Multivariate Studies. Multivariate studies determine the process variation without manipulation of any factors. Data are collected so that several families of variation can be analyzed to determine which family contributes the most variation to the process. The effect of each of these families on the total variation can be described for relative importance using a Pareto chart. One example of a multivariate study is to collect data for critical dimension control using product material. The data can be collected on two wafers out of each lot processed, measuring five sites on a wafer (top, center, bottom, left, and right). The families of variation are site-to-site, wafer-to-wafer, and lot-to-lot. Multivariate data are illustrated in Figure 44.

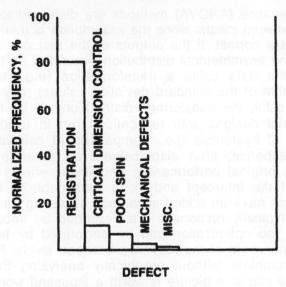

Figure   43.   Example of a Pareto chart.

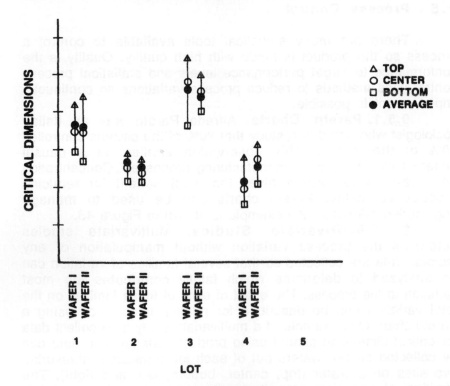

Figure   44.   Example of a multivariate chart.

Nested Variance. The total process variability can be partitioned into major sources of variability using the concept of nested variances:

[71] $$\sigma_{total}^2 = \sigma_l^2 + \sigma_w^2 + \sigma_s^2 .$$

In this example, the data being collected are critical dimension measurements, x_{ijk}, for the i-th lot, the j-th wafer, and the k-th site. The grand average is:

[72] $$\bar{x} = \sum_{i=1}^{L} \sum_{j=1}^{W} \sum_{k=1}^{S} x_{ijk}/(SWL) .$$

where L is the number of lots inspected, W is the number of wafers inspected in the lot and S is the number of inspected sites per wafer. The average of lot i (for i=1,...,L) is:

[73] $$\bar{x}_i = \sum_{j=1}^{W} \sum_{k=1}^{S} x_{ijk}/(SW) .$$

The average of wafer j within lot i (for j=1,...,W and i=1,...,L) is:

[74] $$\bar{x}_{ij} = \sum_{k=1}^{S} x_{ijk}/S .$$

The total variation of lot means is:

[75] $$s_l^2 = \sum_{i=1}^{L} (\bar{x}_i - \bar{x})^2/(L-1)$$

The pooled variation of wafer means within the lots is:

[76] $$s_w^2 = \sum_{i=1}^{L} \sum_{j=1}^{W} (\bar{x}_{ij} - \bar{x}_i)^2/[L(W-1)] .$$

The pooled within-wafer variability is:

$$[77] \quad s_S^2 = \sum_{i=1}^{L} \sum_{j=1}^{W} \sum_{k=1}^{S} (\overline{x}_{ijk} - \overline{x}_{ij})^2 / [SL(W-1)] \quad .$$

The variance components for each family of variation are calculated from their estimates after accounting for nested variation:

$$[78] \qquad\qquad \sigma_S 2 = S_S^2$$

$$[79] \qquad\qquad \sigma_S 2 = S_W^2 - S_S^2/S$$

$$[80] \qquad\qquad \sigma_I 2 = S_I^2 - S_W^2/W \quad .$$

9.5.3 Control Charts. There are two principal types of control charts. The first is a variables control chart, for monitoring outputs such as $\overline{x}$, R, and s (for the mean, range, and standard deviation estimate, respectively). The second is an attributes control chart, for monitoring outputs such as P, C, and U (for proportions or percentages, counts, or uniformity of a process, respectively).

One type of variables control chart for normally distributed outputs has three zones, a green zone, a yellow zone, and a red zone (see Figure 45). About the target center, the green zone is $\pm 1.5\sigma$. The yellow zone is outside the green zone, about the target center, to $\pm 3\sigma$. Outside of the yellow zone is the red zone. The rules for using this control chart are to make no adjustments if the outputs are plotted randomly in the green zone. Adjustments are made to the process if six consecutive points are in the green zone on the same side of the centerline, two consecutive points are in the same yellow zone, or a single point is ever in the red zone.

9.5.4 C_p and C_{pk}. Two process capability indices are C_p and C_{pk}. These indices are used to judge the capability of the process to meet specifications after the process is under statistical control. Standard deviation and mean values are used in the calculations and it is assumed the distributions are normal.

C_p describes the process variation with respect to the existing specification window.

$$[81] \qquad\qquad C_p = \frac{(USL - LSL)}{6\sigma}$$

FACTOR = _____

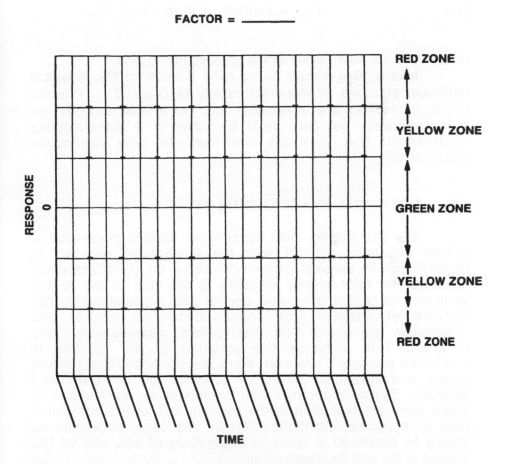

Figure 45. Example of a variables control chart.

where USL and LSL are the upper and lower specification limits, respectively, and σ is the total process variation.

C_{pk} describes how well the process is centered relative to the specification window.

[82] $C_{pk} = \min(USL - \bar{x}, \bar{x} - LSL) / 3\sigma$

where $\bar{x}$ is the mean value of the process distribution.

Ideally, C_{pk} should be as large a value as C_p. A typical minimum standard of manufacturability is $C_{pk} \geq 1.33$. Process capability indices are a valuable tool to ensure designs are manufacturable, but care must be taken in a manufacturing environment that they are not used to slow continuous improvement.

10.0 PRACTICAL IMAGING QUALITY

The total overlay must combine the families of variation of critical dimension control with the families of variation of registration. The simplest method describing the composite effect assumes the total overlay variation is the sum of the nested variances of critical dimension control and registration(67). The families of variation for critical dimension control and registration are across-field, field-to-field, wafer-to-wafer, and lot-to-lot. This assumes the variation of the families is distributed normally . However, there are systematic and random errors, and systematic errors by definition are not distributed normally. Systematic errors can contribute to distributions in which there are less data in the extreme tails of the distribution than in a normal distribution(68). In this case, performances should be described in terms of distributions of 95% and 99.7%, instead of 2σ and 3σ, respectively.

10.1 Field Diameter and Resolution

The design of a lens determines how many pixels, N, can be imaged. The image field and resolution of this design trade off equally according to(53):

[83] $N = \pi[(D \cdot NA)/(2\lambda)]^2$

where D is the field diameter. Chemical support restrains the transition to shorter wavelengths, but for a given number of pixels and no limitations on the quality and availability of lens materials, the use of a shorter exposure wavelength simplifies lens designs and matching because of the lower numerical aperture.

The stitching of multiple fields together allows very large dice to be manufactured, up to wafer scale integration. The lens constrains the resolution and depth of focus latitude but the maximum area becomes a design and wafer fabrication cleanliness issue. Butted fields either employ lines where overlap is made to staggered, enlarged pad areas or with seamless stitching(69). This technique overlays lines which have been vignetted by a defocused edge on the reticle. Superposition of the complementary tapered exposure profiles yields a full exposure, with minimal linewidth variation. Economics will determine when field stitching comes into vogue.

10.2 Exposure-Defocus Diagrams

Process latitude can be quantified by Exposure-Defocus diagrams(70)(71) like Figure 46. The graphical presentation of linewidths varying with defocus, with each curve associated with a different exposure dose, are known as Bossung curves(72). The ED area describing process latitude is derived from these curves, with the acceptable processing latitude defined by the area in which the feature size ±10% is imaged. If the processing latitude is given by a ± value for exposure and defocus, the largest rectangle that will fit in the ED area will be described. Actual construction of an ED curve requires some type of normalization of exposures, to avoid the erroneous appearance of a latitude advantage to high exposure processes. Methods of normalization include taking the natural logarithm of the exposure doses (so a ln (ED) curve is generated) or describing the exposure doses of all features as a percentage relative to a reference feature. The area under the ED window decreases with pitch (the width of the line/space pair).

10.2.1 **Mask Bias.** A mask bias means the linewidth on the mask is larger than the desired nominal linewidth on the wafer. As the resolution nears the limit of a lens, the choice of the magnitude of a mask bias becomes more restricted (by the Rayleigh resolution limit), until zero bias is the only processing choice. Zero bias masks reduce process latitude in an obvious way since a two tailed distribution about the nominal wafer size is impossible. The luxury of a mask bias helps enlarge the latitude

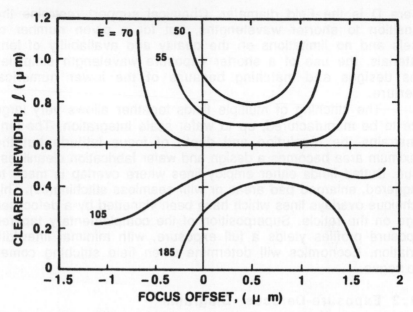

Figure 46. Bossung Exposure-Defocus diagram. Energies are in mJ/cm^2.

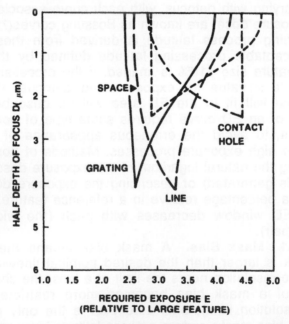

Figure 47. E-D windows for 0.75 micron objects imaged through a 0.35 NA i-line lens with s=0.7. Reprinted with permission of Reference 71.

for both exposure and usable depth of focus(73). The selection of mask bias has an effect on the process latitude as described by the ED curve(71), but it does not expand the ultimate resolution capability(74).

10.2.2 Aspect Ratio of Features. The aspect ratio of features has a dramatic effect on the process latitude window, especially for depth of focus. Isolated small aspect features have the worst latitude, while grating features (i.e., line/space pairs) are the best.

10.2.3 Conjugate Lithography. The exposure dose which yields a flat linewidth response with defocus on a Bossung curve is termed the isofocal, or conjugate, exposure. In Figure 46, the isofocal exposure is between 70 and 105 mJ/cm^2. Generally, the process latitude is not maximum at the isofocal point because the exposure latitude varies so sharply. However, if the exposure control is acceptable, the isofocal point offers maximum depth of focus latitude(75).

10.2.4 Proximity Effects and Degree of Coherence. Proximity effects caused by scattering of light impacts the size and shape of a feature due to the geometry of the feature itself or neighboring features. There are four general classes of proximity effects: linewidth differences between isolated and packed lines or spaces, linewidth differences between clear field line and dark field space, line length shortening of rectangular spaces or lines, and corner rounding(76). Packed lines have more rounded corners and smaller linewidths than isolated lines at zero defocus. Under relatively large defocus, packed lines have more resist loss, accentuated linewidth loss, and more pronounced corner rounding.

As the numerical aperture increases, the size at which proximity effects become apparent decreases(77). Proximity effects are relatively small for features larger than one micron.

The optical proximity effect results in different biases for the different features (e.g., grating, line, space, contact hole, island). This bias variation is sufficient to reduce or eliminate the process latitude overlap of the windows(71) as shown in Figure 47. This means a single exposure dose might not image all the features of interest on the reticle to the desirable feature size ±10% on the wafer. As feature size is reduced, it becomes increasingly difficult to find a single exposure condition which is satisfactory for different features sizes and geometries(27). Selective biasing across the mask or reticle becomes important to maintain equal linesizes across circuits, depending on the aspect ratios (i.e., the ratio of the features' length to width)(76). The process latitude should describe the performance of the most difficult feature.

Proximity effects depend on the partial coherence value and feature size and shape(77). Isolated lines have approximately the correct dimensions in all sizes when the imaging is done with partially coherent light. However, diffraction effects cause deviations from the design size for isolated spaces. For spaces $(0.6-1.0)\lambda/NA$ wide the light intensity in the center of the space is larger than the intensity of the incident light causing the resist to develop faster than in a very large area(74). The undesirable necking and bulging at elbow and pad corners can be reduced greatly by going to a partial coherence, s, of 0.5 and almost eliminated entirely using s=0.7(78). However, the contrast and peak intensity are reduced. For very small, unity aspect ratio contact masks, reducing s to 0.3 from 0.7 will increase the edge slope and give about 50% higher intensity at the center of the contact (but lower total illumination intensity, requiring longer exposures). This higher partial coherence (lower s) improves contact hole image quality and improves the sensitivity to optical defocus aberrations(78)(79). Unfortunately, this means there is a higher sensitivity of the optical transfer process to local defects and dust on the reticle and lens(15).

The general trend is that when the degree of partial coherence is increased (i.e., s becomes smaller), the intensity tolerance becomes wider but the depth of focus becomes shallower (70). However, for features near half micron size, the loss in depth of focus is minimal, but the increase in exposure tolerance is substantial. Illumination uniformity is relatively constant for $0.4 \leq s \leq 0.65$, but worsens quadratically for increased coherence(80).

Proximity effects depend strongly on the resist processing. In particular, the resist thickness is an important factor since the resist contrast depends on thickness(39)(81). High contrast resist processes replicate the aerial image diffraction interference effects more faithfully than lower contrast ones.

10.2.5 Numerical Aperture. There is an optimum numerical aperture at which the depth of focus is greatest for printing features in a particular pitch(73)(82), i.e., for relatively large features the imaging quality under defocused conditions can be higher with a lower numerical aperture lens. There is no benefit to the use of a high resolution lens when only moderate resolution is required because of the loss of depth of focus. Optimizing the depth of focus as a function of numerical aperture for a given feature size requires setting the derivative of depth of focus with respect to numerical aperture equal to zero (using

Equations [45] and [13]).   However, the depth of focus for different feature types (specifying density, grating, isolated line, isolated space, etc.) may not change monotonically with feature size(83), making optimization difficult.   A numerical aperture of 0.45 is close to the ceiling for microfabrication purposes because the depth of focus requirements become increasingly stringent for any higher NA(70).

10.3 Depth Of Focus Issues

The resolution of features always must be qualified by its useable depth of focus. The depth of focus latitude directly affects the critical dimension control. In reality, there are many factors that must be monitored carefully to ensure the highest quality imaging. Resolution and depth of focus issues include substrate topography, substrate reflection, resist thickness and variation, the quality of the resist process, chuck flatness, wafer flatness, the orthogonality of the wafer plane with the optical axis, exposure wavelength, lens aberrations (e.g., field curvature), lens numerical aperture, the feature size being imaged, pellicle flatness, nominal focus selection, and autofocus precision.

The best exposure latitude (or equivalently, develop latitude) corresponds to a nominal focus position near the base of the resist, but moves to the middle of the resist layer as the exposure dose increases and the nominal cleared linewidth decreases(84).

One limit to increasingly higher numerical aperture lenses is that the image focused by a lens with higher NA tends to be slightly broadened by resist absorption(85), attenuating the effective NA. When the angles of the plane waves are below 18° in the resist, which is equivalent to an NA of 0.46 in air, the increase of light absorption can be neglected. Also, the focus effects for high numerical apertures, where the depth of focus becomes comparable to the resist layer thickness, depend upon the oblique direction of propagation of light in the resist(10)(84). These effects include an asymmetry in resist profile on either side of best focus and an asymmetry in the curves on either side of best focus of a Bossung focus-exposure plot, which cannot be explained by lens aberrations(10)(84)(86). This latter asymmetry is shown in Figure 48, which was imaged with a very high numerical aperture, 0.60, g-line lens. These asymmetries worsen with decreasing feature size, increasing numerical aperture, and thickening resist. The resist profile asymmetry on one side of best focus effectively halves the usable depth of focus if subsequent

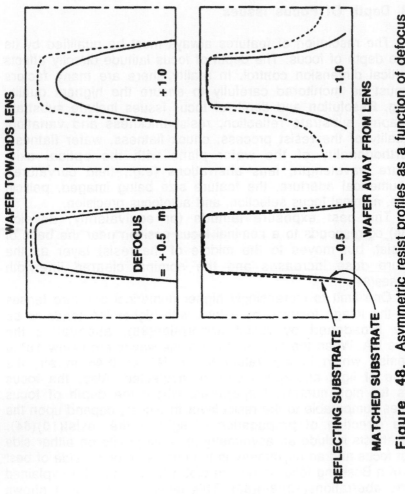

Figure 48. Asymmetric resist profiles as a function of defocus for a 0.60 NA g-line lens with s=0.5. Reprinted with permission of Reference 84.

processes include plasma or reactive ion etching with significant resist erosion or high energy implants.

For thick resist, part of the space within which rays remain adequately converged will be occupied by the resist(73). If DOF is the depth of focus for thin resist (a film of thickness substantially smaller than DOF, where the lateral light distribution is essentially that within a single plane), then the focal range, DOF', for thick layers is(73):

$$[84] \qquad\qquad DOF' = DOF - t/n$$

where t is the resist thickness and n is the index of refraction.

10.3.1 Resolution Versus Defocus Plots. A preferred orientation for a Resolution versus Defocus plot is an X (versus a +) pattern over the lens field, so that an inscribed square field has its corners characterized. Data can be collected by image contrast(43)(44) or linewidth measurement. The data are organized typically with the resolution plotted as the ordinate and the defocus condition as the abscissa (see Figure 49). The data for sagittal and tangential features as a function of defocus each can be regressed to a second order polynomial using the method of least squares. Typical acceptable resolution of linewidths is considered to be ±10% of the nominal target (e.g., $k_1 = 0.8$, referencing Equation [13]). A Resolution versus Defocus plot should reveal the best imagery at the optical axis or a translation error might exist in the platen (i.e., the holder of the reticle) alignment to the center of the optical axis.

Astigmatism. The difference in resolution between sagittal and tangential features at each defocus position indicates an astigmatism error. There is no preferred test structure and line/space pairs, isolated squares and checkerboard patterns are recommended. The difference between the sagittal maximum and tangential maximum at a certain field position (e.g., the right edge of the lens) describes the astigmatism error in terms of defocus units at that field position. Astigmatism can be found anywhere in the lens field, even though the design generally will show this focal error only at the field edges. Figure 49 illustrates an astigmatism error.

Field Curvature. Many radial positions must be characterized to determine the field curvature. At each radial position the curves for the sagittal and tangential resolution as a function of defocus must be found, as well as their respective maxima. Field curvature is the connection of the maxima to form a curve from the lens center to the field edge. Two curves will be

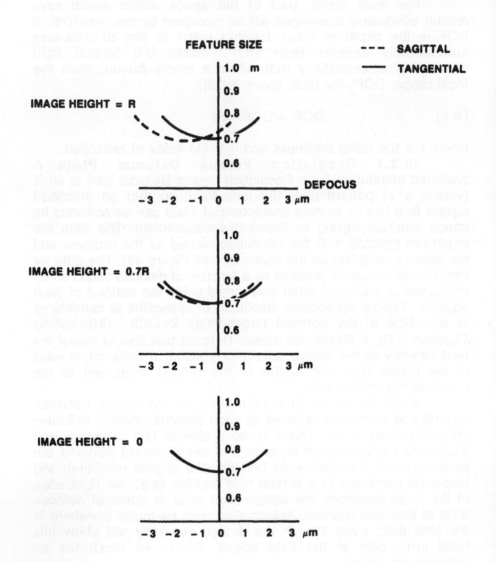

Figure 49. A Resolution versus Defocus plot for different radial image heights.

created, one for the tangential features and the other for sagittal features. Field curvature changes as a function of degree of coherence so MTF testing must be corroborated by testing at the partial coherence the stepper will normally operate at.

Astigmatism and field curvature can degrade the average lens performance by up to one half a Rayleigh unit, of which about 30-40% is in the design and the rest is in the manufacture of the lens(73).

Column Tilt. Resolution versus Defocus plots should characterize the imagery at the lens top, center, bottom, left, and right as well as different radial positions along an axis. Column tilt indicates the wafer plane is not orthogonal with the optical axis and is characterized as a difference in value between the best focus position at the lens top versus bottom or left versus right. Figure 49 illustrates column tilt along one radial arm. The effect of column tilt as feature sizes become more aggressive is there will not be enough depth of focus latitude to resolve the features everywhere over the field. Correction of column tilt is either through adjustment of the chuck defining the plane of the wafer or physical adjustment of the column position. Determining the best focus positions at just the opposite edges of the lens (e.g., top and bottom) allows the column tilt to be found directly by difference. If data are taken at the lens center or other radial positions so data are collected from more than two extreme positions across the field a linear least squares fit will describe the column tilt by the slope of the line.

Coma. It is difficult to separate the effects of true coma from decentered lens elements. From the earlier discussion of lens aberrations, it was disclosed that the tangential coma is three times larger than the sagittal coma(50). These data are apparent from the Resolution versus Defocus plots, where the curve connecting the tangential features will appear to always lag the sagittal features in resolution at each defocus position. Coma causes asymmetrical imagery so a qualitative test structure should be symmetrical on the reticle(52). These test structures should be positioned at several radial positions. One candidate test structure is a long line of width equal to $0.7NA/\lambda$. At the middle of its length on either side of the line are two small squares. The squares cannot be resolved since they have sides equal to $0.3NA/\lambda$. The separation distance of the squares from the line cannot be resolved and is equal to 0.1 micron. These test structures preferably are oriented tangentially. Although it is characteristic that the asymmetric images are present at best focus, the comatic asymmetries show up more clearly in the defocused image.

10.3.2 Autofocus Systems. Autofocus systems are responsible for bringing each field to the nominal focus position. Typically, the autofocus system functions after stepping to a new field prior to exposure.

Types. One type of autofocus system is a grazing incidence optical autofocus system. Usually, multiple wavelengths are used to eliminate interference effects. Infrared light (700-900 nm) is common since there is no resist sensitivity at these wavelengths. A grazing angle (of 2-4 degrees) of reflection from the light source across the wafer surface to the detector is preferable for elimination of substrate reflections since an oblique angle causes greater reflection. The autofocus light is usually incident at 30-45 degrees to the field to avoid interference with topographical patterns.

An air gauge autofocus system(87) emits a pressurized jet of air to sense the wafer surface. Multiple jets are positioned at edge positions of an average field size so die leveling can be sensed. Advantages of air gauge autofocus systems are their high precision and immunity to thin film interference effects. The largest disadvantage is the particulate issue, which is revisited with each decrease in feature size. The air or nitrogen may be filtered to remove particles but the pressurized air may still create an area of turbulence that stirs up local particles. Usually, the pressurized jet is always on to avoid cycling (on again, off again) bursts, which exacerbate the problem.

Leveling. Global leveling of a wafer occurs before alignment or exposure begins. The autofocus system is used to map three (to define the plane) or more sites on a wafer and wafer wedge (i.e., the planes of the polished and back side of a wafer are not parallel) is removed by tilting the chuck upon which the wafer is vacuum clamped. Local leveling improves the usable depth of focus further by compensating for the local focal plane deviation of a wafer. Local leveling(87)(88-89)(90)(91) requires information to determine the local plane of the field. Correction to nominal focus is made by adjusting the two axes determining the image field tilt and the third z axis for height separation of the lens and wafer field. Usually, there is a small penalty in throughput for local leveling and a three-axis laser interferometer is needed to track accurately the wafer's position after local leveling so stage yaw, pitch, and roll don't contribute to misregistration (if there is any post-alignment stage motion before exposure).

Enhancing Focus Latitude. One method used to enhance the depth of focus is called Focus Latitude Enhancement exposure (FLEX). In FLEX, several focal planes are created at different positions along the light axis, and exposures are made using each focal plane at the same field position on the   wafer(92-94). According to SEM measurements of features at the resist/substrate interface, the focus latitude is increased about three times using FLEX, although the edge slope of the features is degraded relative to a conventional, fixed focus method. In FLEX, defocused images in one focal plane are superimposed on the sharply focused images in another plane. Since the image intensity distribution is flatter with defocus, the contribution of this exposure lowers the image contrast of the composite image. Best results seem to be gotten on low aspect ratio features, which is where depth of focus latitude usually is shallowest.

10.3.3 Nominal Focus. Nominal focus position depends on the ambient temperature and pressure since a change in the index of refraction of air causes changes in the magnification and focus of the reduction lens. Temperature is controlled either by enclosure of the stepper inside an environmental chamber where the temperature is controlled to at least $\pm 0.1°C$, or by close monitoring of a thermocouple sensor and compensation to the autofocus system. The amount of focal shift resulting from a $0.1°C$ shift in temperature depends on the glass characteristics of the lens elements (i.e., the change in glass index of refraction with temperature, dn/dT). A barometrically controlled stepper chamber has never been sold commercially, so the ambient pressure is monitored at an environmental station and compensation for the effect on nominal focus is made to the autofocus system. For a 0.3 NA g-line lens, a change in atmospheric pressure of 1 mm Hg causes a focal shift of ~0.15 microns.

A lens is a thin lens if its thickness is much less than the radii of curvature of its two surfaces. Assuming thin lens theory holds, the change in focus, Df, with pressure is(95):

[85] $$\Delta f = \Delta P(k\delta n_l f^2 / P_0)$$

where, ΔP is the difference between the ambient pressure, P, and the standard pressure, $P_0 = 760$ mm (i.e., $\Delta P = P - P_0$), k is a lens constant, δ is a function related to the index of refraction of air, n_l is the index of refraction of the lens, and f is the focal length of the lens. Equation [85] predicts a linear dependence in focal position with ambient barometric pressure.

Nominal best focus is found by evaluating data from a defocus array. The defocus increments are selected based upon the confidence level of the position of best focus and time constraints. Data may be collected from a linewidth defocus array such as

Figure 46, aerial image quality(43)(44), or alignment signal contrast from a latent image(96). Data regression to a quadratic equation allows the best focus to be found after the first derivative is taken, set equal to zero, and is solved. Offsets between the different methods can exist and these must be correlated (preferably to a linewidth defocus array) so the most convenient procedure is available for production use. Convenience should reflect the gauge capability of the method and the time spent on the stepper (versus off-line). The gauge capability is the total variance measured from a multivary study, where families of variation may include test-to-test and operator-to-operator (this family should be small for objective metrology and mathematical analysis of data).

Asymmetric Imaging Effects. As previously mentioned, the focus effects for numerical apertures greater than ~0.3, where the depth of focus becomes comparable to the resist layer thickness, depend upon the oblique direction of propagation of light in the resist(10)(84). These effects include an asymmetry in resist profile on either side of best focus(86). This asymmetry occurs for the top of the resist profile but not its base. The effect this would have on the device performance depends on the resist erosion in the dry etch or the momentum of the dopant during implant. Also, there is an asymmetry in the curves on either side of best focus of a Bossung focus-exposure plot, which cannot be explained by lens aberrations. This effect would evidence itself most strongly in critical dimension control. These asymmetries worsen with decreasing feature size, increasing numerical aperture, and thickening resist.

Asymmetrical resist profiles frequently occur on the outboard lines of arrays. This is the result of an asymmetrical aerial image caused by proximity effects(77).

10.3.4 Depth of Focus Dependence Upon Exposure Duty. Thermal absorption in the lens assembly due to the exposure duty can affect the nominal focal plane and cause defocus errors without a scaling or offset compensation(97). The change in nominal focus with time is an exponential decay(98). Satisfactory modeling can require multiple exponential terms(99). The exposure duty is a function of the transmission of the reticle (i.e., how much of the reticle pattern is glass versus chrome) and exposure dose. The change in glass index of refraction with temperature, dn/dT, causes the effect. Absorption of light by the lens materials at the actinic wavelength becomes important for this reason. Generally, i-line lenses use glasses of higher thermal coefficient of refractive index than g-line(100), so the effect probably would be slightly greater at the shorter wavelength.

However, selection of suitable glasses results in only a 0.2 micron focal shift after a one hour flood exposure of one commercial 0.40 NA i-line lens, which is very competitive.

10.3.5 Wafer Flatness.   Wafers imaged using full field scanners are specified most appropriately for flatness using global criteria. One global flatness metric is the Total Indicated Reading (TIR), which is the difference in elevation between the highest and lowest points on the surface of a wafer (i.e., the range) (101). Wafers imaged using steppers, which print one field at a time, more appropriately are quantified for flatness using local criteria. One local flatness metric is the Local Focal Plane Deviation (LFPD), which is the peak absolute deviation of the wafer surface from the focal plane of a representative size exposure field. Global leveling of the wafer by the stepper prior to exposure compensates for wafer wedge (the frontside and backside planes of the wafer are not parallel). Local leveling of the wafer by the stepper prior to exposure compensates for random polishing errors, bow, and edge effects (e.g., domed or cupped wafers). The flatness variation is most severe at the edge of a wafer(102). The local thickness variation directly affects the usable depth of focus so wafer specifications can be referenced to the minimum feature that must be imaged during construction of the product die.

10.3.6 Chucks.   Wafer chucks are constructed from materials such as aluminum (which can be anodized to improve hardness), stainless steel, metal alloys, or ceramics. There is a strong correlation between average wafer flatness and clamping vacuum(101). Unfortunately, the surface contact with the wafer preferably is minimized so that trapped particles under the wafer do not create a local focal plane deviation. Low contact chucks include designs of raised concentric circles or pins (which are an array of elevated pyramidal structures) which maintain vacuum contact with the backside of the wafer. Chuck flatness is measured with a laser interferometer illumination.

10.4 Illumination

10.4.1 Sources.   Mercury vapor lamps are the illumination sources for broadband and narrow band steppers, as well as the deep ultraviolet (UV) step and scan machine. Mercury lamps contain a starting gas like argon. Hg-Xe lamps emit virtually the same spectrum as mercury lamps with the addition of some infrared radiation and a slight increase in the continuum. A typical spectrum is illustrated in Figure 50. Xenon mainly assists in the lamp ignition and warming quicker to equilibrium. A discharge arc of the high pressure vapor emits a characteristic

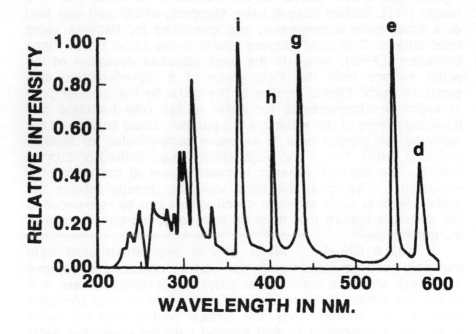

Figure 50. Illumination spectrum of a mercury arc lamp.

spectrum for the element mercury as its electrons relax from an excited state to the ground state. Infrared (IR) radiation is emitted due to the heating of the lamp during its operation. Infrared radiation is filtered with a cold mirror, which passes the IR to a radiator and reflects UV. The mercury spectrum is not of uniform intensity, nor does its intensity vary smoothly as the wavelength changes. As the lamp ages, the relative intensities at each wavelength change. Steppers incorporate an integrator (a light sensor coupled to the shutter servo) to vary the nominal exposure time so that the actual exposure dose remains constant over the life of a lamp. The high intensity peaks at the shorter wavelengths in the ultraviolet and deep ultraviolet have been of the most interest in microlithography since the resolution improves with shorter wavelength.

A cool mercury arc lamp has an internal pressure of 0.3 atmospheres. Intensity is raised by increasing the equilibrium operating pressure, which is in excess of 40 atmospheres(103). This high pressure makes possible lamp explosions, which are both dangerous and expensive. Thermal control of the lamp helps minimize the possibility of a lamp explosion and is related directly to the useful lifetime(104). Short term fluctuations in lamp temperature can cause a variation in illumination uniformity, adversely affecting intrafield linewidth control(105).

Energy output is nonlinear with lamp wattage. Past a certain point, there are diminishing returns for raising the wattage. Stepper lamps of 1000 watts are common and the step and scan uses a 2400 watt lamp(91). The illuminance at the wafer plane for g-line is ~800 mW/cm^2 and for i-line is ~400-500 mW/cm^2 (these values depend on illuminator and reduction lens transmission). Actual exposure times depend on the resist sensitivity and the process. Commercial volume photoresist development has paralleled the emphasis on these peaks, 436 nm (g-line), 405 nm (h-line), 365 nm (i-line), and 250 nm, in chronological order.

An alternative source for deep UV at 248 nm and 193 nm is the use of pulsed gas excimer lasers. These lasers use noble halogen gases such as KrF (at 248 nm) and ArF (at 193 nm). Excimer lasers offer high average power, absence of speckle (i.e., small dots randomly scattered throughout the field), and low spatial coherence. The word "excimer" comes from excited dimer. Excimer molecules or exciplexes exist only in the electronically excited state, since the atoms are unbound or only weakly bound in the lower energy level. The rare gas and halogen atoms reach the upper electronic excited level by chemical reaction in a high voltage (e.g.,15-20kV) pulsed discharge(106).

The average power of a standard commercial excimer laser is above 100 watts, but for microlithography it is as low as 2-5 watts(107)(108). Unnarrowed single pulse energies can be greater than 500 mJ, while narrowed emissions are ~20 mJ/pulse. Spectral narrowing is required to control chromatic aberration and power is reduced appreciably. Spectral narrowing can be accomplished by injection locking, gratings, intracavity prisms, or etalons(107)(108). The last method extends the lasing cavity beyond the excitation region and inserts two etalons (an etalon is used to measure distances in terms of wavelengths of spectral lines). The etalons act as narrow bandpass filters; one has high resolution but has several transmission peaks while the second etalon has lower resolution and is used to select one of the peaks(109).

Maintenance is more complicated for laser sources than mercury arc lamps. The discharge unit can be replaced as easily as a mercury arc lamp. The excimer laser is housed in a separate chamber and the window requires frequent and regular cleaning after ~1×10^8 pulses. Gas refills are needed after ~5×10^6 pulses. A laser head can be completely rebuilt in less than a day, including replacement of the optics, preionization electrodes, and preionizer. This type of major overhaul is performed after ~5×10^8 pulses.

10.4.2 Bandwidth.   Historically, the volume commercial use of optical steppers at shorter and shorter wavelengths has been slowed by lack of chemical support (e.g., high contrast resists, contrast enhancement materials, etc.) since the resolution benefits are indisputable. Generally, lower numerical aperture lenses of shorter wavelengths are easier to design, manufacture, and match than higher NA lenses of longer wavelengths(53). However, the selection and availability of high quality lens materials for shorter wavelength use can be a severe hindrance, especially for correction of image smear due to chromatic aberration(100). Filter technology is becoming more important as the actinic wavelength is reduced since the acceptable bandwidth of illumination is narrowing. High pressure lamps used to increase intensity also raise the continuum and this can significantly degrade the image contrast (by up to 5-10%) without adequate filtration(110). The effective energy is defined as the integral of the product of the spectral intensity, the filter transmittance, and the resist sensitivity.

The bandwidth of commercial g-line lenses of 0.45 NA is ±2.5 nm, compared to ±4 nm for less aggressive 0.35 NA designs. The bandwidth of 0.40 NA i-line lenses is ±1.5 nm, illustrating the stricter bandwidth requirement to control chromatic aberration. It is too difficult to achromatize lenses at 248 nm so the only choice of glass is fused silica. The minimum spectral bandwidth of a 0.35 NA lens using an KrF exclmer laser source is 0.005 nm FWHM and the total spectral bandwidth including the spectral swing is 0.008 nm FWHM. FWHM stands for Full Width at Half Maximum and is the width of the spectral line in hertz at the 50% peak intensity point. The bandwidth is this wide to avoid speckle and still minimize attenuation of the OTF. The required laser bandwidth for fused silica lenses depends on the field size and numerical aperture. The OTF decreases generally and the value added by a higher numerical aperture is less if the bandwidth is not sufficiently narrow. Excimer laser wavelength drift errors are evidenced as focus drifts, changes in distortion values, and astigmatism shifts.

Optimization of the bandwidth by manipulation of the mercury arc actinic narrow-band line filter (e.g., adjusting its type or its physical position) may be done empirically by monitoring the effect on astigmatism of a Resolution versus Defocus plot.

Most of the optical power of the step and scan resides in its achromatic spherical mirrors so average values of optical path differences as a function of field radius are about 0.06 waves for an exposure bandwidth of 235-260 nm(91). This wide bandwidth is favorable for reducing standing wave interference and improving wafer throughput by reducing exposure times.

10.4.3   Wavelength Limitations. As the illuminating wavelength shortens, there are dramatically fewer choices of glasses to construct lenses, and in the deep ultraviolet there are some unique problems(111)(112). The transmission of 248 nm light through fused silica is reasonably efficient at ~92%. However, excimer laser illumination by ArF at 193 nm has a transmission through fused silica of only ~88%. Calcium fluoride and lithium fluoride have slightly more favorable transmission values at 193 nm, but the technology to work with these materials is immature. Lithium fluoride is hygroscopic and is difficult to polish to the exacting surface tolerances required in projection lenses. Unfortunately, the transmission of CaF_2, LiF, and fused silica degrades dramatically after $2 \cdot 10^6$ pulses at 193 nm, worsening as the optical path lengthens. For nonhomogeneous and impure materials, irradiation in the ultraviolet induces formation of absorptive color centers, accompanied by a reduction in transmission. Impurities can play a role, as well as intrinsic point defects. Solarization is highly nonlinear in response and is dose dependent. No reliable acceleration testing is known. At the very least, this limits the refractive lens thickness. Fortunately for photolithography, the damage is nonreciprocal and the low fluxes typical of normal wafer exposures do not cause solarization.

10.4.4 Uniformity. Illumination uniformity commonly is calculated by:

$$[86] \quad U\ (\%) = (\pm 100\%) \cdot (I_{High} - I_{Low})/ (I_{High} + I_{Low})$$

where I is the intensity measurement at different points in the field. The measurements usually are done with an illumination meter and appropriate wavelength sensitive probe, but equivalent information also could be gotten by measuring resist film thickness after exposure and development (this is not preferred since the resist process may confound the results). Typical specifications call for uniformity to be less than ±2.5%.

Uniform illumination is obtained by passing light through a fly's eye integrator or mixing light in a fiber bundle or kaleidoscope. For excimer laser sources, exposures also require multiple pulses. There are variations from pulse-to-pulse in their energy/pulse, so there is a minimum desirable number of pulses for exposing each field(111). Some sources of laser fluctuation noise include the discharge voltage, gas mix recipe, age of the gas fill, pulse repetition rate, head temperature, immediate operating history, and the state of maintenance of the laser discharge system(113). This lower limit is related to the desired tolerance on linewidth control, which is a function of the aerial image quality and the material response of the photoresist. Because of the shorter wavelength of excimer lasers as well as their high peak intensity, a growing number of photo processes have been found to be nonreciprocal (i.e., the linewidth for a given total exposure energy is different for a single pulse of all the energy versus multiple pulses of equal fractions). Exposure control strategies which minimize the number of laser pulses can improve system throughput while allowing simpler, lower repetition rate lasers to be used. More importantly, laser maintenance intervals can be extended and reliability improved(113). The drift of the mean shot energy can be controlled well by feedback control of the laser discharge voltage. Continuous sampling of the pulse energy allows an updated energy-voltage transfer function to be calculated so any exposure dose requirement can be executed with exactly n (10≤n≤20 seems practical) pulses being fired.

It has been proposed that a nonuniform source results in interaction of diffracted orders so any MTF falloff at the field edge(19) can be compensated by a local intensity difference. This is based on the hypothesis that high frequency information generated by the annular area will be gained at the expense of the fundamental order from the center. However, simulations have shown(77) that a source with a central obscuration reduces the image contrast in all cases (i.e., for all radii of obscuration). The best fine line imaging is obtained with a uniform source. Evidently, the loss of the zero-order component is more critical to the image than any higher-order information that might have been obtained. However, for advanced resist processing where MTF values between 0.4 and 0.55 provide useful imaging, annular illumination offers flatter frequency response, much higher exposure latitude for resolution of high spatial frequency patterns, and improved image quality with defocus compared to conventional Kohler illumination with a uniform circular source(82)(113b).

10.4.5 Partial Coherence. The spatial partial coherence for the reduction lens can be determined empirically by measuring the condenser lens spot size at the entrance pupil plane(54). The illuminator is moved aside and a resist covered wafer positioned at the entrance pupil plane is exposed. The developed spot size then can be used, along with the entrance pupil diameter, to calculate the ratio which is defined as the partial coherence for the reduction lens. Most steppers operate with a partial coherence value of 0.5.

Laser sources can emit light in a nearly coherent illumination mode due to the low divergence of the beam while keeping the energy efficiency high. Partial coherence can be obtained by moving the point source in the conjugated plane of the entrance pupil with respect to the condenser(15), e.g., by scanning or by divergence at the source (e.g., by means of a fly's eye element, which is an array of short focusing lenses).

10.4.6 Flare. Flare, or background exposure, is the result of stray reflections and scattering from the reticle, lens, and wafer combinations(23)(41)(114). Flare increases the overall exposure level in low exposure areas and decreases it in high exposure areas. Antireflection coatings help eliminate scattering off surfaces of elements. Antireflection coatings on the reticle and wafer surfaces also improve the image contrast. The internal lens barrel can be painted flat black to reduce reflections. Internal knife-edge stops lining the lens barrel can be used to eliminate internal low-angle reflections that no paint alone can stop. A method of measuring flare uses positive resist as a radiometer(114). Theoretically there should exist no exposure in large opaque regions of the reticle, where large is assumed to mean at least 100 times the size of the Airy disk pattern. A brightfield mask (i.e., a chrome feature surrounded by a clear glass field) is used to create a plot of developed thickness within opaque areas as a function of exposure. The plot should be constant if no flare exists.

10.5 Standing Wave Interference

Standing waves are formed in the resist by the coherent interference effects due to a partially reflecting substrate which result in periodic intensity distributions in the direction perpendicular to the plane of the resist with a period $\lambda_r/2$ where λ_r is the exposure wavelength in the resist(11) (see Figure 51). This results in a sinusoidal dependence of intensity upon the film thickness where the maximum intensity envelope is given at a resist thickness, d, of:

$$[87] \qquad d_{max} = (2m + 1)\lambda_r/4n$$

and the minimum intensity envelope is given at:

$$[88] \qquad d_{min} = m\lambda_r/2n$$

where m is an integer and n is the real part of the resist index of refraction(39)(115). The variation of exposure energy needed to clear a resist film is described by the same equations and since the linewidth of resist patterns is related directly to the exposure dose, linewidths vary with a similar dependence upon resist thickness. The exposure time needed to just clear a resist film (E_0) is at a maximum for a thickness corresponding to d_{max}. The exposure time-to-clear for a resist film is at a minimum for a thickness of d_{min} (interference data are illustrated in Figure 52). The effective light intensity contributing to the exposure of the resist is expressed by:

$$[89] \qquad I = I_0 \cdot (1 - R)$$

where I_0 is the incident intensity and R is the reflection(116). Although the exposure dose requirement is less at a resist thickness corresponding to d_{min}, most processes are controlled

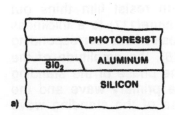
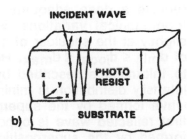

Figure 51. A schematic representation of a typical wafer substrate is shown in (a). A schematic representation of incident and reflected waves during exposure of a photoresist layer deposited on a reflective substrate is shown in (b). Reprinted with permission of Reference 117.

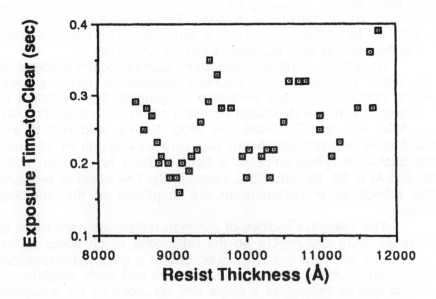

Figure 52. Standing wave interference effect on the exposure time-to-clear.

near a nominal thickness of one micron around a centerpoint of d_{max} ± tens of angstroms. This ensures that all features will receive a dose adequate to clear the film, even with surface topographical variations, where the spun resist film thins out slightly at the shoulder of a nonplanar feature(117) to a thickness of $d_{min} \leq d_{local} \leq d_{max}$. However, for resist films corresponding to thicknesses described by Equation [88], the amplitude of the intensity distribution is minimized since the phase of the standing wave formed by the superposition of the primary wave and the first reflected wave is opposite the phase of the standing wave formed by the superposition of the second and third reflected waves at any resist position z (see Figure 53). Resist thickness variations of $\lambda_r/4n$ from d_{min} can alter the intensity of the standing waves by the ratio I_{max}/I_{min} of 2.5 for a perfect reflector(115). Regardless of the exact choice of resist film thickness, a significant increase in MTF is obtained by thinning the resist(7).

Polychromatic exposure light can result in some wavelength induced intensity smoothing of standing waves, but for dimensions of most interest the benefits are marginal(115). For metal and thin oxide pattern definition steps it is impossible to take advantage of wavelength induced intensity smoothing and the minimum oxide thickness needed for wavelength smoothing (~4000 Å for reflective optics) is large compared to preferred oxide thicknesses with aggressive design rules(115).

Nonplanar surface topography causes local variations in the resist thickness, and spurious reflections, causing local critical dimension control problems. In general, the resolution of the resist pattern is degraded by using highly reflective films like Al, AlCu, Mo, and Ti. Films like SiO_2, Si_3N_4, and poly-Si are much more transparent. At an exposure wavelength of 436 nm, the backward reflection from a flat substrate for silicide, poly-Si, and Al is 33, 23, and 80%, respectively. The effect of reducing film reflectivity is reduction of the amplitude of the standing wave.

The refractive indices of common resists are very close to those of SiO_2 and Al_2O_3 so the reflections which occur at the resist/oxide interface are very weak, and to a good approximation, can be ignored(115). This means the resist and oxide combination of films can be treated as a single film described by the equations for d_{min} and d_{max} above.

The suppression of reflections is critical for maximizing imaging latitude. Variations in reflectivity among substrates affects the MTF with a decrease in resolution caused by diffuse

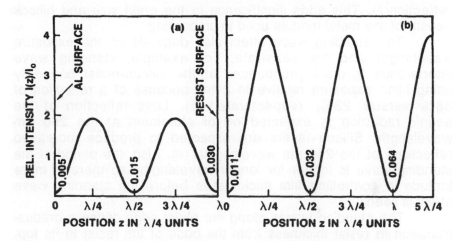

Figure 53. Intensity distribution in photoresist layers (a) of thickness λ and (b) of thickness 5λ/4. The incident light intensity in both cases is I₀. Reprinted with permission of Reference 117.

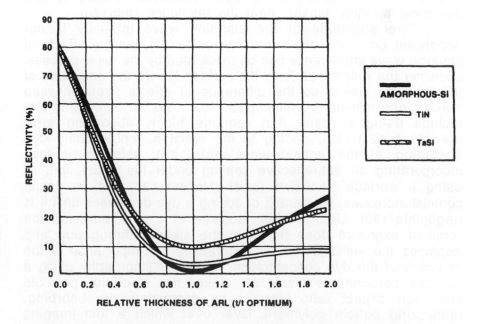

Figure 54. Comparison of the effect on reflectivity for three antireflection layers on aluminum as a function of varying ARL thickness, t. Reprinted with permission of Reference 135.

reflection(7). This adds significance to the grain size and hillock density of the metal module used in processing.

The standing wave effect also depends on the exposure wavelength and the substrate. For example, standing wave interference is more pronounced at the silicon-resist boundary using i-line exposure relative to g-line because of a reflection of 38% versus 23%, respectively(100). Less reflection of the actinic radiation is expected off of aluminum at the 260 nm wavelength. Silicon layers are expected to produce increased reflections at the 260 nm wavelength(118). Also, the period of the standing wave is longer for longer wavelengths so there is more latitude in controlling film thicknesses before the standing wave effect is seen.

The standing waves along the resist edge cause a gradual transition in resist thickness from the base of the resist to its top. This lateral transition zone is proportional to $\lambda/(2NA)$ since the magnitude of the oscillation reflects the extent to which the diffraction pattern penetrates into the geometrical shadow(119). Reduction lenses of high numerical aperture can be used to reduce this transition zone, so high numerical aperture lenses reduce line width fluctuations versus resist thickness relative to low numerical aperture lenses, near the resolution limit(120).

The amplitude of the standing wave intensity is not dependent on z, the depth into the resist(121), but the effects of standing wave interference can be moderated by the resist process. Lowering the reflectance from the surface lowers the amplitude of the standing waves so the difference in effects seen between maxima and minima are minimized. Processes accomplishing this include using a resist that remains highly absorbing after bleaching(121)(122), adding to the resist a nonbleaching dye absorbing at the actinic wavelength(123), using a process incorporating an antireflective coating under the resist film, or using a surface sensitive resist process(124). As the resist contrast increases, the benefit of adding a dye decreases until it is negligible(120). Unfortunately, addition of a dye increases the required exposure dose (lowering the stepper throughput) and degrades the resist profile. This increases the importance of the selection of the dye concentration. Multilayer lithography using a portable conformable mask(125) makes fine resolution possible with high aspect ratio features by combining an absorbing, planarizing bottom polymeric layer over which a thin imaging resist is coated.

There is a marginal improvement in linewidth variations versus resist thickness if contrast enhancement materials(126) are coated over either a low or high contrast resist(120). Contrast enhancement materials (CEM) increase the aerial image contrast photochemically(127). CEM is an optically dense polymer that is extremely photobleachable.

A great advantage is obtained using a thermal image reversal process(128)(129)(130) since there is no change in the exposure time-to-clear as resist thickness changes. This process benefits layers with sharply varying topography.

Other simple processes include minimizing the exposure time (e.g., by increasing the develop time or developer concentration, to substitute chemical etching for photoactivity) or incorporating a post exposure bake(131). If there is sufficient mask bias, overexposure accompanied by over development can reduce standing wave linewidth modulations to lower than 0.05 microns per edge(70). A post exposure bake is most effective at a temperature greater than or equal to the softbake (i.e., the bake immediately after coat and before exposure) temperature with an appropriately long bake time, since the effectiveness of the post exposure bake for reducing the effects of standing wave interference depends upon the diffusion of the photoactive compound (it is the photoactive compound that inhibits unexposed resist dissolution in a developer). Resists that bind the photoactive compound to the resin will not receive optical interference relief from a post exposure bake.

Antireflective layers (ARLs) under the resist film act to minimize standing wave interference. These layers can be either organic coatings or inorganic films. Organic ARLs are convenient since they can be applied during the resist coat process, and then require no special processing other than what the resist receives. The spin-on antireflective material is basically an actinic light-absorbing dye dissolved in a polyimide silane type resin(132-134). Polymeric ARL benefits are limited when topography becomes very severe and undercutting of lines (causing lifting if the adhesion loss is severe enough) is a problem for features near 0.7 micron feature size. Inorganic ARLs include amorphous semiconductors such as silicon and selenium(119), Ta-Si or TiN, and sandwiches of SiO_2 and Si_3N_4(135). For example, a quarter wave coating of either silicon nitride or silicon dioxide is applied to minimize the reflectance and a half wave coating of resist is used because it preserves the low reflectance and is easily controlled to have less than 100Å thickness variation(23). Some data are graphed in Figure 54.

Candidate antireflective layers should have refractive indices n+ik with large ratios of n/k(136). Inorganic ARLs require application of thin films on equipment that is not usually found in a photo bay and this may create manufacturing difficulties.

Compatibility of the ARL over the reflective substrate can be a reliability question (e.g., amorphous silicon over aluminum, raising diffusion and precipitation issues), so Ta-Si or TiN may be preferred over Al films since they need not be removed.

10.6 Vibration

Resolution and depth of focus performance of steppers is affected adversely by excessive floor vibration. Sources of this vibration include building motion, building structural dynamics, acoustic buffeting from clean room air handling, and the dynamic interaction between the stepper and its vibration isolators(22)(137). Stepper resonant frequencies are usually designed around 100Hz so there is no inference from servos.

A vibration isolator is typically a resilient element, such as a metal spring or a rubber mount. However, reduction of machine vibration takes place only above the resonant frequency of the vibration isolator. Most steppers use pneumatic isolators. These isolators can achieve a low resonant frequency (~2-3 Hz), but they do not null out vibration actively, so around the resonance peak, vibrations actually are amplified. Advanced isolators have electronic feedback control using servo control components that operate over an extended frequency range to dampen vibrations actively(137).

Mask-to-wafer vibration has the effect of averaging out the intensity distribution in a given range of vibration amplitude(138). The frequencies of interest are usually between 1-100 Hz. The frequency characteristics of the stepper site can be determined using a very sensitive accelerometer and performing a spectrum analysis of the signal. Measurements must be made for the x and y horizontal axes and the vertical axis. The variability of the amplitude and phase of the frequency spectrum make it difficult to describe vibration concisely. Acceleration measurements are converted to displacements by dividing by the square of the frequency. Alternatively, a vibration histogram can be created which shows the percentage dwell time of the object at each position during the vibration(139). The histogram data are collected from a given time function of the object by summing and normalizing the time the object stays within small position intervals. Vibration displacement also can be monitored with an induction sensor but the sensitivity is limited to about 10 nm.

Simulations(139) show that the depth of focus remains fairly constant for the amplitude of vibration until a threshold is crossed and the depth of focus suddenly ceases to exist. The optimum exposure changes rapidly with vibration amplitude. The depth of focus is highest with a partial coherence of s=0.4. However, s=0.6 is most tolerant to vibrations.

The effects of vibration can be studied directly by observation of the aerial image(44). The real-time nature of the technique allows the monitoring of image placement errors versus time with frequency response limited only by the bandwidth of the detector. It is possible that assignment of an image quality error to astigmatism actually can be due to vibration.

11.0 PRACTICAL IMAGE PLACEMENT

The placement or registration of a reticle die pattern to a previous wafer die pattern is critical to integrated circuit performance. The main sources of registration errors are alignment, stepper induced field errors (only some of which are correctable), mask errors, and process induced errors (typically resulting in a magnification error, caused by a high temperature or stressful film deposition). The results obtained from bench testing a lens do not guarantee necessarily the practical imaging quality obtained when a stepper is used daily.

11.1 Alignment

Alignment is the registration of one field defined by the reticle pattern to the previous fields on the wafer after survey of the locations of alignment targets on the previous layer. Alignment errors include a nonzero mean and a finite distribution from field-to-field of the registration errors. Alignment variation is caused by stage imprecision and signal processing errors. Stage precision is improved by designing the laser interferometer interface to reduce the least count positional inaccuracy, using three axis laser interferometers instead of two(140), making a transition from analog to digital stages, improved software modeling to correct for more of the mechanical and environmental factors affecting stage positioning accuracy, reducing column and stage vibrations, reducing the Abbe offset error (by referencing the stage laser interferometer closer to the actual wafer plane reducing tilt induced translation errors) improving the mechanical stage motion efficiency and measuring the wavelength of the interferometer directly instead of relying on a model expression for the index of refraction of air. Signal processing errors are being improved through different choices of algorithms that determine the alignment target position, digital signal processing(141), and various noise reduction techniques, improved understanding of the statistical error in position measurement, and higher resolution hardware.

11.1.1 Interfield Model. One interfield model describing the overlay of the current wafer grid to a previous one is(142-144):

$$[90] \quad dx_f = dx_W + x_W M_{Wx} - y_W \phi_{Wx} + y_W^2 D_{2x} + R_{Wx}$$

$$[91] \quad dy_f = dy_W + y_W M_{Wy} - x_W \phi_{Wy} + x_W^2 D_{2y} + R_{Wy}$$

$$[92] \quad \phi_{fs} = \phi_f + y_W M_{W\phi} - 2x_W D_{2y} + R_{W\phi}$$

where dx_f and dy_f are field translation values, ϕ_{fs} is intrafield rotation, dx_W and dy_W are wafer to field translation values, x_W and y_W are the wafer coordinate axis system, M_W is wafer scaling, ϕ_W is wafer rotation, D_2 is stage bow, ϕ_f is die rotation, and R_W is the interfield residuals. With a software controlled x-y-θ stage, every systematic grid error is correctable. Figure 55 diagrams some of these errors.

The stages of all the steppers are matched relative to an artifact wafer. A reference stepper is selected and interfield errors are minimized. A first layer pattern is exposed on a wafer, this wafer is rotated 90 degrees, and the wafer is aligned. Software corrections are made to make the stage motions orthogonal (the rotation will double the real orthogonality error) and the x and y steps are made equidistant by adjustment of the y scaling software value. All other stepper stages are referenced to an artifact wafer exposed on this reference stepper.

11.1.2 Types. Alignment system configurations are about as varied as the choice of stepper models(145)(146). The most common alignment types use either brightfield, darkfield, or phase grating methods (see Figure 56). Incident light striking a wafer target scatters the light. The brightfield specular light scattered back in the same spatial dimensions as the incident light is not collected by the darkfield alignment optics. Darkfield light scattered off the target edges outside this cone of illumination is collected by the reduction lens and some is directed toward signal processing(147-149). When diffraction gratings are used(150-152), a beam of light incident on the reticle transmission grating is split into several beams which are collected by the projection lens and reimaged on the wafer reflection grating target. Besides the specular reflection, light will be diffracted in many directions. If a periodic structure (multiple line/space pairs) is used for the alignment target, the light is diffracted only into a few

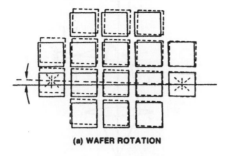

(a) WAFER ROTATION

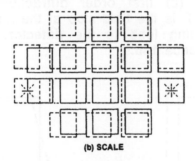

(b) SCALE

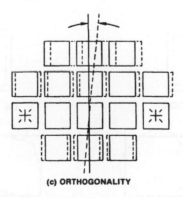

(c) ORTHOGONALITY

Figure 55. Interfield errors.

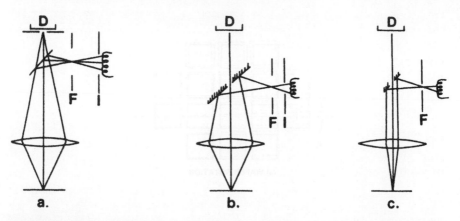

Figure 56. Schematic of (a) partially coherent bright field, (b) darkfield, and (c) first order diffraction interferometric alignment systems. F is the field stop at the reticle/wafer image and I is the imaging lens to the detector. Reprinted with permission of Reference 146.

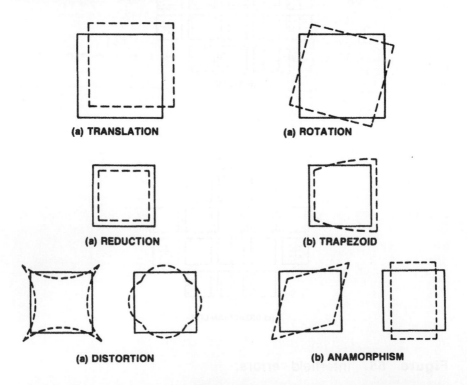

Figure 57. Examples of (a) symmetrical and (b) asymmetrical intrafield errors.

well confined directions, the diffraction orders of the grating. If the zero order and diffracted orders higher than +1 or -1 are blocked, the image always has a sinusoidal shape with 100 percent contrast. If the marks are degraded by planarization or thin film effects or physically impaired the amplitude of the diffracted signal can be lowered but the positional information determined by the period of the grating is the same.

Brightfield alignment can suffer on wafers with topography covered by resist(148). Simulations(146) indicate brightfield alignment systems typically have high intensity, but the signals can be low contrast and brightfield is unlikely to provide the alignment accuracy required by micron and submicron lithography because of relatively large errors caused by small angular asymmetries of the resist coat and adverse sensitivity to uncorrected alignment optical path aberrations. However, dark field alignment can be adversely affected by large, reflective grain boundaries. Dark field alignment errors are relatively small for resist film asymmetries over the target and there is reasonable insensitivity to alignment optical path aberrations. Alignment systems which use the first diffracted orders of the light scattered by a phase grating are relatively insensitive to resist coat asymmetry and defocus and coma effects.

Through the Lens. Most alignment systems pass light through the lens assembly of the stepper. Off axis objectives have a difficult time maintaining the distance to the center of the optical axis so relatively frequent compensation checks must be performed if they are used. Differences in material thermal expansion coefficients and physical displacement are the most common problems for stabilizing the position of the alignment objective.

Direct Reticle Reference. It is desirable for the stepper to reference the reticle directly to the alignment target on the wafer so that reticle translation errors are normalized. Because of symmetrical field errors, the reference should either be on the optical axis (which is rarely practical from a semiconductor manufacturer's perspective since this area is preferentially occupied by a product die) or symmetrical about the field at two or more sites. A single reference point at the reticle field edge can be susceptible to relatively large field errors, causing translation errors in alignment since the center of the reticle position (corresponding to the optical axis) may not be accurately known (minimizing the advantages of direct reticle alignment).

Wavelength. Alignment using the actinic wavelength has an advantage since the exposure optics are corrected chromatically already, which will help stabilize alignment mean shifts.

Alignment with monochromatic light subjects the signal to standing wave interference effects so photoresist thickness and topographical variations exert an influence. Asymmetry of the alignment signal can result in an alignment grid error.

Process optimization can improve the target visibility. On substrates that are neither highly reflective nor that scatter the light much, removal of the resist by local exposure and development or by laser photoablation(106), can improve alignment accuracy and precision. Alternatively, prior to actinic alignment, bleaching of the resist covering the target can improve the signal-to-noise ratio(153). On substrates that are highly reflective or that scatter the alignment light quite a bit, addition of a dye that is absorbing at the alignment wavelength can help absorb scattered light so the signal-to-noise ratio is raised(154). If this dye is not absorbing at the actinic wavelength, there will be no deleterious effects to the resist profile.

To improve target visibility by reducing resist absorption, the alignment wavelength can be lengthened. However, since this alignment wavelength is nonactinic, chromatic aberration and astigmatism will appear and only sagittal aberrations can be corrected(90)(145)(155). The alignment optical focal length must be corrected separately, usually with a separate alignment optical path. Residual chromatic aberration of a wideband alignment source blurs the images for the wavelengths farthest departed from correction in a through-the-lens design. The composite image of blurred and sharp images results in an attenuated edge definition and causes alignment inaccuracies. For nonactinic alignment wavelengths, radial orientation of the alignment targets is preferred since this orientation remains well focused.

The use of multiple wavelengths can improve the composite signal symmetry by minimizing the signal intensity fluctuations from the individual wavelengths caused by optical interference effects(94)(155).

11.1.3 Target Design. Alignment target designs are of almost all imaginable types. The shapes can be crosses, chevrons, bars, islands, concentric circles (e.g., Fresnel zone targets that form a point focus which moves with the wafer), or diamonds. The general shape depends upon the design of the stepper's alignment system that illuminates the target and the method by which diffracted or scattered light is collected for signal processing. In some cases, the density of the target (i.e., mesa or raised features versus trench or depressed features) may not be negotiable (e.g., device structures of deep isolation trenches would preclude mesa targets since these would act like picket fences, interfering with the resist coat) or may depend upon the reflectivity of the deposited substrate and scattering of light off the field.

Aspect Ratio. The aspect ratio is defined here as the ratio of the length to the width of a feature. If it is possible, features of high aspect ratios are favored over small ones because there is more focus and exposure latitude. For example, targets consisting of small trench unity aspect ratio (e.g., contact) features (that are square on the mask) will image as something between a square and a circle under good focus conditions. Under defocused conditions, the images of the contact features will degrade to football or pear shapes. Light scattered off these target features can yield an asymmetric signal which will cause an alignment inaccuracy. However, an advantage of the unity aspect ratio targets is the favorable resist flow characteristics, since there is no long edge to block resist flow and allow asymmetric resist buildup. By comparison, alignment targets of the same density made of dashes (i.e., the length of the target is stretched so the aspect ratio is raised) will have improved focus latitude since a defocused image will yield cigar shaped targets. Although the target ends may not be ideal, the edges will remain straight and light scattered off these edges will continue to yield symmetrical signals. Another benefit is the higher aspect ratio of dash targets relative to contact targets results in more exposure latitude. Dashes permit favorable resist flow also.

Planarization. If resist is coated over a row of mesa features (i.e., features that protrude above the substrate), the resist film thickness is symmetrical with respect to the mesa feature edges for all of the features except the edge ones. The edge features tend to have film thicknesses that are thicker on one side than the other(156). Changes in the phase component of the intensity dominate, and the resulting asymmetric alignment signal causes an alignment shift(146)(157-158). Careful characterization of the resist spin process can yield a planarizing coating over alignment target features(157). Alternatively, planarization structures in proximity to alignment targets act to minimize alignment variation caused by asymmetrical photoresist coating of alignment target features(159). Asymmetric resist pileups on target features are prevented by dummy border pattern placement (which can be a shape identical to one of the target features), helping to stabilize alignment shifts caused by asymmetric alignment light absorption, asymmetric signal reflection, and local variations in the resist's index of refraction after partial bleaching.

11.1.4 Dependence on Field Errors. Alignment data must be collected by the stepper for signal processing so that field errors do not interfere. For this reason, the target site on the field should be blocked (i.e., a scattergram of locations in the field that are aligned should indicate either a single point or symmetrically located misregistration vectors in the field).

Since field errors are minimized at the center of the lens, the center of the field is optimum for placement of alignment targets(160). This may be an option if there are multiple die in the exposure field since there may be a central scribe lane. If it is impossible to position the wafer alignment target at the center of the field, a pair of targets located at the field edge on a diagonal through the optical axis is a good compromise since symmetric field errors will be compensated for so they don't adversely influence the translation value. Alternatively, the field error contribution to wafer y alignment can be minimized if the alignment target for y is positioned at the edge of the field in x but near the origin of the field's y axis(161). The x alignment target is positioned similarly in the field.

Collection of alignment data from across the wafer or across a field allows in-situ correction of some field errors. Wafer scaling, or magnification, caused by high temperature processing or stressful film deposition represents a physical expansion or contraction of the field, also. Isotropic effects can be corrected easily. Field rotation is correctable with data from only two points in the field. With a correction for translation and these two field errors corrected in-situ, total registration is improved.

11.1.5 Mapping and Field-By-Field Alignment. Alignment variation is the sum of the nested variances of stage imprecision and signal processing errors.

$$[93] \qquad \sigma_{align}^2 = \sigma_{stage\ prec.}^2 + \sigma_{signal\ proc.}^2$$

Stage precision (including mechanical, electrical, and metrological noise contributions) is important since most steppers survey only a few sites on the wafer, calculate a wafer grid model, and blind step the pattern. This procedure minimizes the alignment time and helps improve wafer throughput. An advantage of this procedure is the signal processing noise (including algorithm errors, signal asymmetry, bit-wise errors in signal intensity detection and other detector and electrical noise) over the wafer is averaged out. Alignment of every field (field-by-field) is an advantage only if there is no post-alignment motion (so stage errors are zero) and there is low signal processing noise.

For optimum device performance, usually it is desirable to align a current layer directly to the previous layer (e.g., aligning first metal to contact). In this case, alignment to some earlier layer (but not the immediately previous one) would introduce an alignment variation since a translation (of the desired previous pattern) error would be unaccounted for. In circumstances where there are strong interaction effects, it is desirable to align several sequential layers to the same earlier layer.

11.2   Field Errors

Field errors become a very important family of image placement variation for feature sizes near a micron or smaller. The ideal image field is rectangular and described perfectly by the reticle. Under real conditions, field errors contribute to a distortion of the rectangular field(162)(163). Some of the field errors are correctable and others are fixed by the design or manufacture. Some examples of field errors are shown in Figure 57.

11.2.1 Geometric Model.   The analysis of multiple image field placement deviation values using a least squares fit to a geometrical model allows characterization of the individual errors(142)(144)(162)(163). One such model is:

$$[94] \quad dx = T_x - Ry + Mx + K_x x^2 + K_y xy + xr^2 D_3 + xr^4 D_5$$

$$[95] \quad dy = T_y + Rx + My + K_y y^2 + K_x xy + yr^2 D_3 + yr^4 D_5$$

where:

$$r^2 = x^2 + y^2 .$$

In the model, dx and dy are the component image displacement values at various field positions, T is the field translation, R is the field rotation, M is the magnification of the field, K is keystone, D_3 is third order distortion, and D_5 is fifth order distortion. The signs preceding the rotation, R, coefficients in Equations [94] and [95] are opposite for the x and y components since their directions are orthogonal to each other. It is reasonable to expect the translation, rotation, and magnification values after regression would be identical. If Equations [94] and [95] are solved separately, this probably will not result and some type of averaging will be necessary(144). If the equations are coupled (164) then a unique solution will result. The matrix for coupled equations is shown in Table 4.

Lens matching compares the intrafield errors of different lenses and is defined generally as the correctable errors subtracted from the measured errors.

The unity magnification folded catadioptric Hershel-Wynne-Dyson lens(165-167) has advantages in microlithography since the symmetry of its design corrects all odd order Seidel aberrations (i.e., coma, distortion, and lateral chromatic aberration). This makes it possible to achieve excellent correction over a large field with a quarter of the elements of a refractive lens design(167). The design is telecentric on both the image and object side. In reality, there are random field errors, magnification errors, and trapezoid errors. There is no provision for the correction of magnification and trapezoid errors, so these must be treated as uncorrectables. The only correctable field errors are translation and rotation. Distortion of the prism surfaces due to stress induced during the stepper's assembly can be responsible for the magnification and trapezoid errors.

Application of Equations [94] and [95] yields the information necessary to align the reticle exposure field with the symmetric center of the reduction lens assembly.

Reticle Design and Testing. The two tests of interest are the test of a single lens and the test from stepper-to-stepper. The test of a single lens is facilitated with in-situ stepper metrology(43)(44)(144)(168). This in-situ test reticle has an array of alignment targets in the usable field that are imaged and then individually aligned with the stepper's local alignment system. The local displacement in x and y for each target is measured using the laser stage as a yardstick and these data are modeled for intrafield errors. Alternatively, a comparison can be made at selected sites in the field relative to the center of the lens, which is assumed to be ideal. Registration is determined from electrical probe structures(142), optical verniers(169), or box-in-box structures(170). The registration of these structures is found from interlocking two parts, referred to here as the male and female halves (see Figure 58). A test reticle has an array of male structures in the usable field that is imaged. Before the wafer is removed from the chuck, a second exposure pass is made where the reticle is framed down with opaque blades so only the center female structure is exposed, and the laser stage is stepped to interlock the female half with each of the male halves. Stage imprecision shows up in the intrafield model as a residual error since it is a random error. Systematic stage errors such as x and

Table 4. Matrix for Coupled Equations.

OVERLAY ERROR	X INTERCEPT	Y INTERCEPT	R MULTIPLIER	M MULTIPLIER	K_x MULTIPLIER	K_y MULTIPLIER	D_3 MULTIPLIER	D_5 MULTIPLIER	OVERLAY COEFFICIENTS
dx_1	1	0	$-y_1$	x_1	x_1^2	$x_1 y_1$	$x_1 r^2$	$x_1 r^4$	T_x
dx_2	1	0	$-y_2$	x_2	x_2^2	$x_2 y_2$	$x_2 r^2$	$x_2 r^4$	T_y
.	.	.	.	.	.	.	.	.	R
.	.	.	.	.	.	.	.	.	M
dx_n	1	0	$-y_n$	x_n	x_n^2	$x_n y_n$	$x_n r^2$	$x_n r^4$	K_x
dy_1	0	1	x_1	y_1	$x_1 y_1$	y_1^2	$y_1 r^2$	$y_1 r^4$	K_y
dy_2	0	1	x_2	y_2	$x_2 y_2$	y_2^2	$y_2 r^2$	$y_2 r^4$	D_3
.	.	.	.	.	.	.	.	.	D_5
.	.	.	.	.	.	.	.	.	
dy_n	0	1	x_n	y_n	$x_n y_n$	y_n^2	$y_n r^2$	$y_n r^4$	

$$\text{OVERLAY ERROR} = \text{MATRIX} \times \text{OVERLAY COEFFICIENTS}$$

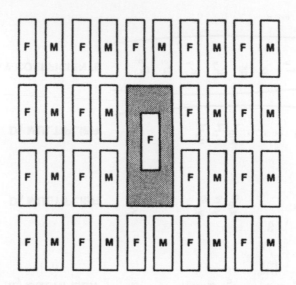

Figure 58. Schematic of an intrafield test reticle showing the layout of the complementary male and female registration halves.

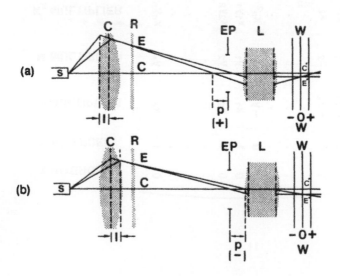

Figure 59. Positive condenser defocus is shown in (a) and negative condenser defocus is shown in (b). Reprinted with permission of Reference 171.

y scaling and orthogonality must be minimized for uniform comparison of results between steppers. Stage scaling errors directly affect field magnification and stage orthogonality shows up as field anamorphism.

Stepper-to-stepper tests may be either in-situ or remote also, but since lens matching is defined as (measured-correctables), stage imprecision must be eliminated as a source of residual error or it will be a confounding effect. Test reticles for inter-stepper matching have adjacent male and female structures in pairs in an array in the usable field. The nominal separation of the structures is known a priori. One stepper exposes fields on a wafer and the second stepper introduces a shift prior to exposure so the male and female structures interlock (for in-situ metrology, a shift is needed to avoid overlay of the alignment targets, so they can be read later).

Testing all random pairs of steppers for matching becomes overwhelming quickly since the number of random pairs increases geometrically as n(n-1)/2.

11.2.2 Symmetric Errors. Symmetric errors are the same value at a given radial distance regardless of position in the field.

Translation. Translation of the field theoretically is not a problem for steppers that reference the reticle directly. Steppers that do not reference the reticle directly have procedures that try to reference the reticle to some fiducial (usually on the stage near the chuck), and then reference the alignment system to the same or a related fiducial, so the alignment system is referenced indirectly to the reticle(161). As overlay requirements tighten, the stability of indirect reticle reference systems becomes a larger issue, since translation instability requires the use of send-ahead wafers to center alignment mean values before the lot is committed for processing.

Magnification. A stepper is image side telecentric if there is no change in magnification as the wafer plane is defocused. Typical telecentric values are ±0.05 μm or less of magnification change for ±2 μm wafer plane defocus. Experimental failure of a test for telecentricity usually is caused by condenser lens defocus, but the effect can be caused by a condenser aberration such as spherical aberration(13). A double telecentric design calls for no magnification change if either the object (i.e., the reticle) or image (i.e., the wafer) is defocused.

In Figure 59(171), the chief ray for a properly focused condenser (C) intersects the optical axis at the entrance pupil (EP) of the lens (L) and exits approximately parallel to the optical axis, for an image side telecentric lens. A ray originating from a defocused condenser intersects the optical axis either in front of or behind the entrance pupil of the lens for positive and negative condenser defocus, respectively. As this ray exits the lens, it is either converging on the optical axis (positive condenser defocus) or diverging from the optical axis (negative condenser defocus). As the wafer (W) focus is varied, the reticle (R) object field CE is magnified or demagnified, affecting the wafer image size C'E'. Telecentric error is inversely proportional to the square of the focal length of the lens.

For steppers that are image side telecentric only, a reticle z stage can be used to correct for magnification errors. Reticle z motions can correct for trapezoid errors quickly, too. Generally, there are six degrees of freedom on a reticle stage: x and y translation, z motion, azimuthal rotation, and front-to-back and left-to-right tilt.   Magnification corrections are made by:

$$[96] \qquad \Delta M = M \cdot \Delta z / EP$$

where ΔM is the desired magnification change, M is the nominal lens magnification, Δz is the reticle z displacement, and EP is the entrance pupil distance on the reticle side of the projection lens(144). It is apparent that reticle z motions become more critical as the reduction ratio of the lens is lowered.

Double telecentric lenses have the disadvantage that magnification and distortion adjustments cannot be made by reticle plane manipulation so more complex adjustments of individual lens elements have to be made for reduction steppers(144). No correction is available for unity magnification double telecentric steppers. Wafer defocus can induce distortion and magnification errors even in cases of wafer side telecentric lenses, but requires severe (i.e., improbable) levels of condenser and wafer plane defocus(170).

Alternatively, the magnification can be adjusted by changing the refractive strength of the lens elements $n_{glass}(\lambda)/n_{medium}(P,T,H)$, where P is the pressure of the gas medium, T is its temperature, and H is its humidity. Typically, the gas is nitrogen and pressure changes of only a couple of psi make a significant (e.g., tens of ppm) magnification change. This low pressurization avoids element flexure, which could change aberration values. Lens assembly pressurization is relatively slow since a change requires equilibration, so it may not lend itself as well as a reticle z stage to in-situ correction of process induced magnification errors (caused by high temperature processing or the deposition of stressful films).

The dependence of magnification on environmental factors is(152):

$$[97] \quad \Delta M_y = 0.5[(LF_L + LF_R) + (LR_L + LR_R)] -10[\alpha_w(T_{m,w} - T_{exp,w}) - \alpha_r(T_{exp,r} - 20) + \alpha_p(760 - P_{exp})]$$

where ΔM_y is the magnification error in the y direction, LF_L and LF_R are the y direction lengths on the left and right side of the field, respectively, LR_L and LR_R are the corresponding lengths on the left and right side of the reticle, respectively, α_w and α_r are the coefficients of thermal expansion for the wafer and reticle, respectively, $T_{m,w}$ is the temperature of the wafer during measurement, $T_{exp,w}$ is the temperature of the wafer during exposure, $T_{exp,r}$ is the temperature of the reticle during exposure, α_p is the magnification correction factor for barometric pressure, and P_{exp} is the barometric pressure during exposure. The coefficients of thermal expansion for silicon, Hoya low expansion LE-30 glass, and fused quartz are $2.5 \cdot 10^6$, $3.7 \cdot 10^6$, and $0.55 \cdot 10^6$ mm/°C, respectively(172).

It is important that the focus walk be small (e.g., ±0.2 μm) for a relatively large change in magnification (±10 ppm). This check is most important in systems that pressurize the reduction lens to achieve magnification changes. Also, the exposure duty should have a negligible impact on magnification change for either the across wafer or wafer-to-wafer families of variation (unless compensation is applied). Although there is a stronger effect on focus, absorption of light in the lens assembly changes the glass index of refraction, dn/dT, causing a magnification shift.

Systematic stage errors such as x and y scaling must be minimized between random stepper pairs for uniform comparison of results in intrafield testing. Stage scaling errors are a confounding effect for field magnification. Also, testing for magnification can be confounded by the feature size of the object so either the test vehicle must be selected carefully or the aerial image should be sampled(98).

Rotation. Field rotation is measured for the current layer being imaged relative to the previous one(169). Systematic sources of field rotation include a rotational misalignment of the reticle on the platen (the platen vacuum clamps the reticle in position at the object plane), non-parallelism between the reticle alignment marks and the stage motions, and asymmetrical keystoning of the system. Random sources of rotation errors include twisting of the optical column from one exposure to the next and stage yaw. Field rotation is an error that is correctable with wafer grid errors during the alignment sequence.

Distortion. Distortion is an uncorrectable error for reduction stepper users. Barrel distortion occurs when the distortion parameters are negative. Pincushion distortion occurs when the distortion parameters are positive. In most practical cases, the third and fifth order parameters will have opposite signs leading to partial barrel and partial pincushion distortion(144).

11.2.3 Asymmetric Errors. Asymmetric errors have different values at the same radial position on opposite ends of a diagonal through the optical axis.

Trapezoid. Trapezoid is a correctable error, except with unity magnification steppers. If the stepper is not telecentric on the object side, a mechanically fine z axis tilt of the reticle will introduce a geometric trapezoidal distortion from an ideal square shape. From the coefficients of the intrafield model, if K_x or $K_y \neq 0$ the trapezoid error is keystone; if $K_x \neq K_y$ and $K_x \neq 0 \neq K_y$, there is an irregular quadrilateral; if $K_x = \pm K_y \neq 0$, there is a kite(163).

Anamorphism. Anamorphic errors are uncorrectable except with the step and scan machine where relative differences in mask and wafer motions allow correction(91). Anamorphism is a non-rotationally symmetrical distortion term which is caused by cylindricity in lens elements(144) or by distortion of the plane of the reticle. Cylindricity of lens elements is due to improper lens manufacture. Anamorphism causes a difference in magnification across the axes. Systematic stage errors for orthogonality must be minimized for uniform comparison of results in intrafield testing. Stage orthogonality is a confounding effect for field anamorphism. Referencing the coefficients of the intrafield model, distortions from an ideal square shape are rectangular if $M_x \neq M_y$ and field orthogonality=0; a rhomboid results if $M_x \neq M_y$ and field orthogonality$\neq$0, and ; there is a rhombus if $M_x = M_y$ and field orthogonality$\neq$0(163).

11.2.4 Illuminator Issues. Condenser lenses are manufactured with certain aberrations that are complementary to those in the reduction lens assembly(13)(54)(171), but control of the magnitude of the aberrations is much less critical, and can be one or two orders of magnitude larger than the reduction lens assembly requires. The cost and quality of a condenser can be much less than the reduction lens assembly.

The z axis position of the condenser lens affects illumination uniformity, the partial coherence value (i.e., the degree to which the pupil of the reduction lens is filled), and the degree of image side telecentricity.

Condenser aberrations are important because they change the directional distribution of illuminating radiation, which affects both the shape and telecentricity of the resulting images. If the wafer is in focus and the projection lens is aberration-free, the mask image (especially for small features) does not depend critically on the shape of the source image(13). Bench testing of the condenser lens is preferred since the interaction of condenser aberrations and projection lens aberrations is complex and difficult to sort out with testing on wafers.

12.0 MASK ISSUES

A few particular issues are treated here since they have special application to optical steppers.

12.1 Particulate Protection

The frequency of particles increases quadratically as their size decreases. The population of particles is related to their size, x, by:

$$[98] \quad \text{Population (area)} \quad \alpha \int_{x=n}^{x=\infty} (1/x) \, dx = \left. \frac{-1}{x} \right|_n^\infty = \frac{1}{n}$$

so the particle population increases inversely linearly. This is important in considering the choice of lens reduction since the chip area of reduction steppers relative to unity magnification increases by the inverse square of the magnification.

12.1.1 Pellicles. Pellicles(173) are thin transparent membranes on a frame offset from the reticle surface that are hermetically sealed dust-free enclosures. The use of pellicles prevents particles from falling onto the reticle surface where it can print as a defect. Particle size immunity to printing is proportional to the standoff distance of the pellicle frame. The standoff distance requirement also depends on the illumination wavelength since scattering of light is inversely proportional to the fourth power of the wavelength.

The most common pellicle films for g-, h-, or i-line exposures are manufactured from nitrocellulose with antireflection coatings. There is little degradation of a nitrocellulose film as a function of lifetime exposure duty for wavelengths at 365 nm or longer if the thickness is selected and manufactured carefully(174). Transmission spectra for nitrocellulose as a function of wavelength exhibit standing wave interference. The oscillatory behavior of reflectivity, R, can be approximated by(175):

[99]   R = r(1-cosX)

where:

$$r = 2[(n-1)/(n+1)]^2$$

and

$$X = 4\pi nt/\lambda$$

where n is the index of refraction of the film (~1.5 for nitrocellulose), and t is the film thickness. The transmission reaches a theoretical maximum of 1.0 when the optical thickness, nt, is an even integer of quarter wavelengths and a minimum when equal to an odd integer. Selection of the pellicle thickness with good transmission depends on the required mechanical strength but affects the periodicity of the waves (at 436 nm, both 0.865 and 2.85 micron thick pellicles have ≥99% transmission, but the thinner, more fragile film has a longer period). Raising the transmission of a single pellicle slows the degradation time of the film. A laser reflectometer can be used to ensure transmission uniformity to less than one fringe in the active area.

Application of an antireflection coating on the nitrocellulose film maintains image contrast and minimizes the amplitude of the waves. Typical antireflection coatings are organic Teflon-like polymers (Teflon is a trademark of E.I. DuPont de Nemours Co.) or inorganic MgF_2. At 436 nm illumination, many antireflective coatings are available that do no degrade with cumulative exposure dose, but at 365 nm the selection of a durable antireflective coating is less trivial(174). The useful lifetime of a pellicle has expired when its transmission varies by more than 1%, corresponding to a shift of the interference fringe to the short wavelength side.

Material selection for pellicles that have acceptable transmission properties, lifetime characteristics, and durability is much more limited for wavelengths shorter than 365 nm, but there are some suitable candidates(176)(174)(177). The heating effect in the pellicle film from deep UV excimer laser exposure is negligible compared to the photochemical change due to oxidation. Selection of a pellicle film with acceptable transmission stability and uniformity still requires matching an antireflective coating material that won't decompose and that has a compatible index of refraction.

Soft defects are defects that are not imaged consistently. In many cases, these printed defects are caused by floating, unsecured particles that attach to, and then release from, the reticle surface. If the mask is pelliclized, the volume of air under the pellicle of a reduction stepper relative to unity magnification increases by the cube of the reduction ratio, so the probability of soft defects increases with increasing ratios of lens reduction.

The focus shift, Δf, resulting from the insertion of a membrane in the optical path is approximately(178):

[100] $$\Delta f = t(n-1)/n$$

where t is the film thickness and n is the refractive index of the pellicle film. The depth of focus at the object plane is relative to the objective lens magnification, M ($M \geq 1$):

[101] $$DOF_{object\ plane} = M \cdot DOF_{image\ plane}$$

so there is less sensitivity to the choice of pellicle thickness for reduction steppers relative to unity magnification steppers.

To ensure less than a 10% variation in exposure, the minimum standoff distance, D (mm) required for a particle size p (μm) is(175):

[102] $$D = (np)/(560NA)$$

where n is the refractive index (1.0 for air and 1.5 for glass) and NA is the numerical aperture. Particles on the glass side of the reticle appear ~33% closer because of the refractive index of the reticle glass, so an equidistant glass standoff is not as effective as a pellicle, but there is an obvious benefit to using 250 mil thick reticles instead of 90 mil thick ones. Common pellicle standoff distances to keep particles out of focus at the object plane range between 3-10 millimeters.

12.1.2 Glass Coverplates. A recent alternative to pellicles is glass coverplates(179). Materials of construction include soda lime, borosilicate, or quartz glass. Advantages of glass coverplates include their greater durability relative to pellicles and chrome damage from electrostatic discharge is eliminated. Like pellicles, addition of a planar glass element to the optical path can introduce optical aberrations.

12.1.3 Voting Lithography. Voting lithography is a technique that superimposes multiple images of nominally identical reticle fields and exposes each with a suitable fraction of the total exposure energy(180). A random defect unique to one of the reticle fields will be averaged out, reducing the influence of the defect on the final image. Typically, voting occurs with two or three reticles. This procedure reduces wafer throughput and assumes precise alignment. Simulations(180) indicate since the image intensity rises roughly linearly over a distance of ~0.3λ/NA, total displacements between the individual images less than this will result in essentially no linewidth change. Increased coherence (i.e., small s) gives a higher intensity slope and yields a smaller linewidth variation. For larger values of s, the interactions with neighboring features are less coherent so there are smooth linewidth variations and defect suppression with voting with results $\geq$300% better than a nonvoted image (since voting effectively reduces the intensity coming through the defect by 1/3). When 3 votes are used, defects which bridge on the mask do not bridge when printed, and the size of defect which can be tolerated for a 10% linewidth variation doubles from 0.24λ/NA to 0.5λ/NA.

12.2 Phase-Shifting Masks and Serifs

Special construction of the reticles can increase the resolution and depth of focus of features being imaged. A conventional transmission mask limits resolution because the electric field corresponding to the intensity pattern has the same phase at every aperture (i.e., opening in the chrome of the reticle), causing a finite intensity in the region between features that ideally should be completely darkened. Constructive interference between waves diffracted by adjacent apertures enhances the field between them. The intensity pattern is proportional to the square of the electric field(181). A phase-shifting mask has a transparent phase-shifting layer covering adjacent apertures so the light coming from one of the features is delayed so that it arrives 180 degrees out of phase. The two diffracted beams will cancel and the desired dark area between images will be obtained(182). The transparent layer of thickness $d = \lambda/2(n-1)$ (where n is the index of refraction of the phase shifter and λ is the wavelength) reverses the sign of the electric field corresponding to the covered aperture. The intensity pattern at the mask is unchanged(181). This is seen in Figures 60 and 61.

The image contrast produced by a phase-shifting mask depends upon the partial coherence value of the illumination as well as the mask spatial frequency and the numerical aperture of the imaging system(181). The benefit of a phase-shifting mask appears as the feature sizes approach the Rayleigh resolution limit, since large features are not susceptible to proximity effects. Phase-shifting masks can increase the exposure latitude significantly(181). The expected improvement is greatest for more coherent light; a partial coherence value of 0.3 gives much greater image contrast for fine features than a value of 0.7(182). Phase-shifting masks can give results equivalent to a numerical aperture increased by ≥33%, but without the large reduction in depth of focus the higher NA lens would yield. Near the Rayleigh resolution limit, the defocus tolerance is less than ±1.1 waves for phase shifting masks if the MTF is to remain ≥60%(183).

Unfortunately, the benefits of greater resolution extend to unwanted defects on the reticle, also. Phase shifts of 180 degrees cause defects twice as small to print. Phase angles less than 180 but greater than 120 degrees give similar improvement in resolution and the lower phase angle reduces the susceptibility of printing defects(182).

An approximation for the spatial frequency (line pairs/mm) at an MTF of 60% is(183):

$$[103] \qquad \nu_{60\%} = [2-(6/7)\sigma]\cdot NA/\lambda$$

where the spatial frequency $\nu=1000/[2\cdot\text{linesize}(\mu m)]$, σ is the partial coherence value, and λ is the wavelength in mm.

A phase-shifting mask has no effect on the imaging of isolated features on a mask without suitable surrounding phase shifted apertures, that are themselves beyond the resolution limit of the system(182)(183).

Serifs allow improvements in terms of reduced radius of corner rounding, reduced area loss, and better preservation of aspect ratio of rectangles(184). Serifs are an optical defect (i.e., their size is well below the resolution limit) placed in proximity to the corners of a feature of interest (e.g., a square contact on a reticle will have subresolution squares placed at the tips of each corner). The slope of the image intensity distribution is lower as the radius of corner rounding is reduced with serifs. However, the dose-related size variability is increased with serifs in the corner area. Serifs help maintain better resist profiles in the

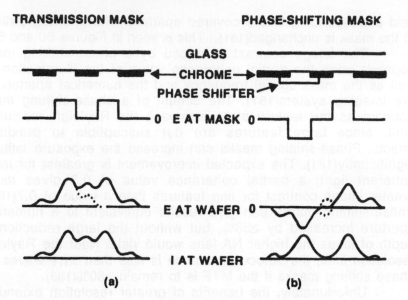

Figure 60. Comparison of the diffraction optics of an ordinary transmission mask with a phase shifting mask. E is the electric field and I is the intensity. Reprinted with permission of Reference 181.

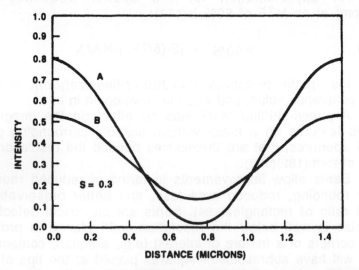

Figure 61. Image intensity for a periodic series of lines and spaces 0.75 microns wide, for no phase shift (B), and 180 degrees phase shift (A). Reprinted with permission of Reference 182.

corner region than rectangles imaged without serifs. This translates into increased depth of focus latitude.

12.3 Excimer Laser Irradiation Damage

When the reticle pattern is formed by excimer laser projection processing, chromium mask damage results for metal films of 70-80 nm thickness(185). Damage of the chromium film on quartz is a result of poor heat transfer following optical absorption causing metal stress and fatigue. Assuming no transmission, chromium film absorbs 79% at 248 nm. The degree of damage is pattern size dependent (smaller features are more desirable) and ranges from erosion of pattern edges to total ablation of chromium film depending on fluencies. At lower fluencies, chromium films crack to various degrees and the damage can be cumulative. This is the major concern for excimer laser wafer steppers, although the energy at the object plane is reduced by the square of the reduction ratio (which presents no relief for unity magnification steppers). Damage thresholds are ~25-50 mJ/cm^2 at 248 nm and total ablation can happen at ~800 mJ/cm^2. Increasing the chromium film thickness appears to improve the heat transfer so damage is avoided. Chromium films of 2 microns thickness offer relief, but these must be etched in reactive ion etchers because the isotropic undercut of wet etching becomes prohibitive. Films of this height are not a problem for reduction lenses with their large objective plane depth of focus. Focused air cooling of the reticle may allow the more conventional chromium masks to continue in use or alternative mask materials may show better resistance to damage.

12.4 Registration Error Contributions

It is desirable that the reticle form as nearly an ideal grid as possible so the image placement errors will be dominated by the distortions of the stepper optical system. For 10X reduction lenses, this may be a reasonable assumption for random errors on the reticle. Unfortunately, 4-5X reduction systems must consider the quality of systematic and random placement errors present on the reticle as registration requirements approach the design limits of the stepper (even though their magnitude at the wafer plane is reduced by the demagnification). Unity magnification steppers processing submicron features absolutely must demand the highest quality reticles since reticle errors degrade registration and critical dimension control capability directly.

It is preferable that all reticles in a reticle set be generated on a single electron beam machine or pattern generator, so that equipment variation is blocked. For a particular device reticle set, absolute errors are not as important as relative ones. In fact, intentional introduction of a magnification offset or "error" on selected reticle layers can be used to compensate for systematic process induced errors (e.g., high temperature deposition of films).

REFERENCES

1.   Tasch, A.F., SPIE, p. 68, Vol. 333, 1982.
2.   Condon, E.U., and Odishaw, H., Handbook of Physics, 2nd Ed., McGraw-Hill, 1967.
3.   Offner, A., "Optical Design and Modulation Transfer Functions", p. 240, Perkin-Elmer Symposium on "The Practical Application of Modulation Transfer Functions", Perkin-Elmer, March 6, 1963.
4.   Williams, C.S., and Becklund, O.A., Introduction to the Optical Transfer Function, John Wiley & Sons, 1989.
5.   Goodman, J.W., Introduction to Fourier Optics, McGraw-Hill, 1968.
6.   Smith, F.G., and Thomson, J.H., Optics, John Wiley & Sons, 1971.
7.   Nakase, M., Photog. Sci. and Eng., p. 254, Vol. 27, No. 6, 1983.
8.   Abbott, F., Optical Spectra, p. 54, March, 1970.
9.   Berkovitz, M.A., SPIE, p. 115, Vol. 13, 1969.
10. Yeung, M.S., SPIE, p. 149, Vol. 922, 1988.
11. Bowden, M.J., J. Elect. Chem. Soc., p. 195C, May, 1981.
12. Yeung, M., Proc. Kodak Microelectronics Seminar, Oct., 1985.
13. Goodman, D.S., and Rosenbluth, A.E., SPIE, p. 108, Vol. 922, 1988.
14. O'Toole, M.M., and Neureuther, A.R., SPIE, p. 22, Vol. 174, 1979.
15. Lacombat, M., and Dubroeucq, G.M., SPIE, p. 28, Vol. 174, 1979.
16. King, M.C., Proc. Kodak Microelectronics Seminar, p. 33, Oct., 1980.
17. Lacombat, M., Dubroeucq, G.M., Massin, J., and Brevignon, M., Solid State Technology, p. 115, Aug., 1980.
18. Lacombat, M., et al., Solid State Technol., p. 115, Vol. 23, Aug., 1980.
19. King, M.C., and Goldrick, M.R., Solid State Technology, p. 37, Feb., 1977.
20. Anon.
21. Offner, A., Optical Eng., p. 294, Vol. 26, No. 4, April 1987.
22. King, M.C., IEEE Trans. Electron Devices, p. 711, Vol. ED-26, 1979.
23. Oldham, W.G., Jain, P., and Neureuther, A.R., Proc. Kodak Microelectronics Seminar, Oct., 1981.

24.   Rosenau, M.D., "Image-Motion Modulation Transfer Functions", p. 252, Perkin-Elmer Symposium on "The Practical Application of Modulation Transfer Functions", Perkin-Elmer, March 6, 1963.

25.   Coltman, J.W., J. Opt. Soc. Am., p. 468, Vol. 44, No. 6, 1954.

26.   Offner, A., Photogr. Sci. Eng., p. 374, Vol. 23, No. 6, 1979.

27.   Tai, K.L., et al., J. Vac. Sci. Technol., p. 1169, Vol. 17, No. 5, Sept./Oct., 1980.

28.   Wake, R.W. and Flanigan, M.C., SPIE, p. 291, Vol. 539, 1985.

29.   Babu, S.V. and Srinivasan, V., J. Imaging Technology, p. 168, Vol. 11, No. 4, 1985.

30.   Babu, S.V. and Srinivasan, V., SPIE, p. 36, Vol. 539, 1985.

31.   Srinivasan, V. and Babu, S.V., JECS, p. 1686, Vol. 133, 1986.

32.   Trefonas, P., and Daniels, B.K., SPIE, p. 194, Vol. 771, 1987.

33a.   Blais, P.D., Solid State Tech., p. 76, Vol. 20, No. 8, 1977.

33b.   Flanigan, M.C. and Wake, R.W., SPIE, p. 44, Vol. 539, 1985.

34.   Arden, W. and Mader, L., SPIE, p. 219, Vol. 539, 1985.

35.   Arnold, W.H. and Levinson, H.J., Proc. Kodak Microelectronics Seminar, p. 80, Nov., 1983.

36.   Daniels, B.K., Trefonas, P., and Woodbrey, J.C., Solid State Tech., p. 105 , Sept., 1988.

37.   Mack, C.A., SPIE, p. 135, Vol. 922, 1988.

38.   Watts, M.P.C., Semiconductor International, p. 124 , April, 1985.

39.   Waldo, W.G. and Helbert, J.N., SPIE, p. 153, Vol. 1088, 1989.

40.   Taylor, G.N., Solid State Tech., p. 105, Vol. 27, No. 6, 1984.

41.   Arden, W., Klose, H., and Krause, A., Proc. Kodak Microelectronics Seminar, p. 11, Oct., 1982.

42.   Scott, F., "Film Modulation Transfer Functions", p. 248, Perkin-Elmer Symposium on "The Practical Application of Modulation Transfer Functions", Perkin-Elmer, March 6, 1963.

43.   Brunner, T.A., and Allen, R.R., SPIE, p. 6, Vol. 565, 1985.

44.   Brunner, T.A., and Allen, R.R., IEEE Elect. Device Letters, p. 329, Vol. EDL-6, No. 7, July, 1985.

45.   Malacara, D., Optical Shop Testing, John Wiley and Sons, 1978.
46.   Jenkins, F.A., and White, H.E., Fundamentals of Optics, 4th Ed., McGraw-Hill, 1976.
47.   Wang, J.Y. and Silva, D.E., Appl. Opt., p. 1510, Vol.19, May, 1980.
48.   Kano, I., SPIE, p. 48, Vol. 174, 1979.
49.   Webb, J.E., SPIE, p. 133, Vol. 480, 1984.
50.   Fincham, W.H.A., and Freeman, M.H., Optics, 9th Ed., Butterworths, 1980.
51.   Van Heel, A.C.S., Advanced Optical Techniques, John Wiley and Sons, 1967.
52.   Toh, K.K.H., and Neureuther, A.R., SPIE, p. 202, Vol. 772, 1987.
53.   Personal communication, J.H. Bruning, President, Tropel.
54.   Peters, D., Proc. Kodak Microelectronics Seminar, p. 66, 1985.
55.   Sewell, H. and Friedman, I., SPIE, p. 328, Vol. 922, 1988.
56.   Goodall, F., and Lawes, R., Proc. KTI Microelectronics Seminar, Nov., 1987.
57.   Subramanian, S., Appl. Opt., p. 1854, Vol. 20, May, 1981.
58.   Hufnagel, R.E., "Random Wavefront Effects", p. 244, Perkin-Elmer Symposium on "The Practical Application of Modulation Transfer Functions", Perkin-Elmer, March 6, 1963.
59.   Chien, P., Liauw, L., and Chen, M., SPIE, p. 197, Vol. 538, 1985.
60.   Hornbeck, R.W., Numerical Methods, Quantum Publishers, 1975.
61.   Averill, E.W., Elements of Statistics, John Wiley and Sons, 1972.
62.   Bedworth, D.D., and Bailey, J.E., Integrated Production Control Systems, John Wiley and Sons, 1982.
63.   Mendenhall, W., Introduction to Probability and Statistics, 4th Ed., Duxberry Press, 1975.
64.   Box, G.E.P., Hunter, W.G., and Hunter, J.S., Statistics for Experimenters, John Wiley and Sons, 1978.
65.   Box, G.E.P., and Draper, N.R., Empirical Model Building and Response Surfaces, John Wiley and Sons, 1987.
66.   Box, G.E.P., and Behnken, D.W., Technometrics, p. 455, Vol. 2, 1960.
67.   Yeung, M., Langston, J., and Sparkes, C., SPIE, p. 32, Vol. 565, 1985.
68.   Arnold, W.H., SPIE, p. 94, Vol. 922, 1988.
69.   Rominger, J.P., SPIE, p. 188, Vol. 922, 1988.

70.   Lin, B.J., IEEE Trans. Electron Devices, p. 931, Vol. ED-27, No. 5, 1980.
71.   Rosenbluth, A.E., Goodman, D., and Lin, B.J., J. Vac. Sci. Technol. B, p. 1190, Vol. 1, No. 4, Oct./Dec., 1983.
72.   Bossung, J.W., Proc. Soc. Photo-Optical Instrum. Eng., p. 80, Vol. 100, April, 1977.
73.   Arnold, W.H., and Levinson, H.J., SPIE, p. 21, Vol. 772, 1987.
74.   Meyerhofer, D., SPIE, p. 174, Vol. 922, 1988.
75.   Lis, S.A., Proc. Kodak Microelectronics Seminar, Nov., 1986.
76.   Chien, P., and Chen, M., SPIE, p. 35, Vol. 772, 1987.
77.   Robertson, P.D., Wise, F.W., Nasr, A.N., Neureuther, A.R., and Ting, C.H., SPIE, p. 37, Vol. 334, 1982.
78.   Flanner, P.D., Subramanian, S., and Neureuther, A.R., SPIE, Vol. 633, 1986.
79.   White, L.K., SPIE, p. 239, Vol. 772, 1987.
80.   Horiuchi, T., and Suzuki, M., Symp. on VLSI Technol., Digest of Technical Papers, May,1985.
81.   Wolf, T.M., Fu, C.C., Eisenberg, J.H., and Fritzinger, L.B., Proc. KTI Microelectronics Seminar, p. 335, Nov., 1989.
82.   Mack, C., Proc. KTI Microelectronics Seminar, p. 209, Nov., 1989.
83.   Feldman, M., Wong, G.G., and Cheng, M., J. Vac. Sci. Technol., p. 241, Vol. B5, No. 1, 1987.
84.   Bernard, D.A., IEEE Trans. Semic. Manuf., p. 85, Vol. 1, No. 3, 1988.
85.   Lin, B.J., IEEE Trans. Electron Devices, p. 419, Vol. ED-25, No. 4, 1978.
86.   Lee, W., Davis, R., Miller, R., and McCoy, J., Proc. KTI Microelectronics Seminar, p. 179, Nov., 1989.
87.   Buckley, J.D., Solid State Tech., Jan., 1987.
88.   Chandra, S., and Wu, F.Y., SPIE, p. 86, Vol. 772, 1987.
89.   Suwa, K., and Ushida, K., SPIE, p. 270, Vol. 922, 1988.
90.   Mayer, H.E., and Loebach, E.W., SPIE, p. 9, Vol. 221, 1980.
91.   Buckley, J.D., Galburt, D.N., and Karatzas, C., J. Vac. Sci. Technol., p. 1607, Vol. B7, No. 6, 1989.
92.   Fukuda, H., Hasegawa, N., Tanaka, T., and Hayashida, T., IEEE Trans. Electron Devices, p. 179, Vol. EDL-8, No. 4, 1987.
93.   Hayashida, T., Fukuda, H., Tanaka, T., and Hasegawa, N., SPIE, p. 66, Vol. 772, 1987.
94.   Sugiyama, S., Tawa, T., Oshida, Y., Kurosaki, T., and Mizuno, F., SPIE, p. 318, Vol. 922, 1988.

95. Hale, K., and Luehrmann, P., Proc. Kodak Microelectronics Seminar, Nov., 1986.
96. Edmark, K.W., and Ausschnitt, C.P., SPIE, p. 91, Vol. 538, 1985.
97. Suzuki, A., Yabu, S., and Ookubo, M., SPIE, p. 58, Vol. 772, 1987.
98. Brunner, T.A., Cheng, S., and Norton, A.E., SPIE, p. 366, Vol. 922, 1988.
99. Bouwhuis, G. and Wittekoek, IEEE Trans. Electron Devices, p. 723, Vol. ED-26, No. 4, 1979.
100. Kano, I., SEMI Technol. Symposium '87, Tokyo, Japan, p. 54, Dec.9-10, 1987.
101. Guidlcl, D.C., SPIE, p. 132, Vol. 174, 1979.
102. Liauw, L., Muray, A., and Chen, M., SPIE, p. 232, Vol. 772, 1987.
103. Bettes, T.C., Semiconductor International, p. 83, April, 1982.
104. Peters, D.W., Proc. Kodak Microelectronics Seminar, Nov., 1986.
105. Gear, G., Proc. Kodak Microelectronics Seminar, p. 104, 1985.
106. Elliot, D.J., Proc. KTI Microelectronics Seminar, Nov., 1987.
107. Ruckle, B., Lokai, P., Rosenkranz, H., Nikolaus, B., Kahlert, H.J., Burghardt, B., Basing, D., and Muckenheim, W., SPIE, p. 450, Vol. 922, 1988.
108. Znotins, T.A., McKee, T.J., Gutz, S.J., Tan, K.O., and Norris, W.B., SPIE, p. 454, Vol. 922, 1988.
109. Pol, V., Bennewitz, J.H., Escher, G.C., Feldman, M., Firtion, V.A., Jewell, T.E., Wilcomb,B.E., and Clemens, J.T., SPIE, p. 6, Vol. 633, 1986.
110. Jain, P.K., Neureuther, A.R., and Oldham, W.G., IEEE Trans. Electron Devices, p. 1410, Vol. ED-28, No. 11, 1981.
111. Rothschild, M., and Ehrlich, D.J., SPIE, p. 466, Vol. 922, 1988.
112. Bobroff, N., Rev. Sci. Instrum., p. 1152, Vol. 57, No. 6, 1986.
113a. Tracy, D.H., and Wu, F.Y., SPIE, p. 437, Vol. 922, 1988.
113b. Fehr, D.L., Lovering, H.B., and Scruton, R.T., Proc. KTI Microelectronics Seminar, p. 217, Nov. 1989.
114. Flagello, D.G., and Pomerence, A.T.S., SPIE, p. 6, Vol. 772, 1987.
115. Cuthbert, J.D., Solid State Tech., p. 59, August, 1977.

116. Ohtsuka, H., and Kanamori, J., OKI Technical Review 124, p. 33, July, 1986.
117. Widmann, D.W., and Binder, H., IEEE Trans. Electron Devices, p. 467, Vol. ED-22, No. 7, 1975.
118. White, L.K., Proc. Kodak Microelectronics Seminar, Oct., 1981.
119. van den Berg, H.A.M., and van den Berg, P.M., IEEE Trans. Electron Devices, p. 1535, Vol. ED-28, No. 12, 1981.
120. Hashimoto, T., Yamanaka, H., Iino, T., and Takahashi, S., SPIE, Vol. 920, 1988.
121. Mack, C.A., Solid State Tech., p. 125, January, 1988.
122. Bolsen, M., Buhr, G., Merrem, H.J., and van Werden, K., Solid State Tech., p. 83, Feb., 1986.
123. Brown, A.V., and Arnold, W.H., SPIE, p. 259, Vol. 539, 1985.
124. Garza, C.M., Misium, G.R., and Doering, R.R., SPIE, Vol. 1086, 1989.
125. Lin, B.J., SPIE, p. 114, Vol. 174, 1979.
126. Griffing, B.F., and West, P.R., Electron Dev. Letters, p. 14, Vol. EDL-4, 1983.
127. Brown, T., and Mack, C.A., SPIE, p. 390, Vol. 920, 1988.
128. Spak, M., Mammoto, D., Jain, S., and Durham, D., Proc. 7th Int. Techn. Conf. Photopolymers, 247, 1985.
129. Balch, E.W., Weaver, S.E., and Saia, R.J., SPIE, p. 387, Vol. 922, 1988.
130. Vollenbroek, F.A., and Geomini, M.J.H.J., SPIE, p. 419, Vol. 920, 1988.
131. Walker, E.J., IEEE Trans. Electron Devices, p. 464, Vol. ED-22, No. 7, 1975.
132. Brewer, T., Carlson, R., and Arnold, J., J. Appl. Photog. Eng., p. 184, Vol. 7, No. 6, 1981.
133. Coyne, R.D., and Brewer, T., Proc. Kodak Microelectronics Seminar, Nov., 1983.
134. Lin, Y., Marriott, V., Orvek, K., and Fuller, G., SPIE, p. 30, Vol. 469, 1984.
135. Nolscher, C., Mader, L., and Schneegans, M., SPIE, p. 242, Vol. 1086, 1989.
136. van den Berg, H.A.M., and van Staden, J.B., J. Appl. Phys., p. 1212, Vol. 50, 1979.
137. Burggraaf, P., Semiconductor International, p. 23, Dec., 1987.
138. Lin, B.J., Solid State Technology, p. 63, Jan., 1987.
139. Lin, B.J., SPIE, p. 106, Vol. 1088, 1989.
140. Sommargren, G.E., SPIE, p. 268, Vol. 1088, 1989.

141. Cote, D.R., Lazo-Wasem, J.E., and Rahmlow, T.D., SPIE, Vol. 921, 1988.
142. Perloff, D.S., IEEE Journ. Sol. St. Circ., p. 436, Vol. SC-13, No. 4, 1978.
143. Arnold, W., SPIE, Vol. 394, 1983.
144. van den Brink, M.A., de Mol, C.G.M., and George, R.A., SPIE, p. 180, Vol. 921, 1988.
145. Heflinger, B., SPIE, p. 70, Vol. 334, 1982.
146. Kirk, C.P., SPIE, p. 134, Vol. 772, 1987.
147. Abraham, G., Kirk, J.P., Tibbetts, R.E., and Wilczynski, J.S., IBM Tech. Disclosure Bull.,p. 3417, Vol. 19, No. 9, Feb., 1977.
148. Wilczynski, J.S., J. Vac. Sci. Technol., p. 1929, Vol. 16, No. 6, Nov./Dec., 1979.
149. Beaulieu, D.R., and Hellebrekers, P.P., SPIE, p. 142, Vol. 772, 1987.
150. Kleinknecht, H.P., SPIE, p. 63, Vol. 174, 1979.
151. Trutna, W.R., and Chen, M., SPIE, p. 62, Vol. 470, 1984.
152. Wittekoek, S., Linders, H., Stover, H., Johnson, G., Gallagher, D., and Fergusson, R., SPIE, p. 22, Vol. 565, 1985.
153. Ohtsuka, H., Funatsu, H., Kushibiki, G., and Koikeda, T., SPIE, p. 70, Vol. 470, 1984.
154. Sawoska, D.A., SanGiacomo, K.D., Jacovich, E.C., and Cordes, W.F., SPIE, p. 217, Vol. 922, 1988.
155. Yao, S., SPIE, p. 118, Vol. 772, 1987.
156. Sheldon, D.J., Gruenschlaeger, C.W., Kammerdiner, L., Henis, N.B., Kelleher, P., and Hayden, IEEE Trans. Semic. Manuf., p. 140, Vol. 1, No. 4, 1988.
157. Chivers, K.A., Proc. Kodak Microelectronics Seminar, p. 44, Oct., 1984.
158. Gallatin, G.M., Webster, J.C., Kintner, E.C., and Wu, F., SPIE, p. 193, Vol. 772, 1987.
159. Waldo, W.G., and Helbert, J.N., Motorola Techn. Devel., p. 15, Vol. 9, 1989.
160. Fujiwara, K., Tokui, A., and Uoya, S., Proc. Kodak Microelectronics Seminar, 1985.
161. Suwa, K., Nakazawa, K., and Yoshida, S., Proc. Kodak Microelectronics Seminar, Oct., 1981.
162. MacMillen, D., and Ryden, W.D., SPIE, p. 78, Vol. 334, 1982.
163. Armitage, J.D., SPIE, p. 207, Vol. 921, 1988.
164. Personal communication, D. Young, Motorola.
165. Hershel, R.S., SPIE, p. 54, Vol. 174, 1979.
166. Dyson, J., J. Opt. Soc. Amer., p. 49, Vol. 713, 1959.

167. Stephanakis, A.C., and Rubin, D.I., SPIE, p. 74, Vol. 772, 1987.
168. Tan, R.V., and Ausschnitt, C.P., SPIE, p. 45, Vol. 565, 1985.
169. Schneider, W.C., SPIE, p. 6, Vol. 174, 1979.
170. Cote, D.R., Clayton, R.H., and Lazo-Wasem, J.E., SPIE, p. 124, Vol. 772, 1987.
171. Dunbrack, S.K., and Langston, J.C., Proc. Kodak Microelectronics Seminar, p. 62, Nov., 1983.
172. Cowan, M.J., Mattis, D.G., and Chapman, E.W., Proc. Kodak Microelectronics Seminar, Oct., 1981.
173. U.S. Patent awarded to IBM, "Pellicle Cover for Projection Printing Systems", 1978.
174. Yan, P., and Gaw, H., Proc. KTI Microelectronics Seminar, p. 261, Nov., 1989.
175. Hershel, R., SPIE, p. 23, Vol. 275, 1981.
176. Ward, I.E., and Duly, D.L., Proc. Kodak Microelectronics Seminar, p. 35, Oct., 1984.
177. Partlo, W.N., Oldham, W.G., and Flynn, S., Proc. KTI Microelectronics Seminar, p. 107, Nov., 1989.
178. Brunner, T.A., Ausschnitt, C.P., and Duly, D.L., Solid State Technology, p. 135, May, 1980.
179. Zavecz, T.E., and Banks, E.L., SPIE, p. 224, Vol. 772, 1987.
180. Toh, K.K.H., Fu, C.C., Zollinger, K.L., Neureuther, A.R., and Pease, R.F.W., SPIE, p. 194, Vol. 922, 1988.
181. Levenson, M.D., Viswanathan, and N.S., Simpson, R.A., IEEE Trans. Electron Devices,p. 1828, Vol. ED-29, No. 12, 1982.
182. Prouty, M.D., and Neureuther, A.R., SPIE, p. 228, Vol. 470, 1984.
183. Levenson, M.D., Goodman, D.S., Lindsey, S., Bayer, P.W., and Santini, H.A.E., IEEE Trans. Electron Devices, p. 753, Vol. ED-31, No. 6, 1984.
184. Starikov, A., SPIE, p. 34, Vol. 1088, 1989.
185. Yeh, J.T.C., SPIE, p. 461, Vol. 922, 1988.

5

ELECTRON BEAM PATTERNING AND DIRECT WRITE

Lee Veneklasen

KLA Instruments, Inc.
San Jose, California

1.0   INTRODUCTION

Lithography is the process of transferring patterns from one media to another. In the VLSI era, electron beam lithography (EBL) is likely to be the most common technique for converting circuit designs into actual spatial patterns. EBL is routinely used to generate master masks and reticles from computer generated design data. These masks are then used in light optical projection printers to replicate the patterns on silicon wafers. EBL is also beginning to find application as a "direct writer", which transfers patterns directly onto the wafer.

Two properties make electrons suitable for pattern generation. Due to the short wavelength of the electron, electron optical image resolution is not limited by diffraction effects. Sharp images of submicron patterns may be formed using either full field or scanning techniques. While resolution has been an important motivation for EBL development, electron beams are also well suited for conversion of patterns from numerical to spatial representation, regardless of resolution. Simple patterns may be drawn by hand and photographically reduced. As die patterns become larger and their individual features become smaller, it becomes necessary to establish a direct, high speed link between computer aided design data bases and the spatial pattern. Optical pattern generators use essentially mechanical means, and are often too slow to transfer today's very large data bases. Since electrons are charged particles, they may be electromagnetically deflected. Elements of a pattern may be both formed and positioned at high speed, so electrons are particularly favored for primary generation of submicron patterns.

While electrons seem ideal for imaging patterns, they are not always ideal for exposing the resists used to record patterns.

Their low mass leads to relatively short and variable penetration with significant lateral scattering. Exposure mechanisms are not always efficient; sometimes leading to high dose requirements and heating phenomenon. The use of high energy charged particles can sometimes damage underlying circuit elements. With its advantages and disadvantages, EBL will play an important role in the VLSI era.

EBL has a dual role. As a generator of masks for optical replication, it must operate with resolution and accuracy that exceeds the printers it supports. As a direct writing vehicle, EBL must process patterns at a rate that competes with alternative submicron replication techniques. To answer these challenges, several EBL "writing strategies" have been developed, and many technological "tactics" have been developed to support the strategies.

This chapter will concentrate upon strategies for optimizing the electron beam lithographic process. By studying these strategies and their fundamental limitations, it becomes easier to use machines more effectively, and to associate different designs with favorable applications. It also becomes possible to decide which technologies will enhance the overall process. There is no universally optimum strategy or machine design for all applications. Instead, there are many application dependent trade-offs that will seem to defy simplification.

It is assumed that the reader is primarily interested in the end result, and how to improve it. Accordingly, this chapter tries to treat EBL as a process rather than machine oriented subject. The first section defines the process and outlines quality criteria that machine designers and users should agree upon. The main features of three basic writing strategies are described and some of the advantages of each are pointed out. Particular attention is paid to the matching of machine strategies with resist properties. The middle section of this chapter is a fairly detailed presentation that attempts to mathematically interconnect the large number of parameters involved in the process. It is hoped that this will interest those who must optimize conditions for a particular process. This section is necessarily complicated and may be omitted if a more general overview is desired.

2.0   THE EBL PROCESS

The semiconductor lithography process begins with a resist coated mask or wafer, and a pattern to be exposed on the resist. The pattern is usually in the form of a CAD data base describing polygons, their locations within dies, and the locations of dies on a wafer. Usually, this compact, standardized representation is first expanded into a machine specific data base. In the expanded representation, polygons are "fractured" into "figures" whose size, type, and boundaries conform to standards that the machine imposes. When the pattern is prepared, the mask or wafer is loaded into a vacuum system and placed upon a "stage" whose location is precisely monitored by interferometry. The machine is automatically calibrated against this distance standard, and pattern exposure begins. When the exposure is complete, the mask or wafer is extracted, and the resist is processed. This leaves either the exposed or unexposed regions of the pattern free of resist coating. Then the pattern is transferred to the substrate by etching, coating, implantation, or other techniques that depend upon the nature of the application.

From the lithographic viewpoint, the EBL process begins with the design data, the substrate and the resist, and ends with the patterned layer of resist upon the substrate. The transitions between the resist coated and clear areas must meet certain conditions of placement, thickness and profile and in turn depend upon previous and subsequent steps in the fabrication process.

2.1 Feature Definition

The three dimensional structure of a resist pattern is important. The length and width of a feature in the resist layer tends to determine the size of fabricated structures. The step height and cross section also influence resist-to-substrate transfer techniques such as etching and liftoff. Feature definition is often discussed in terms of critical dimension control. The term refers to the degree with which the size of the smallest features may be verniered and consistently reproduced. Usually, control of the smallest features is both most important and most difficult because small variations have proportionally large effects. For a process of limited resolution, the smallest features receive the least dose modulation, which leads to difficulty maintaining desired feature profiles in the face of small variations in exposure and process conditions.

Lithographers speak of a "process resolution" as the minimum dimensions of a feature that can be transferred according to certain size error tolerances and process latitudes. Resolution thus refers to the smallest resist structures that are free of unwanted residual resist and have desirable profile. Definition of resolution is necessarily vague and application dependent.

Another important criteria is the roughness of nominally straight edges, and the radius of nominally square corners. Rough edges and round corners, as well as unwanted resist residue, can cause imperfect pattern transfer and in turn, cause open and short circuits, hot spots, and variations in device performance.

Resist thickness tends to be dictated by three dimensional step coverage, desired defect densities, etch resistance and liftoff requirements. Because of electron scattering and penetration effects, lithographic resolution tends to favor thin resist layers, while many pattern transfer processes favor thick layers.

When a contiguous pattern is formed from interlocking figures with artificial boundaries, a "figure butting" requirement is imposed. Nominally contiguous figures must join smoothly without sheer, overlaps or gaps. For submicron lithography, corner radius, size error, edge roughness and figure butting requirements are on the scale of 0.1 μm or below.

2.2   Figure Placement Accuracy

Placement accuracy refers to the location tolerance of the center of a figure with respect to the origin of a die, reticle, or full mask pattern. While the distinction between the size and location of a figure may seem artificial, the two often depend upon different aspects of the process. Figure size can depend upon resist, process, and other scale errors, while placement tends to depend upon beam deflection, stage metrology, registration, substrate flatness, and thermal expansion effects.

A distinction is usually made between absolute and overlay accuracy. Absolute accuracy refers to the position of a figure with respect to an imaginary absolute standard rectilinear grid. This standard is invariant with respect to all conditions and types of machines. Absolute accuracy ultimately measures the repeatability and consistency of feature locations when different technologies are used to perform lithographic steps without cross-calibration. Absolute accuracy is particularly important for mask and reticle writing, where the replicating machines introduce further local distortions.

Overlay errors traditionally refer to the relative placement accuracy of figures when the same machine is used to pattern subsequent layers of a circuit. In most multilevel fabrication a die pattern is "registered" with respect to a "fiducial mark" made upon a previous level. Overlay thus refers to a machine's ability to print different die patterns that are properly registered to each other everywhere within each die. The term "machine to machine" overlay accuracy refers to overlay of levels made on different machines. Within a single die, this term tends to be synonymous with absolute accuracy because both machines require a common measurement standard.

Modern lithography demands both absolute and overlay accuracy within a small fraction of a figure dimension. This requirement is common to all strategies and rates of writing. Specifications of 0.05 - 0.15 μm are typical. As wafers and dies become larger, as writing rates increase, and as features become smaller, accuracy is a continuing technical challenge.

2.3   Throughput and Area Coverage Time

Provided feature definition and placement accuracy criteria are met, the next three quality criteria measure the economic viability of the lithographic process. It would be reasonable to define throughput as the minimum elapsed time to pass a wafer through the lithography process. By this definition, one would include resist/wafer preparation, pre- and post-bakes, resist development, and possibly an amortized overhead involving pattern fracture or mask fabrication. In EBL terminology, throughput usually refers to the load, calibrate, write, and unload cycle of the exposure station. This chapter will also discuss an area coverage time T (sec/cm^2) to actually expose a unit area of pattern. In some situations, wafer handling, self-calibration, and data handling overheads instead of area coverage actually dominate throughput budgets.

The area coverage time is highly dependent upon the writing strategy as well as the exact nature of the pattern. For pattern generators, coverage time tends to correlate with the complexity (information content) of a pattern. This is ultimately true because of basic digital and analog noise/bandwidth limitations.

2.4   Defect Density

A defect is an unwanted lithographic feature that adversely affects the performance of the circuit being fabricated. Insofar as a "fatal defect" can render an entire die or reticle unusable, defects play a significant role in any economics based definition of throughput. Defects are frequently measured in terms of percentage yield of functional circuits. This definition does not isolate their causes to specific fabrication steps. Lithographic defects can be measured and classified by optical or scanning microscope inspection but these techniques are presently too slow to qualify each production wafer.

Lithographic defects can originate in all steps of the process. Causes include particulate contamination, anisoptropic resist or substrate chemistry, and digital or analog noise in pattern generation. As feature sizes decrease, one may expect the size of fatal defects to decrease, and their probability per unit circuit area to increase. Defect reduction is therefore another continuing challenge.

2.5   Automation and Process Integration

An ideal lithography system would probably include integrated resist application, pre- and post-bake, development and quality monitoring capability along with pattern fracture and exposure stations. To date, such an integrated process is unavailable, possibly due to the variety of resist processes, patterns, and operating environments. The exposure station alone requires a high level of automation to operate effectively. Calibration, mechanical and environmental controls act as semi-independent subsystems during exposure. Operator skill requirements tend to distinguish research from production equipment.

2.6   EBL Applications

The quality criteria discussed above apply to lithography for all types of patterns using all types of radiation. One might ask what are the lithographic tasks that particularly favor electron beam lithography. Electrons are fairly easy to produce, to focus, and particularly to deflect at high rates of speed. Electromagnetic deflectability suggests that electrons are particularly suitable for primary pattern generation. Digitally stored data can be converted to analog signals that move the beam according to pattern requirements. Much of the EBL technology evolved directly from a synthesis of CRT, electron microscopy and computer technologies. Early EBL was performed on computer controlled scanning electron microscopes, which soon became tools for high resolution device development and mask fabrication. Most microcircuit production continues to use light optical projection aligners and steppers for wafer lithography, but relies on E-beam generated masks and reticles.

There thus seem to be two basic factors driving direct write applications: resolution and flexibility for maskless patterning. EBL is being applied particularly in the areas of application specific (ASIC) and low volume submicron VLSI wafer fabrication. In these areas, feature size, overlay, or production volume do not always favor masked lithography. Direct write is particularly advantageous during circuit development phases, where a need for masks would slow down iterative design cycles. Flexibility is equally advantageous for ASIC designs, regardless of resolution. Personalized ASICs or gate arrays would otherwise need many unique masks.

Electron beam direct write pattern generators are unlikely to be able to expose area as fast as competing masked techniques. By analogy, one cannot compare a typist and word processor with a copying machine because both are needed to complete a document. In general, the range of EBL application will be determined by the technical and economic constraints of the complete lithographic process rather than by exposure throughput alone.

A third factor influencing the range of EBL application is its capability for metrology, or measurement of lithographic quality. Spot beams can be used to locate and image features on a mask or wafer. Using secondary or backscattered electron images, this can be done at resolution exceeding accuracy specifications. Calibration, registration and self-verification of placement accuracy all rely on this capability. When combined with interferometric position measurements, imaging and scanning techniques are an effective tool for measuring some of the quality parameters.

Electrons are not the optimum choice for all lithography. Where masks are useable, X-rays, ions and possibly very short wavelength light may offer superior penetration and profiles, at comparable resolution. When pattern data can be stored in a mask, it is unnecessary to transmit data during exposure of each pattern. At best, this increases coverage times. It seems likely that several processes will be commonly used, and that different circuit levels will favor different processes.

2.7 Classification of EBL Equipment

2.7.1 Mask and Reticle Makers. One class of EBL pattern generators is specifically dedicated to mask generation. They support masked light optical, X-ray, synchrotron, ion and electron projection lithography. Design effort primarily concentrates upon feature position accuracy and efficient, flexible pattern processing. Accuracy over relatively large distances is important, so that full field mask sets overlay between circuit levels. Overlay is obtained by critical control of absolute accuracy, using self-calibration and environmental control. Feature size and position accuracy is relatively less demanding for reticles that are demagnified, but the need for 1X X-ray and full field optical masks continue to stress both accuracy and feature size capabilities. Accuracy and flexibility are weighted above speed in mask making. Mask makers also tend to use a limited number of sensitive and mature resist processes. Defect prevention, detection, and repair all play major roles in the mask making process, because defects are transferred to each wafer patterned with the defective mask.

2.7.2 Electron Beam Replicators. The distinction between pattern generators and masked replicators has already been discussed. This class of EBL system uses transmission masks, forming an image or shadow upon a wafer surface. In comparison with similar light optical steppers, these systems may offer higher image resolution and comparable throughput. Application of these systems is not yet widespread, possibly because X-rays and deep UV steppers offer attractive alternatives for a similar range of application.

2.7.3 Direct Write Pattern Generators. These systems all write directly on substrates. Based on a combination of image resolution, throughput and degree of automation, they can be loosely classified as research, developmental, or manufacturing designs. Research equipment typically provides small 5 - 50 nm point beams that move over small fields. They accommodate all varieties of substrates from silicon to thin films. To write high resolution patterns using modest pattern electronics and deflection technology, these research machines sacrifice coverage times, but this is not usually a limitation for fabrication of isolated devices and test circuits. This equipment pioneers device technology.

Developmental E-beam systems pattern VLSIs in modest volume. All quality criteria except throughput are emphasized in order to support the fabrication of prototypes. Regardless of the eventual lithographic technique chosen for later chip manufacture, these systems are particularly important as a direct design data to lithography interface. They also set the standards eventually expected of production equipment. Many different machines for in-house prototyping have been designed by semiconductor manufacturers as well as E-beam equipment companies.

Production direct write equipment is intended to produce commercial volumes of IC levels in production environments. Pattern accuracy and feature quality compete with manufacturing economics, emphasizing process throughput as well as accuracy. Reliability and automation become important factors, and level-to-level mixing of lithographic technologies is to be expected.   At present, optical lithographic processes are favored when viable, but production of direct write is beginning to perform tasks that are otherwise not possible.   Production resist processes are evolving simultaneously.

Because production of direct write combines accuracy and speed requirements, these machines are on the limit of both strategy and subsystem technology.   Production throughput particularly drives the evolution of writing strategies, and is forcing their continuing re-evaluation in the context of the complete process.

3.0   EBL STRATEGIES

The term "writing strategy" refers to the way patterns are exposed on a resist.   As a guideline, it should be possible to infer a system's writing strategy only by observing activity at the surface of the resist.   Thus a study of strategy refers to what happens rather than exactly how it is made to happen.

Features are usually located on a mask or wafer as a result of combined stage and beam motions.   Beam motion can be precisely monitored.   Since the goal is to place features with respect to a fixed origin on the wafer, all activity is ultimately calibrated with respect to stage motions.   EBL writing strategies do not require that the stage be stationary or even accurately positioned.   What is required is that stage position be precisely known as it moves about, so that errors in mechanical placement may be compensated by beam motion.   Most EBL strategies employ this trick, and write patterns in small interlocked pieces as though they operated in a quasi-stationary coordinate system.

3.1   Strategies for E-Beam Pattern Replication

An electron beam pattern replicator is functionally similar to its optical or X-ray counterpart.   Patterns are exposed using single or complementary sets of transparent stencil masks that store the pattern information.   An entire die or wafer pattern can be exposed simultaneously by flood beam.   These systems use a "step and repeat" stage motion that exposes one or more dies per step.   The mask is usually a single die reticle, and the stage comes to rest during each exposure.   Interspersed with stage moves is a registration sequence that determines the position, spacing and rotation of the exposure field.   The orientation and focus of the field can be corrected by electronic or mechanical means on a die to die basis.

Several kinds of replicators have been developed. One kind of system uses a back illuminated photocathode mask(1). Photoelectrons emerging from sensitized regions of the mask are accelerated and focused at 1X magnification using a uniform magnetic field. The system exploits the higher resolution of electron imaging, but is otherwise subject to mask fabrication, contamination and registration limitations common to light optical steppers.

Several demagnifying electron beam steppers have been developed(2)(3). These use one or more reticle masks. Since the reticles must be transparent to electrons, various means of forming isolated opaque regions on the mask have been proposed(3), ranging from complementary stencil patterns to grid arrays. With these systems, it is possible to vernier position, magnification, rotation, and focus by electron optical means. Very wide field, low distortion optics are necessary, but high beam currents and fast coverage times are possible. Unique problems include alignment techniques and thermal stability of reticles.

A scanning proximity printer has also been developed(4). This system combines some of the throughput advantages of masked lithography with the calibration and interferometer referenced correction capability of pattern generators. Complementary stencil reticles are spaced one die period apart, and the composite reticle is placed quite close to the wafer surface. A small one mm scale flood beam is scanned over both reticles such that the die pair is illuminated at nominally perpendicular incidence. Offset and scale errors, rotations, and even static and time dependent mask distortions can be corrected by applying small tilts to the illuminating beam as the scan progresses across the mask. The exposed pattern is the shadow of the features on both complementary masks.

This strategy has the advantage of allowing high resolution registration, calibration and self-diagnostics as well as imaging. Grating fiducial marks and holes to form a point beam may be included on the mask. Rocking the illumination allows the use of scanned beam techniques using absorbed wafer current as a video output. Particular challenges include defect free reticle fabrication and handling, and correction of heating effects during scanning of the beam.

Masked EBL is potentially a very fast strategy for submicron lithography because it combines the inherent resolution of electrons with parallel rather than serial exposure of features. It is also inherently a two step process involving the fabrication of unique, special purpose reticles or masks.   Its resolution, accuracy and economy all depend upon the details of the mask process.   It relies on the availability of robust, defect free masks.   One might speculate that electron beam as well as optical, X-ray, and ion projection techniques will find high volume production applications, and that they will rely upon the EBL pattern generation process to provide masks.

3.2 Strategies for E-Beam Pattern Generation

Pattern generators are characterized by the direct transfer of CAD design data to a resist pattern.   Real time processing of large volumes of pattern data is implied.   Pattern generators also include a built-in standard of position, allowing features to be positioned exactly according to vector locations in a data base.   Usually the mask or wafer is rigidly attached to a pair of perpendicular mirror surfaces, which allow any point on the substrate to be positioned with respect to a fixed electron optical axis. The measurement is performed with a laser interferometer with an accuracy of about .01 µm.   Provided the wafer to mirror connection, the laser beam path, the interferometer-to-optics connection, and the optics-to-beam connection are all mechanically and thermally rigid, the position of the beam upon the substrate is precisely known.   To the extent that mechanical positioning errors are detected, these errors may be corrected by deflecting the beam.   Similarly, deflections may be calibrated by matching them to known stage translations.   Thus, in EBL pattern generation each feature position refers back to interferometric distance standards.

There are two common stage motion strategies.   The step-and-repeat strategy moves the stage to a fixed location, corrects small residual errors, and then exposes a "field" of the pattern with beam motion alone.   The "write-on-the-fly" strategy writes while the stage is in motion, and usually modulates its velocity according to fluctuations in pattern density.   Write-on-the-fly strategy depends heavily upon dynamic position error correction, but allows relatively small fields and modest accelerations.   Step-and-repeat strategy implies high accelerations, and fundamentally requires non-writing overhead time to execute steps, but it minimizes calibration dependent position error correction.   Most high speed pattern generators use variations of write-on-the-fly stage strategy to maximize throughput.

Electromagnetic and electrostatic deflection is used to position the beam while writing pattern detail.   The strategies used for this deflection determine both the electron optical design and the data structure used in different systems. Three basic strategies have evolved for writing patterns.

3.2.1 Gaussian Beam Raster Scan (Figure 1). This strategy grew out of television technology and scanning microscopy. The standard raster scan strategy was pioneered by the Bell Labs EBES system(5). The beam is repeatedly scanned along a line while the stage is moved perpendicular to the scan. The resulting area covered is a long narrow stripe that includes segments of die patterns on a wafer. To form the black and white pattern, the beam is turned on and off according to the location of individual pixels along the raster trajectory. Each pixel is defined as a square area equal to the spacing between scans in the raster, so that an arbitrary pattern can be recorded upon an address grid of regular pixel spacings.

The raster scan strategy requires a point beam no larger than the desired resolution of the pattern, because each pixel is equally exposed or not exposed, completely independent of its height. A beam diameter of 1/4 - 1/10 of a minimum feature size is usually considered sufficient for definition of the exposure profile. Raster scan electron optics is similar to a scanning microscope, which demagnifies a crossover formed in the electron gun. The current in the beam is determined by the demagnification and by axial aberrations of the imaging lenses. The optics differ from SEM in that they can accurately scan wider fields, and "blank" the beam on and off in synchronization with the scan.

The raster scan strategy necessarily requires that pattern data be reduced to a bit map that is organized serially along the raster scan trajectory. CAD polygon representations are usually first reduced to primitive figures that conform to stripe and die boundaries. In this reasonably compact form, they are presented to the exposure system which reduces the intermediate format to a bit map in real time. This data handling strategy is dictated by the enormous volume of data supplied to the blanker. [Up to 10^{12} bits on a 0.1 μm grid].

The throughput of a raster scan system is controlled by two factors: the time between bits transmitted to the beam blanker, and the exposure time that the beam must dwell upon each pixel. Under different conditions, either factor can limit writing speed. Given a sensitive resist and bright electron gun, the fundamental limit is the bit rate and timing jitter that digital logic can support. For example, at 100 MHz, 10^8 pixels per second can be exposed. If a pattern requires a 0.1 μm grid to properly locate all feature boundaries, it takes 100 seconds to cover each square cm of the pattern.

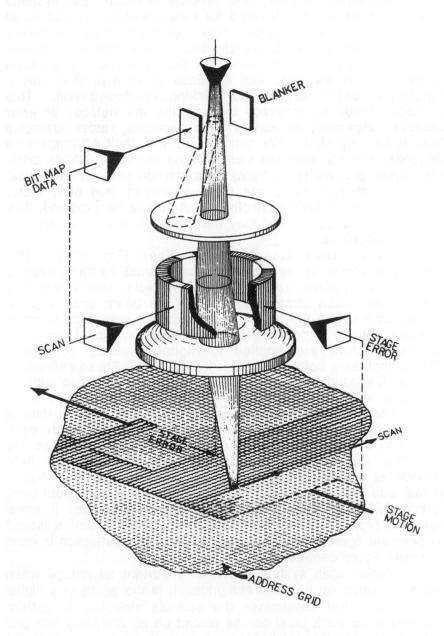

Figure 1. Raster scan pattern generation strategy.

The basic simplicity and flexibility of raster scan systems make them particularly favored for mask making application, at least if past history is a guide. This is perhaps because accuracy rather than throughput is usually the primary concern. Raster scan is relatively simple because only the stage and a single level of deflection is involved, and the dose of each pixel is only a function of beam current and blanking synchronization. This simplicity leads to accuracy by reducing the number of error sources. However, as feature sizes decrease, raster strategies tend to be very slow. For constant bit rate, their coverage time depends inversely upon the square of the pixel size. While faster electronics and higher capacity data storage can partially offset this disadvantage, there are other strategies that offer faster coverage using today's technologies. As might be expected, they are more complicated, and they have taken longer to evolve into accurate machines.

3.2.2 Fixed Beam Vector Scan (Figure 2). This strategy is similar to raster scan in that a small beam is used to expose pattern pixels serially. The term vector scan refers to a different deflection strategy for placing the beam upon a pixel. Instead of blanking the beam over unexposed areas, the beam is directed only to exposed areas. To do this, digital words or position vectors are applied to digital-to-analog converters (DACs) that drive both X and Y deflection axes. These deflections are superimposed upon stage position error corrections to address any point within a writing field.

For the fixed beam vector (FBV) strategy, pattern data is reduced to a list of position vector words that specify where each pixel is to be exposed. Pattern data is usually pre-processed into primitive non-overlapping figures that conform to field boundaries but do not specify each vector location within a figure. A real time "figure generator" is used to step the beam from point to point within the figure according to hardwired recipes. In some systems, high speed, smaller field electrostatic deflection issued to generate figures while lower speed wide field deflection is used to locate figure centers in the writing field.

Vector scan systems have an important advantage when writing patterns on fine address grids. It is the <u>width</u> of a digital vector word that determines the address structure of pattern placement, so each pixel can be placed on an arbitrarily fine grid without requiring that the beam move to <u>each</u> exposed pixel on the grid. Early FBV systems did not exploit this opportunity, and required many bits to expose each pixel in a pattern. In comparison with raster scan, more data had to be generated for

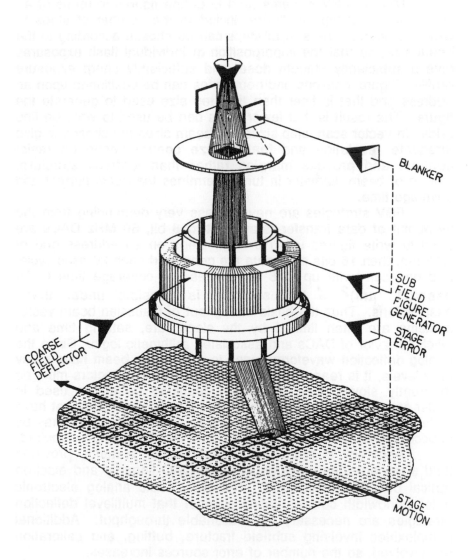

Figure 2. Vector scan strategy using a beam formed by the image of a Gaussian crossover or fixed aperture.

each beam move, and coverage times were much slower than for raster scan strategies.

Modern FBV systems tend to define figures in terms of an origin, size and type of figure, including the number of steps to write the figure. The size of steps can be chosen according to the beam size, so that the superposition of individual flash exposures give a sufficiently smooth dose, and sufficiently sharp exposure profile. Figure locations and boundaries can be positioned upon an address grid that is finer than the step size used to generate the figure. The result is that fewer steps can be used to write on fine grids. In vector scan, step sizes and beam sizes are chosen to give adequate resolution and feature size control based on resist contrast and process margins rather than address structure. Choice of beam diameter in turn determines the beam current and coverage time.

FBV strategies are nevertheless very demanding from the viewpoint of data transfer rates. If two 8 bit, 50 MHz DACs are used to write figures using 0.1 μm steps on an address grid of 1/32 μm, then 16 bits of of data are generated each 20 nsec cycle, and one can write up to 8 μm figures. A coverage time of 20 nsec/(.1 μm)^2 = 200 sec/cm^2 is possible under these assumptions. Thus the basic throughput of Gaussian beam vector systems are often limited by the clock rate, settling time and glitch behavior of DACs and associated arithmetic logic. Since the analog deflection waveforms used to deflect the beam have many grey levels, it is reasonable to expect that FBV deflectors must be inherently slower than the binary blanking deflectors used in raster scan strategy. However, potential throughput does not have to fall quadratically with address structure, so vector scan may be expected to become more favorable as finer addressing is required.

Accuracy is potentially high in these systems, provided field sizes are small enough to avoid DAC linearity and electron optical distortion limitations. However, basic analog electronic noise bandwidth considerations suggest that multilevel deflection strategies are necessary for reasonable throughput. Additional complexities involving subfield fracture, butting, and calibration are involved, so the number of error sources increases.

Both raster and FBV strategies also tend to favor sensitive resists because of electron optical limitations. High currents must be focused into small spots. For example, to sustain 200 sec/cm^2 coverage with 0.1 μm exposure edge profiles, one must concentrate 5 nA per μC/cm^2 of resist dose into a 0.1 μm beam diameter. For sensitive resists, this is possible using modern thermionic electron guns, but higher dose resists demand more advanced field emission technology to sustain coverage times.

Fixed beam vector strategy has widespread applications in research and developmental direct write systems, particularly in high resolution development applications. Such systems have been historically favored for exploring the limits of device dimensions. They will probably play an important role in submicron mask generation, for which sensitive resists are available.

3.2.3 Variable Shaped Beam Vector Scan (Figure 3). A shaped beam strategy departs from fixed beam strategy by exposing many pixels of a figure simultaneously. Provided most features in a pattern are composed of many contiguous resolution or address elements, this leads to faster coverage times. For example, a checkerboard pattern can be composed faster by exposing square areas, while a dot array must necessarily expose each dot separately. In a shaped beam system, each figure is further fractured into smaller shaped "flashes" that interlock to dose the figure uniformly. If the electron optics can form these shaped beams, e.g., variable sized rectangles, triangles, etc., then many pixels of a figure can be exposed simultaneously without compromising the ability to place figure edges on an arbitrarily fine grid.

The higher throughput potential of the variable shaped beam (VSB) strategy(6)(7)(8) arises from two characteristics. If a larger area can be exposed before moving the beam, then it can expose more area in a minimum dwell time. This advantage applies when electronic DAC and data processing limitations predominate. The second advantage arises from the fact that gun brightness and optical aberrations tend to limit the current that can be placed within a resolution element rather than an extended image, so it is possible to place more current in a shaped beam. This advantage applies when using higher dose resists. In this case, the resist dose divided by available current is the factor that limits coverage time. For these reasons, VSB strategy is favored for fastest coverage of rectilinear patterns on less sensitive resists.

VSB strategies employ several levels of beam deflection to locate and shape the beam. In addition to the interferometer, stage and electromagnetic tracking system, a fairly slow wide field coarse deflection is used to locate the center of smaller minor "subfields". Usually the subfield deflection is used to position the center of figures, and also to step the shaped beam to points within the figure. Some systems separate subfield and figure generation deflection functions. Another level of deflection is necessary to

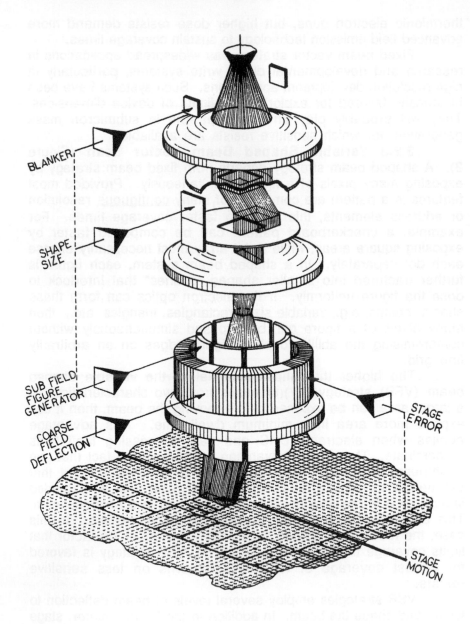

Figure 3.   Variable shaped beam vector scan strategy. Shape sizes are varied to conform to feature dimensions.

blank the beam and to vary the size and/or type of shape. This multilevel deflection hierarchy allows independent optimization of noise/bandwidth and dynamic range for each level. Each level is calibrated against stage motions so that ideally, all activity occurs on a common address grid and all fields overlay precisely.

Accuracy potential is also high in VSB systems, but calibration is demanding because of the many additive sources of placement error. In addition to several levels of position calibrations, shape size and dose calibration is also necessary. Both the size and net area of shapes are particularly important to control if they are to properly interlock and dose the figure. Since variably shaped areas imply variable current within the shapes, small shapes can involve large dose error unless area is well calibrated.

VSB systems fracture CAD design data according to the deflection hierarchy described above. Continuous polygons are fractured along artificial field boundaries where figures must be butted together, and figures are ordered to obtain efficient stage and coarse deflection activity. The machine specific intermediate format is usually an ordered list of figure words. These describe the flash size and dwell time, the figure subfield origin, and the nature of the step trajectory to be used to compose the figure from a single shape. From this compacted data, a real time "figure generator" creates the sequence of subfield deflection vectors that step the shape within a figure.

VSB pattern fracture appears to be more difficult than for raster strategy, but it also offers more flexibility to high resolution lithography. Raster scan bit maps allow very simple overlap removal, feature size scaling, contrast reversal and other techniques used to tune the lithographic process. Vector strategy retains individual polygon representation throughout, making it inherently difficult to establish relationships between features. Trigonometric relationships are involved at each of many subfield boundaries. On the other hand, raster strategy is unfriendly to proximity correction, i.e., the technique of varying dose to compensate for electron scattering. This is because a uniform scan forces each pixel to be equally exposed. The proximity correction technique seems to be necessary for submicron direct write applications that require thick resists, while micron scale masks using thin resists need less correction. This is another important distinction to consider when matching strategy with application. In general, vector shaped beam strategy tends to tradeoff fracture time and effort for higher throughput potential in production applications.

Shaped beam systems involve a rather different set of electron optical challenges. Rather than demagnifying a source, the optics must image a shape with comparable resolution and higher total current. The basic goal is to provide maximum current density into a shape that is comparable to the minimum feature sizes, while maintaining exposure edge slope comparable to desired resolution and corner radius. As will be discussed later, it is not necessary to concentrate as much current into each resolution element because many are exposed simultaneously. This shifts the electron optical emphasis from gun brightness to illumination uniformity and minimization of beam interactions. In general, VSB optics are more complex but contain less exotic technology.

Coverage times are difficult to specify for VSB systems because they depend heavily upon the details of the pattern and resist. For dot arrays at high resolution, they can be no faster than other strategies, but for more contiguous circuit patterns, parallel pixel exposure is always faster if subject to comparable technological limitations. Area coverage times may be estimated if average allowed shape (flash) size, current density, minimum flash cycle time, and dose requirement are all known. For example, if 1 μm^2 area flashes can be used to expose a submicron pattern using 100 nsec dwell at 100 A/cm^2 current density, then coverage times of 10 sec/cm^2 can be achieved using 10 mC/cm^2 resist. This kind of calculation leads to throughput estimates in the range of 5 - 10 wafer levels per hour.

There are several VSB strategy systems addressed to development and direct write production applications. IBM(9) pioneered the strategy for in-house development, and several manufacturers now offer systems(10) with varying feature size, accuracy, throughput and resist process targets.

3.3 Multi-beam and Hybrid Strategies

Throughput potential of the strategies above can be related to the degree of parallelism. Masked replicators are fast because they expose many features simultaneously. Raster and fixed beam vector strategy is slower because it is only possible to expose one pattern element at a time. Shaped beams are faster because many elements of single features can be exposed simultaneously. It is also possible to obtain parallelism using multiple beams. These can be controlled independently to write different parts of the same die pattern, or controlled in parallel to write identical patterns at several die sites. It is also possible to conceive of a system that writes identical patterns on several wafers. In these multiple beam systems, raster, vector or VSB strategies are used to write features with each individual beam.

Several prototype multi-beam systems have been developed. Veeco Corp. devised a VSB system that shapes and deflects the beam in a single beam column, but accelerates and images the shape onto multiple chip sites using an array of electrostatic lenses(11). Microbit Corp. has developed small independent electrostatic columns that each use an array of electrostatic "fly's eye" lenses(12). Another proposal uses an array of beamlets within a single lens to expose adjacent flash sites, minimizing beam interactions by separating the ray paths for the beam(13).

In multiple beam systems, coverage times decrease in proportion to the number of beams, provided that individual beam currents, and electronics-limited flash times are maintained. In the past, this has proved difficult because optical aberrations tend to increase in small bore, low excitation lenses. To date, multiple wafer systems have not been tried. There is room for much innovation.

A hybrid strategy might be defined as one that uses different strategies in different levels of deflection. For example, raster figure generation is not incompatible with vector figure location on finer grids. Shaped beams can also be formed by high speed analog deflection, or by dot matrix blankers or cathodes. There are probably many new combinations to explore, based on quantitative evaluation of their reward.

3.4 Summary

This section reviews EBL writing strategies from the viewpoint of an observer watching the activity on the wafer. They are classified according to their capability to rapidly transfer pattern data to resist. Accuracy is an ongoing quest for all strategies, and the difficulty in achieving accuracy is loosely related to the number of positioning variables and their dynamic range and bandwidth requirements. The accuracy and throughput potential trade-offs associated with various strategies are generally correlated with their application.

Coverage times are shown to be associated with the degree of parallel vs. serial exposure of pattern elements. The concept of a flash as a single size and position state of the beam is developed. Strategies are characterized by the number of resolution elements that one flash exposes simultaneously. This tends to establish electronic and electron optical technological requirements.

Replication strategies are very fast because they expose up to one die or wafer per flash, while pattern generators are slower because they expose much smaller areas in a serial manner. Pattern generator coverage times are limited either by available current and resist dose requirements (sensitivity/current = sec/cm^2) or by electronics-limited flash time per flash area (also sec/cm^2). The following section develops a quantitative formulation of writing strategy.

4.0   THEORY OF WRITING STRATEGY

This section will attempt to identify and associate the many parameters involved in the lithography process. There are three general groups of parameters involved: pattern variables, resist variables, and machine variables. They combine to determine the lithographic quality criteria discussed in Chapter 1. Depending upon the reader's viewpoint, some variables are given and some may be controlled, but their interplay is important to optimize the entire process.

4.1   Throughput

In the context of this discussion, throughput is measured by the time to complete the patterning of a single wafer level using a single beam patterning machine. This time has two components: writing time and non-writing overhead. Non-writing overhead includes mechanical wafer handling and vacuum activity, registration, and machine self-calibration. These overheads are difficult to discuss because they depend upon machine details. They will not be covered in detail even though non-writing overheads can exceed writing times for some systems. As feature sizes decrease, writing time for pattern generation must increase unless new technology is brought to bear, so this discussion will focus upon writing time.

Writing time may be further separated into exposure time and writing overhead time. To disassociate wafer and die sizes, writing time will be discussed in terms of "area coverage time" T in units of sec/cm^2 of pattern, where writing time $T = T_E + T_{WO}$ (exposure + write overhead). The total time to cycle a wafer of coverage area A through a system is:

Cycle time = non-write overhead + $A(T_E + T_{WO})$,   (sec);
Throughput = 3600 sec/cycle time,   (wafer levels/hr).

4.2   Pattern and Dose Parameters

Figures 4a,b identify the parameters used to discuss writing strategies. In the figure, the wafer is covered with individual dies each containing "patterns". The total area of all dies is A (cm^2), where a percentage c(%), of the area is exposed to form a pattern. Patterns are composed of "figures" whose edges lie upon a regular address grid of period Δ (μm). The origin of the grid may be shifted by die-to-die registration, but remains fixed with a die. Each figure with index i is composed of flashes of area a_i(μm) whose centers are defined by the vector or raster deflection used to write the figure, and is not necessarily equal to the beam area a_i (μm^2) used to expose each flash. The subfield size γ_{sf} (μm) is the size of the deflection field within which figures may be written by <u>vector</u> (i.e., pattern controlled) deflection. If a coarse deflection is also used, its field is γ_{cf}.

Figure 4 shows one possible way of composing a figure from individual flashes. In this analysis, it is assumed that all flashes and dwell times are identical within a figure. The figure can be uniquely described by a data word that is either a bit map or vector representation of the figure and its necessary exposure. In practice this tends to constrain figures to rectangles, parallelograms, etc., although more complicated possibilities may be considered.

Figure 4b analyzes the dose of an individual flash. Figure doses may be considered a linear superposition of flashes. Each flash is exposed by a beam of area a_{bi} which may be Gaussian or a flat-topped shaped beam. In general, the current density profile of a flash J(x,y) (Amps/cm^2) is the convolution of an ideally shaped image with an electron optical point spread function of width δ_b. Due to optical imaging imperfections, shaped images have diffuse edges of width δ_b.

As the beam enters the resist, scattering and other non-local resist effects cause the actual dose to differ from the incident exposure. The incident exposure J is further convolved with a resist point spread function of width δ_r, resulting in a dose distribution D(x,y) whose edges are even more diffuse. It is this three dimensional dose distribution and the solubility vs. dose characteristics of the resist that actually determine the developed resist pattern(14).

In order to include possible thermal, charging and resist scattering effects, the flash dose profile D(x,y) is explicitly

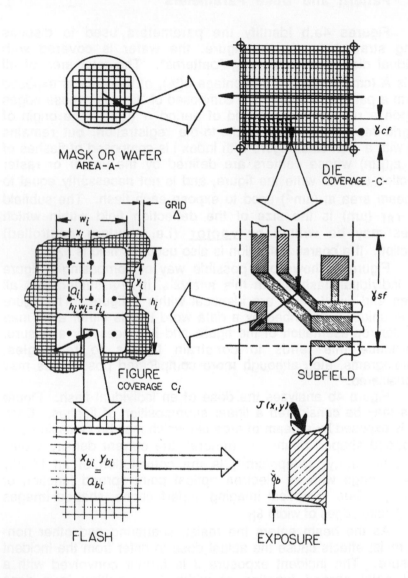

MASK OR WAFER
AREA -A-

DIE
COVERAGE -C-

γ_{cf}

GRID
Δ

w_i

x_i

Q_i

y_i

h_i

$h_i w_i = f_i$

FIGURE
COVERAGE C_i

SUBFIELD

γ_{sf}

$\begin{matrix} x_{bi} & y_{bi} \\ & = \\ & a_{bi} \end{matrix}$

FLASH

$J(x,y)$

δ_b

EXPOSURE

Figure 4a. Parameters associated with an analysis of EBL strategy.

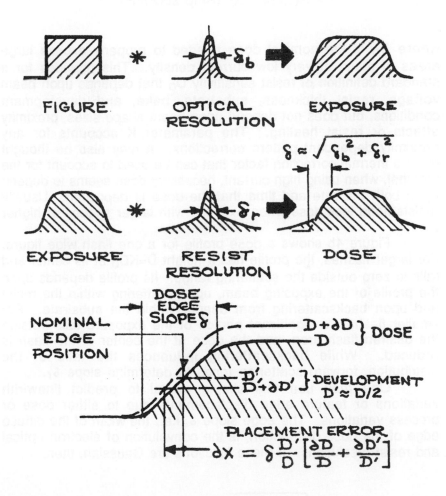

Figure 4b. Resist dose profile, and the concept of dose edge slope.

assumed to be a function of beam area, current density and dwell time:

[1] $D = K(J, a_b, t)\, D_0$ (Amp sec/cm^2)

where D_0 is the nominal dose needed to properly expose large areas of resist at very low current density. This provides for a standard definition of resist sensitivity D_0 that depends upon beam voltage, resist thickness, substrate, bake, and development conditions, but does not depend upon beam shape sizes, proximity effects or resist heating. The parameter K accounts for any proximity dependent pattern corrections. It may also be thought of as a thermal correction factor that can be used to account for the fact that, when using high current, necessary dose seems to depend upon both the rate and time that the dose is deposited. Usually resists need less dose when exposed with larger shapes or higher J.

Figure 4b shows a dose profile for a one flash wide figure. For larger flashes, the profile has a height $D=KD_0$ at its center and falls to zero outside the scattering range. Its profile depends upon the profile of the exposing beam, upon scattering within the resist and upon backscattering from deeper within the substrate. For smaller flashes, a significant portion of the exposure falls outside the desired flash area, and the dose at the center of the flash is reduced. While backscattering influences the tails of the distribution, forward scattering tends to determine slope δ_r.

One may use this simple model to predict linewidth variations or feature edge displacements due to either dose or process variations. The edge slope δ, i.e., the width of the diffuse edge of the dose distribution, is the convolution of electron optical and resist resolutions. If these functions are Gaussian, then:

[2] $\delta = \sqrt{\delta_b^2 + \delta_r^2}$ (μm)

This dose edge slope is a measure of feature size control. Suppose that all resist that receives a dose greater than D' is removed by the developer. For large features, a dose of $D \sim 2D'$ will reproduce the desired feature. If the percentage dose error is $\partial D / D$ and the percentage development level error is $\partial D' / D'$, then the position of the developed edge moves a distance:

[3]
$$\partial x = \delta\ \frac{D'}{D}\left[\frac{\partial D}{D} + \frac{\partial D'}{D'}\right]$$

where δ is the local edge slope at the development threshold. The dose edge slope can also be considered an empirical quantity that can be measured by intentionally varying dose or development conditions. It is a convenient measure of CD size control that is possible when exposure resolution, resists, and processes are being optimized for a particular lithography task. If the exposure edge slope δ_b is known, then experimental values for δ_r may be used to classify resist performance.

One of the purposes of this discussion is to point out that electron optical resolution and resist resolution effects are largely interchangeable when feature size control and process margins are the end goal. Both resist sensitivity vs. resolution and electron beam intensity vs. image resolution are competing parameters, so a variety of trade-offs are possible in the quest for lithographic quality.

It is also important to realize that the address grid Δ is not related to beam and resist parameters δ_b and δ_r. The address grid refers to the placement of ideal exposure distributions, and does not imply that features can be placed or processed with comparable accuracy.

4.3 Pattern Fracture

In the CAD data base, patterns are usually expressed as overlapping polygons with arbitrary angles and sizes that do not correspond to the writing strategy of a pattern generation machine. The first task of pattern fracture is to impose artificial boundaries within a die pattern. These boundaries reflect the limits of electron optical deflection fields and stage motions. In vector scan systems, it is also necessary to remove figure overlaps so that regions will not be doubly dosed. In raster scan systems, overlap removal can occur within the machine, after a bit map becomes available. During this fracture, curves, non-orthogonal lines, and complex polygons are fractured into primitive figures according to machine specific recipes. The intermediate fracture output is an expanded list of primitive figures, each of which can be generated by the real time electronics of the pattern generator. This list is sorted and organized in an efficient writing sequence. Usually all figures in a subfield are completed before a slower coarse deflection or stage move is requested.

The next fracture step usually composes actual figure words by assigning doses, flash (shape) sizes, step increments and subfield origins as needed to specify machine activity. At this point, dose corrections and other process machine dependent variables can be merged with the desired pattern data. For compactness, the resulting data almost always remains in a machine specific vector format that can be stored on disk or magnetic tape. This part of the pattern fracture is often performed on large computers operating independent of the exposure process.

Seen at the data path to electron optics interface, the writing data consists of many individual flash commands interspersed with occasional coarse deflection and stage commands. For submicron patterns, the volume of data passing through this interface is very large, e.g., up to 10^{12} bits in a few hundred seconds. In modern systems, this flash data is almost always generated from more compact figure data using specialized logic. This final fracture step is characteristic of the writing strategy. Since most figures are composed of multiple flashes, the largest data volume and highest processing rates are encountered here. The structure of the flash word tells much about system strategy.

4.4   Flash Word Content

In general, the system needs to be told where to place a flash, the size and shape of the flash, and how long to wait before the next flash. For a system writing on an address grid Δ using flashes of maximum size $x_{max}\ y_{max}$, and subfield size γ_{sf}, the flash word contains at least:

[4] $$\log_2 \left[\gamma_{sf}^2 / \Delta^2 + x_{max}\ y_{max} / \Delta^2 + n_d \right]$$

bits of pattern dependent data. The expression in brackets represents the number of possible different states of the system during one flash. The first term, γ_{sf}^2 / Δ^2, is flash position, the second is shape size, and the third term, n_d, is the number of possible dose levels. For example, a blanker with on and off stages needs $\log_2 2 = 1$ bit. In principle each bit can change for the next flash, although most systems hold dose and shape constant while writing a figure.

This expression is correct for both replication and pattern generator strategies. In a replicating stepper that prints one die per flash, $x_{max} y_{max}$ is the die area, and $\log_2 x_{max} y_{max} / \Delta_{sf}^2$ is the number of bits that define all possible mask patterns. This flash data is stored on the mask, so electronic storage and transmission is avoided. The stage vector location subfield γ_{sf}^2 is the wafer area A, so $\log_2 A / \Delta^2$ bits stage data are needed to locate each of A $/x_{max} y_{max}$ die locations. The rate of active data transfer is low, and the system is simple and fast because the data is stored on the mask.

The mask pattern generator supporting this stepper was probably a raster scan system. Raster deflection is entirely periodic, and passes over each grid square for the same time, regardless of pattern detail. There are no vector subfields, so $\gamma_{sf}^2 = \Delta^2$, and no pattern dependent position bits are needed. Similarly, the only available flash size is $x_{max} y_{max} = \Delta^2$, and no variation in beam or flash size is involved. A total of A $/ \Delta^2$ flashes of one bit each are used, and the data structure is a bit map of the pattern. The strategy uses a very large number of one bit flashes transferred at high speed.

4.5 System Accuracy

The location of a feature within a die may be thought of as the vector sum of several field vectors. With respect to a chip center, a feature position is reached via a stage vector, a position error vector, a coarse field vector, and one or more subfield vectors. The position of a figure edge is additionally influenced by the shape size and dose vectors. Each of these vectors operate on a grid that is the size of the field, and is discrete to a quantized address structure equal or less than the pattern address. It is the task of system registration and self-calibration to make sure that each level of deflection operates on superimposed address grids. During writing, calibrations must remain stable so that these field grids remain superimposed. To the extent that calibrations are inaccurate or unstable, these grids will be offset, stretched, rotated or distorted, resulting in figure edge displacements.

In the general system, the positioning vectors are listed in Table 1.

In the table, the number of steps per field is a measure of the dynamic range over which the field acts. It is closely associated with the stability requirements of the hardware. The number of moves per wafer is a rough measure of the positioning bandwidth necessary to avoid excessive overhead. To assure that the vector sum of all positioning vectors locates a figure to within a net error, e, it is necessary to calibrate each field to within a small fraction of e. It is also necessary to choose field sizes so that realistic noise/bandwidth criteria can be met.

TABLE 1. POSITIONING VECTORS OF VSB STRATEGY

Vector	Stage Position	Error Correct.	Coarse Deflect.	Subfield Deflect.	Flash Size
Steps/field edge	$\sqrt{a}/\Delta$	$\gamma err/\Delta$	$\gamma cf/\Delta$	$\gamma sf/\Delta$	x_{max}/Δ
Moves per wafer	$>A/\gamma_{cf}^2$	$\leq A/\Delta^2$	$A/\gamma sf^2$	$>A/a_{max}$	$>A/f$

The net error for any given vector field, for example the coarse field, has several components. An error in grid overlay can be discussed in terms of offset, size, rotation, and nonlinear distortion. An offset error displaces the entire field. A gain or rotation error causes displacements that are proportional to the field vector, and distortions depend upon higher powers of the field vector. All errors of a large field become offset errors in the origin of the next smaller field.

In most EBL systems, each vector field is independently calibrated against the stage interferometer system. For example, the measured stage location of several fiducial marks in a die defines a size and rotation for the local coordinate system of a die. These measurements are made with all other deflection levels centered to avoid interaction with other calibration errors and drifts. Similarly, coarse and subfield deflections are calibrated by countering motions of the stage, resulting in local offset, gain and rotation numbers for electron deflections. To the extent that a subfield offset is the result of a coarse field distortion, these errors are corrected by applying "dynamic correction" vectors to the coarse field vector. Each calibration is the result of several deflections and associated position measurements.

The accuracy and stability of each vector can usually be traced back to the sum of the following errors: beam position measurement error during calibration, e_c, quantization error, e_q, and random noise/drift error, e_n. The sum $e_c + e_q + e_n$ is the uncertainty of the point of origin (offset) of the vector, and (x/γ) $(e_{cgain} + e_{crot} + 2\ e_q)$ is the net result of gain and rotation and dynamic corrections that are often applied using digitally truncated addition or multiplication.

For an arbitrary position on the wafer, positioning error vectors add statistically. In general, the largest errors occur when all levels of deflection are simultaneously at the extremes of their allowed fields. Thus the worst case estimate of feature displacement would be the linear sum of all possible errors for all positioning levels listed above. Fortunately, one can show that the worst case error is very improbable because the pattern area written with all vectors at maximum is very small. Most analyses apply statistical addition where e_{max} represents the 1σ deviation of the normal distribution that arises from the statistical convolution of many approximately equal error contributions of arbitrary functional form, and e_{co} is chip registration error. Overlay errors are usually given for the 3σ width $3e_{max}$ of the position error distribution.

$$[5] \qquad e_{max} = e_{co} + [\underset{Levels}{\Sigma} \, e_c^2 + e^2_{cgain} + e^2_{crot} + 3e_q^2]^{1/2} \, .$$

From Eq. 5 it is clear that the total error depends upon the number of levels of positioning, and the number of position measurement calibrations needed to calibrate the system. The writing strategy plays an important role in accuracy as well as speed. In general, speed is obtained using shaped beams and multiple deflection levels, whereas accuracy favors simple deflection and stage strategies. Insofar as raster scan only involves stage and a single analog coarse field calibration, it tends to be an accurate strategy. More complicated VSB strategies tend to require finer digital grids and more accurate individual calibrations to achieve similar accuracy. This trade-off is an interesting area for further study.

4.6 Coverage Time Model for Serial Pattern Generation

The goal of this section is to derive a general expression for the "beam on" exposure time in a pattern generator. In general, this is difficult because the exposure time involves a complex interplay of parameters that depend upon the writing strategy, the pattern, the machine limitations, the resist, and the lithographic quality required. For this model, it is assumed that CAD patterns are already fractured into a primitive figure representation that allows only rectangles or parallelograms. It is also assumed that each figure is composed using either a fixed or variable beam shape whose size and dose are not varied during exposure of the figure.

The pattern is described by sets of figures of height, h_i, and width, w_i, and area, $h_i w_i = f$. The figures are composed of flashes by moving the beam within the figure in such a way that n_i flashes of dimension x_i, y_i and area, a_i, uniformly dose the figure. There are n_{ix} flashes along the width dimension of the figure, and n_{iy} flashes along the height dimension of the figure, so the figure is fractured into square, rectangular or possibly parallelogram shaped flashes.

The statistics of specific patterns is described by an area coverage distribution, c_i (%), which is the fraction of the total exposed area that contains figures in the set i, so $\Sigma_i c_i = 1$. The mean coverage c (%/cm^2 of pattern) is the exposed area of one cm^2 of pattern, so that set i covers cc_i cm^2 per sq. cm of pattern.

Each figure is to be fractured into as few or many flashes as is desirable to minimize the total time to expose the pattern. This seems to be a simple problem until it is realized that exposure time rather than flash count is the quantity to be minimized. The fracture of figures into flashes should be optimized while obeying constraints imposed by not only the die pattern and writing strategy, but also the system electronics, electron optics and resist process. Further constraints are imposed by the address structure upon which figures are placed and sized, and by the dose edge slope δ needed for feature definition.

The beam used to expose each flash may be of fixed or variable size, depending upon the strategy. If small enough, the beam may be round and use an approximately Gaussian intensity distribution, or it may be shaped rectangles with diffuse edges of slope δ_b. It is not assumed that the exposing beam is the same size as a flash. Flash size is defined strictly according to the deflection trajectory, regardless of beam size or shape. A typical VSB strategy uses beam shapes equal to flash shapes, but Gaussian and small fixed shape strategies may use flash overlap to advantage. Beam size parameters will be distinguished by the subscript b, so one flash of area a_i is exposed using a beam of area a_{bi}. Where flash overlap is used, the figure dose distribution (Figure 4b) uses an effective currently density $J_{eff} = J_{abi}/a_i$, where on the average each flash receives an additional dose a_{bi}/a_i from adjacent flashes.

The dose, $K_i D_o$ Ampsec/cm^2, required by figure f_i that uses n_i flashes of time t_i is $I_i n_i t_i / f_i$ (Ampsec/cm^2). Each flash area $a_i = f_i / n_i$. The current used is the beam current density, J, times the beam shape shape area, a_{bi}, so $I_i = Ja_{bi}$. Thus the dwell time for each flash in figure set i is:

[6] $t_i = K_i D_0 / J \ [a_i / a_{bi}]$.

      The parameter, J, in Eq. 6 is the constant current density in the beam, and is selected before exposure. The selection of current density can be limited by three factors, depending upon machine and resist constraints. The deflection system, drive electronics, and pattern generator impose a minimum flash time $t_i \geq t_{min}$ for all figures, during which the DACs and blanker must increment and remain stable during a flash. The electron optics impose a current density limit $J \leq J_{max} \ (\delta_b)$, above which exposure edge slope requirement δ_b cannot be met. The third basic limitation is the maximum current $I_{max} = Ja_{bmax}$ that can be used to expose the largest flash $a_i = a_{max}$ used in any figure. This current limitation can be imposed by electron optical beam interactions, or by resist heating. Beam interactions diffuse and defocus the image, causing excessive exposure edge slope δ_b. Resist heating can cause distorted figures, and a strong dependence of dose $K_i D_0$ upon shape current Ja_{bi}. It can also cause unwanted edge roughness and loss of resolution.
      The operating current density J must be chosen to be the lowest value that is consistent with all three constraints, t_{min}, J_{max}, I_{max}. If the exposure is flash time limited, then the largest shape a_{bmax} and the largest flash area a_{max} will call for the shortest flash, because the thermal enhancement factor K (a_{bmax}) is lowest. Thus from Eq. 6 with $t_i = t_{min}$:

[7] $J \leq K(J, a_{bmax}) D_0 \ a_{max} / t_{min} \ a_{bmax}$.

The optical current density limit is controlled by electron gun brightness and aberrations, such that:

[8] $J \leq J_{max} \ (\delta_b)$.

This limit is independent of shape size and flash size. The beam current limit occurs when the largest shape area a_{bmax} is used, so:

[9] $J \leq I_{max} / a_{bmax}$.

By inserting the above limiting values for J into Equation 6, multiplying t_i by the number of flashes cc_i/a_i in set i, and summing over i, three limiting expressions for coverage time are obtained:

$$[10] \qquad T_E \geq c \left\{ \begin{array}{c} I_{min} \; a_{bmax}/K(J,a_{max})a_{max} \\ D_0/J_{max} \\ D_0 a_{bmax}/I_{max} \end{array} \right\} \sum_i c_i K_i/a_{bi}$$

In Equation 10, the beam and flash size parameters a_{max} and a_{bmax} may be chosen to minimize the exposure time T_E. They are free parameters so long as dose edge slope criteria δ are not violated, and the rules of the machine strategy are obeyed. This optimization technique might be called "fracture optimization"(15). The technique varies according to writing strategy, but the goal is always to minimize T_E within the pattern, machine and resist constraints that have been isolated in Equation 10.

4.7   Raster Scan Fracture Optimization

The raster scan strategy operates on the fixed address grid on which the pattern is defined. The beam is deflected to each pixel, so c = 1, regardless of pattern. The largest flash area is $a_{max} = \Delta^2$. The beam is a fixed diameter Gaussian distribution, so that $a_{bi} = a_{bmax}$, and $K_i = K(a_{bmax})$. Since all figures use the same beam area a_{bmax}, and all pixels are addressed, pattern statistics play no role in the exposure time. Optimization reduces to choosing the beam area a_{bmax} to be as large as possible without causing excessive dose edge slope. If the overall process requires a dose edge slope δ to assure accuracy of feature sizes, and the resist has an inherent scattering-limited resolution δ_r, then the tolerable exposure edge slope is $\delta_b \leq (\delta^2 - \delta_r^2)^{1/2}$. The beam area a_{bmax} may be chosen based on these criteria. In raster scan, the Gaussian beam is in continuous motion, and is unblanked upon entry to a figure. Superposition of elementary Gaussians with a diameter $\sqrt{a_{bmax}}$ at half height forms an exposure profile with edge slope $\delta_b \sim \sqrt{a_{bmax}}$. If the beam diameter is chosen to be $a_{bmax} = \delta_b$, then the dose profile requirements are met. Use of beam diameters much smaller than resist resolution δ_r results in marginal improvement in accuracy but can result in much longer exposure times if coverage is current density limited. $D_0/(\delta^2 - \delta_r^2)$ is a better criteria than sensitivity D_0 for choosing raster scan resists, because less sensitive, high resolution resists may allow larger beam diameters. The resulting increase in current tends to cancel increases in required dose. For raster scan strategy, optimized coverage times are:

$$[11] \quad T_E \geq \begin{cases} t_{min}/\Delta^2 & \text{if } J = KD_0\Delta^2/t_{min}\delta_b^2 \\ K(J_{max},\delta_b)D_0/J_{max}\delta_b^2 & \text{if } J = J_{max} \\ K(I_{max})D_0/I_{max} & \text{if } J = I_{max}/\delta_b^2 \end{cases}$$

where, $\delta_b^2 = \delta^2 - \delta_r^2$ and $a_{max} = \Delta^2$.

4.8   Fixed Beam Vector Optimization

A FBV strategy allows the beam to be stepped to predetermined positions on the pattern, rather than scanned over the entire pattern. It exploits pattern coverage c(%) by addressing only areas to be exposed. Since beam addressing is by vector word rather than bit map, it is also possible to locate figures on a fine address grid without generating a correspondingly high flash count. In principle, flashes larger than Δ^2 can be used to generate figures. Flash overlapping can be used to further relieve current density constraints. There are, therefore both figure fracture and beam size trade-offs that can be used to optimize coverage time.

In the context of FBV strategy, the beam can be either a demagnified image of the source, or a small fixed-shape beam with diffuse edges(16). It differs from the VSB in that beam size stays fixed, so a_{bi} always equals a_{bmax} in Equation 10. Accordingly, pattern statistics do not play a direct role in the coverage time. However, since shape size is fixed, the flash size a_{max} is closely related to the beam size a_{bmax} if excessive overlap is to be avoided and small figures are to be possible. The beam can never be larger than a minimum feature dimension, nor can it be much larger than the step used to generate multiple flash figures.

There are two basic methods of forming figures with a fixed vector strategy. The first method uses a fixed figure generator step $\Delta_s = \sqrt{a_{max}} \geq \Delta$, and exposes figures using flashes of area $a_i = \Delta_s^2$. This composes figures on a grid of period Δ_s. To the extent that figures can be designed on a coarser size (as opposed to position) grid, the flash count can be reduced and the strategy can be faster. The second method uses a variable step grid, allowing the beam to overlap. So long as the figure generation address grid is very small, and flash times are chosen to compensate varying overlap, it is possible to minimize flash count without compromising address structure.

In either case, the net exposure distribution must satisfy edge slope conditions. If a Gaussian image of a point with half height diameter $\sqrt{a_{bmax}}$ is used, then the edge slope of the figure exposure depends upon the step size used. If the step $\Delta_s << \delta_b$, then the exposure distribution will have an edge slope $\delta_b \sim \sqrt{a_{bmax}}$, and the beam must be approximately the diameter of the desired edge resolution. If the steps are approximately equal to the beam diameter $\sqrt{a_{bmax}}$, then the figure dose convolves somewhat differently, and $\delta_b \sim 0.5 \sqrt{a_{bmax}}$. This means that if larger steps are possible, one can also use a larger beam $a_{bmax} \sim 4 \delta_b^2$, with corresponding reduction in flash count and current density limitations. If a small shaped beam of approximately one step size $\sqrt{a_{max}}$ is used, then a_{bmax} is independent of δ_b, and still larger flashes may be used. However, in contrast to VSB strategy, these flashes, must remain comparable to the feature size increment rather than the feature size itself. (The Bell Lab EBES IV system seems to exploit some of these ideas)(17)(18).

To summarize this discussion in terms of Equation 10, the beam size a_{bmax} may be chosen approximately according to the formula:

$$[12] \qquad a_{bi} = a_{bmax} = \delta_b^2 + a_{max}$$

where a_{max} is the flash area used to write figures. In the face of flash time and current density limitations, it is desirable to minimize flash count, $c_i a_{bmax}/a_{bi} a_{max} = c_i/a_{max}$, and to maximize flash area, so the figure fracture strategy described above will result in minimum coverage as a function of tolerable flash area a_{max}.

$$[13] \quad T_E \geq c \begin{cases} t_{min}/a_{max} & \text{if } J = KD_o/t_{min} \\ \{K(J_{max}, \delta_b)D_o/J_{max}(\delta_b^2 + a_{max})\} & \text{if } J = J_{max} \\ K(I_{max})D_o/I_{max} & \text{if } J = I_{max}/(\delta_b^2 + a_{max}). \end{cases}$$

It should be noted that this FBV vector coverage time is independent of figure area distribution, but depends critically upon tolerable flash size. Tolerable flash size in turn, depends upon the desired variability of small features. To some extent, dose can be used instead of position to vernier feature sizes, so the implications of Equation 2 should be accounted for in choosing the step increments $\Delta_s = \sqrt{a_{max}}$. In most cases, the beam and flash must be considerably smaller than any figure, but not necessarily as small as intuition might suggest.

4.9   VSB Fracture Optimization

A variable shaped beam vector strategy can use shape areas a_{bi} that are much larger than a resolution element δ_b^2. This is why the strategy is usually faster for rectilinear features $f_i >> \delta_b^2$. Fewer flashes of lower current density are needed, which relieves flash time and current density constraints. Since the shape size is variable, flash overlap is not used. Thus $a_{max} = a_{bmax}$, and the shaped beam is chosen to exactly cover the figure using a rectangular flash, $a_{bi} = a_i = f_i/n_i$.

A figure fracture algorithm may in principle choose any flash size up to the machine imposed limits x_{max}, y_{max}. This simple algorithm is not generally desirable. The following scheme is one possible technique that accounts for all machine and resist limitations.   For reasons that will become clear later, an additional flash area constraint $a_{max} = a_{bmax}$ will be imposed along with size constraints x_{max}, y_{max}. This area is not necessarily equal to the product of maximum linear shape dimensions because it will be used as free parameter to limit the total current, $J \, a_{max}$, in the largest shape. Given a figure of shorter dimension, n_i, one might first fracture h_i into n_{iy} pieces where $n_{iy} = [h_i/y_{max}]$(rounded up to the nearest integer). The shorter dimension uses as few flash dimensions $y_i = h_i/n_{iy}$ as possible. To obey the maximum area requirement, the x flash dimension must be $x_i' \leq a_{max}/y_i = a_{max} n_{iy}/h_i$ to obtain an integer number of flashes along the longer axis w_i so, $n_{ix} = [w_i/x_i] = [w_i h_i/n_{iy} a_{max}]$. The number of flashes in the figure is:

$$[14] \qquad n_i = n_{ix} n_{iy} = [w_i h_i/[h_i/y_{max}] \, a_{max}] \, [h_i/y_{max}] \,.$$

If we make the simplifying assumption that the smaller figure dimension h_i is less than y_{max}, then when rounded upwards, $[h_i/y_{max}] = 1$. This assumption seems reasonable for submicron interconnect, gate, and window layers. It allows subsequent discussion to be based on figure area $f_i = h_i w_i$ alone. The number and area of flashes in figure set i is:

[15] $n_i = [f_i/a_{max}]$, and $a_i = a_{bi} = f_i/ [f_i/a_{max}]$

where $a_{bi} < a_{bmax} = a_{max}$ if flashes do not overlap.

The effective flash count in Equation 10 becomes:

[16] $\sum_i c_i K_i(J \cdot a_i) / f_i[f_i/a_{max}]$.

In VSB strategy, the rounded off term causes the coverage time to be pattern dependent, because f_i does not cancel out when figure dimensions are comparable to flash dimensions. For some figures, smaller than optimum flashes a_{max} must be used. These flashes contain less than the maximum available current J, a_{max}, and are exposing figures inefficiently. While this effect is an inherent flaw in VSB strategy, the advantages of large flashes usually compensate.

To compute accurate coverage times for VSB strategy, it is necessary to choose a_{max}, compute Equation 16, and then compute coverage times per Equation 10. However, to gain physical insight, define a mean figure area $f' = \sum_i c_i f_i/\sum_i c_i$ and imagine that the pattern is entirely composed of these figures. On the average, flashes of area $a' = f'/[f'/a_{max}]$ are being used. With this simplification, the general equation x for VSB strategy becomes:

$$[17] \quad T_E \geq cK(J,a') \left\{ \begin{array}{ll} t_{min}/K(J,a_{max}) & \text{if } J=KD_0/t_{min} \\ \text{and} & \\ D_0/J_{max} \\ \text{and} & \\ D_0a_{max}/I_{max} & \text{if } J=I_{max}/a_{max}. \end{array} \right\} [f'/a_{max}]/f' \quad \begin{array}{l} \\ \text{if } J=J_{max} \\ \end{array}$$

The fracture optimization problem may now be stated: choose a maximum flash area so as to minimize the area coverage time T_E.

This exercise will be left to Figure 5c, where a specific example is considered.

4.10 Example of Strategy Model Results

Figures 5 a, b, and c are examples of the use of this strategy model under a specific set of assumptions. Coverage time per unit area T_E is plotted versus the maximum flash area a_{max}. Figures a,b,c, show raster, FBV, and VSB strategies evaluated by Equation 11, 13 and 14 respectively. Each expression generates three limiting curves corresponding to flash time, current density and max beam current constraints. The longest coverage time always applies.

In this example, all three strategies are plotted for the same pattern, machine and resist assumptions. This is not to imply that all strategies involve the same machine constraints. In fact, t_{min}, J_{max} and I_{max} vary widely according to the writing strategy and technology employed in different machines. The numbers are only a starting point for discussion, and do not represent any specific machine. The curves are plotted logarithmically, and functional dependences are labeled, so that the curves may be scaled up or down as multiplicative constraints are changed.

A coverage c = 50% of f' = $1\mu m^2$ figures is chosen to represent a dense submicron pattern. The assumed resist requires a dose D_O = 10 $\mu C/cm^2$, and can tolerate a maximum current of I_{max} = 0.5 μA without showing signs of thermal enhancement (K = 1.0). The limiting electron optical parameters J_{max} <100A/cm^2 may be obtained at up to I_{max} = 0.5 μA with an exposure edge slope of δ_b=0.1 μm in a modern column(8). A minimum flash time, t_{min} <20 nsec (50 MHz), is assumed for all strategies, although it is optimistic for VSB systems, and somewhat longer than flash times used in raster scan systems.

Figure 5 may be used in two ways. If machine and process limitations are fully characterized, then the figure can be used to optimize pattern design and fracture to maximize throughput. Figure 5 can also be used as a machine design tool to determine specifications and expected performance for different patterning processes.

In the VSB example, Figure 5c, optimum coverage occurs when the pattern is fractured into flashes no larger than a_{max} = 0.5 μm^2, i.e., 0.7 μm squares or 0.25 x 2 μm rectangles. This point occurs at the intersection of the current density and current

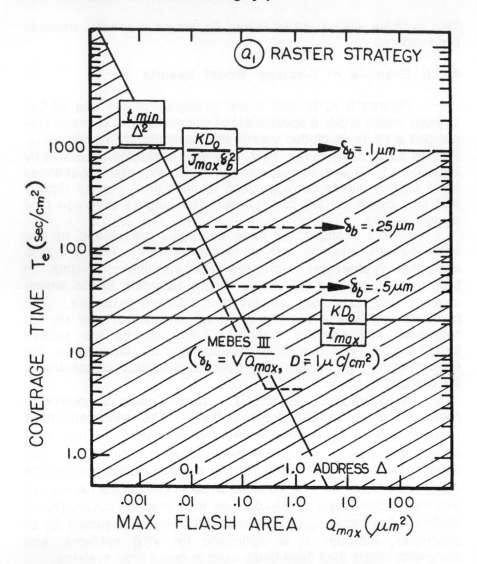

Figure 5. Coverage time constraints for three different writing strategies. Minimum flash time, maximum current density and maximum tolerable current can each limit coverage time under specific conditions. The clear areas above all curves show the coverage time domain that is available if patterns are fractured with maximum flash area a_{max}. Depending upon the strategy, the address structure, step size or flash size are related to flash area, and are also shown along the x axis. These curves are appropriate only for the conditions given in the text.

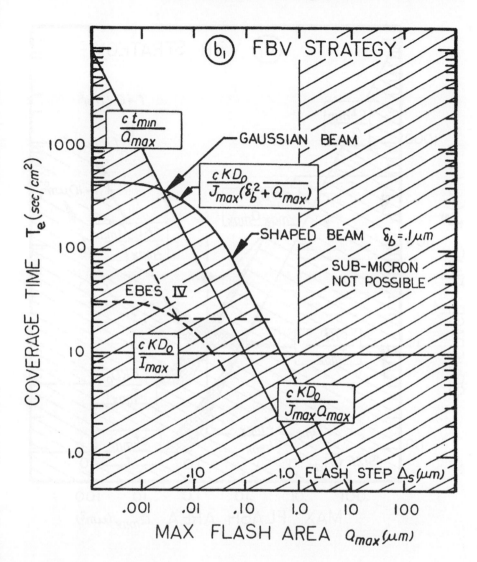

$\dfrac{c\,t_{min}}{Q_{max}}$

GAUSSIAN BEAM

$\dfrac{c\,KD_0}{J_{max}}(\delta_b^2 + Q_{max})$

SHAPED BEAM $\delta_b = .1\,\mu m$

SUB-MICRON
NOT POSSIBLE

EBES $\underline{\text{IV}}$

$\dfrac{c\,KD_0}{I_{max}}$

$\dfrac{c\,KD_0}{J_{max}\,Q_{max}}$

FLASH STEP $\Delta_s\,(\mu m)$

MAX FLASH AREA $Q_{max}\,(\mu m)$

COVERAGE TIME $T_e\,(sec/cm^2)$

Figure 5. (continued)

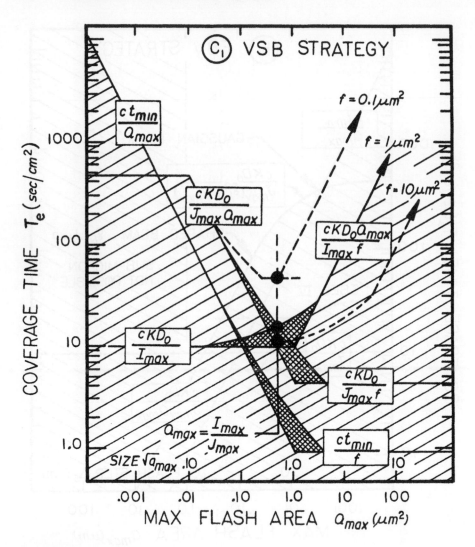

Figure 5. (continued)

limited curves where $a_{max} = I_{max}/J_{max}$. Coverage times are quite sensitive to deviations from this optimum, so considerable economy results from careful optimization of the fracture. In the example, a flash time $D_o/J_{max} = 100$ nsec would actually be used, so the assumed $t_{min} = 20$ nsec did not play a role. Figure area f' may be varied by translating those portions of the curve that depend upon mean figure size. The optimum a_{max} proves to be remarkably independent of mean feature area for f' = 0.1 - 10 μm^2, which means that there exists approximate fracture guidelines for VSB strategy that depend primarily upon resist current limit and machine parameters. These guidelines are $a_{max} = I_{max}/J_{max}$ for insensitive resists, and $a_{max} = t_{min} I_{max}/D_o$ for sensitive resists. Both guidelines are relatively independent of f' because flash time and current density limitations have similar functional dependence.

A second important conclusion may be drawn from Figure 5. All strategies have an intrinsic coverage time limit cKD_o/I_{max}, regardless of machine and fracture sophistication. At this limit, the parameters K, D_o, I_{max} are resist properties. Along with scattering parameter δ_r, KD_o/I_{max} seems to be a figure of merit for resists used in EBL machines. It accounts for thermal effects as well as sensitivity. A 40 $\mu C/cm^2$ resist that could accommodate $I_{max} = 2\mu A$ would have shown the same coverage time if current density and flash time requirements were met. This emphasizes the role that resist design plays in fracture optimization and system design. In this example, 10 sec/cm^2 is the ultimate coverage time allowed by the resist.

In the limit of very small features ($a_{max} < \delta_b^2$) written on sensitive resists, the strategies obey almost the same equations. Coverage times are controlled by minimum flash time, and VSB loses its advantage because multipixel flashes are not allowed. To the extent that raster and fixed beam vector systems typically allow a smaller t_{min}, they are faster for some patterns. Raster scan strategy is probably the fastest alternative for a fine dot matrix pattern that might be used for archival memory or diffraction gratings. Under these conditions raster scan is desirable because bit map blanking electronics is usually faster than vector word/DAC electronics.

A comparison of FBV and VSB strategies yields interesting observations. At first, it would seem that the two strategies are similar, and in fact FBV seems somewhat faster because it doesn't depend upon pattern statistics $[f'/a_{max}]/f'$. For most applications, this comparison is not valid because FBV cannot use large flash steps and still compose submicron features of variable size. However, there are two specific applications where this similar strategy may excel. The first application is for very periodic patterns that can be composed using a limited repertoire of fixed shapes on a coarse grid. The second application is for 4-10x reticle writing, where features may be composed on a coarser grid, but accuracy requirements still favor the small edge slopes and fine feature positioning offered by a shaped beam vector strategy.

The FBV strategy can also be significantly improved by higher current density and faster figure generation. The EBES IV system composes figures at up to t_{min} = 2nsec using steps of 1/8 μm(17)(18). A Schottky emitter allows J_{max} = 1600 A/cm^2 in a ab_{max} = 0.0156 μm^2 (1/8 μm dia) beam containing I_{max} = 0.25 μA. The data is shown in dotted lines in Figure 5b, where coverage times of about 20 sec/cm^2 are predicted. Figure 5b shows that high current density optics and very fast figure generation are the technological focus of fixed beam vector strategy.

Coverage times for raster scan strategy are controlled by address grid structure and resist sensitivity. In Figure 5a, flash areas in the range of a_{max} = 0.0025 - 0.025 μm^2 (0.05 - 0.15 μm address structure) would be necessary for submicron microcircuit patterns. With this assumption, the predicted coverage time is 1000-350 sec/cm^2. With increasing address structure, coverage times drop to 20 sec/cm^2 for patterns that can be written on a 0.7 μm grid. Raster scan technology is therefore quite favorable for mask and reticle generation. Modern mask making processes also use more favorable parameters than Figure 5a suggests. The MEBES III system is usually used to make 1X optical masks with down to 1 μm minimum features, or to make 4 - 10x reticles requiring features in the range of 2 - 5 μm. These patterns allow flash areas in the range of a_{max} = 0.01 - 0.25 μm^2 (0.01 - 0.5 μm address sizes). Exposure conditions are usually J_{max} = 100 A/cm^2 at D_0 = 1.25 μC/cm^2 (10 KeV) and I_{max} < 2 μA. Minimum flash time is 12.5 nsec (80 MHz). These constraints are plotted on Figure 5a in dotted lines. In this favorable range of application, coverage times range from 100 to 5 sec/cm^2. While the strategy is very slow for small address structures, it rapidly becomes favorable for mask making applications. High bit rates, high current density, and sensitive, higher resolution resists are the technological focus of raster scan development.

4.11 Writing Overheads

The coverage time accounts only for the time the beam spends in a position to expose resist. It does not account for electronic and mechanical activity during which exposure is not allowed to occur. Writing overheads occur during coarse and subfield vector slews, and possibly during stage activity. They are accounted for as T_{wo} = non-exposing time per cm^2 of pattern area.

The flash overhead t_0 can occur as part of each flash cycle, while the beam is being slewed to the next site within a figure. If the figure is continuous, it is not absolutely necessary to blank the beam during the slew, provided the slew time is somewhat shorter than the flash time $t_0 < t_i$. If t_0 is comparable to t_i then considerable overhead is incurred because it is multiplied by the flash count.

The figure overhead t_{fo} occurs between each figure, while the beam is being slewed to the origin of that figure. Since this slew is over a longer distance, and can expose unwanted regions, it is necessary to blank until the beam has settled at the location of the first flash. The value of t_{fo} is multiplied by the figure count for a vector strategy.

The coarse field deflection overhead t_{co} occurs each time the coarse field deflection slews to a new origin of subfield activity. If two electromagnetic levels of deflection are not involved, then $t_{co} = 0$. Otherwise, t_{co} is multiplied by the number of subfields $1/\gamma_{sf}^2$ in a sq. cm. of pattern.

With a step-and-repeat stage trajectory, the stage must slew to a new location each time a coarse deflection field is written. If this settling time is t_{so} and the coarse field area is γ_{cf}, then t_{so} is multiplied by $1/\gamma_{cf}^2$ to obtain the overhead per cm^2 of pattern. It is this overhead that forces step-and-repeat systems to use large coarse deflection fields. If a write-on-the-fly stage trajectory is used, t_{so} can be zero if the stage position error can always be kept within the range of deflection corrections. This is possible provided pattern coverage density fluctuations do not demand excessive stage acceleration or velocity. This is usually possible within individual chips, even using modest correction and coarse deflection fields, but additional overhead is incurred at chip boundaries. In summary, overheads may be included in the coverage time model as follows:

[18] $T_{wo} \geq t_o \sum_i cc_i[f_i/a_{max}]/f_i + t_{fo} \sum_i cc_i/f_i + t_{co}/\gamma sf^2 + t_{so}/\gamma cf^2$

where t_o, t_{co} and t_{so} can be zero for some systems.

The transfer of figure data from memory or disk to the active data path can also be a source of overhead. Figure data is usually stored in large buffer or FIFO memories that are being loaded during writing. Both load and access time can generate bottlenecks in the writing process, particularly if figures are small. If each figure is represented by m Bytes of data, then $m\Sigma_i$ cc_i/f_i bytes/cm^2 must be loaded during writing coverage time $T_E + T_{WO}$ (sec/cm^2). The average loading rate needed to incur no additional overhead is:

[19] $m \sum_i cc_i/f_i / (T_E + T_{WO}) \sim mc/f' (T_E + T_{WO})$

(bytes/sec) .

If net coverage times are to be in the range of 10-20 sec/cm^2 for submicron figures of f' = 1-5 μm^2 at c = 50%, and m = 10 Bytes/figure then data transfer rates range from 50 to 5 megabytes/sec and must be sustained for 500 to 100 megabytes of data. Fortunately, most of today's patterns contain larger figures, and compaction techniques are available to minimize the transfer of repetitive pattern areas. However, today's disks only sustain 2 - 20 megabytes/sec peak rates, so data delivery bottlenecks may appear in the future. Fast EBL is a very data intensive process.

4.12 Electron Optical Optimization

Preceding sections have shown that beam current and current density are critical parameters in coverage time optimization. The electron optical system is limited by the brightness of its electron gun, the imaging defects of its lenses, and beam interactions along its beam path. The relative importance of these limitations depends upon the writing strategy used. In the absence of beam interactions, an electron optical system with an aperture half angle ß (rad) can provide a current density $J_{max} = \pi \beta^2/B$ (A/cm^2), where B is the gun brightness

(A/cm^2/sr).   Including beam interactions, the resolution disk
diameter or exposure edge slope δ_b is approximately:

[20] $\delta_b \sim [\{C_s \beta^3/2\}^2 + \{C_c \beta \Delta V/V\}^2 + \{PI^{2/3} L^{2/3}/V^{4/3} \beta^{4/3}\}^2]^{1/2}$

where C_s and C_c are spherical and chromatic aberration
coefficients (cm) of the final lens, $\Delta V/V$ is the energy spread in
the beam, including "Borsch" effects, and where the last term is
an expression for transverse beam interactions(19). P is a
proportionality constant, I is the beam current, and L is the length
of the imaging optics.

From Equation 20, one may see that there exists an
optimum angle β_{opt} that will minimize resolution δ_b at a current
I.   This optimum angle may be related to the necessary gun
brightness using the maximum shape area discussed in the
preceding section. For a given choice of a$_{bmax}$, an <u>optimum</u> choice
of brightness is implied:

[21] $B_{opt} = J/\pi \beta^2_{opt} = I_{max}/\pi \beta^2_{opt} a_{bmax}$.

If resolution requirements are to be met using the highest possible
current in a largest shape, then it is necessary to use gun
brightness as free parameter in Equations 20 and 21.   Extremely
high brightness is desirable only when a$_{bmax}$ is small, i.e., for
raster and FBV strategies.   In some unapertured shaped beam
systems, the gun brightness actually determines the optimum
aperture angle directly, and is chosen according to design goals
a$_{bmax}$ and δ_b. Figure 2 of reference 8 treats these matters in
greater detail. This discussion points out that optical design is
closely linked to broader questions of strategy, resist and pattern
goals. It is the task of the optics designer to anticipate these goals
and optimize the optics accordingly.

5.0   THE RECORDING MEDIUM

The writing strategy sections show that resist properties
profoundly influence both the design and performance of EBL
equipment. Feature definition, throughput and defect densities all
depend upon both the machine and the resist process.   For

example, feature definition and placement accuracy are the combined result of exposure profile and resist resolution. It has been shown that one may be traded against the other to optimize coverage times. Similarly, resist sensitivity may be traded against tolerance to maximum current to optimize process performance with acceptable coverage. This section discusses the role of the resist in the overall process, emphasizing physical rather than chemical effects that directly influence machine design.

An electron beam resist is usually a thin polymer layer. Sometimes sensitizers are used to create local reactive sites. Pattern transfer depends upon differential solubility caused by electron exposure. The dose profile $D (x,y,z)$ can be thought of as a three dimensional _latent_ solubility distribution. Along with the actual exposure, other factors such as beam heating, pre-and post-baking, aging, oxidation, and pre- and post-irradiation effects can influence this latent solubility distribution, so the strategy related parameters D_0, K, δ_r are characteristic of exposure conditions _and_ the entire resist process; not just the resist material. In most processes, the differential solubility is the result of chemical and structural rearrangements that change the molecular weight. Intervening mechanisms are often only empirically understood. Pre- and post-exposure processing techniques are receiving increased attention as a means of optimizing the electron beam exposure process. Resists are classified as positive or negative depending upon whether exposure increases or decreases solubility in developer. There are cases where both exposure level and process can reverse the polarity of the developed pattern.

Electrons can penetrate several microns into resist or substrate material. Both the width and depth of a dose distribution depend upon beam voltage. Electron optical and deflection optimization tends to favor fixed voltages for specific machines, while fabrication processes will favor various resist thicknesses and beam voltages. It seems that there is not one ideal beam voltage, so there will probably be machines with various voltages.

5.1 Sensitivity

The sensitivity or required dose D is the number of electrons per cm^2 needed to expose the pattern. Due to scattering and beam-induced thermal effects, the dose varies according to the pattern and exposure conditions. For example, isolated submicron lines made with small flashes usually need greater doses than large areas exposed with large flashes. The nominal dose D_0 can be defined as that dose required to obtain equal lines and spaces for large lines written at very low beam current. Under these conditions, scattering and resist heating do not effect the dose at the center of a line. Referring to Figure 4b, equal lines and spaces imply that the nominal dose is twice the development threshold D_0', so that the steepest region of the dose distribution is used to define edge positions. By measuring linewidth vs. dose, it is possible to measure the dose edge slope δ in Equation 2. While this is not a unique definition of D_0 or δ, it is a useful calibration tool.

Sensitivity fundamentally depends upon the effectiveness of electrons in altering the solubility of a resist. This in turn depends upon the amount of electron energy transferred to the resist via those mechanisms that alter solubility, and not via other competing mechanisms. Studies of electron/resist behavior try to determine this energy loss distribution (vs. x, y,z) as a function of incident exposure dose $J(x,y)$ t. Given the dose distribution and development rate vs. dose data for the resist, figure shapes and wall profiles can be predicted(20).

5.2   Contrast

Resist contrast is usually defined in terms of a relative change in developed resist thickness vs. dose. While this definition gives insight only for large features, one can say that high contrast resists are generally desirable because they tend to develop along 3-dimensional contours of constant dose. Contrast is related to practical definitions of "resolution" because small features necessarily involve less modulation of dose. By varying the dose and sometimes by adjusting the exposure distribution to be different than the desired dose distribution, it is possible to obtain correct feature sizes with various wall profiles(21). In practice, these procedures are largely empirical. In general, smaller features favor a narrower exposure and larger dose to obtain vertical wall profiles.

5.3   Resolution

Resist resolution is equally difficult to define. In light optical lithography, typical definitions involve the highest spatial frequency of lines and spaces that can be obtained subject to thickness and process margin requirements. It is possible to define a single valued modulation transfer function, but it is difficult to isolate resist resolution from exposure resolution because some resists may be able to support higher spatial frequencies than light optics can provide. Using electrons, it is possible to provide an essentially square wave exposure profile, so resist resolution should be a measurable quantity. However, it has been pointed out that a definition of resolution cannot be based on a simple contrast transfer function(22). The dose profile is the convolution of the exposure profile with a more complicated point spread (resolution) function that combines the effects of interaction range, forward scattering and backscattering. Several Gaussian functions of various radii and relative amplitude are frequently used. The combined effect of non-local energy transfer and forward scattering has a typical range of 0.02 - 0.2 μm, while the broad backscatter distribution may have a 2 - 5 μm radius. With increasing beam voltage, the forward scatter radius decreases while the backscatter range increases. As linewidths decrease, the net effect of this convolution with a more complicated point spread function is first to decrease the amplitude but not the edge slope of dose distribution. Extremely high resolution has been demonstrated using very thin resist on even thinner substrates(23). Under these conditions backscatter is negligible, and forward scatter is minimized. Possibly molecular size, nonlocal interactions, and quantum noise are ultimate limits.

One practical measure of resist resolution is the radius of an inside corner of feature. If the resist resolution δ_r is higher than the exposure resolution δ_b, then the bottom inside corner radius of the pattern should measure the resist resolution. Figure 6 compares two resist processes that were both exposed with 1.5 μm shapes at $J = 5A/cm^2$ using a beam edge slope $\delta_b < 0.07$ μm. Figure 6a was exposed using 0.5 μm thick PMMA resist. It shows <0.1 μm corner radii. Figure 6b was exposed using 1.5 μm thick MP 2427. Corner radii are about 0.5 μm. Thickness, contrast, scattering and thermal effects combine to produce a large difference in the pattern acuity that can be obtained using the two resist systems.

5.4   Edge Profile

Resist wall profiles play an important role in fabrication processes. Chrome masks use a simple wet etch of a thin layer, so sloped walls do not directly effect pattern accuracy. The details of the immediate interface define the etch boundary. However, the low contact angles that are common in the PBS mask process may adversely effect dose and process latitude because a low angle

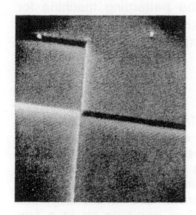 

a. .5 μm PMMA/GaAs
   D = 60 μC/cm²
   3:1 MIBK:IPA 180 sec

b. 1.5 μm MP2427/.6 μm SiO₂/Si
   D = 42 μC/cm²
   1:4 MP2401:H₂O 180 sec

□
1 μm SQ.

Figure 6. Corner radii and wall profiles for two different resist processes.

results in more edge movement per unit of removed resist. for lift-off processes, thick resists with vertical or reentrant sidewalls are favored. For aggressive dry etch processes, thicker resists and vertical sidewalls are favored, and chemical resistance to plasma etch becomes an important factor in the choice of resist. All of these considerations influence the parameters, D_0, K, and δ_r, that specify the performance of the patterning machine for particular fabrication steps. They also determine the optimum pattern fracture (flash sizes and exposure times) that best exploits the machine and resist.

5.5   Proximity   Effects

The term proximity effect refers to the tendency for adjacent features to influence each other's dose distribution. Backscattered electrons from the substrate are those electrons that scatter from lower lying orbitals of the resist molecules, returning up through the resist at high energy. Since the electron to atom mass ratio is low, these electrons lose little energy but are scattered through wide angles. They originate up to a few microns deep in the substrate, so they can pass upwards through the resist up to several microns from their point of entry. They can expose adjacent features. Up to half the total dose can be due to backscattering. The scattering properties of electrons make proximity effects more serious in EBL than other lithographic techniques that use lower energy or heavier particles.

For registration, backscattered electrons are exploited to locate fiducial marks underneath the resist layer. During exposure, they are a disadvantage because they cause shifts in feature boundaries. Referring to Figure 4b, they cause a shift in the dose profile ∂D, causing a shift in feature edges. Proximity effects depend mostly upon the fraction and range of backscatter dose, but their potential for shifting edges also depends upon the dose edge slope δ.

EBL frequently applies figure dose correction to minimize proximity effects. Figure 7 schematically illustrates two kinds of proximity correction. The desired pattern is two narrow lines near a wide line. Figure 7a shows exposure and uncorrected dose distribution. If some of the net dose is due to a diffuse distribution, and some is due to a sharper distribution, then the net dose is a sharper profile on top of a diffuse background. There is no development threshold that gives correct dimensions for all three features. Figure 7b shows the "self consistent method" (24) that intentionally over or underexposes figures to give a more

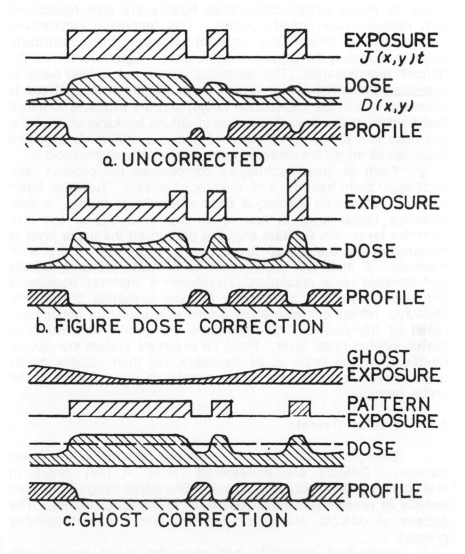

a. UNCORRECTED

b. FIGURE DOSE CORRECTION

c. GHOST CORRECTION

Figure 7. Schematic representation of two different techniques for correction of proximity effects.

uniform net dose. This is done by varying flash dwell times on a figure by figure basis. Sometimes figures are also refractured with narrow overexposed outlines. This technique emphasizes edges while minimizing interior doses that contribute disproportionately to adjacent figures(25). Figure 7c shows the "Ghost" technique(26). This technique uses a defocussed beam to pre-dose the resist with a reverse tone image. If the defocus is chosen to match the backscatter range, and the exposure to match that fraction of the dose that is due to diffuse backscatter, then the net dose has the features elevated above a uniform background dose, but all edges are centered upon a development threshold.

Each of these techniques complicates the process, and increases both fracture and coverage times. The third basic proximity correction technique uses a multilevel resist. A thin, sensitive upper layer is coated on top of a thicker but less sensitive layer. An E-Beam exposed pattern on the upper layer is transferred to the lower layer by light optical or dry etch methods. A multilevel resist tends to alleviate proximity effects and increase resist resolution. Resolution is improved because a thin upper layer has a narrower forward scattering distribution. Multilayer resists can also allow a favorable choice of upper layer resist by transferring etch resistance and profile constraints to the underlying resist layer. From an exposure system standpoint, multilevel resists seem to be desirable, but their viability needs to be viewed in the context of the economics and reliability of the entire process.

5.6   Resist Defects

Lithographic defects are unwanted features in the resist pattern. Defects are considered "fatal" if they result in malfunction of the fabricated circuit. Standards range from zero defects in reticles to several per layer in multichip wafers. The causes of defects are distributed throughout the lithography process.

Particulate matter in and upon the resist surface can scatter electrons and influence the dose distribution. This causes "pinholes" and "pin spots" in the pattern. Particulate control upon both masks and wafers is an important aspect of the EBL machine and environment. Since electrons use evacuated chambers, contamination control considerations extend beyond the laminar flow techniques used in optical lithography.

Defects can also be caused by random bit errors in pattern data. Raster scan systems tend to suppress isolated bit map errors by integrating over adjacent pixels. In vector scan systems, a one bit error can misplace an entire flash or figure, guaranteeing a larger defect. Elimination of digital noise is another aspect of defect reduction.

Resist chemistry can also play a role in defect generation. Trace chemicals in the resist or on the substrate surface can nucleate reactions that locally alter the sensitivity or latent solubility of the resist layer(27). One experiment suggests that many chrome mask defects may be due to this mechanism(28). In the future, these chemical effects may place additional constraints on the design of electron resist processes.

5.7 Charging Effects

Most resists are insulators that can charge up during electron exposure. As the pattern is exposed, charge is inhomogeneously distributed on the resist. Resist surfaces may discharge to the underlying substrate either during the flash (via induced conductivity) or may slowly decay after exposure. Since charging is nonuniform, residual charging can cause beam deflections that result in pattern dependent overlay errors that are difficult to distinguish from other sources of feature placement error. Some processes use a thin conductive layer over the resist. Another successful trilayer resist process uses an ion shower to chemically alter the resistivity of the resist previous to exposure(29). These techniques may significantly improve the overall process.

5.8 Thermal Effects

In the strategy section, it has been suggested that heating of resists can place a lower limit on coverage times. Serial pattern generation requires intense localized beams whose current depends upon resist dose. When writing small features, these simple Gaussian or shaped beams cannot be larger than feature dimensions, so relatively high power density is necessary. Any associated local heating of the resist can alter the latent dose distribution.

Thermal effects become more important as throughput increases and feature sizes decrease. Traditional Gaussian beam strategies tend to use relatively low currents. Using a shaped beam strategy, lower coverage rates become possible because higher current can be used. So long as feature sizes remain significantly larger than the resist thickness and electron penetration depth, heat dissipation is a more efficient planar process that depends upon current density. However, as flash dimensions become smaller than the characteristic range of thermal diffusion, heating depends upon total current rather than current density. Since coverage ultimately depends upon current, area coverage is directly limited. Calculations from ref. 30 show that temperature rises of several hundred degrees are possible using micron scale flashes containing microamps of current. At 20 KeV, thermal profiles during a flash seem to have a "resolution" of one or two microns(31). On SiO_2, even at 10 A/cm^2, some combinations of flash size and exposure time exceed the glass transition temperature of the resist, so one may expect both physical and chemical properties of the resist to change. Noticeable thermal effects in developed patterns are frequently signaled by flow patterns or thickness modulations that can be observed before development.

Figure 8 shows several manifestations of thermal effects in insensitive resists on a silicon wafer. While they are particularly dramatic examples, they serve to illustrate the problems. Each 16 μm square was exposed using 40 A/cm^2, 1.5 μm square shapes containing 0.85 μA. The top row was exposed using a high resolution 0.5 μm thick PMMA resist that is overdeveloped to obtain a nominal D_0 = 60 μC/cm^2 sensitivity. The bottom row was exposed using 1.5 μm thick MP 2427 process with a nominal D_0 = 45 μC/cm^2 sensitivity. These nominal doses were measured from other sites on the two wafers that were exposed at 5 A/cm^2 using 0.25 μm shapes. They represent the required dose in the limit of minimal temperature rise. The high current flashes were measured to have less than δ_b < 0.15 μm edge slope, and were stepped in increments of 1.5 μm, so the flash and figure exposure distributions were uniform and sharply defined. The left hand column shows the figure exposed at the correct dose as judged by correct location sizes in the well defined image on the upper surface of the resist. The right column shows 50% underdosed images that show the modulation of the effective dose distribution deeper within the resist.

From this data, one first notes that much less than nominal dose was needed to expose figures using high current shapes. In the case of thick MP resist, the enhancement factor was about K = 0.50, so the net dose KD_0 = 20 μC/cm^2 delivered at a rate of 0.85 μA was less than half of that needed when delivered at a low rate. Compared with the 5 A/cm^2, 3 nA control experiment, a coverage time improvement of 410X rather than the expected 273X was realized.

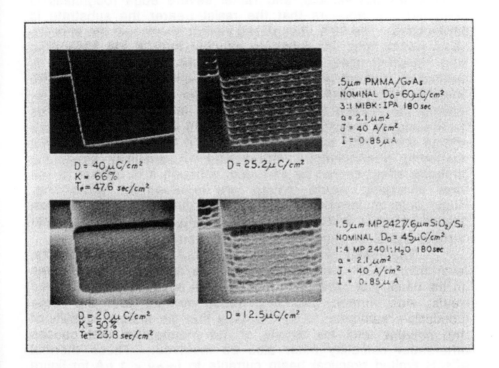

.5μm PMMA/GaAs
NOMINAL D₀=60μC/cm²
3:1 MIBK:IPA 180 sec
a = 2.1 μm²
J = 40 A/cm²
I = 0.85 μA

D = 40 μC/cm²
K = 66%
Tₑ = 47.6 sec/cm²

D = 25.2 μC/cm²

1.5 μm MP2427/.6μm SiO₂/Si
NOMINAL D₀ = 45μC/cm²
1:4 MP 2401:H₂O 180 sec
a = 2.1 μm²
J = 40 A/cm²
I = 0.85 μA

D = 20 μC/cm²
K = 50%
Tₑ = 23.8 sec/cm²

D = 12.5 μC/cm²

Figure 8. Thermal effects in two different resist/substrate systems. On the left are shown patterns whose dose was chosen to give approximately correct feature sizes. Resist heating may be recognized by the low ratio $K = D/D_0$, and by the rough edges and missing corners. The under-exposed patterns on the right suggest the temperature profiles during exposure. The thick resist profile suggests that long time constant effects also occur. Lower current exposures on the same wafers did not show.

However, the sensitivity enhancement is accompanied by degeneration in figure quality. Although a well resolved figure pattern appears in the upper surface of the resist layer, the wall profiles are not vertical, and rather severe edge roughness is observed. It appears that the resist nearer the substrate is underdosed. The 50% underdosed pattern shows that the effective dose profile near the substrate strongly reflects the serpentine step trajectory used to generate the figure, and the first flash is particularly underdosed. To cause the trajectory dependence, the effect seems to have decayed slowly after the flash is complete, leaving surrounding area more sensitive after the beam doubles back along its trajectory. At lower current density, the modulated effects were much weaker, so this cannot be a rate independent scattering phenomenon.   Some kind of sensitivity enhancing charging effect cannot be excluded, although it is difficult to see how charging would cause pre-development patterning. Regardless of detailed interpretation, feature quality degradation placed an upper limit of approximately I_{max} <0.2 µA on the beam current useful for exposing this resist system.

The PMMA results show similar but less dramatic effects, even though flash dwell times were longer and local temperatures in the resist may have been higher. This is possibly because the resist was thinner, providing better contact with the more conductive substrate. It seems likely that the thermal stability of the polymer and the details of the mechanism that causes differential solubility must also be involved.   These thermal effects limited practical beam currents to I_{max} < 1 µA for figure degradation similar to the MP resist, so in practice, the practical coverage times KD_0/I_{max} were lower in spite of PMMA's higher dose.

Figure 9 shows the dose $K(a_b, J) D_0$ plotted against beam current used to expose the 16 µm test figure with correct edge placement. Both current density and shape size were varied to gather this data set for several resist thicknesses and resist processes. In this experiment, beam current $I = a_b J$ proved to be the parameter that controlled the correct dosage $D = K(I) D_0$. This result would be consistent with the idea that the shape size was not as important as the net power input. While the normal dose requirement D_0 varied with the type of resist and developer conditions, the shape of the curves seems to depend primarily upon thickness. If trajectory effects are ignored, it should be possible to modify dwell times on the basis of flash area, thus correcting doses for thermal changes in resist sensitivity. The first order correction does not remove edge roughness and profile

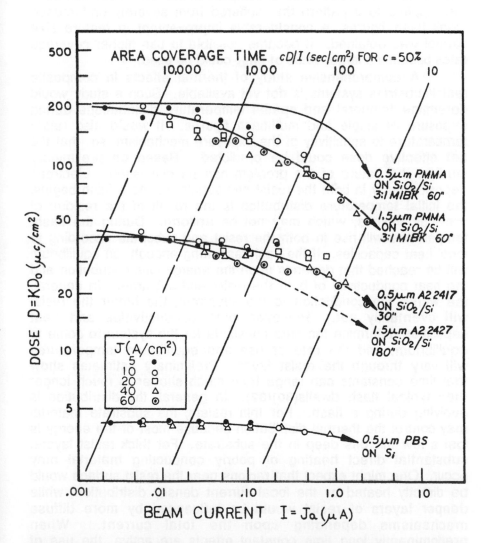

Figure 9. Reduction of dose as a function of beam current for several resist/substrate systems. In this graph, the dose required for correct sizing of 16 μm square pads was determined for many combinations of current density and square flash area. Results correlate with beam current J_a and resist thickness. Corresponding coverage times are shown above.

problems, but it can insure that figures received optimum dose regardless of flash size used to compose them. When this approach was applied to a pattern that suffered from severely underdosed small flash figures, a considerable improvement in feature size control was obtained. It became possible to use higher coverage rates because a larger value of I_{max} could be used.

A comprehensive study of thermal effects in composite resist/substrate systems is not yet available. Such a study would determine temporal and spatial temperature distributions during exposure of single and multiflash figures. It would also relate temperature to sensitivity of the exposure mechanism, so that the net effective dose could be predicted. Based on preliminary studies, the nature of the problem can be discussed. Electrons deposit energy in both the resist and substrate. As a flash begins, the initial temperature distribution is the result of the residue of previous flashes, which may not be uniform. During the flash, temperature will rise in both the resist and substrate according to local heat capacities. If the flash lasts long enough, an equilibrium will be reached that depends upon the energy loss distribution and the heat conductivity of both the resist and substrate. In general, the worse the conduction to the substrate, the hotter the resist will eventually get. However, both conductivities and heat capacities determine the time constants for the system to come to equilibrium, and the rate of rise and equilibrium temperatures will vary through the resist layer. Preliminary estimates show that time constants can range from much shorter to much longer than typical flash dwells(30)(32). In general, the distribution is evolving during a flash. For thin resists, the substrate material may control the thermal distribution because most of the energy is lost a few microns deep in the substrate. For thick resist layers, substantial direct heating of poorly conducting material may occur. One might expect that regions near the resist surface would be directly heated by the local current density distribution, while deeper layers of resist would be influenced by more diffuse mechanisms depending upon the total current. When predominantly long time constant effects are active, the use of more sensitive resist should minimize temperature rise by allowing shorter flashes. As well as reducing K D_0, one would expect the maximum tolerable current to increase, so that the coverage rate factor K D_0/I_{max} would be much more favorable. However, estimates of silicon surface time constants are in the 20 nsec range, so short time constant effects may sometimes be shorter than minimum flash times(32). Clearly, the problem is complex and highly dependent upon resist and machine process parameters. It is not surprising that informal estimates of the severity of thermal effects vary with the writing strategy, the process, and the patterning goals. What seems to be clear is that the thermal behavior of resist/substrate systems will receive increased attention. Writing strategies, beam voltages, fracture parameters and machine tradeoffs will probably be influenced by deeper understanding of the physics behind resist behavior. The speculation in this section is designed to encourage research rather than provide answers.

5.9 Fracture Optimization for Resist Limited Conditions

It has been shown in the strategy section that coverage times for a specific resist process depend upon a series of choices. In particular, maximum flash area and current density may be chosen, within limits, to optimize coverage times(15). In many cases, thermal effects in the resist limit the usable beam current to values well below the machine's capability. It remains to chose J and a_{max} to use this current most efficiently.

Figure 10 shows selected results of a study aimed at optimizing the free parameters for several common resist systems(33). The pattern is a mixture of 0.5., 1.0 and 2.0 μm CD proximity test patterns. The goal is to minimize coverage times while maintaining acceptable edge roughness, feature size control, and submicron feature profile. No proximity or flash time corrections are used, but many other combinations of current density and flash sizes were also evaluated. Although resist thickness and proximity effects cause the feature quality to vary between resists, the examples shown generally represent the fastest coverage obtained before thermal degradation became evident.

This experiment illustrates an important strategy that may be used to optimize the pattern fracture. At this phase of its evolution, the AEBLE-150 imposed limits were $J_{max} = 40$ A/cm^2, h_{max} or $w_{max} = 2.0$ μm, and $t_{min} = 200$ nsec. For insensitive resists in Figure 5c, the optimum choice of maximum flash area is I_{max}/J_{max}. To fracture these patterns, the resist limited current, I_{max}, was first determined by separate experiments using feature quality as a criteria. Then a_{max} was chosen according to Figure 5c.

In these cases, the optimum maximum flash area is considerably smaller than the largest available flash. In general, it proved desirable to reduce beam currents to tolerable levels by

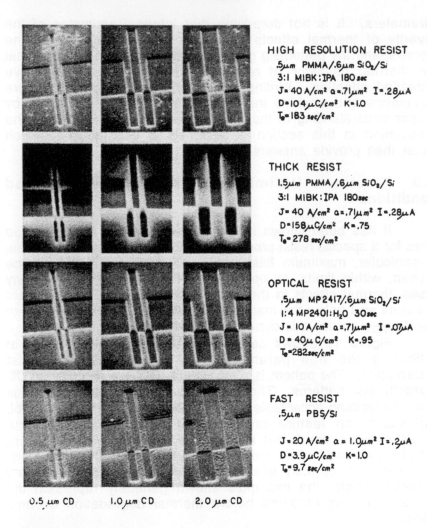

HIGH RESOLUTION RESIST
.5μm PMMA/.6μm SiO₂/Si
3:1 MIBK:IPA 180 sec
J=40 A/cm² a=.71μm² I=.28μA
D=104μC/cm² K=1.0
T₀=183 sec/cm²

THICK RESIST
1.5μm PMMA/.6μm SiO₂/Si
3:1 MIBK:IPA 180 sec
J=40 A/cm² a=.71μm² I=.28μA
D=158μC/cm² K=.75
T₀=278 sec/cm²

OPTICAL RESIST
.5μm MP2417/.6μm SiO₂/Si
1:4 MP2401:H₂O 30 sec
J=10 A/cm² a=.71μm² I=.07μA
D=40μC/cm² K=.95
T₀=282 sec/cm²

FAST RESIST
.5μm PBS/Si

J=20 A/cm² a=1.0μm² I=.2μA
D=3.9μC/cm² K=1.0
T₀=9.7 sec/cm²

0.5 μm CD   1.0 μm CD   2.0 μm CD

Figure 10. A comparative study of submicron patterns in different resist/substrate systems. The patterns simulate the mixture of features that might be found in a circuit pattern. In each case, the combination of shape area, current density and dwell time that gave acceptable patterns at minimum coverage time was selected for presentation. In all cases except PBS, resist behavior similar to Figure 8 prevented the use of higher current and the demonstration of faster coverage.

reducing flash area rather than current density. In this experiment, the fastest fracture used the same flash area, a_{max} = 0.7 μm^2, for all three test features, even though it would be possible to compose the larger features from larger, lower current density flashes. Instead of varying flash area, the aspect ratio h_i/w_i was varied, so that each feature was written using the same current $l_{max} = Jh_iw_i$.

This kind of fracture has several advantages. Although there was considerable thermal enhancement of the required dose in some cases, the required dose did not vary with feature size because the current did not vary. The fracture avoided overdosing large features, which is otherwise typical of thermally limited situations. This constant area fracture also tended to maximize the time average current, so that it was close to the maximum tolerable. This led to minimum coverage times regardless of pattern statistics. In general, the experiments suggested that when using insensitive resists with limited tolerance to current, it is desirable to use relatively small, high current density flashes, even though more flashes are needed.

One will also notice that coverage times in Figure 10 are much longer than one might wish. The machine parameters would suggest that it should have been possible to write much faster. Using present AEBLE 150-imposed limits of J_{max} = 100 A/cm^2, h_{max}, w_{max} = 2 μm, t_{min} = 100 nsec. and beam interactions-imposed limits of l_{max} = 1.0 μA at δ_b = 0.15 μm, the 50% coverage times would be 52, 79, 20 and 5 sec/cm^2 from top to bottom. In the actual experiment, resist heating effects prevented this performance in all resists except PBS. Less favorable fracture strategies yield even slower results. The experiment emphasizes the importance of a careful determination of l_{max} and K D_0, followed by a fracture and machine setup specifically chosen for the process. It also emphasizes the need for more sensitive, thermally robust resists.

6.0   AN EBL DIRECT WRITE PATTERN GENERATOR

This section will discuss a modern direct writer more specifically. The AEBLE 150 is manufactured by Perkin-Elmer Corp.(8)(34). It is an outgrowth of the VHSIC I program in cooperation with Hughes Research Labs. The system was specifically designed for direct writing of features averaging one micron and below on a variety of resist systems. It combines submicron absolute and overlay accuracy with moderately high coverage rates on relatively insensitive resists. The system differs from a mask maker not so much in its accuracy goals, but rather in potential coverage rate using the less sensitive resists that are frequently necessary in submicron device fabrication.

The system uses variable shaped beam strategy instead of raster or fixed beam vector scan alternatives. Vector scan was chosen because very fine address structure is necessary for submicron lithography. An equivalent 1/128 µm raster scan address structure would demand a 1560 GHz flash rate! Variable shaped beam was chosen because relatively high coverage rates are possible without resorting to more sophisticated electron gun, deflection and figure generation technology.

The system is designed for submicron direct write lithography with mean feature areas $f' = 1 - 10 \ \mu m^2$ using maximum rectangular shape dimensions of 2.0 µm. The electron optical design uses a lanthanum hexaboride cathode and large aperture, moderate scan field optics that favor high current density, small shapes that are stepped at high speed to compose figures. The optics-limited edge slope $\delta_b \geq 0.07$ µm, and increases with current density and shape area according to the formula $\delta_b = [0.072 + (0.0009J_{ab})^2]^{1/2}$ (µm)(8). The beam is stepped at $t_{min} = 100$ nsec without blanking between flashes within a figure. According to Figure 5c, coverage times of 2 - 10 sec/cm^2 are possible with submicron patterns. Fracture optimization is possible in the job preparation software by imposing upper limits on current or shape size.

Figure 11 shows the major functional blocks of the system. The job preparation software operates on a large super-mini computer. It accepts several CAD format data bases. These are eventually converted into lists of figure words and, where possible, into "microcell references". The software uses a hierarchical scheme that allows redundant frames and subfields to be fractured once and used repeatedly. For patterns that are composed of standard circuit elements, this provides economy in both fracture and input data handling.

The data path subsystem operates during writing. Data expansion logic stores figure, microcell and subfield origin data in a large buffer memory. Each figure word is accessed from either microcell or individual figure sections of this memory, and passed onto the pattern generator. The figure words set up origin, shape and exposure data, and initiate the figure writing sequence described previously. Coarse field moves, frame center and

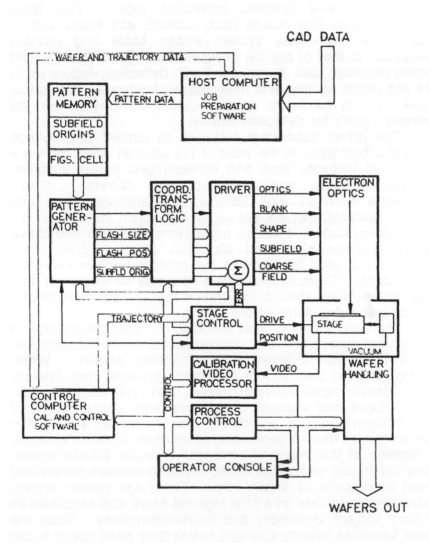

Figure 11. Functional block diagram of the AEBLE 150 direct write lithography system.

timing data are also transmitted in the form of special purpose words.

All positioning data passes through coordinate transformation and dynamic correction logic. This logic ultimately insures that coarse field, subfield, and shape activity occur in a coordinate system whose scale and rotation correspond to that of the die being written. Gain, rotation, and dynamic distortion data is combined with deflection vectors using RAM and digital addition and multiplication hardware. The output of this logic is hardware specific digital words that are the command signals for deflection activity.

The driver subsystem includes all current and voltage drivers that contribute to the state of the electron optics during a flash. This includes lens and electron gun power supplies, deflection drivers, and shaper/blanker drivers. Static adjustments such as high voltage, gun emission, lens and alignment currents, etc. are set up by a multiport interface to the control minicomputer. Position measurements gather coordinate transform, shape size, and distortion data. Curve fitting techniques are used to minimize the random error in individual measurements. The sum total of this data may be thought of as a pattern-independent state of calibration that is applied to drivers and data path subsystems before writing.

The wafer handling, load and vacuum, video processor, and stage control subsystems are microprocessor controlled units that operate somewhat independently of writing activity. Wafer handling removes wafers from cassettes, positions them laterally and rotationally upon transport pallets, and replaces them after writing. Load and vacuum control oversees transportation of pallets through a vacuum airlock and onto the moving stage. The video processor uses a backscattered electron signal to measure the location of the beam with respect to wafer fiducial marks. During calibration and setup cycle, it also processes transmitted current from scans of a grid edge. The stage control system receives trajectory data on a chip segment basis, and calculates its trajectory subject to velocity and acceleration limits. Since the system knows of velocity changes before they must occur, it can smooth the trajectory for minimum overhead. Writing occurs at a pace dictated by these stage trajectories.

System setup and wafer registration are performed primarily using line scan techniques. When a shaped beam is passed over an edge, the transmitted current can be used to determine the position, size, shape and uniformity of the beam. Shape size and position measurements are used to load lookup table memory with data that will be presented to the electron optics each time the shape is called.   Position and uniformity measurements are made under varying lens conditions to set up focus, alignment, stigmation, and Kohler imaging conditions. Deflection gain, rotation and distortion measurements are made by simultaneously moving the stage and beam to several locations in the field. When completed, these adjustments should overlay the coarse and subfield deflection with the interferometer controlled stage coordinate system.   Wafer dependent adjustments are executed separately.

This system uses wafer registration techniques that resemble those used for mask writing.   For mask or reticle generator, the pattern must conform to an orthogonal grid whose scale and linearity refer to an absolute standard. Mask making places special emphasis upon accuracy over distances comparable to the mask size. Control of the magnetic and thermal environment is necessary. This system employs similar techniques for wafer writing.   Before writing, a "coarse wafer map" is taken. This locates the die position, size and rotation, and is also capable of measuring their individual heights with respect to the beam focus. In total, this data set allows the system to write each wafer in a local coordinate system defined by fiducial marks. In-plane wafer distortions, thermal expansion and flatness effects are calibrated before writing. During writing, position is controlled entirely by the interferometer system, but the beam offset and coarse field calibration are occasionally monitored. Between calibrations, the system relies on environmental control to maintain stability.

This general calibration strategy is dictated partly by the need to "mix and match" different lithography techniques. In the case where EBL is used to make masks or first level wafer patterns, there are no fiducial marks to locate patterns. Levels must be written in conformance with absolute standards that can be matched with the light optical, X-ray or other techniques that may be economically favored for other process steps. In a mix and match situation, there is no substitute for absolute accuracy over relatively large distances. Thus the absolute accuracy of the machine may also be exploited for writing registered layers.

While much of the preceding discussion emphasizes writing speed limitations, this and other VSB systems have proven to be much faster than raster and Gaussian beam vector systems. The coverage rates demonstrated in Figure 10 may be compared to the MEBES III raster scan system.   The thin PMMA and PBS results are 57X and 41X faster than possible with a 0.1 μm address, 80 MHz pixel rate raster scan system.

On the other hand, the direct write EBL systems have been slower than expected in finding their place in the manufacturing environment. In particular, intense concentration upon strictly machine technology has overshadowed some of the process problems that are unique to EBL. Paramount among these are resist heating which has limited realistic throughput, and scattering in thick resists, which cause profile and proximity problems. Increased speed has also made accuracy goals more illusive, due to both calibration complexity and resist charging effects. In the author's opinion, this situation will change with time, as electron specific processes and optimized fracture strategies come into more general use.

7.0 THE FUTURE

Electron beam lithography is an evolving technology. Undoubtedly both its range of application and its equipment performance will increase. As patterns become smaller and more complex, both accuracy and coverage rates must increase. If process performance is ultimately measured in units of cost and turnover time per viable circuit element, then progress will extend traditional writing strategies and hardware configurations. In some areas, present equipment already approaches fundamental limits of electron optical, analog electronic, data processing, and data storage technology. One might speculate that in the electronic subsystems, progress may be a "bootstrap" process, where traditional machines contribute to the development and manufacture of components for advanced EBL designs. Particularly in the areas of writing strategy and resist processes, there seems to be room for innovation.

EBL might be expected to make an especially critical contribution in the development and manufacture of new, unique, or relatively low volume IC designs. It is an effective tool for an iterative design process because it can modify a pattern upon command. When comparing EBL with light optical or other replication techniques, this fundamental difference in role should be emphasized quite separately from issues involving circuit dimensions. EBL will probably evolve in directions that particularly favor primary pattern generation, partly as an ASIC manufacturing tool and partly as a primary high resolution mask and reticle generator.

EBL pattern generation is a very data intensive process because VLSI circuit patterns contain large quantities of unique data(35). It is the author's opinion that a closer coordination between CAD design standards and EBL writing strategies would considerably enhance the entire process. Relatively straightforward coordination such as matching circuit cells with available field sizes would reduce fracture time, overlay problems, and data storage requirements. The use of a common, non-overlapping figure repertoire would allow lithography to more accurately reproduce the circuit designer's intent.   If libraries of fractured, dose corrected, and pre-tested circuit pattern elements were employed at the CAD station for assembling complex VLSI's, then lithographic optimization could proceed in advance of the prototyping of large scale designs. Similarly, some of the techniques of fracture optimization discussed in Section 4.0 should prove to be important in optimizing throughput in the face of process limitations.   Evidently, each resist process requires a study of optimum fracture and exposure parameters before it can be fully exploited.

As the EBL process matures, the further evolution of resist materials and processes that are tuned to specific EBL mask and wafer fabrication steps, seems to be a high priority task.   The development of high resolution/high throughput equipment has placed additional demands upon the resist process. The analysis of writing strategies shows that the ratio of required dose to maximum tolerable current sets an upper limit to potential coverage rates.   As features shrink and coverage rate increases, both power and power density within the writing beam must increase.   As well as being specifically sensitive to electron radiation, resists need to be tolerant to the thermal variations that occur as a result of changes in beam sizes and pattern variations of underlying materials.   In particular, the physical/chemical mechanisms responsible for solubility changes, proximity effects, resolution limitations, thermal effects and defects all seem to require study at the level of electron/material interaction.   These effects seem to have a profound influence upon the overall viability of EBL processes.

The data processing, absolute accuracy and in-vacuum aspects of EBL equipment design seem to be particularly important in determining the character of machines.   Perhaps electron optics will slowly recede as a pacing technology in EBL, because overall process performance seems to be equally limited by resists, electronic and calibration technologies.   However, the use of electrons in the context of primary pattern generation implies hardware that is very different from machines that use photons in the context of replications.   In particular, pattern preparation, environmental control and in-vacuum wafer handling each generate equipment design challenges that are different from those encountered in photolithography.   These differences are presently reflected in the cost of equipment.   While simpler, less expensive

REFERENCES

1.   Ward, R., "Developments in Electron Image Projection", Proc. 15th Symp. Electron, Ion, Photon Beam Technology pp. 1830-1833 (1980).

2.   Heritage, M., "Electron Projection Microfabrication System", Proc. 13th Symp. Electron, Ion, Photon Beam Technology pp 1135-1140 (1975).

3.   Frosien, J., Lischke, B., Anger, K., "Aligned Multilayered Structures Generated by Electron Microprojection," Proc. 15th Symp. Electron, Ion, Photon Beam Technology, pp. 1827-1829 (1980).

4.   Nehmiz, P., et al., "E-beam Proximity Printing, "J. Vac. Sci. Technol. B3 (1) pg. 136 Jan/Feb (1985).

5.   Herriot, P., Collier, R., Alles, D., Stafford, J., "EBES, A Practical Electron Lithography System" IEEE Transactions on Electron Devices. Vol. ED-22 No. 7, pg. 385 (1975).

6.   Pfeiffer, H.C., "Recent Advances in EBL for High Volume Production of VLSI Devices," IEEE Transaction of Electron Devices. Vol. ED-26 No. 4, pg. 663 (1979).

7.   Goto, E., Soma, T., Ickesawa, M., "Design of Variable Aperture Projection and Scanning System for Electron Beams". J. Vac. Sci. Technol. 15(3) May/June (1978).

8.   Veneklasen, L., "A High Speed EBL Column," J. Vac. Sci. Technol. B3(1), Jan/Feb (1985).

9.   Moore, R.D., et al. "EL3, a High Throughput, High Resolution EBL Tool," J. Vac. Sci. Technol. 19(4) pg. 950 (1981).

10. For example JEOL, Hitachi and Perkin-Elmer.

11. Brodie, I., et al. "A Multiple Beam Exposure System for Submicron Lithography," IEEE Transactions on Electron Devices. Vol. ED 28 No. 11 (1981).

12. Smith, D.O., Harte, K. "Electron Beam Array Lithography". J. Vac. Sci. Technol. 19(4) pg. 953 Nov/Dec (1981).

13. Van der Maast, K., Jansen, G., Barth, J. "The Shower Beam Concept". Microcircuit Eng. 3 pp. 43-51 (1985).

14. Phang, J.C.H., Amed, H. J. Vac. Sci. Technol. 16 pg. 1754 (1979).

15. Takamoto, K. "Optimization of System Parameters for Throughput in a VSB Exposure System". J. Vac. Sci. Technol. B5 (3) May/June (1987).

16. Weber, E.V., Yourke, H.S. "Scanning E-beam System Turns Out Wafers Fast," Electronics, Vol. 50 No. 23 pg 96 (1977).

17. Alles, D. et al., "EBES 4, A New E-beam Exposure System," J. Vac. Sci. Technol. B5(1) pg. 47 Jan/Feb (1987).

18. Thompson, M.G.R. et al, "The EBES 4 Electron Beam Column," J. Vac. Sci. Technol. B5 (1) pg 53 Jan/Feb (1987).

19. Jansen, G.H., Stickel, W. "Trajectory Displacement and Space Charge Effect in E-Beams", Microcircuit Eng. 4 pp. 167-176 (1984).(Eq. 6 used with authors advice, based on later work. Note: also that ΔV should include "Borsch effect.")

20. Chen, A., Neureuther, A., Pavkovich, J. "Proximity Correction in VSB Lithography" J. Vac. Sci. Technol. B3(1) Jan/Feb (1985).

21. Rosenfield, M.G., Neureuther, A., Ting, CH. "The Use of Bias in EBL for Improved Profile and Linewidth Control". J. Vac. Sci. Technol. 19(4) pg. 1242 Nov/Dec (1981).

22. Stickel, W., Langner, O. "Edge Contrast, a New Definition for Comparative Lithography Tool Characterization". J. Vac. Sci. Technol. B1(4) Oct/Dec (1983).

23. Broers, A.N. "High Resolution E-beam Fabrication Using STEM". Proc. Ninth Int. Cong. on Electron Microscopy, Vol. 3., Toronto (1987).

24. Parikh, M. "Self Consistent Method for Proximity Correction". J. Vac. Sci. Technol. 15, pg. 931 (1978).

25. Otto, O., Griffith, A. "Proximity correction on the AEBLE 150". J. Vac. Sci. Technol. B6(1) Jan/Feb (1988).

26. Owen, G., Rissman, P. "Proximity Effect Correction by Equalization of Background Dose". J. Appl. Phys. 54 No. 6 pg. 3573 (1983).

27. Dean, R. "Oxidation as a Major Cause of Defects in E-beam and Optical Resists". Proc. SPIE Conf. Sec. 773 pp. 92-102 (1987).

28. Milner, K. MEBES Users Group. Jan 1986 (unpublished).

29. Nomura, N. et al. "An E-beam Direct Write Process for 1/2 μm DRAM" Proc SPIE Conf. Sec 932 pp. 30-38 (1988).

30. Ralph, H.I., Duggen, G., Elliot, R. "Resist Heating in High Speed E-beam Pattern Generator". Proc. 12 Int. Symp. Electron, Ion, Photon Beam Technology. 1982 Electrochemical Soc. Proc. Vol 83-2 (1982).

31. Morai, F., Okazaki, S., Saito, N., Dan, M. "The Effects of Accelerating Voltage on Linewidth Control with a Variable Shaped E-beam System". J. Vac. Sci. Technol. B5(1) Jan/Feb (1987).

32.   Young, L.J.   Perkin-Elmer Corp. (unpublished).
33.   Veneklasen, L., Humphreys, M., Howard, G. (unpublished).
34.   King, H. et al. "An EBL System for Submicron VHSIC Device
       Fabrication".   J. Vac. Sci. Technol. B3(1) Jan/Feb
       (1988).
35.   Veneklasen, L. "EBL and Information Transfer".
       Microcircuit Eng. 3 pp. 33-42 (1985).

6

X-RAY LITHOGRAPHY

William B. Glendinning

U.S. Army ETDL*
Ft. Monmouth, New Jersey

Franco Cerrina

Director, Center for X-ray Lithography
University of Wisconsin
Madison, Wisconsin

1.0 INTRODUCTION

The inquisition of well established optical (visible) IC chip lithography practice and principles led to the X-ray-radiation-based alternative. The pervasive scrutiny of the optical-mask shadow details at the mask's opaque pattern edges revealed perplexing questions(1). Conclusions were made that visible (400 nm to 600 nm) proximity and projection optical printing was diffraction limited and would not provide high quality delineation at IC device minimum feature size (MFS) dimensions of <2.0 µm(2). Conversely, X-ray-mask-produced shadows with only small Fresnel diffraction limitations (wavelengths of 0.4 nm to 1.5 nm) would extend high quality IC production lithography to 1.0 µm MFS and less. X-ray flux could provide high volume, high density IC device production as well as meet the necessary favorable throughput, yield, and processing cost criteria. With time, the X-ray lithography method, although lacking even few sizable production-use examples, retained support and still does by virtue of it's fairly unencumbered diffraction feature. In addition, X-ray's associated large depth of focus, wide process latitude, and excellent linewidth control, furthered arguments favoring the X-ray alternative. The relative dust-and-defect immunity of X-ray printing, an elegant and additional plus for high quality/high yield IC processing, has not as yet been fully established(3).

*Presently retired and acting as a consultant.

The text in Parts 1.0-10.0 of this chapter will present the growth and development of the present day X-ray lithography printing methods mostly from a process and user viewpoint. The X-ray printing method will be defined and described as a system in terms of the components of source, mask, aligner, and resist. The important component design error budget factors are detailed to provide both insight and clarity regarding the X-ray system complexity. The strengths and weaknesses of the major X-ray printing system sources and aligners are cited. Some emphasis is then given to mask technology in terms of fabrication, membrane and absorber materials, and absorber patterning. Finally, the extent of IC device damage from X-ray lithography process steps is shown. The chapter ends with a few brief conclusions and directs attention towards the subject of Part 11.0, namely, the synchrotron as a unique and technically ideal X-ray source.

2.0   X-RAY PRINTING METHOD - SYSTEM APPROACH

The X-ray method of lithography, like its spawning predecessor (visible optical proximity printing), can be defined in a system approach manner. The X-ray system's major components are: source, mask, aligner, and resist as seen in Figure 1a. In Figure 1b, X-ray photons emitted from the radiation source of finite dimensions d travel through distance D to impinge on the mask of field size F. The X-ray flux continues it's path through the mask membrane of thickness t. The photons either pass through clear mask areas including the smallest feature width W, or are absorbed in the absorber areas of height h. The exiting aerial image of flux from the bottom of the mask travels the gap spacing distance s to be finally absorbed in the resist creating photoelectrons. A latent image in the resist corresponding to the geometric mask absorber pattern instantly forms by the photoelectron-induced changes to the molecular weights within the resist layer (see Chapter 2). This simple physical model is applicable to all X-ray lithography systems regardless of size, power, fixed or scanned flux, path direction of flux (vertical to horizontal), or stepped or fixed (full wafer) field.

2.1   X-ray System Definitions

Since the X-ray system is employed to perform key lithography process steps in the IC chip manufacturing line, severe lithography constraints are levied in context of the

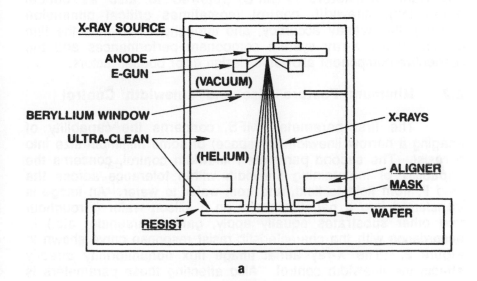

a

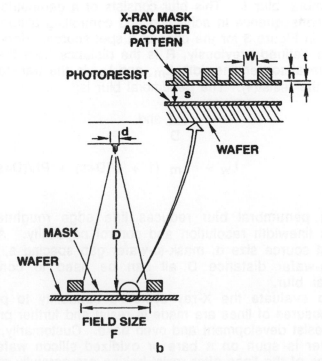

b

Figure 1. a.   X-ray lithography system concept; b.   X-ray exposure physical model.

complexity of the particular circuit device being fabricated. Four important parameters -- MFS (referred to also as critical dimension), linewidth control (sometimes critical dimension control), the overlay accuracy, and throughput -- dictate the fine details of the X-ray system component performances and the respective component allowable design error budget factors.

2.2   Minimum Feature Size and Linewidth Control

The first parameter, MFS, concerns the capability of imaging a narrow linewidth (or space) of some minimum size into a resist. The second parameter, linewidth control, concerns the capability of maintaining linewidth within tolerance across the field F, from field to field, and from wafer to wafer. An image is transferred into a negative resist on a silicon wafer (throughout text other substrates equally apply, gallium arsenide, etc.) in accordance with the characteristic resist response curve shown in Figure 2. The X-ray aerial image flux nonuniformity directly affects the linewidth control. Also affecting these parameters is the penumbral blur δ. This blur consists of a geometric shadow factor of consequence in achieving and controlling dimensions as illustrated in Figure 3 for the case of a spot source. Here D, d, h, and s are defined previously, R is the distance from the source axis to the absorber linewidth L_m, and L_w is the transferred Lm image in the resist[4]. The penumbral blur is:

[1]
$$\delta = \frac{sd}{D}, \text{ and,}$$

[2]
$$L_w = L_m \, (1 + s/(D+s) + Rh/(D+s)).$$

Reducing penumbral blur reduces line edge roughness and increases linewidth resolution and control capability. As seen, equivalent source size d, mask-to-wafer gap spacing s, and the source-to-wafer distance D all can be used to control the penumbral blur.

To evaluate the X-ray system's capability to print fine lines, exposures of lines are made in resist and further processed through resist development and oven bake. Customarily, the thin resist layer is spun on a bare or oxidized silicon wafer. The dimensions of the lines after resist baking are carefully measured by special optical microscopes or by electron-optics (scanning electron microscope [SEM]) methods. As an example, in high

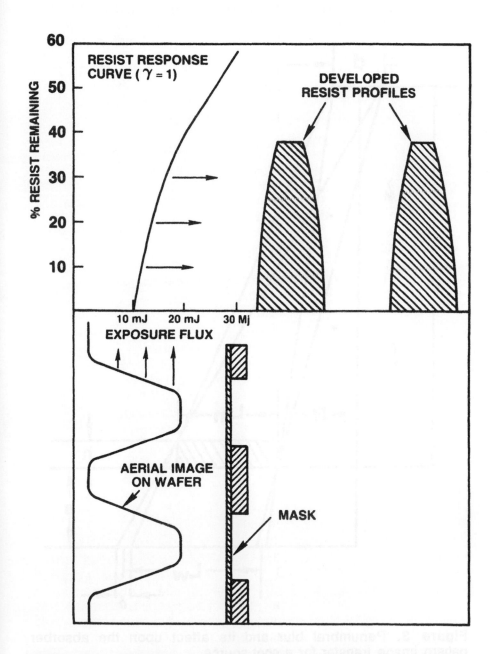

Figure 2. Transfer of mask absorber pattern into negative resist in accordance with the resist's transfer curve.

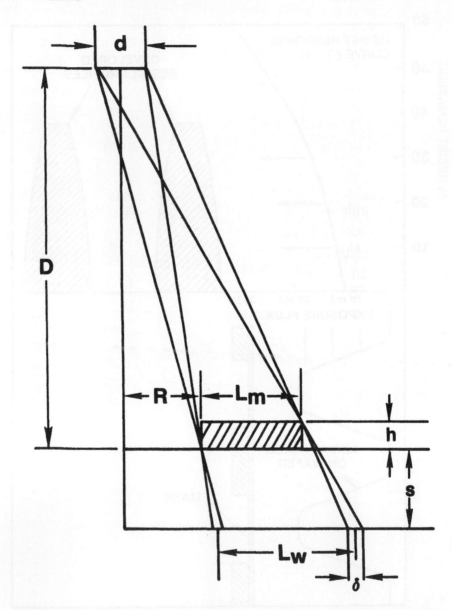

Figure 3. Penumbral blur and its affect upon the absorber pattern image transfer for a spot source.

density IC device work such lines would be made at a minimum feature size of 0.5 μm(5). High quality linewidth control of 75 nm (3σ) is expected. To fully qualify the X-ray system, x- and y-directed lines must be evaluated throughout the entire exposure field F, across the entire wafer, and from wafer to wafer. The line edge quality is also of importance and is measured as part of the linewidth artifacts. An edge roughness per line side of 75 nm (3σ) is acceptable. Table 1 shows the design error budget numerical values levied against a spot source, aligner, mask, and resist of an example X-ray system in order to achieve the required MFS, and both the MFS and the edge roughness controls (2σ range is compatible with the tolerances cited above).

2.3   Overlay Accuracy

The next parameter for judging the capability of an X-ray system is the overlay accuracy. The manufacturing of IC devices requires many process steps involving the lithographic delineation of geometries. To make IC chips to exacting electrical performance specifications and achieve production yields, each lithographic mask level must be aligned perfectly to all of the other lithographic levels and vice versa. In terms of the device MFS, acceptable production yields occur if the x- and y- direction overlay accuracy value is at least 1/5 of the MFS. An MFS of 0.5 μm requires that the overlay accuracy be <±0.1 μm (2σ). Overlay accuracy values represent measured x- and y- directed data sample populations with spreads of 2 or 3 sigma from the mean. Sample data is obtained by the fabrication and test of simple electrical devices, optically read verniers, and opto-electronically read bench marks (e.g., matrix of market crosses). See Figure 4 (6)(7). Table 2 shows the design error budget numerical values levied against the source, aligner, and mask components in order to achieve the desired overlay accuracy for a spot source X-ray system example. Mask-to-mask (±80 nm) and aligner (±60 nm) are the major error components. The wafer contributes to the overlay error (distortion) and is therefore included in the error budget.

2.4   Throughput

The last parameter for judging the X-ray system performance capability is throughput. The throughput directly bears on the essential value of any lithography method, i.e., the throughput value must favorably affect the manufacturing cost per

TABLE 1.   Elements of the linewidth control design error budget for a spot source X-ray system example.   MFS = 0.5 μm; Linewidth control = ±50 nm (2σ); Penumbral blur = MFS/3 = 170 nm.

Component	Error Element	Design Budget Allowance (± nm)
Source	Source-to-wafer (40 μm to 50 μm)	1 0
	X-ray flux uniformity: Across field (± 2%)	5
	Exposure to exposure (± 2%)	5
Aligner	Mask-to-wafer stability	1 0
Mask Edge profile <150 nm/side Roughness <50 nm/side	Absorber projected angle Linewidth uniformity: E-beam pattern Across field	1 0

1 0 |
	Mask-to-mask	1 0
	Processing: Across field	1 0
	Mask-to-mask	1 0
	X-ray transmission Across mask (± 2%)	4
Resist Thickness >300 nm Edge roughness <50 nm	Processing development control	3 0

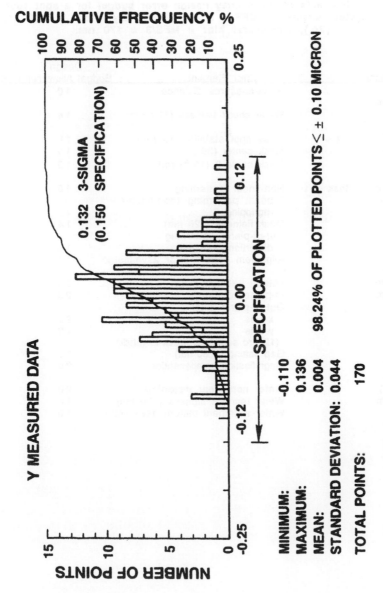

Figure 4. Y - direction overlay error data -- two different mask levels printed on the same wafer. Courtesy Perkin-Elmer.

TABLE 2.   Elements of the overlay design error budget for a spot source X-ray system example.   MFS = 0.5 nm; Total x,y errors = ±100 nm (2σ); Penumbral blur = MFS/3 = 170 nm.

Component	Error Element	Design Budget Allowance (± nm)
Source ±20 nm	Mask-to-source distance	10
	Wafer chuck flatness (120 nm)	14
	X-ray spot stability (30 μm)	10
	Mask center (30 μm)	10
	Mask angle 3 (10⁻⁴) rad	10
Mask (Single Mask ±60 nm)	Non-linear patterning	40
	E-beam patterning (non-linear/ regripping	40
	Mask flatness (120 nm)	14
	Out-of-plane patterning distortion (120 nm)	14
	Alignment pattern error asymmetric	10
Alignment ±60 nm	Lateral translation x,y	20
	In-plane rotation	20
	Out-of-plane tilts:	
	x-axis	28
	y-axis	28
	(12 arc seconds, mask-to-mask)	
	Gap setting (± 60 nm)	20
	Magnification compensation	20
(Wafer) ±30 nm	Wafer non-linear distortion	20
	Wafer rechuck flatness (120 nm)	14
	Wafer alignment pattern asymmetry	10

chip equation. Hence, the throughput should be as high as possible. It is evaluated by lithographically printing a series of wafer levels for a measured period of time. The time interval must be long enough to include the prospects and consequences of downtime (equipment breakdown) should it occur(8). Table 3 shows the design error budget numerical values levied against a spot source, planar-motor driven aligner stage with physical optoelectronic alignment detection/positioning control (zone-plate/grating method). Source, mask, aligner, and resist components are all throughput determinants. The source and resist especially are shown in other chapter sections as the salient throughput determining factors.   A throughput of forty-five 100 mm (diameter) wafer levels per hour for 0.5 μm MFS lithography is acceptable, or, an equivalent exposed area value (mm^2/hour) for the case of using larger diameter wafers (e.g., 150 mm, or 200 mm).

3.0   X-RAY SYSTEM COMPONENTS

To understand the X-ray lithography method requires familiarity with the features, materials, configurations, physics, electronics, and the principles of operation of the X-ray system components of source, mask, aligner, and resist. Broad views of these components can only be presented here. Some depth in mask technology is provided later in the chapter. The reader is directed to Chapter 2 and Part 11.0 of this chapter for in depth discussions of photoresist and X-ray synchrotron source subject matter, respectively.

3.1   Sources For X-ray Flux

The selection of a source type levies several important design constraints: source size, power emitted, mask-to-source distance, characteristic and broadband radiation, flux scanning, aligner orientation, mask membrane and absorber material, and photoresist spectral sensitivity and contrast. In practice, three categories of sources have shown lithographic quality to varying degrees as noted in Table 4. Electron impact (fixed, rotating), plasma (laser, gas-puff), and the electron storage ring or synchrotron (warm magnet, cold [superconducting] magnet). Ideally, the X-ray emitter should be of minimal size <1 mm, provide collimated flux (approximated by large source-to-mask distance [2m to 4m], or, grazing-angle scanning), emit copious

TABLE 3: Elements of the design throughput budget for a spot source X-ray system example. 50 wafer levels (100 mm diameter) per hour, 7 fields (3 cm x 3 cm) per wafer level.

Component	Error Element	Design Budget Allowance
Source	Power input	10 kW
	Source-to-wafer distance	18 cm
	Penumbral blur	170 nm
	Beryllium window	15 μm
Mask	X-ray transparency	50%
Aligner	Overhead:	20.5 seconds/wafer
	Transport	2.5 seconds
	Gap prealign	2.8 seconds
	Transport to/from exposure	15.2 seconds
	Subfield time:	12.6 seconds/wafer
	Accelerate	0.1 seconds
	Move	0.5 seconds
	Decelerate	0.2 seconds
	Fine lateral align	0.5 seconds
	Gap adjust	0.5 seconds
Resist	Exposure time/field (resist sensitivity = 8 mJ/cm^2)	5.5 seconds/field

TABLE 4. X-ray source types and related parameters.

Parameter	Electron Impact		Plasma		Storage ring[a]	
	Stationary	Rotating	Z-Pinch	Laser	Warm	Superconducting
Material	Palladium 5 - 6 kW	Tungsten 10 kW	Krypton Ipk 400-500 kA	Steel,Cu Nd: Glass	1.0 GeV	0.7 GeV
Distance Source/ Mask (cm)	40	18.0	18.0	10	1500	1500
Source size (d) (nm)	4	1.5	1	0.1 - 0.2	x: 0.5 z: 0.075	x: 0.7 z: 0.7
Penumbral blur (nm)	400	170	170	300	x: 5-10 z: 0.05	x: 1.3 z: 1.3
Wavelength (nm)	0.44	0.7	0.6 - 0.8	0.4-2.2 1.2-1.6	0.4-2.0 λc: 1.0	λc: 8.45
Irradiance at resist mW/cm^2	0.3	1.5	2.5 - 12.5	5.0-25[b]	50	100
Isolation window (μm)	Be, 50	Be, 15	Be, 15 - 25	no Be, 0	Be, 25	Be, 25

a. See Table 8, this chapter.
b. With 100 J/pulse laser output.

photons of complementary energy for a per field exposure time of a few seconds duration. The storage ring is near to an ideal emitter source. Its awkward features are its physical and cost size which will be mentioned later (also see Chapter 1).

3.1.1 Electron Impact Source.
Electron impact X-ray sources emit the smallest quantities of photons. Electrons bombarding a solid target produce both a characteristic line (stripping of inner nucleus electrons) and continuous Bremsstrahlung (deceleration of electrons) X-rays(9). Emitted power, for example, from a 10kW rotating tungsten (M- line) anode is absorbed enroute to the wafer/resist by: a beryllium window (25 μm, T=0.61), ambient gas (He, 18 cm, T=0.96), boron nitride mask (3.0 μm, T=0.72), and a polyimide layer (2 μm, T=0.88), where T is the flux fraction transmitted through the medium. The total transmission coefficient is T_T = 0.38. For the optimum X-ray system performance the mask, resist, and anode materials are customarily selected for compatibility with respect to the characteristic radiation. If a resist with a sensitivity of D_i = 8 mJ/cm^2 is used, and an assumed field exposure time t_{exp} = 5.0 sec applied, the incident on-resist X-ray power irradiance ϕ must be:

[3] $$\phi = D_i/t = \frac{8mJ/cm^2}{5.0sec} = 1.6 \ mW/cm^2 \cdot$$

A tungsten anode X-ray generation efficiency which relates irradiance to input power P(10)(11) is:

[4] $\eta = \phi d^2 /P \ T_T$
$$= [1.6 \times 10^{-3} \ W/cm^2 \cdot (18)^2] / [10 \ kW \cdot 0.38]$$

$$\eta = 136 \ \mu W/W\text{-}Sr.$$

The maximum power dissipation of a rotating anode is:

[5] $$P = (\Pi /4)^2 (\rho C_v k d^3 v)^{1/2} \Delta T, \ where,$$

ρ is the density of the material, C_v is the specific heat, k is the thermal conductivity, d is the spot size, v is the surface velocity, and ΔT is the maximum allowable temperature rise. For tungsten,

$$P = (\pi/4)^2 \left[\underbrace{(19.3)}_{cm^3} g(0.155) \underbrace{W-s}_{g/°C} 1.99 \underbrace{W}_{cm/°C} 0.1^3 \ cm^3 2.2(10^3) \underbrace{cm}_{s} \right]^{1/2} 3345°C,$$

and,   P = 7.5 kW.

The incorporation of water cooling enables a P value increase from 10 kW to 20 kW to be realized(12)(13).   High density, high melting point metals such as molybdenum and tungsten have high figures of merit for high power anodes. Rotation velocities of 7000 rpm to 8000 rpm are required.   The vacuum condition of electron impact anodes are usually maintained via high transmission beryllium windows which absorb very few of the emitted photons enroute to the resist/wafer.

   3.1.2 Plasma Sources. Those who work with plasma physics workers have known for some time that plasmas of multimillion-degree temperatures can emit intense X-ray pulses(14)(15)(16). The average pulsed power of such sources (beam and electron-discharge heated types) ranges from several to ten times more irradiance (5 mW/cm^2 to 30 mW/cm^2) on the resist wafer than the electron impact anode type.   As is desirable, characteristic X-ray lines (0.4 nm to 2.0 nm) can be generated through selection of target and gas materials, and the laser or electron discharge energy and properties.   Some of the plasma sources tend to be relatively bulky and higher in cost, however, their potential of 3 to 5 times higher wafer level throughput make them economically attractive. The laser plasma source spot is a fraction of the source spot of either the electron discharge plasma or the impact anode types.

   Laser-Beam Heated Plasma Source. In the most advanced laser plasma source system a Nd+3:glass slab laser, pumped by an array of Xenon flashlamps, pulses at up to a 2 Hz rate during exposures and delivers pulse energy of 25 Joules at a 1.054 μm wavelength. The focused laser beam impinges upon and vaporizes a thin iron alloy film target. The targets are indexed by a tape cassette drive to enable a supply of fresh metal for each pulse impact. The focused laser beam further couples its energy (1·10^{13} W/cm^2) into a spatial metal vapor creating dense million degree microplasma temperatures. About 10% of the laser pulse energy is converted to a soft X-ray energy spectrum producing copious photons (λ=1.2 nm to 1.6 nm) suitable for photoresist absorption from a plasma spot size of only 150 μm to 200 μm diameter. The small spot size reduces the penumbral blurring factor and provides a design tradeoff for the source-to-wafer distance factor. For a practical laser plasma X-ray system: source-to-wafer distance D=10 cm, gap=20 μm, pulse repetition rate R=2 Hz (pulses/s), and the total windowless (no beryllium membrane) transmission fraction T_T which includes both helium ambient and mask (1.0 μm silicon) absorption at a wavelength peak of λ=1.2 nm, T_t = 0.59. The source X-ray energy release per pulse is E_p=2.5 J/pulse. For a spot-type X-ray source flux distribution, at the resist/wafer plane the incident energy intensity per pulse, E_w, is:

$$[6] \quad E_w = E_p T_T \cdot \omega / 4\pi D^2 = \frac{2.5 J/pulse(0.59)}{4\pi(10)^2 \ cm^2} = 1.17\omega \ mJ/cm^2/pulse$$

where ω is a factor to compensate for a non-uniform atmospheric helium distribution. For ω=1, a uniform distribution, E_w = 1.17 mJ/cm²/pulse.

For a resist with a sensitivity S=24 mJ/cm², the number of high intensity laser pulses N_p to achieve proper exposure is:

$$[7] \quad N_p = \frac{S}{E_w} = \frac{24mJ/cm^2}{1.17mJ/pulses/cm^2} = 20.51 \ pulses \ .$$

Practically speaking N_p=20 pulses. Since the laser pulsing rate R_p=2Hz, the time for exposure t_{exp} would be:

$$[8] \quad t_{exp} = N_p/R_p = \frac{20 \ pulses}{2 \ pulses/sec} = 10 \ sec.$$

The average irradiance φ on the resist is:

$$[9] \quad \phi = E_w N_p/t_{exp} = \frac{24mJ/cm^2}{10 \ sec} = 2.4 \ mW/cm^2 \ .$$

To achieve an average irradiance φ of 15 mW/cm², necessitates multiple burst pulsing and/or increased laser power. With an irradiance of 15 mW/cm² an exposure time t_{exp} of less than 2 seconds per printing field occurs and conforms to the criterion of high throughput. The windowless nature of the X-ray transmission path described above is unique and makes possible the delivery of almost half of the entire soft-type flux spectrum to the resist. For various resists, e.g., novolak, a relatively high

(50%) of the soft X-rays. The X-ray output energy spectra and yield is controllable via target composition, laser pulse shape, peak power, focused spot size, target geometry, and the background gas(17).

Notice the appearance of an extremely small penumbral blur δ (Equation 1) value of:

$$\delta = \frac{sd}{D} = 25 \times 10^{-4} cm(0.01)\ cm/(12)cm = 21\ nm.$$

Electron Discharge Heated Plasma Source. Two types of electron discharge heated plasma sources have produced high density X-ray photon energies (1.0 kV to 2.0 kV). These photons are compatible with the component materials necessary to accomplish IC chip lithography(18)(19)(20). These imploding plasma discharges (X-ray pulse duration=20 ns) take place in large work chambers (0.1 mm^3 to 0.2 mm^3) coupled via special transmission lines to a very large oil-cooled inductance-capacitance tank circuit having shaping, compression, and output stages.

The gas-puff Z-pinch configuration shown in Figure 5 has been driven in a repetitive mode (1 Hz to 10 Hz) at 400 kA-650 kA discharge currents of about 1.0 ms duration. The puff diode gas breakdown releasing the narrow X-ray photon pulse occurs in the time during the upper leading edge of the high tank output current pulse. The krypton or neon gas "puffed" into the diode electrode region builds up and forms a hollow gas cylinder being synchronized with the tank current discharge pulse as described above. After discharge, the residual gas is pumped from the chamber to allow the implosion cycle to repeat. The cylindrical plasma region with axial spot size of 1 mm to 2 mm emits X-ray pulses of 100 J to 200 J amplitude of 20 ns width at repetition rates of 3 Hz to 10 Hz (21)(22). The photon energy depends upon the noble gas: krypton emits 1.6 kV to 2.0 kV at 0.6 nm to 0.8 nm wavelength, and neon gives 1.4 kV to 0.9 kV at 0.9 nm to 1.4 nm wavelength. Such transmission, through the confining beryllium window (10 μm to 15 μm), 18 cm helium, silicon nitride or boron nitride mask (1 μm to 2 μm), and polyimide electron stopper (2 μm), delivers 2.0 to 4.0 mJ/cm^2/pulse energy into the resist. For a resist sensitivity of 8 to 80 mJ/cm^2, the resist exposure is achieved by as few as 2 to 40 pulses. The printing time for a one mask field, therefore requires a total exposure time ranging from 0.2 sec to 4.0 sec. By using a longer series of pulses, the undesirable effects arising from pulse to pulse non-

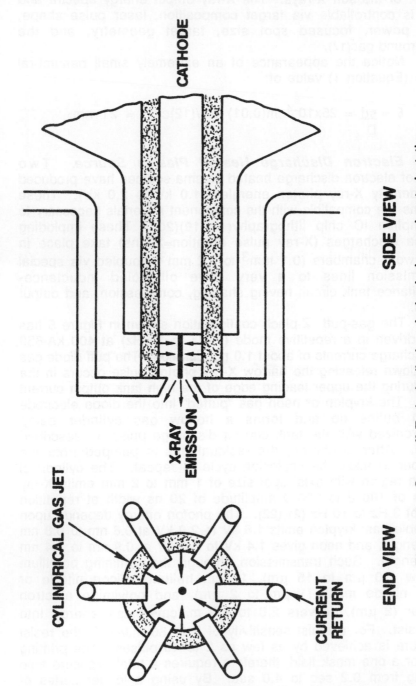

Figure 5. Gas-puff Z-pinch diode; pinched plasma located at the center of the diode measures about 1 mm diameter and 2 mm axial length.

pulses, the undesirable effects arising from pulse to pulse non-uniformities can be statistically compensated. Hence, exposure times of a few seconds or more are required. The best imaging work has been reported at repetition rates of several Hz.

Although some high quality resist imaging has been demonstrated, because of the size, cost, transmission tank and line durability, overall performance stability and the debris factor cause these intermediate-power sources to require engineering enhancements to become IC production viable.

4.0   MASK TECHNOLOGY

In general, the X-ray lithography method requires a sharp photon aerial image to be formed at the resist using the irradiance from the X-ray system's source(23)(24). Large process latitude can result from masks with high contrast, steep absorber walls, and smooth absorber line edges. For overlay accuracy the mask needs: accurate pattern placement, accurate alignment marks, high laser or light transmission for alignment, minimal distortion, and planarity. In Figure 6a the quantities d, D, s, h, and R are as defined previously. A magnification factor $\Delta R/R = s/D$, and penumbral blur varies as a function of R:

$$[10] \qquad \delta\,(R) = sd/D + h/D \cdot (R\text{-}d/2)$$

Some flux transmission actually takes place through the top and bottom corners of the absorber pattern edges. Figure 6b shows that out-of-place distortions δ_s of the mask or the wafer produce lateral xy shifts of the transferred mask pattern data on the wafer/resist by the amount δ_r. An alignment error δ_{align} results such that:

$$[11] \qquad \delta_{align} = R/D \cdot \delta_s$$

For an R = 2.0 cm, D = 18 cm, and δ_s = 100 nm, then δ_{align} = 11 nm.

Absorber defect density must be ≤ 1 defect/cm^2 (including mask repair). In present day thinking, for an MFS of <0.5 μm, defect densities should be <0.1 defect/cm^2. Through the life of the mask, the extent of in-plane dimensional distortions and the changes in membrane transmission coefficients must remain within close tolerance.

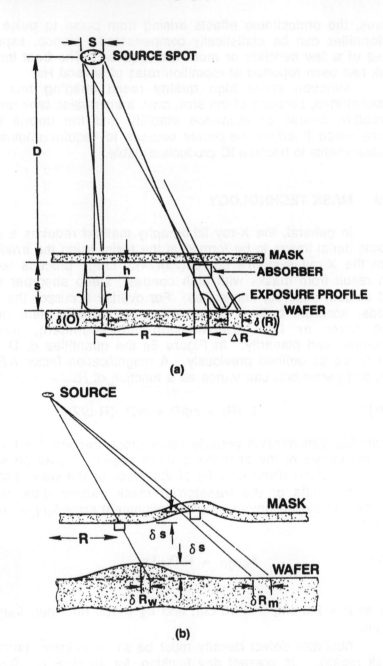

Figure 6. a. Magnification and penumbral blur variation with position R in the mask field; b. Alignment errors from out-of-plane mask or wafer distortions.

4.1   Minimum Linewidth and Control

The major mask factors affecting minimum linewidth and control are: contrast, mask transmission and uniformity, source energy spectra and uniformity, absorber line edge profile, pattern generation accuracy, distortion, geometric magnification, and indirectly the resist properties including secondary electron range. Sharp aerial image (line edge) profiles depend chiefly upon the absorber line edge profile in Figure 7. In addition both the absorber line edge profile location and its direction within the exposure field causes small linewidth variations. Not only can linewidths increase toward peripheral field locations, but a tilting of the resist walls occurs, i.e., resist towers lean radially inward towards the source(25). The absorber wall angles should be greater than 70° to keep the wall-smearing blur effect as small as possible and much smaller than the simple penumbral blur δ defined earlier in the chapter. Achieving high mask contrast and high uniform mask transmission depend directly upon material, absorption coefficients, and the physical thickness of the mask membrane, absorber, and electron stopper (if used). Absorption coefficients for some of the typical X-ray mask materials are shown in Figure 8. Since the X-ray absorption in a material is proportional to its electron density, X-ray mask membranes need low atomic numbers. However, such low density elements/compounds lack other properties needed (Young's modulus, tensile strength, etc.) and hence intermediate density materials are made to suffice (silicon, silicon nitride, boron nitride, etc.). Window material for the source vacuum integrity and protection are low density/low atomic number elements (beryllium). Absorbers are high density/high atomic number elements (tungsten, gold, tantalum).

The mask contrast factor C_m, can be thought of simply as the ratio of photon transmission through the clear mask membrane (non-absorber area), T_{clear}, to the transmission through the absorber area, $T_{absorber}$. Usually, the ratio is calculated on the basis of absorption coefficients corresponding to the X-ray source's characteristic radiation. Here:

[1 2] $C_m = T_{clear}/T_{absorber}$

and assuming that:

$$\lambda = \lambda_{wL} = 6.98\text{Å}, \ t_{BN} = 2.5 \ \mu m, \ t_{Au} = 0.7 \ \mu m,$$

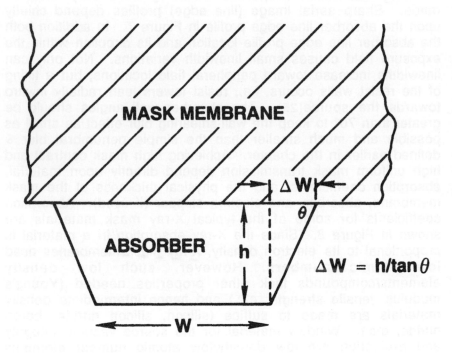

Figure 7. Absorber line profile; h=height, W-MFS (e.g.), and θ is the angle of the wall's slope.

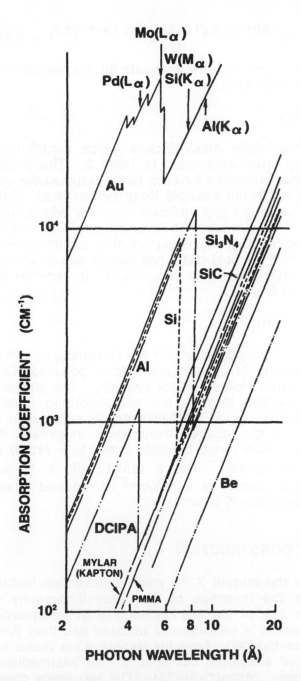

Figure 8. Absorption coefficient vs. photon wavelength for various X-ray mask window materials.

then:

$$C_m = e^{-800(2.5\times10^{-4})}/e^{-5\times10^{-4}} (0.7) = 9.97$$

Mask contrast values ≥10 are adequate for the resolution of MFS and linewidth control(26).

4.2   Overlay

The five major mask factors which contribute to the system overlay error were listed in Table 2. The design error budget allowance presented for each factor assumed the applicable masks were used in the example X-ray system cited. Differential runout error due to the gap variation from one printing level to a subsequent level, produce an overlay error in level-to-level printing. However, capacitive gap-sensing alignment/positioning control, on a field by field basis has shown gap positioning to a tolerance of ±100 nm. The runout error is reduced by such precision to ±11.0 nm.

4.3   Throughput

The mask component affects the throughput by virtue of the absorption property of its membrane (plus polyimide layer if it is used for electron stopper and/or strength). The photon energy must be transmitted through the non-absorbing mask areas (membrane) to provide an irradiance at the resist/wafer surface sufficient to yield the required throughput. In general the field exposure time should range between 2 sec to 5 sec to achieve economical throughput. For a resist with a 10 mJ/cm^2 sensitivity, an irradiance of 2 mW/cm^2 is required to complete proper exposure within 5 seconds.

5.0   MASK CONSTRUCTION

One of the current X-ray mask construction features has survived from the inception of the X-ray lithography method itself, i.e., the use of the starting material of an epitaxial n-n+ silicon wafer which is subsequently oxidized and then thinned by an etch back method(27). From the beginning the choice for X-ray membranes and absorbers has been to use intermediate-Z and high-Z materials, respectively(28). The two major divisions of mask fabrication are the blank, and the absorber/patterning. The absorber/patterning is performed either as a subtractive (etch) or additive (plating) process.

The blank assembly typically starts with a silicon wafer which is oxidized, epitaxially layered, or with a boron nitride deposition layer as shown in Figure 9. The silicon wafer is etched back and epoxy mounted to a pyrex or glass ring. After depositing a tantalum/gold plating base (additive), a thick stencil resist is spun and baked followed by a thin chromium etch mask layer. The blank is then completed by spinning on and baking an imaging resist poly(butene-1-sulphone) (PBS). Variations from this particular procedure and materials have included: stretching a transparent membrane over a silica or silicon mounting frame, and the use of mylar(29)(30), kapton(30), and titanium(31) as membrane materials, respectively. Other membrane forms include: aluminum/oxide(32), silicon nitride(33), silicon nitride/silicon dioxide/silicon nitride(34)(35), silicon oxynitride(36), and triple layer membranes (tensile silicon-doped boron nitride with compressive layers of boron nitride)(37).

It is imperative to consider the interactions of the thermal and mechanical properties of the membrane and its support ring(38)(39). Considering Table 5 and the blank assembly described above: the thick pyrex support ring with an expansion coefficient of $3.3 \times 10^{-6}/°C$ matches the silicon substrate closely ($2.6 \times 10^{-6}/°C$). Matching these coefficients with the membrane coefficient itself assures that the preset membrane tension will change only a small amount as temperature fluctuates. Membrane coefficients for silicon nitride, silicon, silicon carbide, and boron nitride hydride indicate compatibility with the pyrex/silicon combination, and therefore, should be used together. Since membranes with high elastic modules can reduce local pattern distortions from high absorber stress, silicon carbide appears as the leading membrane candidate.

The additive absorber patterning process starts with the exposure and development of patterns in the PBS resist as in Figure 10. The thin chromium layer is etched to become a mask for reactive ion etching (RIE) of the stencil resist. After stripping the chromium layer, gold is plated onto the stencil resist mask. The resist is then stripped followed by removal of the tantalum/gold plating base. The resulting gold X-ray mask absorber patterns have steep line edge wall profiles (Figure 11). Numerous references exist which describe the electroplating process(40-48).

The subtractive absorber/patterning process shown in Figure 12 starts with a blank prepared in a manner similar to

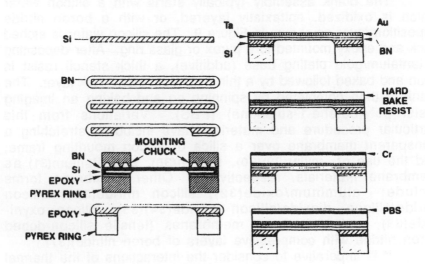

Figure 9. Mask blank fabrication process.

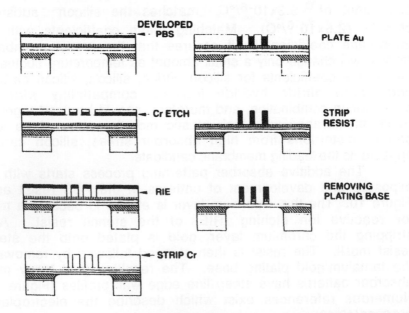

Figure 10. Additive mask absorber patterning process via gold electroplating. Courtesy Perkin-Elmer.

TABLE 5: X-ray mask and window material properties

Materials	Coefficient of Thermal Expansion (°C^{-1})	Young's Modulus (dynes/cm^2)
BN	$1 \cdot 10^{-6}$	$1.33 \cdot 10^{12}$
Si$_3$N$_4$	$2.7 \cdot 10^{-6}$	$1.55 \cdot 10^{12}$
Ai$_2$O$_3$	$9 \cdot 10^{-6}$	$3.73 \cdot 10^{12}$
SiO$_2$	$0.4 \cdot 10^{-6}$	$7.38 \cdot 10^{12}$
SiC	$4.7 \cdot 10^{-6}$	$4.57 \cdot 10^{11}$
Si	$2.6 \cdot 10^{-6}$	$1.62 \cdot 10^{12}$
Be	$12.3 \cdot 10^{-6}$	$3.02 \cdot 10^{12}$
Pyrex	$3.3 \cdot 10^{-6}$	

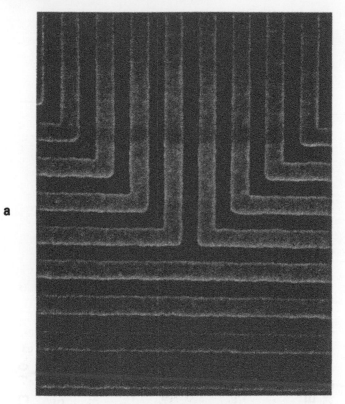

DENSE LINE/SPACE ARRAYS(0.8 micron pitch)

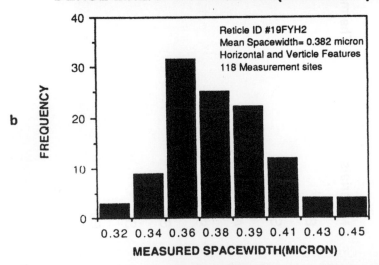

Figure 11.   Electroplated gold absorber lines: a. 0.4 μm line/space on 1 μm boron doped silicon; b. Histogram of horizontal and vertical features on the reticle in a. Courtesy Hampshire

0.5 Micron

c

0.3 Micron

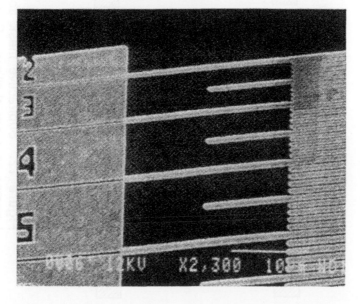

d

Instruments. Line/space (0.7 μm gold) on 5 μm boron nitride: c. 0.5 μm, and d. 0.3 μm. Deviation 0.040 μm from mean (3σ). Courtesy Perkin-Elmer ALO.

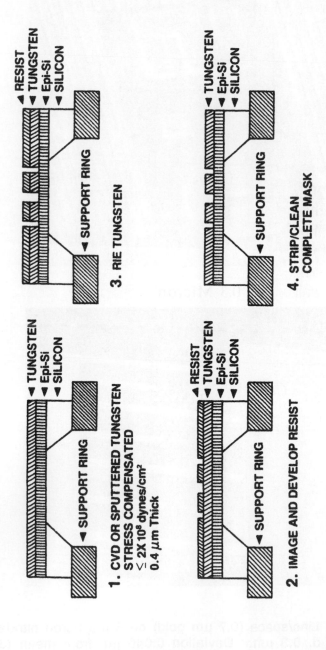

Figure 12. Subtractive mask absorber patterning process via tungsten on epitaxial silicon. Courtesy Hampshire Instruments.

that of preparing an additive absorber blank. Only one resist layer is exposed, developed, and used to delineate absorber geometries in three etching steps: RIE of tantalum, sputter etch of gold, and RIE of tantalum(49). A polymer is finally deposited over the entire mask area to act as a photoelectron-Auger electron suppressor. Electron beam lithography is commonly used to write the submicron MFS absorber patterns. The RIE and sputter etch combination has been used to pattern a variety of multilayers such as tantalum/gold/tantalum(50), titanium/gold/titanium(51), and chromium/gold/tantalum(52). The sputter etch patterning suffers from redeposition of gold pattern feature on the sidewalls producing wall angles of 65°-73°(45)(53). Vertical walls in 7000Å gold were reported from reactive ion milling(54). The additive process in general has the advantage of producing vertical wall features and less local mask pattern distortions as a result of the low stress electroplated gold.

5.1   Mechanical and Optical Distortions

Local membrane distortions occur within the mask's exposure field from the stresses caused by the placement of resist and absorber layers and the subsequent patterning of these layers(47)(55). Adjusting membrane residual tensile stress between 5×10^8 and 1×10^9 dynes/cm^2 gives flat membranes ≤ 1 μm and provides a control mechanism for compensating absorber induced tensile stresses of 2×10^8 dynes/cm^2(56). The low pressure chemical vapor deposition method (LPCVD) and its plasma enhanced counterpart, enable great control of the boron nitride hydride and silicon nitride membrane layer properties to be exercised. Varying reactant gas ratios and substrate temperature affect absorption, density, stoichiometry, residual stress, and refractive index. In Figure 13, the residual stress of boron nitride: hydride deposited by LPCVD is made to vary between 1×10^9 dynes/cm^2 tension to 8×10^8 dynes/cm^2 compression by changing the diborane-to-ammonia flow ratio at various temperatures(57). Refractive index increases (2.0 to 3.0) and optical absorption increases resulted. Using CVD methods, membranes of silicon nitride have been made with tensile stresses ranging from 2.5×10^9 to 1×10^{10} dynes/cm^2(35)(44). An accepted way to measure film stress on a wafer is to measure the amount of curvature before and after removing the film coating from only one side of a wafer (see Figure 14)(45)(58)(59). Measuring the bulge displacement of the membrane center point under differential pressure conditions also enables the calculation

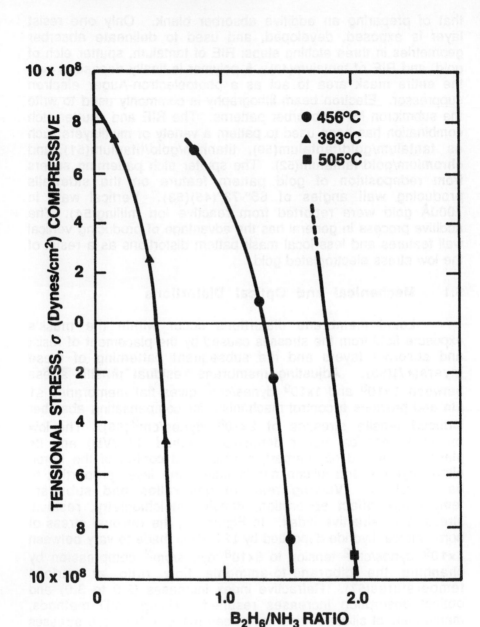

Figure 13. Mask membrane stress vs. B_2H_6/NH_3 ratio as a function of temperature in an LPCVD process.

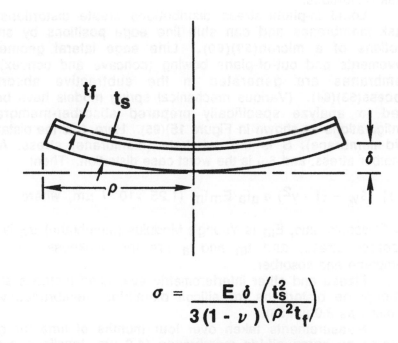

$$\sigma = \frac{E \, \delta}{3 \, (1 - \nu)} \left(\frac{t_s^2}{\rho^2 t_f}\right)$$

WHERE ν = **POISSON'S RATIO**
E = **YOUNG'S MODULUS**
E/(1 - ν) = 2.3 × 10^{12} dynes/cm^2
for 1:1:1 silicon

Figure 14. Mask membrane stress measurement using the radius of curvature technique +13.44 µm (convex) bow with 4.55 µm film of BN on both sides; after BN removal from one side curvature became +16.32 (convex). Courtesy Westinghouse.

of stress.   For least distortion of mask patterns the membrane stress should be the dominant stress(60).   Larger membrane-to-film thickness ratios also provide a means of minimizing local mask distortions.

Local in-plane stress distributions create distortions of mask membranes and can shift line edge positions by small fractions of a micron(59)(60).   Line edge lateral geometric movements and out-of-plane bowing (concave and convex) of membranes are generated in the subtractive absorber process(63)(64).   (Various mechanical spring models have been used to analyze specifically prepared absorber-membrane configurations as shown in Figure 15)(65). Here, L is the distance (mid-membrane), S is the substrate (membrane) stress, A is absorber stress, and δ_w is the worst case distortion. Then:

$$[13] \quad \delta_w = (1 - v^2)\, \sigma_a t_a / E_m t_m\ (1.25 \times 10^{-3})\ \mu m, \text{ where,}$$

v = Poisson's ratio, E_m is Young's Modulus (membrane) σ_a is the absorber stress, and t_m and t_a are the thicknesses of the membrane and absorber.

Fizeau and other interferometric evaluation methods show out-of-plane distortions for silicon oxynitride membranes vary as much as 2.0 μm(36).

Measurements taken over four months of time for gold patterns on boron nitride membranes (4.0 μm, tensile stress = 1.2×10^9 dynes/cm^2) indicate a mask stability with overlay error of σ=30 nm.   Thermal and stress cycling of these bonded boron nitride membranes show no fatigue, loss of tension, or bond slippage(66). High integrated synchrotron fluxes of the order of 130 kJ/cm^3 show in-plane xy stripe (1 cm wide absorber) pattern distortions of hydrogenated boron nitride masks ranging to 1 μm. At doses of 200 J/cm^2, the hydrogenated membrane's (4 μm thick) optical absorption coefficient increases and transmission (λ=632.8 nm) falls from 65% to 40%(67). However, hydrogen-free mask membranes of BN/B$_3$N/BN suffer only slight distortions of 10 nm, and transmission (λ=632.8 nm) changes of less than a few percent for doses of 2000 kJ/cm^3(68).

5.2   Defects

An attractive feature of the X-ray lithography method is the possibility of immunity from dust particles(69) and certain defects (Figure 16). Energy absorbing particles and defects are

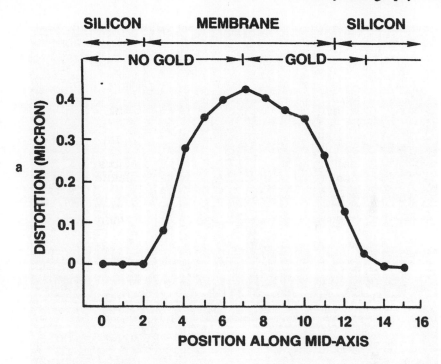

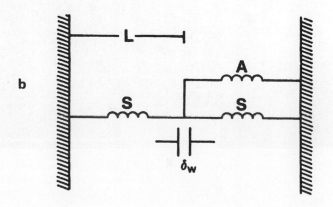

Figure 15. a. Distortion due to stress in gold covering half of the mask membrane; b. Simple spring model used to estimate distortion due to worst-case pattern loading.

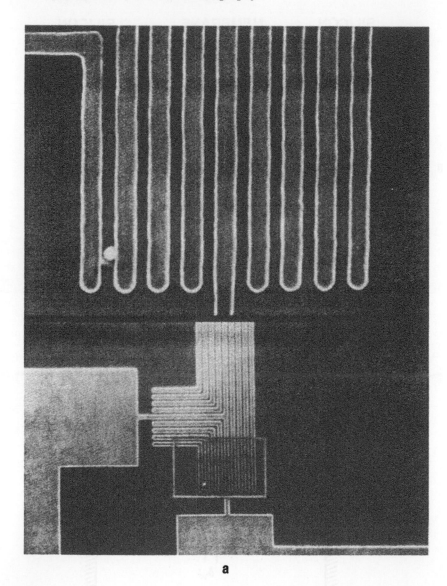

a

Figure 16. a. Reticle mask with 0.4 μm line/space array in 0.4 μm thick gold absorber on 1.0 μm silicon membrane, a 0.4 μm soft (latex) defect partially obscures a space; b. Wafer printed without showing the defect into 0.81 μm thick Hitachi RE5000P resist. Courtesy Hampshire Instruments.

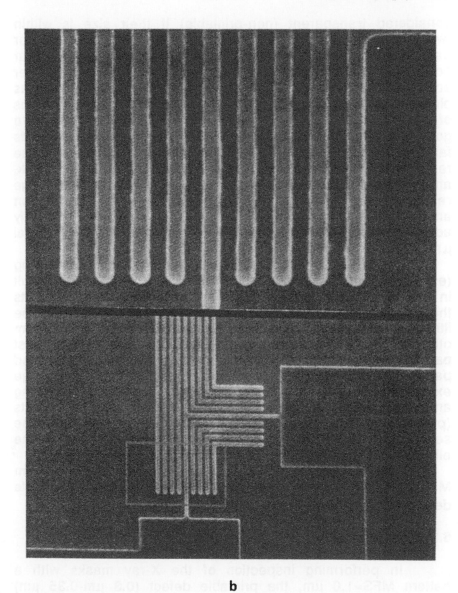

b

Figure 16. (continued)

considered transparent (non-printable) if their size is within two times the size of the penumbral blur (e.g., 160 nm) or particle/defect size $\leq$ 320nm. This only applies to the clear portions of the mask including clear areas adjacent to absorber edges. In the opposite case of nonabsorbing defects, such as pinholes in absorber areas, immunity does not exist and pinholes of penumbral size remain printable defects. In the subtractive gold patterning process of boron nitride hydride printable defects fall into three types: missing absorber gold from pinholes in the resist, unetched gold absorber from inadequate patterning mask, and opaque silicon residues from incomplete silicon backside edge removal(70). Since dust particle (low-Z) immunity is a photon energy dependent effect, higher energy sources are relatively advantageous in ignoring even the larger sized particles (several μm).

In addition to the lithographic process, wafer processing (e.g., deposition, ion implant, and etching) and handling can result in defects that also reduce device manufacturing yield. Defects that repeat in the exposure field are undoubtedly due to the lithographic process. Randomly distributed defects result from other process steps or from wafer handling. The airborne dust particles are found to be predominantly low density carbon-based particles that emanate from people and clothing. The large exposure latitude for the X-ray lithography method allows using an overexposure to reduce the impact of lower contrast defects (particles) without appreciable linewidth change. In fact, the selection of exposure, resist, and processing can result in the elimination of both soft and certain hard defect printing errors.

The argument has risen that for the production of 0.5 μm MFS IC chips, such as a 16 MBit DRAM, will require printable defect densities of <0.1 defect/cm^2 (71).

5.3   Inspection

In performing inspection of the X-ray masks with a pattern MFS=1.0 μm, the printable defect (0.3 μm-0.35 μm) density and location were first evaluated manually via highly magnified optical means. The expensive and time consuming inspections, on a sample basis made at rates of 1.0 cm^2/45 min, were too costly to be extended into the X-ray printing realm at 0.5 μm MFS. Comparison microscopes with synchronously driven stages were used for the comparison of the mask and the printed wafer in order to determine printable defects attributed to the mask(70). Commercial optical instruments, which compare the data base input pattern to the absorber mask pattern, exhibit enhanced and broadened defect analysis capabilities with the detectable defects ranging down in size to slightly below 0.5 µm.

Although defect detection, sorting, and mapping with respect to any particular patterned mask set allowed the prediction of lithography process limits and yield to be used for IC product quality control, manufacturing technology was enhanced even more as mask repair techniques emerged. Laser ablation was adapted for removing isolated high-Z opaque defects. Focused ion beam milling provides high resolution (100 nm) trimming of excess gold from absorber features(70). Pinholes in opaque absorber areas can be filled selectively with photochemical deposition by laser CVD and laser enhanced plating techniques(72-73). In the vein of standardization, tape and disk recorded defect map data taken via optical inspection machine has been made compatible with the input requirements for the drive control of an ion beam repair tool(74).

5.4   Pattern Generation

The generation of high accuracy mask absorber patterns is absolutely critical to high resolution (0.5 µm) MFS X-ray printing. Special electron beam addressing (100 to 50 nm), electron spot size (150 nm) and intensity, and substrate handling/positioning are required to convert a standard commercial electron beam pattern generator (see Chapter 5)(75) for meeting the requirements of 0.5 µm MFS X-ray mask patterning with ±50 nm to 74 nm placement accuracy, and ±40 nm to 80 nm linewidth control. The most difficult criterion to meet is placement accuracy. In addition, for pattern transfer, tri-level resist processing is essential (described in Chapter 2) in which a 400 nm NPR imaging resist layer, a 100 nm silicon dioxide and a 1.4 µm HPR204 resist are layered on the absorber. It is imperative to use proximity correction software to reduce the pattern edge smearing by the effects of back and front scattered electrons caused by the impinging high energy (20 kV-30 kV) writing beam. Since a low intensity writing beam is used in order to achieve minimal spot size the pattern exposure time suffers considerably, in fact, in some cases two-pass superimposed exposures have been attempted to achieve the required absorbed resist energy density.

6.0   ALIGNMENT

The performances of the X-ray system aligner and mask components essentially determine the overlay capability (e.g., ±100 nm, 2σ) of the overall X-ray system.   Other relatively small overlay errors are contributed by the wafer and source/alignment interface.   The alignment task if performed ideally would position the mask-to-wafer exactly in parallel plane positions at the specified design wafer-mask separation distances, and the parallel planes would be normal to the axis of radiation and contribute an aligner overlay error of ±60 nm maximum.   Figure 17 shows the aligner coordinate system related to the six degrees of freedom of movement required to compensate for the major misalignments of: lateral shift (ϵ_x, ϵ_y), inplane rotation ($\epsilon_{\theta z}$), out of plane tilt ($\epsilon_{\theta x}$), out of plane tilt ($\epsilon_{\theta y}$), gap setting (ϵ_z), and magnification.   A further description in Figures 18 and 19 shows how all misalignment errors resolve into ΔX and ΔY quantities. The aligner's closed-loop servo features automatically compensate (during the X-ray mask field exposure) for these misalignments.   Both X-ray, and visible-optical (coherent and non-coherent) radiation have been used for sensing misalignment.   Highly sensitive subsystems detect the relative position of the mask with respect to the wafer(76). All receive optoelectronic signals from matched pairs of alignment marks placed at three or four orthogonal peripheral positions outside of the mask field F, and at the corresponding locations on the wafer(77).   X-ray flux used in an alignment sensing scheme required special back etching of the silicon wafer for transmission of the signal flux into a null X-ray detector to control servo positioning to 100 nm(23).   Later versions of alignment detection schemes shifted more towards coherent or incoherent light for the generation of alignment signals.

6.1   Interferometric   Schemes

The Fresnel zone plate and grating method using laser radiation appeared first in a linear form as shown in Figure 20a(78).   The relative positions of linear zone-plate patterns on the mask are measured with respect to line diffracting gratings on the wafer (Figure 20b).   The one-dimensional Fresnel type lens focuses incident laser radiation into a line focus at the surface of the wafer.   When the focused line aligns with the fine linear grating etched into the wafer surface, diffraction of the laser beam occurs from the grating causing negative and positive orders to emanate off the normal to the wafer surface (Figure 21).   The diffracted laser energy from the 1st+ and 1st- orders is absorbed

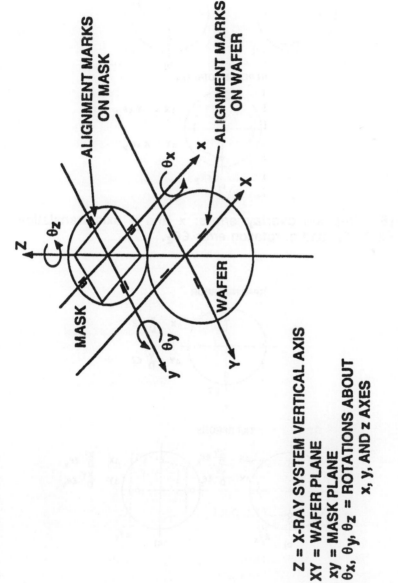

Z = X-RAY SYSTEM VERTICAL AXIS
XY = WAFER PLANE
xy = MASK PLANE
θ_x, θ_y, θ_z = ROTATIONS ABOUT x, y, AND z AXES

Figure 17. Aligner coordinate system related to the six degrees of freedom of movement -- translations occur in the x, y, and z directions and rotations occur about the x, y, and z axes.

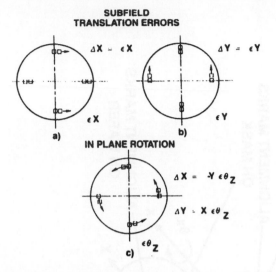

Figure 18. In-plane overlay errors: x and y axis translation errors a. ϵx, b. ϵy, and c. rotation error $\epsilon\theta_z$.

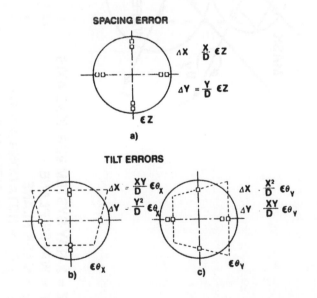

Figure 19. Out-of-plane overlay errors: a. mask-to-wafer gap ϵz, and tilts about the: b. x-axis $\epsilon\theta_z$, and c. y-axis $\epsilon\theta_y$.

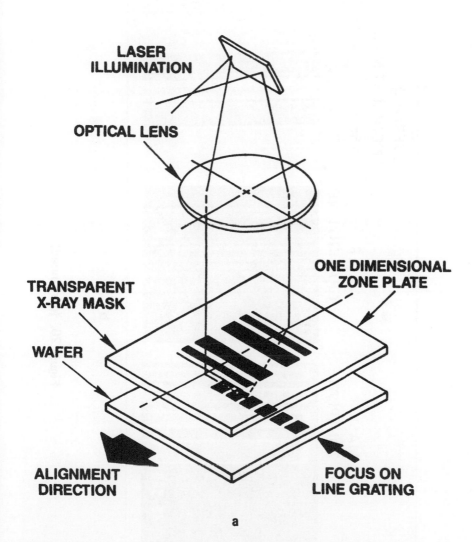

LASER ILLUMINATION

OPTICAL LENS

TRANSPARENT X-RAY MASK

WAFER

ONE DIMENSIONAL ZONE PLATE

ALIGNMENT DIRECTION

FOCUS ON LINE GRATING

a

Figure 20.   a.   Linear zone-plate-grating alignment model; b. Lateral alignment marks on the wafer and the mask.   Courtesy Perkin-Elmer.

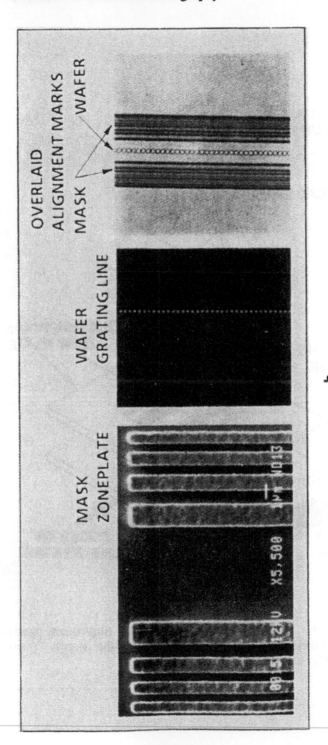

Figure 20. (continued)

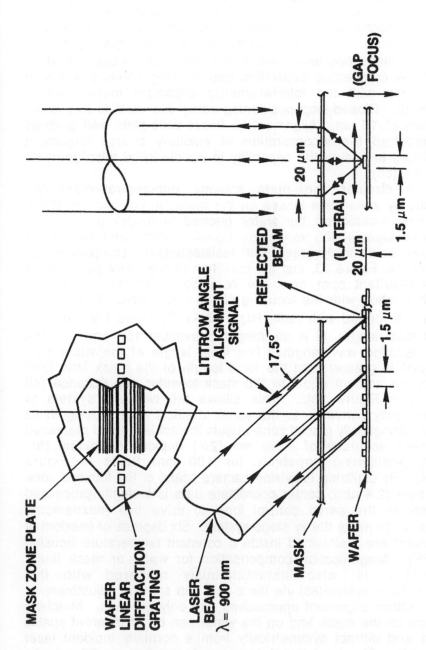

Figure 21. Schematic of linear zone-plate-grating; alignment signals are diffracted from grating at the Littrow angle.

and sensed by a photodetector. When the maximum amount of diffracted light is sensed by the photodetector exact alignment takes place, i.e., the grating is directly beneath the zone-plate. Three orthogonal zone-plate grating pairs are used for inplane and θ_z positioning. A fourth orthogonal zone-plate grating pair is added for measuring expansion or contraction of the same mask or wafer. A compatible capacitive gap sensing scheme works in conjunction with the interferometric alignment method as in Figure 22. Closed-loop positioning using this method has given accuracy of 15 nm. Alignment by linear zone-plate and gratings has required the incorporation of auxiliary coarse alignment techniques to assure the capability to handle initial misalignments to ±1 mm(80).

A circular zone-plate scheme demonstrated(81) that focusing by zone-plate lenses on the mask, in conjunction with a third lens located on the wafer (etched in oxide) (Figure 23), produced signals with very large signal-to-noise ratio for various surfaces, oxide thicknesses, and resists(82)(83). The principle is, as shown in Figure 23, that the condition of alignment generates a single resultant spot from the coincidence of the mask lens focusing action with the focusing of the wafer lens. The circular zones are formed with radii, $R_N = (fN\lambda)^{1/2}$, where f is the zone target focal length, N is an integer (maximum 15), and λ is the laser radiation wavelength. The focal length of the wafer lens (240 μm) is greater than the focal length of the mask lens (200 μm) by an amount equal to the mask-to-wafer gap distance (40 μm) as in Figure 23b. This allows the two-lens system to concentrically converge to the same identical spot. Using three sets of orthogonally placed zone targets the achieved and measured alignment accuracy of ±250 nm (2σ) depicts the global (full wafer) positioning capability for 100 mm fields in Figure 23c(84). By digitizing television camera scans of the focused zone plate spot (2.4 μm), center coordinate data is instantly processed and enters the servo control loop to drive five piezoelectric motors which move the xy stage system. Six degrees of freedom of movement are maintained inside a constant temperature housing (±0.2°F). Magnification compensation for wafer or mask linear distortions is also instantaneously rendered with this aligner/stage system(85) via the z-axis gap spacing adjustment.

Other alignment approaches use only gratings. Matching gratings on the mask and on the wafer can be of different spatial period and diffract symmetrically from a normally incident laser beam(86). The experimental step-and-repeat X-ray lithography system aligns mask-to-wafer when the symmetrically diffracted

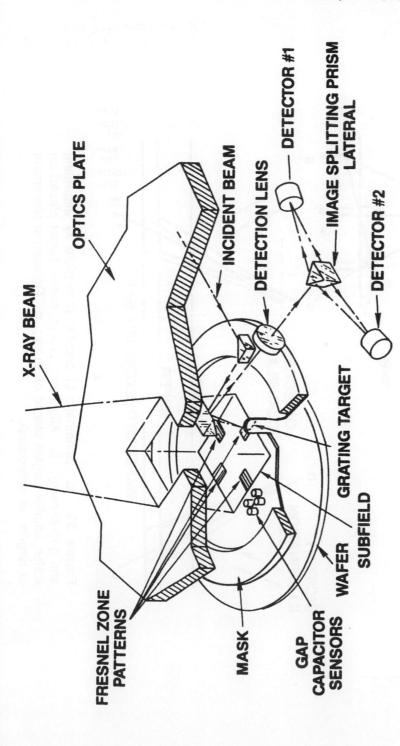

Figure 22. Incorporation of a linear zone-plate-grating alignment method with a capacitance gap-sensing scheme. Courtesy Perkin-Elmer.

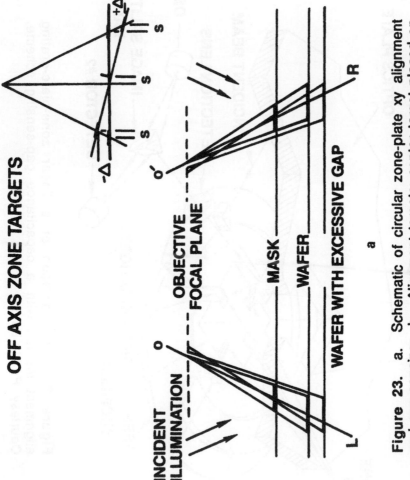

Figure 23. a. Schematic of circular zone-plate xy alignment and gap-sensing; b. Alignment targets: center target placed on wafer, smaller targets placed on mask; c. Orthogonal placement of targets on wafer/mask.

ALIGNMENT TARGET DESIGN

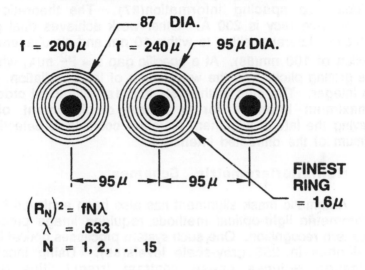

$$\left(R_N\right)^2 = fN\lambda$$
$$\lambda = .633$$
$$N = 1, 2, \ldots 15$$

(b)

ALIGNMENT TARGET LAYOUT

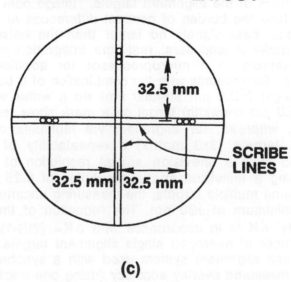

(c)

Figure 23. (continued)

beam intensities are equal. The initial coarse alignment ±1mm achieved by Moire fringe patterns requires operator control and provides gap spacing information(87). The theoretical fine alignment accuracy is 200 Å. Other work achieves dual grating detection of lateral position to within 10 nm and simultaneous gap detection of 100 nm(88). At a specific gap $s = P^2 \, m/\lambda$, where, P is the grating pitch, λ is the wavelength of laser radiation, and m is an integer. The sum of the 1st+ and 1st- diffracted orders are at maximum intensity for a lateral displacement of P/2. Achieving the lateral alignment therefore consists of detecting the maximum of the diffracted intensities.

6.2   Non-Interferometric Schemes

Automatic mask alignment has also been achieved by non-interferometric light-optical methods requiring image processing and pattern recognition. One such system processes optical images to halftones in 256 gray-scale levels(89). Using incoherent illumination reduces phase contrast irregularities due to photoresist thickness or gap spacing variations. However, sufficiently sharp optical images of alignment marks in various planes (mask or wafer) must be realized. Automatic position detection requires the detection of straight and frequently low-contrast edges of the alignment targets. Image converters (video cameras) face the burden of contrast differences at edges which might supply data signals no larger than the noise amplitude. Using operational amplifiers, real time integration provides line-by-line integrals to a microprocessor for additional halftone processing. Experiments with the combination of a square-shaped annular target (~2.0 μm annular rim) on a wafer with a cross-shaped (5.0 μm linewidth) target on a mask show repeatability of ±125 nm, whereas, for alignment via multiples of the same alignment targets (3x3 matrix) a repeatability of ±40 nm is determined (with a television spatial resolution of 0.5 μm per line). Using a television spatial resolution of 0.25 μm per line and the same multiple targets, the measuring uncertainty is found to be a minimum of ±30 nm. The reduction of the measuring uncertainty ΔK is in accordance with $\Delta K = [2t(t-1)]^{-1}$, where t is the number of averaged single alignment targets. This light-optical-based alignment system used with a synchrotron source yielded a measured overlay accuracy (using one mask) of ±40 nm (3σ) in x and y. The alignment mark targets consisted of 0.5 μm silicon dioxide, 2 μm of pyrolytic silicon dioxide, and 1.5 μm resist thickness(90).

Yet another light-optical non-interferometric alignment system similar to the alignment system above incorporates: television camera image detection of the matching alignment mark pairs (wafer,mask) (Figure 24) via bifocal objectives, light path converters, differentiation of the video scan across the image line edges to derive control signals for a null (zero-detection), and autofocusing and autopositioning(11). The distance between the double focal points is set with high accuracy to equal the gap spacing s (10 μm). Setting the specified s value at the three orthogonal alignment target locations almost eliminates wafer-flatness-caused alignment errors. The alignment system with multiple servo controls for multiple-axis alignment is driven by stepping the xy wafer-stage motor and electromagnetic force operated elastic plates for micromovement positioning of both wafer and mask stages. The Θ and Z positioning is performed at the mask stage, and xy lateral positioning at the wafer stage. Figure 25 shows lateral positioning data, Δx, obtained from signal processing of the combined mask-wafer mark image according to $\Delta x = k\Sigma_n$ (a-b) where, k is a conversion constant, n is the number of averaged scan lines, and a and b represent bi-level signal widths. Fine xy lateral displacement control of the stage to 7 nm, combined with an alignment servo stability of 22 nm enabled step-and-repeat overlay accuracy of ±150 nm (2σ) to be achieved over a 100 mm wafer (two different masks were used). The total alignment time under microprocessor/computer control is 0.5 seconds.

7.0   RESIST

The photoresist component of the X-ray lithography method is of great importance in determining the X-ray system's achievable capability: minimum feature size (resolution), linewidth and edge control, and maximum throughout. The reader is directed to Chapter 2 pertaining to photoresist materials and processing concepts. The photoresist, flux intensity, and mask are related in the following way: High contrast positive resists make good linewidth control possible with lower contrast mask (i.e., the profile slopes of the X-ray flux intensity aerial image within the gap space can be less). These lower contrast masks are less difficult to manufacture. Such resists also can lead to a smaller source-mask distance with a resultant flux intensity increase since the penumbral blur increase would be compensated. The use of single level resists of appreciable thickness for more complete

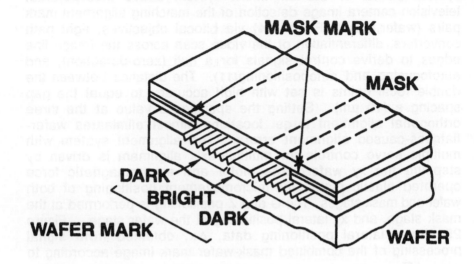

Figure 24. Single-lined marks are on the wafer surface and double-lined marks are on the mask surface.

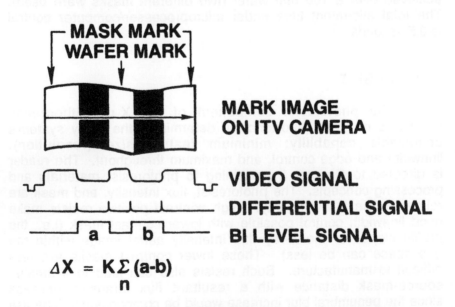

$$\Delta X = \frac{K\Sigma(a-b)}{n}$$

Figure 25. Schematic of mark positioning error detection.

wafer topology coverage is a tradeoff also. Positive resists in general are ten or more times less sensitive than the negative resists, hence, throughout becomes a deciding factor in the selection of a resist for production use.

If the relatively fast negative resists are used, then the aerial image profile slopes must be made steeper. Hence, the mask absorber wall angles must be sharper and the mask fabrication technology once again becomes difficult. The penumbral blur should be made smaller either by extending the source mask distance, or by reducing the gap spacing. The flux reduction can be tolerated because of the compensation made through the high resist sensitivity. The customary swelling of negative resists requires caution in lithography practice at MFS $\leq$0.5 μm, and as a result complex multiple layer resist (MLR) processing is considered a worthy alternative.

Besides being judged by its sensitivity and resolution, an X-ray resist must be judged for its dry etch resistance to the reactive ion etching and other dry active ion etches used in delineating the layers (silicon dioxide, poly-silicon, silicon nitride, aluminum, etc.) that are deposited or grown on silicon wafer surfaces during IC fabrication processes. Resist contrast, as defined in Chapter 2, is a measurable resist parameter which indicates the resolution capability of a resist. A comparison of various X-ray resists is shown in Table 6. Although the absorption of X-ray resists can be very uniform the deposited energy density profile may not be. This non-uniformity can be caused by electron escape from the top resist surface and from backward energy dispersion into the resist from the silicon substrate. Higher energy X-rays such as from the Bremsstrahlung range can produce more than five times the effective dose in the bottom of the resist as compared with the resist's surface(91). The unique resist tilting effect described previously which arises at the peripheral field from point sources (electron impact, plasma) can be controlled or minimized by the use of MLR processing.   Placing the impinging resist layer in the MLR structure at an appreciable distance (i.e., increase bottom layer thickness) from the silicon substrate remedies the substrate backscattered dispersion effect. The MRL method also achieves steep resist profiles and MFS's < 0.5 μm via a uniformly thin and planarized exposing resist layer(92).

8.0   METROLOGY

TABLE 6: Resists used in X-ray Lithography

Resist	Symbol	Sensitivity (mJ/cm^2)	Resolution (μm)	Contrast (t)	Dry Etch Resistance	Comments
	PBS (+)	167	<1	1.27	Poor	Poor, adhesion Mead Tech., KTI Chem.
	COP (-)	35-200	<2	1.07	Poor	Mead Tech., Hunt
	PMMA (+)	5600-8500	<0.1	1.7-2.0	Poor	Dupont, Tokyo Ohka
	PCMS (-)	8	0.5	1.8	Good	Toyo Soda, HP, Mead Tech.
Ray-PF	- (+)	50-200	<0.3	-	Good	Hoechst
WX-214	- (+)	500	<1	-	Good	Olin-Hunt
RX-242	- (+)	200	0.3	-	Good	Olin-Hunt (BESSY)
		400	4	-	Good	Olin-Hunt Perkin Elmer
HPR-204	- (+)	8300	0.1	-	Good	Olin-Hunt
ECX-1029	- (-)	200-250	0.3-0.4	-	Good	Rohm & Haas Novolak
ECX-1092	- (-)	10-20	0.2	Low	Good	Rohm & Haas Novolak
EK-88, (771)	- (-)	9	1.0	1.3	Good	Kodak
Kodak 571 -			0.75	High	Poor	-
	DCOPA(-)	14	1	-	-	Trilevel, BTL 0.3% O_2/N_2 amb.
	DCPA+ +BABTDS (-)	4	0.5	Low	-	BTL, Plasma dry develop
(Novolak)	IBM (+)	750	0.5	-	-	NMOS devices IBM
RD-2000N Hitachi	-(-)	1200	0.5	-	-	NMOS devices IBM
PR1024MB	-	150	0.5-0.75	-	-	MacDermid, Hampshire
RE-5000P Hitachi	-	14-38	-	-	-	

The progressive development and production of IC chips into the submicron range via X-ray lithography (as well as electron beam and optical) required the extension and enhancement of the concomitant metrology base. The reader is directed to Chapter 3 for an excellent description of the related issues affecting the growth of finer line metrology with respect to linewidth measurement. The overlay accuracy and line edge roughness also have importance in determining the full capability of any lithography system. The quality factors linewidth control, overlay accuracy, and edge roughness can be measured in terms of the printed resist image or in terms of it's subsequent transferred likeness etched into a silicon surface layer (silicon dioxide, polysilicon, etc.). Obviously the latter method places some constraints upon the resist material and its processing.

At submicron dimensions, edge roughness measurement (≤ 50 nm) lends itself more to the scanning electron microscope (SEM) type measuring tool. In some cases, expensive and extremely precise electron beam pattern generators (MEBES II, III) have been used for the assessment of the three quality factors. Overlay accuracy is also accessible through optical vernier methods(93)(94). The reading of verniers via microscope viewing is slow, operator dependent, and necessitates high image magnification and double interpolation to achieve a measuring resolution of 25 nm (vernier increments of 100 nm). The optical vernier technique has some advantage as a non-destructive procedure.

Overlay accuracy and linewidth can each be measured via simple electrical test devices (see Chapter 3). These fast and fully automated electrical techniques involve the deposition and delineation of single or double layer conductive films which add some complexity but enable accurate determination of lithography system performance status(95)(96).

9.0 X-RAY SYSTEM

The characteristics of the components of a near ideal X-ray system for submicron ≤ 0.5 µm MFS are: Source: (1) high intensity, uniform, collimated, and stable photon flux, and (2) energy compatible with both high mask contrast and resist absorption; Masks: (1) high contrast and abundant clear area transmission, (2) vertical and smooth absorber pattern walls, (3) high absorber pattern accuracy, (4) long term freedom from spatial or transmission distortions (from exposure radiation, mechanical creep, or otherwise), and (5) near zero opaque and/or pinhole defects; Aligner: (1) perform rapid step-and-repeat movement, (2) align and maintain mask and wafer parallel to each other and normal to radiation flux, (3) maintain gap spacing within close tolerances, (4) align in the xy directions and to close tolerance to the corresponding pattern geometries of the mask and wafer; Resist: (1) high contrast and high sensitivity at source energy, (2) low defects, (3) very good adhesion, (4) very good dry etch resistance, and (5) compatible development and bake processes.

The source collimation criterion is almost realized via the synchrotron storage ring features coupled with the use of favorable radiation reflective material properties (scanning mirror scheme) at the 0.6 nm to 1.5 nm wavelengths. Achieving some degree of collimation lessens some of the X-ray system design constraints. This relaxation simplifies: gap spacing control, mask-pattern absorber wall-edge profile, source-to-mask distance, and the size of mask field. These simplifications accrue merely by reducing source-caused penumbral blur and the combination run out and gap-variation wafer-to-mask misalignments. The noticeable disadvantage of the collimated source is the loss of magnification compensation control of wafer and/or mask linear distortions(97). Also, some Fresnel diffraction effects have been noticed(98). The photon flux from electron storage rings (synchrotron) is unique in terms of its low dispersion. Such flux renders all of the above advantages, besides providing an intense photon density. Hence, although not completely collimated, the storage ring appears as the outstanding source candidate for achieving very high quality IC chip patterning(99) with unmatched throughput(100). Table 7 summarizes various X-ray system development and manufacturer efforts. Examples of resist exposures are shown in Figure 26.

9.1   X-ray Radiation Damage to IC Devices

Various works have evaluated the radiation sensitivity of MOS capacitors and MOSFET device structures(101-103). The IC device fabrication processes in general have long been questioned as to their role in inducing or enhancing the device's radiation sensitivity. The energy of the X-ray lithography process has likewise been suspected enough to provoke a study in which n- and p- channel MOSFET transistors and MOS capacitors were exposed to 10 kW, AlK$_\alpha$ radiation(104). The devices were made using only photolithographic processes and subsequently dosed with various levels (0.375 J/cm^2 to 8.24 J/cm^2) of accumulated X-ray flux.

TABLE 7. X-ray systems for IC chip lithography

University and National Laboratories

Component	Univ. Wisconsin	Brookhaven NL	Stanford Univ.
Sources: Electron Impact: Stationary Rotating			
Plasma: Laser:			
Storage Ring: Conventional (warm)	0.8-1.0 GeV 125 mA Opt. & Mech. Scan	0.6-1.0 GeV Opt. Scan[b]	2.0 GeV 10-20 mA
Superconductor (cold)			
Aligner: Orientation xy detection Gap detection	Vert. LZP Capacitance	Vert.[b] Opt.[b]	Vert.
Masks: Full field	$Bn/B_3N/Bn,B_3N$[d] $B_4NH/W(-)$[g]		
Step & repeat	$B_4NH/W(-)$[g] Si, SiN	B-Si/Au(+),BN	
Resists: Positive	MMA and, c,d	Prop. Novolak[b]	PMMA,PBS and, c
Negative	c,d	RD-2000N[b]	

TABLE 7.   X-ray systems for IC chip lithography (cont.)

IC Industrial Companies - X-ray System Developers

Component	IBM	BTL	Hughes	Hewlett Packard
Sources:				
Electron Impact:				
Stationary	Al,0.83 nm	Pd, 0.44 nm		Pd, 0.44 nm
Rotating			Al, 0.83 nm	
Plasma:				
Laser:				
Storage Ring:				
Conventional				
(warm)				
Superconductor	GeV^e			
(cold)				
Aligner:				
Orientation:	Horiz./Vert.	Horiz.	Horiz.	Horiz.
xy detection	Opt.	CZP	Opt.	Opt.
Gap detection		CZP	Opt.	Opt.
Masks: Full Field		B$_3$NH	Mylar	B$_4$NH
Step & repeat	B-Si,BN	BN/B$_3$N/BN		
Resists: Positive	Novolak		PMMA	
Negative	RD-2000N	COP,DCOPA	COP	PCMS

TABLE 7.   X-ray systems for IC chip lithography (cont.)

EQUIPMENT MANUFACTURERS

Component	Hampshire	Perkin-Elmer	Micronix[a]
Sources: Electron Impact:			
Stationary			Pd,6kW,0.44 nm
Rotating		W,10kW,0.7 nm	
Plasma:			
Laser:	Nd: glass, iron alloy, 1.2-1.4 (nm)		
Storage Ring:			
Conventional (warm)[f]			
Superconductor (cold)			
Aligner: Orientation	Horiz.	Horiz.	Horiz.
Xy detection	Optical	LZP[h]	CZP[h]
Gap detection		Capacit.(Dyn.)	CZP
Masks: Full field		Ti,SiC,BN/Au(-/+)	BN/Au(+/-)
Step & Repeat	B-Si/W(-),Au(+)	B-Si,BN/Au(+)	BN/Au(+)
Resists: Positive		ECX-1029	
Negative	AZ1350	EK-88	

a.   No longer manufacturing
b.   IBM leased line
c.   Intel
d.   BTL
e.   Oxford Instruments contract for cold magnet ring for IBM site.
f.   Normal warm synchrotron ring being built by Maxwell-Brobeck for installation at University of Louisiana
g.   Hewlett Packard
h.   CZP and LZP are circular and linear zone plate, respectively.

a

Figure 26. Resist exposures via: a. Electron impact source tungsten M-line, (1.0 μm ECX1029 resist) b. Laser plasma, 0.45 μm lines/spaces, ±0.05 μm (3σ), 1.1 μm Hoechst RAY-PF over 1.0 μm Al topography; and c. Synchrotron 0.5 μm lines ECX 1125, 38 mJ/cm^2. Courtesy (1) Perkin-Elmer, (2) Micronix, (3) Hampshire Instruments, and (4) Rohm and Haas, with Center for X-ray Lithography (Univ. Wisconsin), respectively.

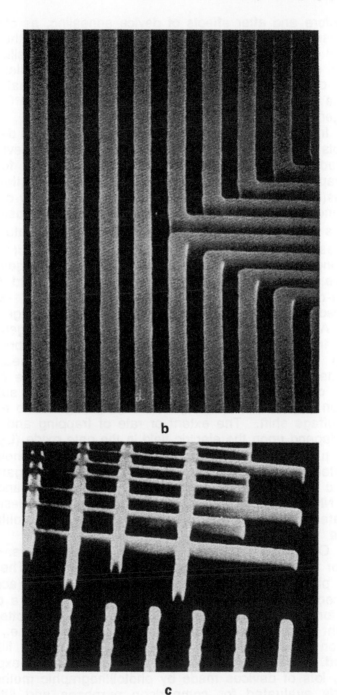

b

c

Figure 26. (continued)

The before and after effects of device annealing, as obtained by capacitance voltage (CV) plots(105), indicated that the radiation induced damage (fixed oxide charge and interface [silicon/silicon dioxide] states) could be removed by annealing. Similarly plots of source-drain current vs. gate voltage revealed that radiation damage was present but curable by annealing. A 400°C anneal in hydrogen/nitrogen gas for thirty minutes was used.

In other work, trapped charge damage has been detected via CV plots on MOS capacitor and p-channel MOSFET devices which received an accumulated X-ray dose from the four X-ray lithography exposure steps used in the fabrication of the devices(8). By annealing, device damage caused by copper (0.9 keV) and aluminum (1.5 keV) radiation, immobile positive oxide charges (5×10^{11} charges/cm^2) and fast surface states (7×10^{11} charges/cm^2 eV at mid bandgap) were removed(106). The annealing required ten minutes exposure to a hydrogen/nitrogen, or pure nitrogen atmosphere. The gate threshold (onset of source-drain current) shift from -20 volts to -3.2 volts produced by annealing agreed with the CV plotted voltage shift.

Another work evaluated the damage from the generation of neutral traps in the silicon dioxide of MOS devices fabricated with trilevel resist and X-ray lithography. No annealing was performed. Such neutral traps, if present within the completed device oxides, can capture holes evolved under subsequent radiation environment (e.g., gamma-ray) giving rise to a negative gate voltage shift. The extent or rate of trapping and threshold shifts depend upon the electric field in the gate oxide (L_{eff} 0.3 μm to 5.5 μm)(107). In Figure 27a, the NMOS threshold shift is presented as a function of γ-ray dose in rads with gate voltage (electric field in oxide) a parameter. The performance of the above NMOS devices were compared with identically designed and fabricated devices made using trilevel resist photolithography (Figure 27b).

One might expect the threshold voltage shift vs. γ dose in rads for these identical devices to be very similar. The disparity in the plots shown can be attributed to the slight (accidentally produced) oxide thickness difference. Notice that if a correction factor for oxide thickness difference is applied and plotted against the right hand ordinate, a near unity value emerges, i.e., an equal radiation sensitivity exists whether X-ray or optical lithography is used. In the first mentioned radiation damage experiments above, lots of devices made by photolithographic methods were similarly evaluated for comparison purposes and little or no difference in the radiation extent of damage was observed. Other

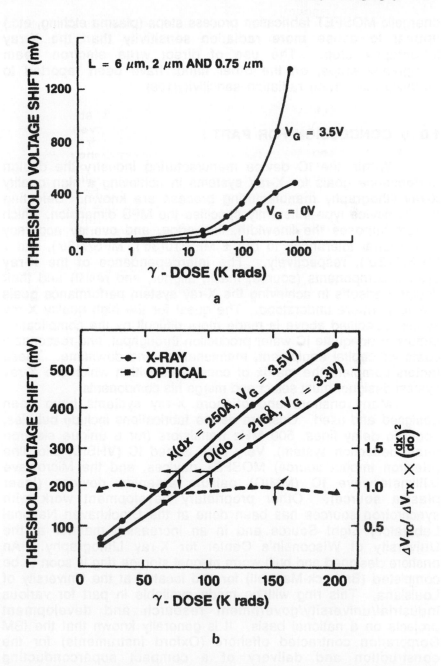

Figure 27. Threshold shift of NMOS devices versus gamma dose: a. With gate voltage as parameter; and b. Comparison of X-ray versus optical lithography damage.

energetic MOSFET fabrication process steps (plasma etching, etc.) appear to cause more radiation sensitivity than the X-ray lithography step.   The use of direct write electron beam lithography steps, on the other hand, have been reported to enhance the device radiation sensitivity(108).

10.0 CONCLUSION FOR PART I

Within the IC device manufacturing industry the design performance goals for X-ray systems in achieving a high quality X-ray lithography manufacturing process are known.   Selecting the IC device type inherently specifies the MFS dimension, which in turn imposes the linewidth, line edge, and overlay accuracy dimensional tolerances to be: $\pm$ MFS/10(3σ), MFS/7(3σ), and $\pm$ MFS/5(3σ), respectively.   The interdependence of the X-ray system components (source, mask, aligner, and resist) and their design tradeoffs in achieving the X-ray system performance goals are furthermore understood.   The quest for the high quality X-ray system specified above is made more difficult by the complicating factors of adequate IC wafer production throughput, and reasonable costs for capital equipment, maintenance, and downtime.   These factors complete the matrix of constraints within which the X-ray system designer must select and merge his components.

Many onshore and offshore X-ray systems have been designed and used.   Onshore device fabrications include bubbles, acoustic delay lines, 500 MHz transistors (for a unique electron beam deflection system), Very High Speed IC (VHSIC) pilot line (electron impact source) MOSFET devices, and the Microwave Millimeterwave IC (MMIC) gallium arsenide devices (laser plasma source).   Other proprietary development work with synchrotron sources has been done at the Brookhaven National Laboratory Light Source and in an increasing manner at the University of Wisconsin's Center for X-ray Lithography.   An onshore designed and built worm magnet storage ring is soon to be completed (Brobeck-Maxwell) for and located at the University of Louisiana.   This ring will be made available in part for various industrial/university/government research and development projects on a national basis.   It is generally known that the IBM Corporation contracted offshore (Oxford Instruments) for the construction and delivery of a compact superconducting synchrotron to be a captive part of an X-ray lithography oriented IC device development/production center.   These actions and others (either partially or totally offshore) will guide, define, and pace the future utilization of X-ray lithography.

The IC device designers conventional demand for rectangular device geometries (vias, source/drain regions, gates, etc.) may help determine the preferable lithography method (X-ray, optical, electron beam, ion beam) as the IC device MFS is extended below 0.5 μm. In this realm, the corners of geometric rectangles are difficult to maintain as right-angles (90°). Perhaps geometric corners with finite radii might even remedy some device performance instability and such devices could work just as well with π/4- sized smaller vias, etc. Some expectancy exists also regarding wafer alignment mark stability and degradation under varied and hostile IC device process environments.

Lithography methods for 0.5 μm MFS IC device production (16 Mbit DRAM) are required now; lithography methods for, and IC device processes for 0.35 μm MFS (64 Mbit DRAM) are both under development now.   An assessment of present non-conventional IC device research, such as quantum-well structures, points out that the need for MFS resolution < 0.35 μm is to be expected and will also include the 256 MBit DRAM (0.25 μm, cir.   1995-7).

The fascinating high-powered synchrotron source text follows in the second part of this chapter. The electron physics oriented subject matter provides a scientific awareness of the full technical prospects and the potential limits of X-ray lithography.

The author regrets that the above text does not include the very important data processing subject matter concerning the X-ray printing tools (source, wafer input/output, and aligner integration) automated control networking: data flow electronics, software, data storage, diagnostics, etc.

11.0 SYNCHROTRON RADIATION SOURCES

11.1 Introduction

This part of Chapter 6 is organized as follows. We first review the basic principles of the emission of X-rays from electron storage rings. The properties of the radiation which are most relevant to X-ray lithography are then presented, and finally, we discuss the system used to relay the radiation from the storage ring to the exposure station.

Without entering in any detailed discussion, it is necessary to mention synchrotrons are considered to be by some as the best sources to power the X-ray lithography steppers needed for advanced lithographies. These new technologies are needed for the manufacturing of the high-density and high-volume semiconductor integrated circuits based on devices with dimension of 0.35 µm and smaller.

11.2 Properties of Synchrotron Radiation

11.2.1 A Basic Accelerator.

Synchrotrons and Electron Storage Rings (ESR) are accelerators capable of producing stable beams of particles of very high energy, in the million (MeV) to billion (or giga, GeV) electron volt range. The machines were first designed for basic high energy and nuclear physics research, but found later application in the industrial and medical field. Yet another application is that in the area of semiconductor processing, where other types of accelerators are commonly employed (e.g., ion implanters). Figure 28 shows the conception of an accelerator, from which several beamlines extract the radiation and direct it to the exposure stations. With some impropriety, we commonly refer to ESRs as synchrotrons. Their basic structure is quite different but the properties of the radiation is the same — hence the terminology.

11.2.2 Radiation Emission: Basic Process.

An electron storage ring performs the function of capturing and storing electrons which are generated by an injector system. The electrons are kept in stable orbits that close upon themselves; these orbits are defined by the action of a series of magnets that are used to (a) define the orbit and (b) focus the electrons. Let us at first consider the simple case of a perfectly circular orbit, of an electron moving in a completely uniform magnetic field B. No work is performed by the magnetic field on the electronics, since the Lorentz force is perpendicular to the velocity, and the electrons would be moving at constant speed in a circular path. However, accelerated charges radiate electromagnetic energy at the expense of their kinetic energy(109). This is indeed what happens in the case of electrons moving in a storage ring, where the losses must be compensated in order to achieve a stable system.

The radiation emitted by low-energy electrons captured in a circular motion can be decomposed in two linear oscillations with angular frequency ω. The two oscillations are exactly 90° out of phase. Each one will give rise to a dipole emission pattern described in the far field by the classical equation:

$$\frac{dP}{d\Omega} = e^2 R^2 \omega^4 / 4\pi c^3 (\sin^2\theta)^{-1}$$

[14]

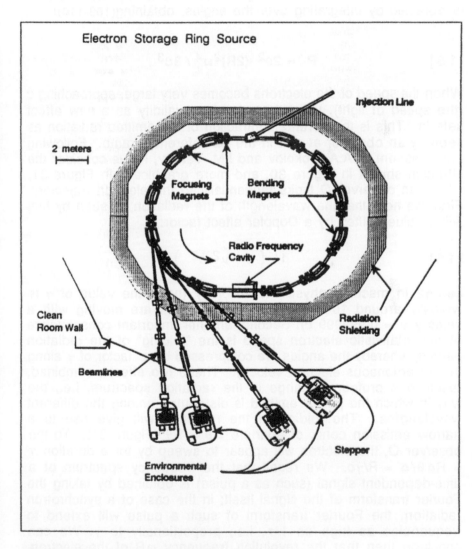

Figure 28. Basic accelerator structure.

where dP is the power radiated in the solid angle $d\Omega$ at an angle θ from the dipole axis[109]. In other words, electrons orbiting in a circular path are equivalent to a dipole antenna, which generates the familiar pattern shown in Figure 29. The total power radiated is obtained by integrating over the angles, obtaining(109-110):

[15] $P = 2e^2 (2R)^2 \omega^4 / 3c^3$.

When the speed of the electrons becomes very large, approaching c (the speed of light), these formulae lose validity as a new effect sets in. This is the Lorentz contraction of the emitted radiation as seen by an observer at rest in the lab reference frame. Following the treatment of A.A. Sokolov and I.M. Ternov, let us consider the situation shown in Figure 30, and more graphically in Figure 31, where an observer O looks tangentially to the electron trajectory. First, we note that the wavelength of the radiation as seen by him will be blue shifted by a Doppler effect factor:

[16] $1/ \left(1- v^2/c^2 \right)$

i.e., $\gamma 2$ in machine physics terminology (note: the value of γ is typically around 1000, so that the electrons are moving with a velocity $v = 0.999999$ c). Second, another important consequence of the relativistic electron speed is the "folding" of the radiation pattern, whereby the angles are compressed by a factor of γ along the instantaneous electron velocity. These two effects, combined, result in a profound change in the radiation spectrum, i.e., the way in which the power emitted is distributed among the different wavelengths. The folding of the radiation will give rise to a narrow emission cone, of aperture $\Delta\theta = 1/\gamma$ (Figure 31). To the observer O, the electron will appear to sweep by for a duration $\tau = R\Delta\theta/\Delta = R/\gamma c$. We recall that the frequency spectrum of a time-dependent signal (such as a pulse) is obtained by taking the Fourier transform of the signal itself; in the case of a synchrotron radiation, the Fourier transform of such a pulse will extend to frequencies as high as $1/\tau$, i.e., proportionally to γ. We can conclude then that the revolution frequency ωR of the electron, c/R, is shifted toward the higher frequencies by a factor τ^3, coming about because of the narrow angle of emission (contributing a factor of γ) and of the blue shift (contributing γ^2). The radiation becomes a continuum (the pulse's Fourier

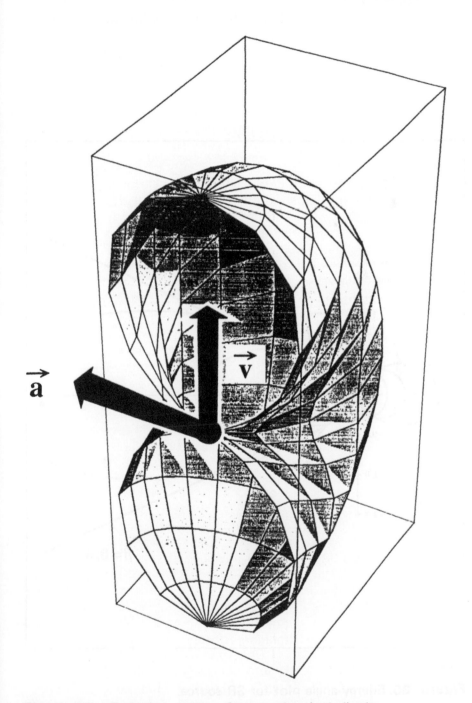

Figure 29. Emission pattern from a classical dipole.

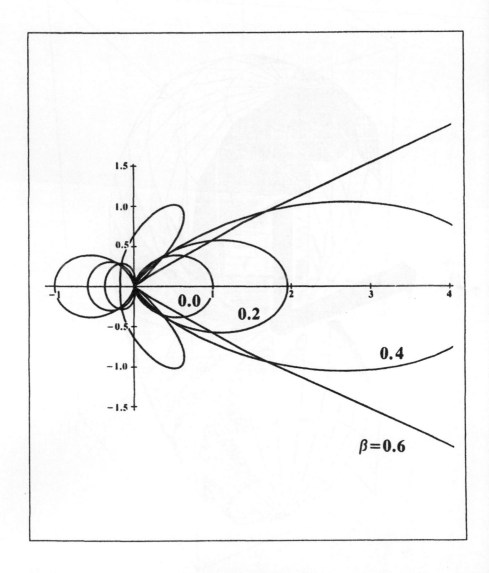

Figure 30. Energy-angle plot for SR source.

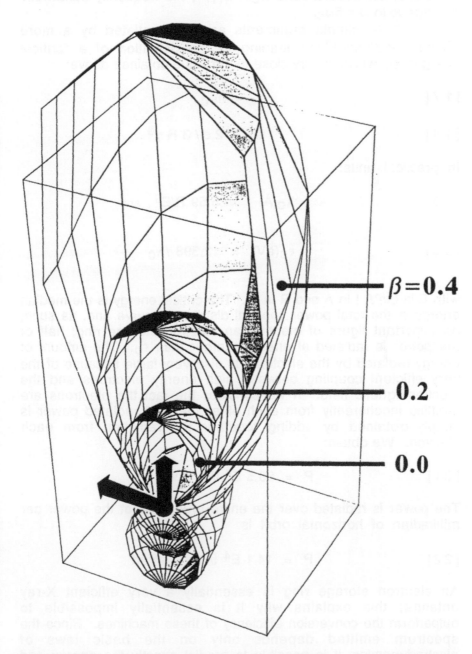

$\beta = 0.4$

0.2

0.0

Figure 31. Synchrotron radiation emission cone. Emission pattern from a high relativistic electron.

transform) centered around $\omega_C \approx \omega_R \gamma^3$; the frequency bandwidth extends up to $\omega \approx 5\omega_C$.

These simple arguments are substantiated by a more formal analysis,[110] leading to the definition of a "critical energy", ε_C, which is very close to the value obtained above:

$$[17] \qquad\qquad \varepsilon_C = h\omega_C$$

$$[18] \qquad\qquad \omega_C = 2c / 3R\gamma^3 .$$

In practical units:

$$[19] \qquad\qquad \lambda_C(Å) \;\; = \; 5.59 \, R/E^3 , \text{ and}$$

$$[20] \qquad\qquad \varepsilon_C(eV) \;\; = \;\; 12,398 \, /\lambda_C$$

with E in GeV, I in A and R in m. The critical energy is the median energy in the total power spectral distribution curve and, as such, an important figure of merit of an electron storage ring; half of the power is radiated at energies larger than ε_C. The amount of energy radiated by the electrons is also very large because of the very efficient coupling between high energy electrons and the electromagnetic field. If we consider the fact the electrons are emitting incoherently from each other the total radiated power is simply obtained by adding together the radiation from each electron. We obtain:

$$[21] \qquad\qquad P \; = \; 88.5 \, E^4 \, I/R \;\; kW .$$

The power is radiated over the entire orbit, so that the power per milliradian of horizontal orbit is:

$$[22] \qquad\qquad P \; = \; 14.1 \, E^4 \, I/R \;\;\; W/mrad .$$

An electron storage ring is essentially a very efficient X-ray antenna; this explains why it is essentially impossible to outperform the conversion efficiency of these machines. Since the spectrum emitted depends only on the basic laws of electrodynamics, it is possible to predict exactly the energy and photon flux from these sources. (Note: ESRs are used as primary standards for the calibration of detectors in the XUV. The National

Institute of Standards and Technology maintains an ESR, SURF II, primarily for calibration purposes). By inspecting the above equations, it can be seen that the properties of the radiation depend only on any two of the three variables (E,B,R), since:

[23]   R = 33.35 E/B m

for relativistic electrons, with B in kG. The spectral distribution of the radiation can be obtained analytically and is a fairly complex expression involving a Bessel function of fractional order(110). However, it is possible to rewrite the emission spectrum in terms of the reduced variables λ/λ_c and, by normalizing the power with the electron energy, to obtain a universal curve. For example, we can rewrite Equation 21 as:

[24]   P = 15.83 ε_cEI   kW

with the usual meaning of the variables. It is important to note how it is possible to obtain the same radiation (both in spectrum and in total power) from two very different machines, as long as ε_c and the product EI are the same. Thanks to these scaling properties, it is then possible to compute the flux only once and then rescale it for the particular case being considered. Computer programs have been developed to exactly model the source and to easily generate tables and plots(111).

A reduced spectral distribution is shown in Figure 32 and can be used to compute the actual flux from any machine. The power radiated is given per milliradian of orbit. For example, let us consider a machine with E = 1 GeV, R = 2 m and I = 0.1 A. Such a ring will have ε_c = 1109 eV, λ_c = 11.18 Å and will radiate 0.705 Watts/mrad. If we keep in mind that an exposure beamline will accept around 30 mrads of radiation, more than 21 Watts will be delivered to the beamline. It is important to notice that while the median energy is at the wavelength 11.18 Å, the spectrum extends well into the visible (and even infrared) on the longer wavelength side. The synchrotron radiation spectrum falls off more rapidly towards shorter wavelengths (higher energies) than it does toward the longer. It can be shown(110) that the power emitted behaves as $\omega^{1/3}$ for $\omega \to 0$ and $\omega^{1/2} e^{-\omega}$ for $\omega \to \infty$. A question of importance in X-ray Lithography is the average number of photons that are generated in a given window of energy. It can be shown that for the whole spectrum, the verge photon energy is given by(112):

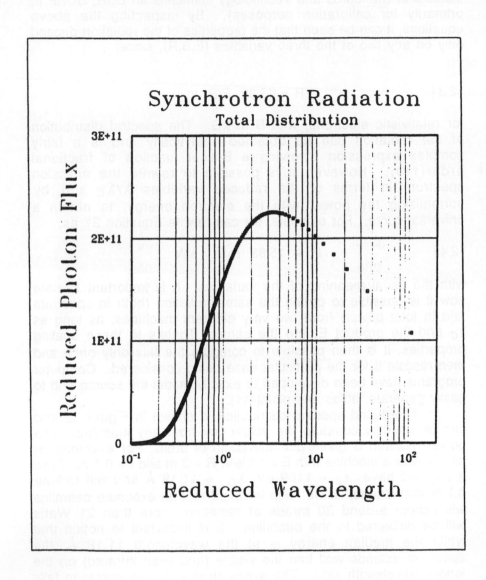

Figure 32. Radiation spectrum emitted by a storage ring.

[25] $<\theta> = 8\,\epsilon_c / 15\,\sqrt{3}$

[26] $<N> = 2.03 \times 10^{19}\ P /<\epsilon>$

with P in watts and ϵ in eV. For the above case, we would then have 1.3×10^{17} photons per second of average energy 341 eV. In general the average number of photons depends on the bandwidth selected.

In summary, the spectral distribution of the radiation emitted by a synchrotron or by an electron storage ring is characterized by a very wide frequency distribution, extending from the infrared to the X-rays. The power radiated is also large, making the electron storage ring a very efficient X-ray source.

11.2.3 Angle Distribution. Another important property of the radiation emitted by ESRs is its angular distribution. As we have briefly discussed above, the Lorentz transformation contracts the "Figure 8" shape typical of the dipole pattern in a narrow beam of aperture $1/\gamma$. The motion of the electrons along the orbit makes the narrow cone "sweep" horizontally, thus leading to a very uniform horizontal distribution. Vertically, the beam is instead characterized by a distribution width $\sim 1/\lambda$, as shown for our typical machine in Figure 33, where the results of a Monte-Carlo simulation of the radiation generation are compared with an experimentally measured power profile. The shape closely resembles that of a Gaussian function. The distribution in energy of the photons generated in the process is not uniform as well, as illustrated in Figure 34, which shows in more detail the relation existing between photon energy and emission angle. From this model we can note: a) how many more photons are generated at low energies, b) how close they are to the orbit plane, c) that the ones of lower energy subtend a wider vertical angle. The strong chromaticity of the source is typical of the synchrotron radiation process, and in first approximation we can write that photons of wavelength λ will be emitted with a distribution of standard deviation given by(110,113):

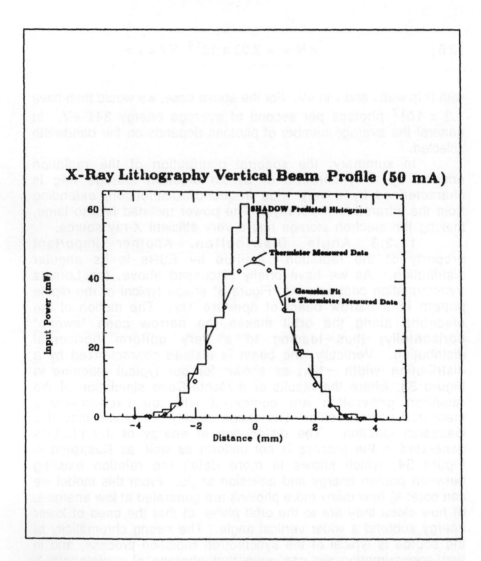

Figure 33. Vertical beam (50 mA) power distribution as observed at a target located at 4 m. The histogram shows experimental data and the smooth curve results from a theoretical calculation.

[27]
$$\theta = \frac{1}{\gamma}\left(\frac{\lambda}{\lambda_c}\right)^{0.435}$$

The narrow vertical emission angles have led to considering synchrotron radiation as being "collimated". This statement is an approximation, as shown above. It is however, true that the opening angles are very small (a fraction of a milliradian), particularly in comparison with other sources.

Horizontally, these effects exist as well but are averaged out by the motion of the electrons. Scaling relationships exist as well for the angle distribution, and the dependence of the photon energy spectrum on the observation angle is quite apparent in Figure 35, that shows the power emitted at different (reduced) angles from the orbit plane in function of the photon energy, while Figure 34 shows the same data but as a function of the emission angle for different photon energies. The curves can be scaled to any machine by making the appropriate substitutions in the units. The most important observation is that harder radiation is always emitted closer to the orbit plane.

We want to briefly notice the strong linear polarization of the radiation. To an observer located exactly in the orbit plane, the radiation would appear to be 100% polarized horizontally. This can be understood if we recall the separation of the circular motion in two harmonic oscillations exactly 90 degrees out of phase: to the observer exactly in the plane the second oscillation (vertical from his point of view) will be all but invisible. Formally, $\theta = 0$ for this perpendicular component (cf. Figure 29) so that no emission is observed on axis. As soon as the observer moves off-axis, the second oscillation becomes "visible", so that a perpendicular component is detected; however, because of the narrow width of the overall opening angle, the radiation quickly fades. By integrating over all the spectrum(110) we find that the degree of polarization can be obtained by noticing the ratio between parallel emission and perpendicular emission:

[28]
$$\alpha = W\pi / W\sigma = 1/7 .$$

This would give a polarization degree of exactly 75%. Although polarization effects are important in several instances they do not appear to play any significant role in XRL, so that we will not consider the argument further.

11.2.4 Equivalent Source. The emission of radiation is a random process governed by Poisson statistics, controlled by the

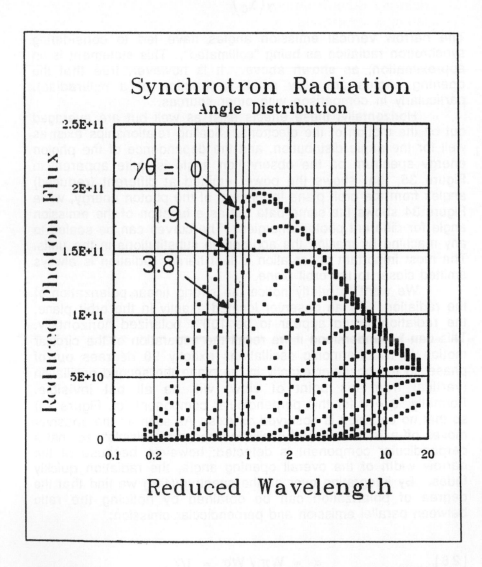

Figure 34. Photon flux as a function of elevation angle for different wavelengths.

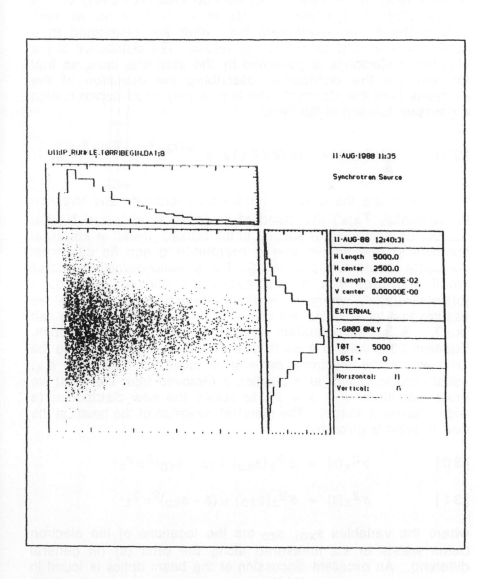

Figure 35. Angle of emission vs. photon energy for a synchrotron radiation source.

deterministic laws described above. The results we have so far derived apply to the case of an isolated electron moving on the central orbit. The electrons stored in an ESR do not all have exactly the same parameters but rather are distributed in a random way around some average values. The distribution of the electrons trajectories is governed by the statistical laws, so that for example the distribution describing the deviation of the electrons from the standard orbit is to a very good approximation a gaussian function of the form:

[29] $N(x,x') = N_0 /2\pi \sigma_x \sigma_{x'} e^{-x^2/2\sigma^2_x -x'^2/2\sigma^2_{x'}}$

where (x,x') are the electron transverse coordinate and direction in horizontal, $\sigma_x(x')$ the standard deviation in $x(x')$ and N_0 the number of electrons per second in the bunch. (Note: in statistical mechanics, one should use a coordinate q and its canonical conjugate momentum $p = \partial H/\partial q$). For a monoenergetic beam of small aperture, if x is the coordinate chosen, then $p_x = |p|\sin(x')\sim|(p)|x'$, where x' is the cartesian angle. Since we assume that $|p|$ = const., we can drop it and consider only the pair (x,x'). A similar equation applies in the vertical direction. Equation 29 is valid at a waist location, i.e., at a position along the orbit where the electrons come to a focus (of dimension σ_x, usually called waist); at a position s removed from the waist we must substitute $x \rightarrow x + x's$ to obtain the new distribution (a wider Gaussian shape). The standard deviation of the beam at the new position is given by:

[30] $\sigma^2_x(s) = \sigma^2_x(s_{xo}) + (s - s_{xo})^2\sigma^2_{x'}$

[31] $\sigma^2_z(s) = \sigma^2_z(s_{zo}) + (s - s_{zo})^2\sigma^2_{z'}$

where the variables s_{xo}, s_{zo} are the locations of the electron beam waists in x,z measured along the orbit (s) (in general different). An excellent discussion of the beam optics is found in the work of G.K. Green[114]. If we assume that the X-rays are emitted by the electrons exactly along their instantaneous velocity, then from the point of view of an observer located at some distance D from the tangent point, the X-rays will appear to form a patch with extension σ_x in horizontal and σ_z in vertical exactly in Equation 31, with s = D. In general we can use a form similar to Equation 29 to represent as well the propagation of the

X-ray beam. Equation 30 can then be used to describe the shape of the X-ray beam during its propagation.

In a real machine, the photons are not generated exactly along the electron trajectory but rather at some random angle whose distribution is consistent with the radiation distribution (e.g., Figure 34)(112,115). The radiation angles always refer to each individual electron instantaneous orbit plane(115), so that if the electron orbit forms an angle α with the reference or central orbit at the emission point, then the radiation lobe is oriented along the orbit at the same angle α. The radiation pattern is then the convolution of the radiation fan and of the electron directions at that orbit location. If we approximate the radiation with a Gaussian distribution we can make use of the fact that the convolution of two Gaussians is still a Gaussian of standard deviation given by:

[32]
$$\sigma^2_T = \sigma^2_R + \sigma^2_{x'}$$

where σ_R is the standard deviation of the radiation. The angle distribution used in Equation 31 is thus augmented by the photon angular distribution. For machines of low beam energy, the photon energy spread normally dominates.

In conclusion, the physical aspect of the synchrotron radiation source is that of a Gaussian distribution of point sources, with a distribution given by Equation 31 an with an angle aperture following Equation 32. An example is illustrated in Figure 36.

A very important figure of merit of any optical source is brightness, the density in phase space of the ensemble of rays generated by the source. In general, the higher the brightness the "better" the source. For example, both point ($\sigma_x = 0$) and collimated sources ($\sigma_{x'} = 0$) have zero phase space, granted for opposite reasons. For a given beam of rays propagating through an optical system, the well known Smith-Helmholtz invariance holds (in the paraxial approximation $x' \sim \sin(x')$)(116):

[33]
$$x\, x' = \text{const}.$$

When extended to an ensemble of N_0 rays, this reduces to the Liouville theorem stating the conservation of the phase space density of trajectories(117). If we now define the brightness:

[34]
$$B(x,x') = \frac{N(x,x')}{2\pi\sigma_x\sigma_{x'}}$$

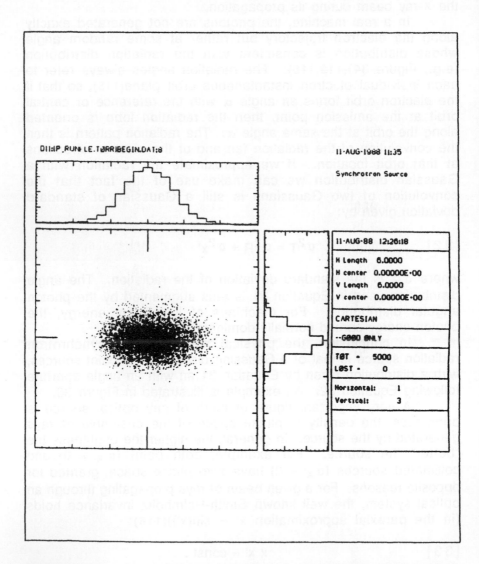

Figure 36. A typical X-ray source, illustrating the original cross-section and the phase space at the origin and after a 10 m propagation.

where N is the photon flux (photons/s), Equation 29, it is easy to convince ourselves that B (x,x') must be conserved following the Liouville theorem, since it is nothing else than the density of the photons in the space of the coordinates (x,x'). The flux N (x,x') is proportional to the electron beam density in phase space and includes the source radiation angle standard deviation. These effects are easily observed in Figure 36, where the evolution of phase space is easy to follow. The ellipses describing the (x,z) spaces are changed after propagation but the total area is constant. The radiation emitted by the storage rings has very high brightness, in comparison to other sources because of both a) small source size and b) small included angle. This effect is further enhanced in wiggler and undulator sources(113).

Another important issue is that of the source coherence. The degree of coherence essentially measures the capabilities of two radiation beams to interfere; a fully coherent source will exhibit strong modulations in the interference region that will be absent in the incoherent case. The details are somewhat complicated and beyond this discussion, but we would like to note that because of the broad spectrum the radiation is very temporally incoherent. Conversely, the small source size reduces the spatial coherence only slightly because of the long optical distances involved; for all practical purposes, the radiation can be considered as being emitted from a small volume containing independent radiators. A typical X-ray optics calculation would then start with the computation of electric field due to a nonochromatic point source to obtain the monochromatic intensity, including diffraction and physical optics effects; this calculation must then be repeated for all the wavelengths involved and the intensities added together. Finally, the calculations should be repeated for all the points defining the source, a quite formidable task. Since the source is formed by an array of independent radiators (thus incoherent) we can use the convolution theorem to add the processes together. The convolution theorem allows a separate calculation of the diffracted image and of the geometrical source effect, followed by a simple convolution integral, thus greatly simplifying the computational steps necessary. These steps are implemented in existing codes that produce a full 2-d or 3-d image calculation(118).

11.2.5 Radiation Distribution at the Mask. On the basis of the above discussion, we can define exactly the influence that the peculiarities of the SR source may have on the formation of the image in the lithographic process. It is important, because of the very high resolution involved, to verify the extent of these effects. For example, a pattern feature must be defined to within a

1/10 of a critical dimension overall(119). Several error sources contribute to this value so that the amount allocated to each component is much less than the typical 1/10 of the smallest feature size. For example, at 0.25 μm linewidths, only 25 nm form the budget allocated to the overall error. One-fifth (5 nm) may be the fraction allocated to the exposure. The simple optical system sketched in Figure 37 can be used to describe in detail the effects of the electron beam on the image formation(120). The system we discuss is of the non-focusing type but any optical system can be accommodated with few changes. An optically active system (i.e., focusing or defocusing) will result in a different value for the spatial and angle beam extent, but the discussion remains otherwise valid. We recall that in order to achieve a uniform intensity at the mask position some form of scanning must be provided. This can be achieved by keeping the mask-wafer stationary and rastering the X-ray beam or by keeping the radiation fixed and mechanically scanning the mask-wafer assembly. Other schemes, based on oscillating the electron beam itself, are unlikely to gain widespread acceptance. The optical system shown can indeed be used to study the effects of any type of scanning by choosing suitable values for the parameters. For simplicity we limit the discussion to the case of a plane scanning mirror, although this does not affect the conclusions. If we consider the photon beam to be a Gaussian beam, the propagation from the source point to the mask via the reflection from the mirror gives rise to a beam with:

$$[35] \qquad \sigma^2_x = \sigma_x(0)^2 + D^2\sigma^2_{x'}$$

$$[36] \qquad \sigma^2_z = \sigma_z(0)^2 + D^2\sigma^2_{z'} .$$

The beam waist is located exactly at the tangent point; the size will be independent of whether the (flat) mirror is being scanned or not. In this equation D(P) is the distance source-mask (mirror-mask). We can safely neglect the effect of the small reflection angle 2θ. The effect of the "painting" of the beam vertically over the mask will be that of introducing a "bias" in the vertical velocities distributions because of the linear scanning rate. From the point of view of XRL what is important is the penumbra which is created by the finite size of the electron beam. It can be shown that the penumbra is obtained by:

$$[37] \qquad \sigma_{g,x} = g / D \, (\sigma_x)$$

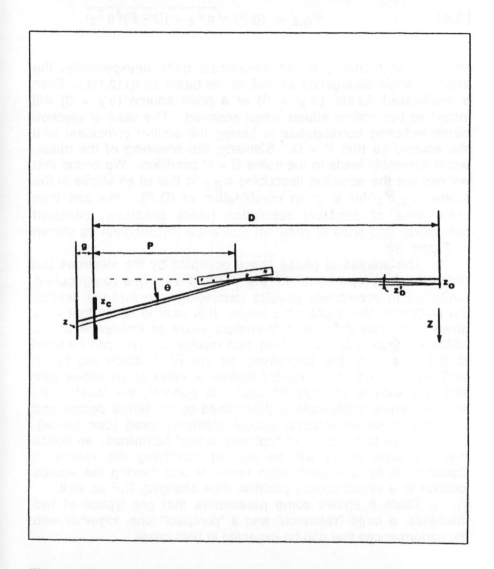

Figure 37. Model optical system for resolution studies.

[38] $$\sigma_{g,z} = (g/P) \sqrt{\sigma^2_z + (D - P)^2 \sigma^2_{z'}}$$

so that the scanning action introduces, quite unexpectedly, the effect of angle divergence as well as the beam size(112,120). Even a collimated beam ($\sigma_{z'} = 0$) or a point source ($\sigma_z = 0$) will introduce penumbral effects when scanned. The case of electron beam wobbling corresponds to having the emittor coincident with the source, so that P = D. Similarly, the scanning of the mask-wafer assembly leads to the same D = P condition. We notice that we can put the equation describing $\sigma_{g,z}$ in that of an ellipse in the plane ($\sigma_z, \sigma_{z'}$) for a given combination of (D,P). We can thus draw lines of constant resolution (more precisely, constant penumbra) and lines of constant emittance (hyperbolas) as shown in Figure 38.

The volume of phase space occupied by the electrons (no radiation) is proportional to the product $\epsilon_x = \pi \sigma_x \sigma_{x'}$, called emittance in accelerator physics terminology, so that the smaller the emittance, the brighter the beam. It is easy to notice that for a given resolution there is a maximum value of emittance above which it is impossible to achieve said resolution. This can be used to define exactly the tolerances on the ESR, since the beam emittance is one of the typical figures of merit of an accelerator and may include strongly its cost. In general, the "best" (i.e., smallest) value achievable is determined by the lattice design and by the number of electron optical elements used (see below). Coming back to the case of "optically active" beamlines, we notice that the main effect will be that of modifying the values in Equation 38 by a magnification factor M and moving the source location to a virtual source position, thus changing D,P as well.

Table 8 shows some parameters that are typical of two machines, a large "research" and a "compact" one, together with the performances that can be expected in both cases.

12.0 TYPES OF MACHINES

The first type of accelerator useful for spectroscopical study was the electron synchrotron. All accelerators share the same block structure, illustrated in Figure 28.

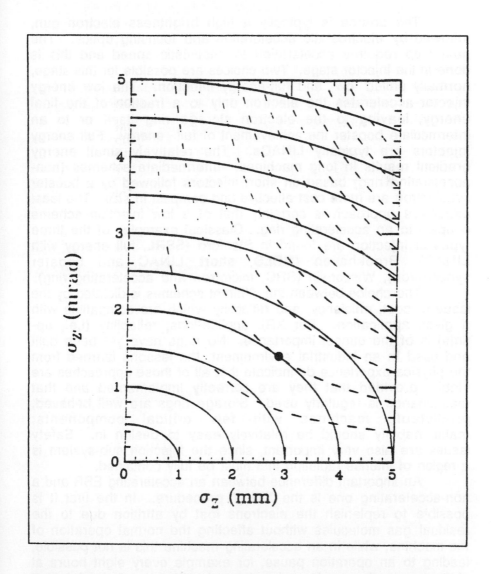

Figure 38. σ_z vs. $\sigma_z{}'$ The dashed lines refer to lines of constant emittance while the ellipses show lines of constant resolution.

Source → Pre Accel → Injector → Booster → Machine .

The source is typically a high brightness electron gun, followed by electrostatic acceleration and focusing optics. The next step requires acceleration to relativistic speed and this is done in the injector stage. Two choices are possible for this stage, normally called low- and full-energy injection. The low energy injector accelerates the electron only to a fraction of the final energy, leaving to the electron storage ring itself or to an intermediate booster the achievement of full energy. Full energy injectors are typically LINACs. The relatively small energy gradient results in long machines. Intermediate schemes (non-accelerating ring) based on short injectors followed by a booster synchrotron are more cost effective and compact in size. The least expensive approach is certainly that of a low injection scheme coupled to an accelerating ring. Classical examples of the three types of injection are found in Stanford (SSRL, full energy with LINAC), Brookhaven (NSLS, short LINAC and booster synchrotron), Wisconsin (SRC, microtron and accelerating ring).

The choice between the different schemes is dictated by the issue of cost, efficiency, and reliability which are compatible with a given application. For XRL applications, reliability (i.e., up-time) is of the utmost importance. No rings have yet been built and used in an industrial environment; the lessons learned from the physics experience do indicate that all of those approaches are viable, provided that they are correctly implemented and that maintenance is regularly used. Storage rings are well behaved, predictable machines with few critical components; maintainability should be relatively easy to design in. Safety issues are also very important, since the injection sub-system is a region of intense radiation that must be fully contained.

An important difference between an accelerating ESR and a non-accelerating one is the "filling" procedure. In the first it is possible to replenish the electrons lost by attrition due to the residual gas molecules without affecting the normal operation of the machine, while in an accelerating machine this is not possible, leading to an operation pause, for example every eight hours at shift change times. Here is where the lifetime of the injected current plays a critical role; the (almost) exponential decay of the current should be such that the average beam current is large enough to sustain the steppers' throughput. When calculating production figures, one should always use the average current:

[39] $<I> = I_{inj}\ (1/T)\ {}_0\!\int^T e^{-t/\tau}\ dt$

[40] $= I_{inj}\ (\tau/T)\ (1 - e^{-T/\tau})$

[41] $\sim I_{inj}\ (1 - T / 2\tau)$

for long lifetimes τ. We can easily verify that if we require that $<I> = 0.8\ I_{inj}$ then we need a lifetime in excess of 20 hours for an 8 hour shift. The "filling-up" process normally lasts 15-20 minutes for a well operated machine.

TABLE 8. Research and Compact Synchrotron Parameters

Parameter	Units	Research	Compact
Beam Energy	GeV	1.0	650
γ		1950	1270
Radius	m	2.0	0.5
Magnetic Field	kG	16	43
Critical Energy	Å	10	10
Radiation Angle	mrads	0.5	0.8
Current	mA	100	200
σ_R	mrads	1	2
σ_x	mm	1.2	2
$\sigma_{x'}$	mrads		
σ_z	mm	0.3	1
$\sigma_{z'}$	mrad	0.5	0.6
Penumbra x	nm	1.6	2.7
Penumbra z	nm	0.4	1.3
Power/mrad	W/mrad	0.7	1.0
Power(Beamline)	W	21	30

The ESR itself is essentially a lattice of magnetic fields used to steer and control the electron beam in its path through the vacuum chamber. An essential component is the radio frequency cavity which is used to provide the energy lost by the synchrotron radiation emission process. The RF cavity is a key component and the subject of extensive studies(121). Dipole magnets are used to define a closed orbit, while quadrupoles, exapoles, and octupoles are used as active elements in controlling the beam position and shape(122). Because of the steady state requirement conditions, only some beam orbits and configurations are stable (like in an optical resonator). The electrons are characterized by an Hermite-Gaussian distribution of trajectories(113-114). The intrinsic circularity of the orbit is broken down in sectors, equivalent to each other. Each sector contains one bending magnet and some other focusing magnets; several arrangements are possible leading to the different type of lattice design. The type and the complexity of the lattice has a direct impact on the quality of the electron beam, i.e., its emittance. Basically, a simple lattice without higher order elements will make it more difficult to achieve a beam of small dimensions and good collimation. Conversely, sophisticated schemes allow the achievement of extremely compact electron beams. Fortunately, the requirements on the beam size and divergence of XRL are quite relaxed, greatly simplifying the ring design and construction(123). This has lead to the proposal of "compact" ESRs, i.e., scaled down versions of the large accelerators used in high energy physics and in spectroscopy. Besides the distinction between "large" and "small" rings, there is also another distinction between "warm" and "cold" rings, depending on the use of resistive or superconducting magnets. The rationale of the two choices is the following. For XRL applications, a small footprint is highly desirable. The minimum footprint is, in ultimate analysis, determined by the maximum magnetic field achievable and by the distance between bending magnets (straight sections). For example, our 1 GeV ring will require a magnetic field of 1.6 Tesla to bend the electrons into a 2 m radius circle. This gives a minimum size of some 6-8 m when we take in account straight sections. Normal magnets are made by water cooled copper coils with a laminated iron yoke used to shape and confine the magnetic field. The technology is simple and robust, and the system limitations come (a) from the maximum field that can be achieved and (b) from the large power that is dissipated, leading to a sizable addition to the ring operating costs. Conversely, the superconducting magnets allow high fields and by definition use essentially no power. If we had 4 Tesla magnets available, our 1 GeV machine could have now a bending radius of 0.5 m, considerably smaller than the normal magnet version. The superconducting magnets are however of complex design and should be considered still a high-risk option. The cost of the helium system can also add considerably to the cost of operation. The focusing magnets are always at room temperature.

The main advantage of cold rings is smaller footprints at equivalent energy. An inspection of Equation 24 shows that trade-offs are possible between the parameters that define the main aspects of a storage ring. As a rule, large currents are the most difficult to obtain and to keep in an ESR. It is then convenient to trade beam energy vs. current in order to achieve a given throughput for a fixed radius and this is clearly where one of the advantages of the cold ring is. Against cold rings are several engineering issues, in particular the non-triviality of magnet construction as well as the difficulty of locating focusing (electron) optics and the injection line in a small area.

The warm ESRs are the ones traditionally used in SR work. Their design and construction is well established. Several commercial vendors are actively working in this area, particularly in Japan(124). In the United States, Maxwell Laboratories is designing and constructing a warm storage ring. At the time of this writing several projects were underway for cold rings, the most notable ones being COSY (Germany), Oxford (England) and Sumitomo (Japan) in the area of commercial vendors and Brookhaven National Laboratory in the area of U.S. based efforts. The German and British efforts have very close ties with their respective national laboratories.

13.0 BEAM TRANSPORT SYSTEMS

The problem of delivering a high intensity radiation beam to the mask requires a careful engineering study in order to ensure cost-effectiveness, reliability and safety. The system used to accomplish this is called a beamline. It can be divided into subsystems performing the different tasks required for the overall functionality, that is, the relay of the X-ray generated by the ESR to the mask-wafer assembly.

First and foremost, the beamline must deliver the X-rays to the mask-wafer assembly in a controlled and convenient way. This is dictated by the exposure tool being used (stepper) and by the mask-resist combination. An exposure window can be easily defined by forming the product of the mask carrier transmission times the resist absorption:

$$[42] \qquad W(hw) = T_{carrier}(hw) \, A_{resist}(hw) \, .$$

The physical meaning can be easily grasped, since it represents the part of the spectrum that will have useful photons, i.e., photons that will be absorbed in the photoresist. Figure 39 shows how the window peaks around 10 Å. Softer radiation is absorbed by the carrier, creating heat and possibly radiation damage; harder radiation is not absorbed by the photoresist and is useless. A well designed beamline will then reject photons emitted outside of the exposure window, delivering radiation that is well matched to the system in use.

Second, the beamline must provide a bridge between the different environmental conditions at the source and exposure station. The environment of the exposure station and of the storage ring are very different, with the first being kept at reduced vacuum (p ~ 20 - 200 torr) or at atmospheric pressure, while the ring is always in ultra-high vacuum, i.e., at pressures of 10^{-9} torr or less. These low pressures are obtained only after prolonged outgassing of the ring and conditioning, since the radiation itself desorbs absorbed gases from the ring walls: it takes a fairly long time (weeks) before the walls are scrubbed clean and a satisfactory pressure is achieved. Exposure of the ring to high pressures in an uncontrolled way may result in long down times due to the need for reconditioning the vacuum. Thus, means of maintaining the pressure differential without affecting the delivery of the radiation must be devised; at the same time, the safety of the ring must be maintained even in case of catastrophic failures.

Third, the beamline must provide a data channel for communications between the stepper and the storage ring. Finally, the hard radiation generated in the ring during injection must be fully contained for the operators' safety. A large body of literature exists on this subject(125-126) and we will not dwell on it, save for noticing that safety radiation is achieved by carefully placed shields located along the beamlines and by heavier, fixed shielding, around the ring itself. The shielding along the beamlines is not needed because of the X-rays, which are efficiently stopped by the vacuum vessel walls, but rather because of the harder gamma rays that may be generated by the collision between the high energy electrons and the residual gas molecules. This bremsstrahlung may happen to be oriented along the beamline if by chance a region of (relatively) large gas trapping exists at the tangent point.

Breaking down the beam transport system into subsystems, we find that we can define the following blocks: vacuum, optics, communication, safeties. We will address these points in turn.

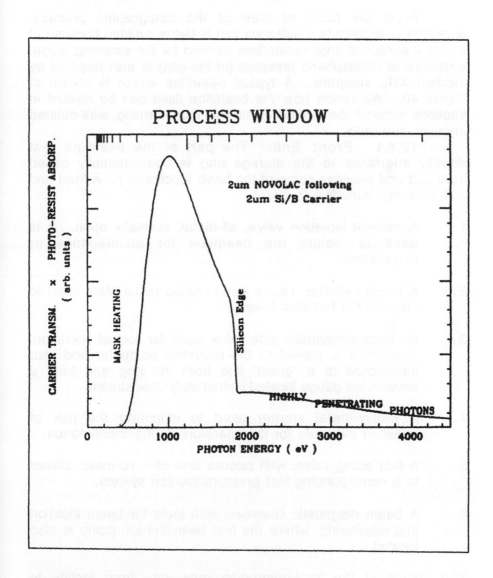

Figure 39. Process exposure window.

13.1 Vacuum Requirements

From the point of view of the lithographic process, operation in vacuum is a nuisance that is better avoided because of all the mechanical implementations needed for the scanning stage. Exposure at atmospheric pressure (of He gas) is also required by modern XRL steppers. A typical beamline layout is shown in Figure 40. We notice how the beamline itself can be divided in sections around the optical systems, thus forming well-defined vacuum sub-units.

13.1.1   **Front End.** The part of the beamline that directly interfaces to the storage ring is conventionally called front end and provides some of the basic functionality. A front end will typically include:

1.   A manual isolation valve, all-metal, normally open. It is used to isolate the beamline for maintenance or installation.

2.   A photon shutter, i.e., a water-cooled metal plate used to shut out the radiation beam.

3.   An electropneumatic gate valve, used for normal operation. This valve is slaved to the beamline computer and also interlocked to a "grant" line from the ring and from a vacuum ion gauge located immediately downstream.

4.   A heavy-metal shutter used to eliminate the risk of radiation exposure for the operators during maintenance.

5.   A fast acting valve, with closure time of ~ 10 msec, slaved to a corresponding fast pressure monitor system.

6.   A beam diagnostic chamber, with tools for beam location and monitoring, where the first beamline ion pump is also located.

The details of the implementation may vary from facility to facility, but the basic functions are those performed above. The front-end should be kept simple in order to increase reliability. As discussed below, the power densities typical of a lithography ring are relatively modest and do not require excessive precaution from a cooling point of view.

13.1.2 **Extraction Windows.** Located at the other extremity, near the exposure station, the exit (or extraction)

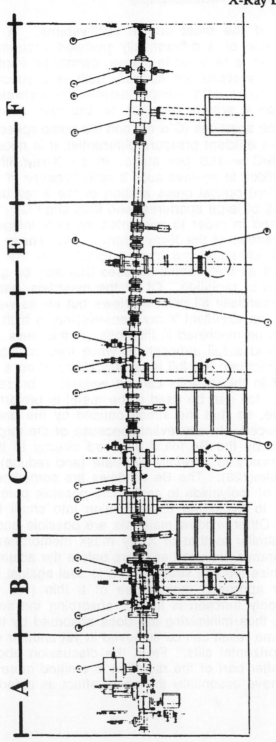

Figure 40. Layout of an XRL beamline.

window is one of the most critical subsystems. A possible alternative, the use of a differentially pumped vacuum system where the pressure is reduced in stages, cannot be implemented efficiently. These systems are normally built as a succession of apertures or pinholes with pumps installed in between. The efficiency of such a system depends on the ratio between the conductance of the apertures ($C \propto d^2$) and the pump speeds S(128). In order to get an efficient pressure differential, it is necessary to have at least S/C = 100 per stage. In an X-ray lithography beamline it is difficult to achieve such a ratio because of the large (relatively speaking) optical cross section of the X-ray beam that mandates the use of large apertures (and thus large C). Windows must then be used in order to guarantee vacuum integrity while allowing the extraction of the X-ray beam in air. These windows must be able to withstand the force originated by the pressure differential as well as the thermal stresses that may be generated by the absorption of radiation. Of all the materials, beryllium is the best from essentially all point of views but for its well-known toxicity. Be has an excellent X-ray transmission, a high Young's modulus and can be machined in thin foils. A thickness of 25 μm is sufficient to hold a full atmosphere while transmitting a good fraction of the spectrum; the foil is usually clamped on a metal or Viton gasket, while thicker foils can be welded or brazed to the support. Thinner foil can be used if the metal is preformed in a convenient shape, so that the stress caused by the pressure is considerably reduced. In any event, because of the large power density absorbed by the Be foil if exposed directly to the white beam, it is necessary to carefully evaluate (and reduce) thermal stresses in the film(128). The Be windows are sometime coated with a thin layer of polyimide to seal some possible porosity and, more important, to avoid the disintegration into small flakes in case of failure. Other window materials are possible and the list would be very similar to that of X-ray mask membranes. Most often Si:B membranes are used as filters before the actual window in order to minimize thermal stresses and to seal against low level leaks. Another approach makes use of a thin (1-2 mm) Be prefilter whose only function is that of absorbing the softer part of the spectrum, thus minimizing the dose absorbed by the metal window. The same result can be achieved in yet another way, that is the use of horizontal slits. From the discussion above, it is clear that the softer part of the radiation is emitted more off-axis; thus, a slit will have essentially the same effect as a (adjustable) filter.

13.1.3 Valves and Delay Lines. The vacuum beamline can be sectioned by valves located at several positions; these valves are automatic and interlocked to vacuum gauges against possible wrong openings. The closing time of these valves is normally 1-2 sec, so that they are not suitable as safety valves since the catastrophic failure of a component creates a pressure shock wave moving at the speed of sound. In helium, this is 965 m/s, giving a progression time of about 1 msec/m of beamline. A typical beamline would be 10-13 m long, so that a fast-acting valve (typically with a closure time of 10-15 msec) would have to be located next to the storage ring itself. In order to improve the odds of successfully safeguarding the ring, an "acoustic delay line" can also be installed(129). The acoustic delay line slows the progression of the front, providing the time needed by the fast valve to close and intercept the shock wave. Nothing can be done about the tail of fast molecules that are in direct line of sight with the beamline. Outside of the fast-closing valve, the other components are Viton-sealed gate valves for the sectioning valves, and all-metal angle valves for other purposes (forevacuum connections, etc.).

13.1.4 Pumps and Construction. The construction of the beamlines follows standard UHV practices, based on the use of all-metal flanges and ion pumps all the way to the Be window itself. This construction normally gives highly reliable systems with little maintenance; other pumps, such as turbomoleculars, should not be used for these applications but can be used for pumping down. An excellent choice is the air-bearing-based oil-less pumps manufactured by Alcatel; they can be started at atmospheric pressure and be used to provide the pumping during the bake out procedure. After this necessary stage, the beamline must be conditioned to be X-ray beam; no matter how careful the bake out procedure, the energetic X-rays will always force desorption from chemisorbed species. When initially opened to the ESR, the initial pressure raise may force the shutdown of the front valve (interlocked to the vacuum gauges), so that a gradual conditioning may prove necessary. This may require careful planning in an industrial environment. Particular attention must be dedicated to the risk of contaminating the surfaces of the mirrors used in the beamline during any of the preparation and installation steps. A base vacuum of 1×10^{-10} torr or better must be maintained at all times when in operation to avoid residual gases cracking at the mirror surface and buildup of a carbonaceous layer that may degrade significantly the mirror reflectively.

Vacuum of this quality is possible with standard practices, if strictly enforced.

Downstream of the Be window, the requirements are much more relaxed. An important consideration is the protection of the window from the atmosphere. No oxygen or water vapor should be allowed near the Be window, since under the intense X-ray beam ozone and other active species are formed and they corrode quickly the metal of the window, with potentially catastrophic results. Another issue that has not been fully developed is that of the window needed for a scanning beamline; existing solutions are not fully satisfactory(130).

13.2 Optical

The optical system, if installed, performs the function of distributing uniformly and in a controlled way the radiation across the mask. There are several reasons for which one or more mirrors need to be installed in a beamline. First, the presence of one mirror provides a break in the machine-stepper line of sight, allowing one to locate a radiation stop for full containment. Second, a collecting mirror system may increase the amount of radiation delivered to the mask-wafer assembly, lightening the requirements on the machine itself. Third, mirrors act as low-pass filters that remove the harder radiation that would not be absorbed by the resist and would end up in radiation damage to the underlying wafer. A careful assessment is necessary in order to avoid design pitfalls while delivering maximum system performance. While discussing alternatives, one must keep in mind the exposure window defined above as well as the speed of the exposure tool itself. An important parameter is the horizontal width of the field that must be filled by the beamline. A full field exposure is not practical so that one must resort to forming a uniform (horizontal) line image and to a scanning system to generate the required uniformity in the other direction (vertical). It is always possible to use suitably shaped apertures to improve the beam uniformity, but this option should be used with caution because it may prove too expensive in terms of power loss if the beam profile is strongly non-uniform to start with. Aperturing should be used for fine-tuning the beam rather than for providing the main shaping. In the following discussion we assume a field of 50 mm (H) by 25 mm (V) as our exposure area, with a requirement of ±3% uniformity.

13.2.1 One Mirror: Collecting Radiation. A single mirror can be used to increase the amount of radiation delivered to the mask-wafer assembly by increasing the beamline aperture(131). For example, for a beamline 15 m long and a horizontal field of 50 mm (f/300), a 100 mm wide collecting mirror located at 5 m from the tangent can provide a gain of 6 (from f/300 to f/50) in comparison to a straight-through system. However, the image formed by such a mirror will not be well suited to the requirements of the exposure system; the image is not a line, but rather forms the characteristic "smile" of glancing optics. Furthermore, the variation in incident angle across the sagittal direction will force a non-uniform distribution at the image. If the system design can tolerate these shortcomings (usually much larger than the typical ±3%) then the option is a good one. The mirror can be optimized to form a focus at the mask or past it; typically the use of a toroidal mirror will allow one to set the tangential (vertical) focus at the mask and the sagittal (horizontal) one quite past it in order to fill the required width. The main problem here is that the optical designer has not enough free parameters to adjust in order to fulfill the requirements of line image and uniformity, while increasing the beamline aperture. For example, one can form an excellent line image by using a tangentially cylindrical mirror (i.e., non focusing in the sagittal), but with no gain in luminosity.

   13.2.2   Two Mirrors: Solacing and Shaping. The use of two mirrors(132) allows increased flexibility to the optical design, since now one can compensate aberrations. Several designs have been proposed in the past for monochromator beamlines(133), but the requirements of an XRL system are different to the point that new designs are necessary. If one considers the combinations offered by the use of two toroidal mirrors there are several possible choices that may provide the required line image.   Great care must be exercised in the selection, since the uniformity may be irreparably compromised by the reflectivity changes across the surface of the mirrors, both from a power and from a spectral content point of view. Proprietary designs overcome this shortcoming by providing an excellent line image and beam uniformity(134). In general the luminosity can be easily increased to f/20 without compromising the uniformity, both power and spectral. From several point of views, such a beamline is ideal.   All the requirements (uniformity, power, image modularity) are fulfilled or exceeded. The extra cost of the optical system is well compensated by the increased performances.

   A beamline based on two mirrors requires a more complex vacuum system, since it must be possible, for example, to replace one of the mirrors without affecting the other.   This can be achieved by a judicious use of sectioning valves, modular design and flexible optical design.

13.2.3 Three Mirrors: Scanning. The vertical scanning action can be provided by a mirror. On the plus side, a scanning mirror provides a well controlled beam and allows the design of a fixed mask-wafer assembly, with dynamic alignment during exposure. This greatly simplifies the design of the stepper (as in the Perkin Elmer approach). On the minimum side, the scanning mirror adds one more reflection and may harm the beam uniformity. For example, a single mirror beamline where the mirror is used for collection and scanning will have unacceptable uniformity. This is true also for a two-mirror beamline, where the change in incidence angle due to the scanning action appreciably changes the system magnification. At a glancing angle of 2° or 1° the scanning range of 25 mm/8 m = 3.1 mrads is definitely not negligible. The best approach appears then to be a three-mirror system, where the scanning action is fully separated from the beam-shaping (or image-forming) action. A flat mirror following the couple of focusing mirrors above will provide the ideal solution, since it will raster the beam across the field without affecting the image shape and the spectral power. This is true if the scanner is operated at a larger incidence angle, since it is the mirror at the smaller incidence angle that determines the overall system lower pass frequency. The use of a flat scanner has another appealing facet: upgradability. For example, the two-mirror beamline designed for a Karl Suss stepper (mechanical rastering, fixed beam) can be implemented for use with a Perkin Elmer stepper by simply adding one more stage, the scanning option. Everything else remains the same. This considerably reduces complexity, inventory and maintenance costs.

13.2.4 Optical Components Specifications. The mirrors used in an XRL beamline must deliver the maximum power in the prescribed image. There are two main sources of potential problems: figure errors and finish. The first causes the rays not to be focused in the correct position, thus degrading the image quality. Because of the reduced requirements of the optics (after all, the beamline is a condenser, not an image forming system) the tolerances on the figure are quite relaxed. An error of one wavelength (He-Ne testing laser) across a mirror 0.25 m long gives a slope error of 1.3 µrad, leading to an error of 13 µm after a 10 m long beamline, quite acceptable. The requirement on the surface finish is more stringent. The reflectively of a mirror at glancing angles is affected by the surface roughness of standard deviation σ via an exponential term:

[43] $R(\lambda;\sigma) = R(\lambda;0)\exp\ (4\pi\sigma\cos(\theta)\ /\ \lambda^2)$

hence, at $\theta = 88°$ and $\lambda = 10$ Å, it is necessary to keep the roughness to less than about 0.74λ for the reflectivity not to degrade more than 10%. This sets it at about 7.4 Å for X-ray lithography, a value not at all unreasonable by today's optical industry standards.

  The mirrors used for focusing and scanning are typically manufactured from fused silica. The material is well known to the optical industry and can be polished with high figure and finish, well within the tolerances described above (1 wavelength figure, 20 Å finish). The power density encountered on a typical XRL beamline does not warrant the use of cooled optics. If the power per milliradian P_0 is of the order of 1 W, on the basis of simple, not an imaging forming system, geometrical arguments, we can verify that the power density at the first mirror will be of the order of:

[44] $P = \dfrac{(1000\ P_0)\cos\theta}{\theta_v D^2} = \dfrac{(1000\ P_0)\ \gamma\cos\theta}{D^2}$

where D is the source-mirror distance (m), θ_v the radiation opening angle (radians) and the factor of 1000 takes in account the conversion from mrads to rads. For $D = 5$ m, $\gamma = 1000$ and $\theta = 88°$, we obtain a power density of about 140 mW/cm^2. Hopefully, the mirror will reflect about 80% of this, leading to an absorbed power density of 28 mW/cm^2 that can easily be dissipated without any special techniques. For example, SiO_2 (e.g., versus Al) has a thermal conductivity of 0.013 (2.37) W/cm^2/°K so that a 5 cm thick mirror will have at most (neglecting radiation and lateral conduction) a ΔT of the order of 10 (0.06) °C between front and back surfaces. Clearly higher beam energy machines (larger γ) will present more problems.

  The reflectivity of the mirrors depends strongly on the coating material. Reflective optics in the X-rays is based on the fact that the index of refraction is less than 1.0 in that energy region, so that total external reflection is possible. However, the reflectivity is strongly affected by the material characteristics within a skin depth (of the order of a few λ) so that the surface conditions play a very important role. A few monolayers of

Carbon (Z = 6) can significantly reduce the reflectivity of a gold mirror (Z = 72). As discussed above, roughness may decrease the reflectivity as well. Care must be exercised when depositing the material in order to have a smooth surface; Au/Cr coatings (500 Å/50 Å) are often used and can be deposited from thermal sources, electron guns or sputter guns. Since the materials are all strongly absorbing, in X-ray we do not observe a well defined critical angle as we do for (non-absorbing) dielectrics in the visible. This means that the selection of the design reflection angle is dictated by a system design based on the increase in cost with larger incidence angles (longer mirrors) weighted against the reflectivity increase. The decrease in reflectivity may often be compensated by an increased acceptance angle (wider mirror). Once more, a system design based on a full ray-tracing(135) is necessary in order to balance the various requirements. Practically speaking, angles less than 88° make for "easy", 88 - 89° difficult and more than 89° for very difficult mirrors. An initial decrease in reflectivity of a newly installed mirror is to be expected, because of the formation of an initial graphic carbon monolayer on the mirror surface. The decrease should quickly saturate at a value of about 10% less than the original (fresh surface) value. This is where the requirements of UHV are the most stringent. Mirrors kept at vacuums of the order of 1.0×10^{-10} have shown essentially no degradation after more than two years of continuous operation at SRC. If a mirror shows signs of degradation, there are only a few choices. The first is that of stripping, cleaning and recoating; most of the times the original values can be restored. Cleaning schemes (both in-situ and ex-situ) based on the use of plasma discharges have been proposed, with mixed success(136). The idea is obviously very attractive but needs further development.

In summary, mirrors must be manufactured in a clean environment and kept clean during use. Standard optical shop techniques, augmented by UHV procedures, can provide satisfactory results.

13.3 Data Communication

The beamline must provide a communication bridge between the stepper and the ESR. From an operational point of view, there is no reason for a stepper to communicate directly with the ring. The beamline replaces, or actually is, a "light bulb": the stepper may request the beamline to deliver a dose to expose a given field, and the beamline system must either grant or refuse the request depending on the system's status. In the same way the ring should not interrogate the stepper, but simply grant (or refuse) the request of the beamline to open the front end valves. Of course, essentially infinite variations are possible on the implementation of the hardware and software. In general, the most important requirement is that of a judicious implementation of a system based on a high-speed link and hardware interrupts, as well as hardware backup systems. Watchdog timers are an absolute necessity. In one implementation, an Ethernet backbone is used for fast data transfer while handshake lines are used to verify the granting of valves and shutter opening or closing actions.

13.4 Safety Issues

As briefly mentioned above, two different types of safety issues must be considered around a storage ring source: personnel and machine. When speaking of safety, the radiation we speak of is the hard radiation generated by the high-energy electrons when they interact with the vacuum chamber walls and/or residual gas. The radiation is typically formed by gamma rays, although neutrons generated by nuclear radiations induced by the electrons are also present. A storage ring is a very "clean" machine, with very little (if any) residual radioactivity; most of the radiation is generated (a) at well defined locations or (b) by the attrition of the gas molecules. The injection procedure is always a source of higher radiation levels (because of less than 100% efficiency). From a personnel point of view, the shielding of radiation must be complete. Specifications vary from laboratory to laboratory and are as a rule much more stringent than the federally approved minimum radiation levels(125). In the case of compact industrial rings, the reduced size of the machine simplifies the problem by reducing the size (and weight) of the main shielding. In a laboratory environment, where skilled personnel are normally at work, procedures based on non-full containment are acceptable as long as the "hot" areas are clearly marked and/or interlocked. In an industrial setting this is not acceptable and full radiation shielding must be implemented. This essentially forces the use of a beamline-break mirror to allow one to put a radiation stop. The machine itself is typically fully enclosed in a concrete vault so that no radiation may leak out. The beamlines have a lead shielding arranged in a telescopic fashion so that nowhere can an operator be in a direct path to the machine. The shielding design must always be reviewed by an authorized medical physics team before implementation(125).

The questions of instrument safety follow quite different guidelines. In a synchrotron radiation X-ray lithography system, several beamlines share the use of a single light source, the synchrotron. Any accident happening on any beamline should not be capable of bringing the ring down or affecting the other beamlines. This means that the beamline systems must be capable of fully containing the accident. The main concern is that of an accidental catastrophic break of the final vacuum isolation window. As discussed above, acoustic delay lines may be used to provide time for the safety valves to shut down isolating the faulty beamline. Experience has shown that catastrophic failures are extremely rare on a "mature" system, where the safety subsystems have been thoroughly developed. An XRL source will require extra protection in the areas where breakdowns are otherwise most likely to occur: electrical feedthroughs, windows (should be avoided unless absolutely necessary), and bellows. Most of the times these breakages do not lead to catastrophic vacuum failure. Residual gas analyzers must be used permanently to monitor the vacuum composition by checking mass 32 (O^+_2, indicative of a beamline leak) and 4 (He^+, likely to indicate a window leak). The safety interlocks on the beamlines follow standard approaches, with a microcomputer managing the opening and closing of valves as well as the pumping system. Manual overrides allow non-standard operations and maintenance. The vacuum gauges used throughout the beamline must be capable of generating interrupt signals to the valves' interlock system independent of the computer operation. While a completely foolproof system does not exist, a careful engineering study will reduce the likelihood of non-containable accidents to an acceptable minimum.

13.5 Machines and Lithography

A synchrotron-based X-ray lithography system comprises three main subsystems: the source, the beamline and the exposure station. We have described the source and the beamline, while the exposure tools are covered in the first part of this chapter. At this point, we want to emphasize that the design of a successful lithography process requires a careful engineering systems study. There are many trade-offs which are possible, delivering performances which are approximately the same but with widely different costs. For example, beamline acceptance may be used to compensate for a low-power machine; modularity is essential in order to accommodate tools with different scanning requirements, and so on.

From a lithography point of view, there are two parameters that will determine the viability of the approach. Resolution is certainly not an issue, at least until features of less than 0.1 μm are needed. Power density and cost are. The economics have been addressed in several papers(137); here we limit ourselves to the discussion of the lithographic parameters excluding cost. We have chosen this position because of the relative meaning of the cost of a process. The absolute cost is not a very significant figure of merit; cost effectiveness is. Cost-effectiveness must not be confused with net cost. To make a trivial example , a large bulldozer is certainly more expensive than a garden spade, but few would argue that the spade is more cost effective tool to build highways. Synchrotron based X-ray lithography is certainly expensive, but is also capable of exceedingly large economies of scale.

A well-designed X-ray lithography synchrotron radiation system can easily deliver in excess of 50 mW/cm^2 over a field of 50 mm (H) by 25 mm (V) of collimated radiation in the lithographically useful exposure window. (We notice, en passant, that most photoresist systems have a sensitivity which is quoted by referring to a reading from a reference instrument [such as a calorimeter]; these readings do not measure the lithographically useful dose since they respond to the whole incoming spectrum. A tailored beamline will deliver only useful photons so that an efficiency factor must be considered). Resist systems with sensitivities in the range of 50-100 mJ/cm^2 are available in experimental formulations and should be reaching a stable formulation in the market in early 1991. This makes it easily possible to achieve the goal of 1 sec/field exposure; assuming an overhead of 1 sec for stepping and alignment, a 6 inch diameter wafer would be exposed in 30 sec. Including wafer loading/unloading the system should be capable of delivering more than 60 6-inch diameter wafers per hour and will most likely be limited by the exposure tool overhead. For a manufacturing plant using 13 beamlines, this would be equivalent to 60 x 23 x 13 = 18940 wafer starts a day. A storage ring would support a very large silicon operation. Clearly, even a large initial capital cost can be amortized quickly by such a production rate.

The technology to build efficient sources and beam transport systems is here today, and work is progressing on the implementation of several such sources in an industrial environment. Time will tell if the synchrotrons do provide an economical, as well as a technical, answer to the challenges of the lithography for the end of the century.

14.0 ACKNOWLEDGEMENT

This chapter is based on the outcome of many years of activity and it would be difficult to acknowledge all the contributions to its content. In particular, Barry Lai wrote a large section of SHADOW and F. Baszler of TRANSMIT, codes on which many of the conclusions are based. Among CXrL staff, R.K. Cole provided many ideas in the beamline optical design and other areas; C. Welnak wrote most of the documentation of SHADOW. The continuous support of the University of Wisconsin, of the Synchrotron Radiation Center and Center for X-ray Lithography staff is also acknowledged. This work would have been impossible without the support of DARPA/NRL to the Center for X-ray Lithography. The operation of the Synchrotron Radiation Center is supported by the National Science Foundation.

REFERENCES

1. Spears, D.L., Smith, H.I., High resolution pattern replication using soft X-rays. Electron. Lett., 8: p. 102-104 (1972).
2. Gordon, E., Herriott, D.R., Pathways in Device Lithography. IEEE Trans. Electron Devices, 22: p. 371-375 (1975).
3. Plotnik, I., Metrology applied to X-ray lithography. Solid State Tech. J., 32: p. 102 (1989).
4. Jaeger, R.P., Hefflinger, B.L., Linewidth control in X-ray lithography: The influence of the penumbral shadow. Proc. SPIE - Int. Soc. Opt. Eng., 471: p. 110-126 (1984).
5. Viswanathan, R., Acosta, R.E., Seeger, S.D., Voelker, H., Wilson, A.D., Babich, I., Maldonado, J., Warlaumont, J., Vladimersky, O., Hohn, F., Crockatt, D., Fair, R., Fully scaled 0.5 μm metal-oxide semiconductor circuits by synchrotron X-ray lithography: Mask fabrication and characterization. J. Vac. Sci. Technol., 6B: p. 2196-2201 (1988).
6. Harrell, S., Alexander, D., Characterization techniques for X-ray lithography submicron metrology. Proc. SPIE - Int. Soc. Opt. Eng., 471: p. 103-109 (1984).
7. Fay, B., Hasan, T., Electrical measurement techniques for the characterization of X-ray lithography systems. Solid State Tech. J., 29: p. 239-243 (1986).
8. Wilson, A.D., X-ray lithography: Can it be justified? Solid State Tech. J., 29: p. 239-243 (1986).
9. Bernacki, S.E., Smith, H.I., Characteristic and Bremsstrahlung X-ray radiation damage. IEEE Trans. Electron Devices, 22: p. 421-428 (1975).
10. Blais, P.D., A practical system for X-ray lithography. Proc. Tech. Symp., Tokyo: 1-7 to 1-11 (1982).
11. Ostercamp, W.J., Efficient X-ray Generation. Phillips Rev. Rpt., 3: p. 303 (1948).
12. Hayasaka, T., Ishihara, S., Kinoshita, H., Takeuchi, N., A step-and-repeat X-ray exposure system for 0.5 μm pattern replication. J. Vac. Sci. Technol., B3: p. 1581-1586 (1985).
13. Kreuzer, J.L., Hughes, G.P., LaFiandra, C., Precision alignment for X-ray lithography. Proc. SPIE - Int. Soc. Opt. Eng., 471: p. 84-89 (1984).
14. Nagel, D.J., Whitlock, R.R., Greig, J.R., Pechacek, R.E., Peckerar, M.C. Developments in semiconductor

microlithography. Proc. SPIE - Int. Soc. Opt. Eng., 135: p. 46-53 (1978).

15.   Matthews, S.M., Cooper, R., Plasma sources for X-ray lithography. Proc. SPIE - Int. Soc. Opt. Eng., 333: p. 136-139 (1982).

16.   Gutchek, R.A., Murray, J.J., High resolution soft X-ray optics. Proc. SPIE - Int. Soc. Opt. Eng., 316: p. 196-202 (1982).

17.   Frankel, R.D., et al., Proc. Kodak Microel. Seminar, p. 82,(1986).

18.   Burkhalter, P.G., Shiloh, J., Fisher, A., Cowan, R.D., X-ray spectra from gas-puff Z-pinch device. J. Appl. Phys., 50: p. 4532-4550 (1979).

19.   Peacock, N.J., Speer, R.J., Hobby, M.G., J. Phys., B2: p. 798 (1969).

20.   Nagel, D.J., Dozier, C.M., Klein, B.M., Mather, J.W., Bul. Am. Phys. Soc., 18: p. 1363 (1973).

21.   Pearlman, J.S., Riordan, J.C., X-ray lithography using a pulsed plasma source. J. Vac. Sci. Technol., 19: p. 1190-1193 (1981).

22.   Okada, I., Saitoh, Y., Itabashi, S., Yoshihara, T., A plasma X-ray source for X-ray lithography. J. Vac. Sci. Technol., B4: p. 243-247 (1986).

23.   Smith, H.I., Spears, D.L., Bernaki, S.E., X-ray lithography: A complementary technique to electron beam lithography. J. Vac. Sci. Technol., 10: p. 913-917 (1973).

24.   Feder, R., Spiller, E., Topolian, J., Replication of 0.1 μm geometry with X-ray lithography. J. Vac. Sci. Technol., 12: p. 1332-1335 (1975).

25.   Neukermans, A.P., Current status of X-ray lithography, Part II. Solid State Tech. J., 27: p. 213-219 (1984).

26.   Fencil, C.R., Hughes, G.P., Submicron Lithography. Proc. SPIE - Int. Soc. Opt. Eng., 333: p. 100-110 (1982).

27.   Spears, D.L., Smith, H.I., X-ray lithography - A new high resolution replication process. Solid State Tech. J., 15: p. 21-26 (1972).

28.   Shimkunas, A.R., Advances in X-ray mask technology. Solid State Tech. J., 27: p. 192-199 (1984).

29.   Maydan, D., Coguin, G.A., Maldonado, J.R., Somekh, S., Lou, D.Y., Taylor, G.N., High speed replication of submicron features on large areas by X-ray lithography. IEEE trans. Electron Devices, 22: p. 429-433 (1975).

30.   Greeneich, J.S., X-ray lithography: Part I - design criteria for optimizing resist energy absorption; Part II -

pattern replication with polymer masks. IEEE Trans. Electron Devices, 22: p. 434-439 (1975).

31. Buckley, W.D., Nester, J.F., Windischmann, X-ray lithogrpahy mask technology. J. Electrochem. Soc., 128: p. 1116-1120 (1981).

32. Funayama, T., Takayama, Y., Inagaki, T., Nakamura, M., New X-ray mask of Al-AlO structure. J. Vac. Sci. Technol., 12: p. 1324 (1975).

33. Spears, D.L., Smith, H.I., Stern, E., X-ray replication of scanning electron microscope generated patterns. Proc. SPIE - Int. Soc. Opt. Eng., : p. (1972/3).

34. Suzuki, K., Matsui, J., Kadota, T., Ono, T., Preparation of X-ray lithography masks with large area sandwich structure membrane. Jap. J. Appl. Phys., 17: p. 1447-1448 (1978).

35. Ebata, T., Sekimoto, M., Ono, T., Suzuki, K., Matsui, J., Ulitchi, C., Nakayama, S., Transparent X-ray lithography masks. Jap. J. Appl. Phys., 21: p. 762-767 (1982).

36. Csepregi, L., Heuberger, A., Fabrication of silicon oxynitride masks for X-ray lithography. J. Vac. Sci. Technol., 16: p. 1962-1964 (1979).

37. Adams, A.C., Capio, C.D., Levinstein, H.J., Sinha, A.K., Wang, D.N., U.S. Patent 4,171, 489; 1979; assigned to Bell Telephone Laboratories.

38. Hofer, D., Powers, J., Grobman, W.D., X-ray lithographic patterning of magnetic bubble circuits with submicron dimensions. J. Vac. Sci. Technol., 16: p. 1968-1972 (1979).

39. Blais, P.D., O'Keefe, T., Tremere, D., Cresswell, M., A practical system for X-ray lithography. Semicon-West Tech. Prog. Proc., May 28, 1982.

40. Parrens, P., Tabouret, E., Tacussel, M.C., Preparation of X-ray lithography masks with 0.1 μm structures. J. Vac. Sci. Technol., 16: p. 1965-1968 (1979).

41. Georgiou, G.E., Jankoski, C.A., Palumbo, T.A., DC electroplating of submicron gold patterns on X-ray masks. Proc. SPIE - Int. Soc. Opt. Eng., 471: p. 96-99 (1984).

42. Yamagishi, F., Kimura, Y., Furukawa, Y., Fabrication of silicon polyimide complex X-ray masks. Fujitsu Sci. Tech. J., p. 85 (1980).

43. Brors, D.L., X-ray mask fabrication. Proc. SPIE - Int. Soc. Opt. Eng., 333: p. 111-112 (1982).

44. Suzuki, K., Matsui, J., SiN membrane masks for X-ray lithography. J. Vac. Sci. Technol., 20: p. 191-194 (1982).

45.   Acosta, R.E., Maldonado, J.R., Towart, L.K., Warlaumont, J.R., B-Si masks for storage ring X-ray lithography. Proc. SPIE - Int. Soc. Opt. Eng., 448: p. 114-116 (1983).
46.   Gong, B.M., Ye, Y.D., Fabrication of polyimide masks for X-ray lithography. J. Vac. Sci. Technol., 19: p. 1204 (1981).
47.   Ono, T., Ozawa, A., High contrast X-ray mask preparation. J. Vac. Sci. Technol., B2: p. 68-72 (1984).
48.   Bassous, E., Feder, R., Spiller, E., Totalian, J., High transmission X-ray masks for lithographic applications. Solid State Tech. J., 19: p. 55-58 (1976).
49.   Fraser, D.B., Lou, D.Y.K., U.S. Patent 3,975,252, 1976.
50.   Maydan, D., Coguin, G.A., Levinstein H.J., Sinha, A.K., Wang, D.N.K., Boron nitride mask structure for X-ray lithography. J. Vac. Sci. Technol., 16: p. 1959-1961 (1979).
51.   Garrettson, G., Neukermans, A.R., HP gives peek at X-ray aligner. Semi. Intl. p. 17 (1983).
52.   Triplett, B.B., Hollman, R.F., X-ray lithography for VLSI. IEEE Proc., 71: p. 585-588 (1983).
53.   Neukermans, A.P., Status of X-ray lithography at HP. Proc. SPIE - Int. Soc. Opt. Eng., 393: p. 93-98 (1983).
54.   Bartelt, J.L., Slayman, C.W., Wood, J.E., Chen, J.Y., McKenna, C.M., Minning, C.P., Coakley, J.F., Hollman, R.E., Perrygo, C.M., Mask ion-beam lithography: A feasibility demonstration for submicrometer device fabrication. J. Vac. Sci. Technol., 19: p. 1166-1171 (1981).
55.   Lepselter, M.P., Alles, D.A., Levinstein, H.J., Smith, G.E., Watson, J.A., A systems approach to 1-μm NMOS. IEEE Proc., 71: p. 640-656 (1983).
56.   Plotnik, I., Porter, M.E., Toth, M., Akhtar, S., Smith, H.I., Ion-implant compensation of tensile stress in tungsten absorber for low distortion X-ray masks. Microel. Engrg., 5: p. 51-59 (1986).
57.   Adams, A.C., Capio, C.D., The chemical deposition of boron-nitrogen films. J. Electrochem. Soc., 127: p. 399 (1980).
58.   Bromley, E.I., Randall, J.N., Flanders, D.C., Mountain, R.W., A technique for the determination of stress in thin films. J. Vac. Sci. Technol., B1: p. 1364-1366 (1983).
59.   Klokholm, E., An apparatus for measuring stress in thin films. Rev. Sci. Instrum., 40: p. 1054-1058 (1969).

60. Garrettson, G., Neukermans, A.R., Mckt. Engrg. Academic, London, p. 247 (1983).

61. Glang, R., Holmwood, R.A., Rosenfield, R.L., Determination of stress in films on single crystalline silicon substrates. Rev. Sci. Instrum., 36: p. 7-10 (1965).

62. Yanof, A.W., Resnick, D.J., Jankoski, C.A., Johnson, W.A., X-ray mask distortion: Process and pattern dependence. Proc. SPIE - Int. Soc. Opt. Eng., 632: p. 118-132 (1986).

63. Karnezos, M., X-ray mask distortions. Solid State Tech. J., 30: p. 151-156 (198

64. Frankel, R.D., Peters, D.W., Engineering of reticles for laser-based-plasma sources. Microel. Manuf. Test., 10: p. 8-9 (1987).

65. Ruby, R., Baldwin, D., Karnezos, M., The use of diffraction techniques for the study of in-plane distortions of X-ray masks. J. Vac. Sci. Technol., B5: p. 272-277 (1987).

66. Karnezos, M., Effects of stress on the stability of X-ray masks. J. Vac. Sci. Technol., B4: p. 226-229 (1986).

67. Johnson, W.A., Levy, R.A., Resnick, D.J., Saunders, T.E., Yanof, A.W., Radiation damage effects in boron nitride mask membranes subjected to synchrotron X-ray exposure. J. Vac. Sci. Technol., B5: p. 257-261 (1987).

68. Levy, R.A., Resnick, D.J., Frye, R.C., Yanof, A.W., Wells, G.M., Cerrina, F., An improved boron nitride technology for synchrotron X-ray masks. J. Vac. Sci. technol., B6: p. 154-161 (1988).

69. Peters, D.W., Dardzinski, B.J., Frankel, R.D., Defect printability for soft X-ray microlithography. Proc. SPIE - Int. Soc. Opt. Eng., 1263: p. 99-109 (1990).

70. Atwood, D.K., Fisanick, G.J., Johnson, W.A., Wagner, A., Defect repair techniques for X-ray masks. Proc. SPIE - Int. Soc. Opt. Eng.,471: p. 127-134 (1984).

71. Burggraaf, P.S., X-ray lithography and mask technology. Semi. Intl., (1985).

72. Ehrlich, D.J., Osgood, R.M., Silversmith, D.J., Deutsch, T.F., One-step repair of transparent defects in hard-surface photolithographic masks via photodeposition. Electron Device Lett., 1: p. 101-109 (1980).

73. Weigmann, U., et al., Repair of electroplated Au masks for X-ray lithography. J. Vac. Sci. Technol., B6: p. 2170-2173 (1988).

74. Econonou, N.P., Cambria, T.D., Mask and circuit repair with focussed-ion beams. State State Tech. J., 30: p. 133-136 (1987).

75. Herriott, D.R., Collier, R.J., Alles, D.S., Stafford, J.W., EBES: A practical electron lithographic system. IEEE Trans. Electron Devices, 22: p. 385-392 (1975).

76. Flanders, D.C., Smith, H.I., A new interferometric alignment technique. Appl. Phys. Lett., 31: p. 426-428 (1977).

77. Austin, S., Smith, H.I., Flanders, D.C., Alignment of X-ray lithography masks using a new interferometric technique - Experimental results. J. Vac. Sci. Technol., 15: p. 984-986 (1978).

78. Fay, B., Trotel, J., Frichet, A., Optical alignment system for submicron X-ray lithography. J. Vac. Sci. Technol., 16: p. 1954-1958 (1979).

79. Kouno, E., Tanaka, Y., Iwata J., Tasaki, Y., Kakimoto, E., Okada, K., Suzuki, K., Fujii, K., Nomura, E., An X-ray stepper for synchrotron radiation lithography. J. Vac. Sci. Technol., B6: p. 2135-2138 (1988).

80. Nelson, D.A., diMilia, Warlaumont, J.M., A wide range alignment system for X-ray lithography. J. Vac. Sci. Technol., 19: p. 1219-1223 (1981).

81. Feldman, M., White, A.D., White, D.L., Application of zone plates to alignment in microlithography. J. Vac. Sci. Technol., 19: p. 1224-1228 (1981).

82. Feldman, M., White, A.D., White, D.L., Application of zone plates to alignment in X-ray lithography. Proc. SPIE - Int. Soc. Opt. Eng., 333: p. 124-130 (1982).

83. Lavine, J.M., Mason, M.T., Beaulieu, D.R., The effect of semiconductor processing upon the focussing properties of Fresnel zone plates used as alignment targets. Proc. SPIE - Int. Soc. Opt. Eng., 470: p. 122-135 (1984).

84. Fay, B., Novak, W.T., Automatic X-ray alignment system for submicron VLSI lithography. State State Tech. J., 28: p. 175-179 (1985).

85. Novak, W.T., A lithography system for X-ray process development. Proc. SPIE - Int. Soc. Opt. Eng., 393: p. 106-113 (1983).

86. Kleinknecht, H.P., Diffraction gratings as keys for automatic alignment in proximity and projection printing. Proc. SPIE - Int. Soc. Opt. Eng., 174: 63-69 (1979).

87. Lyszczarz, T.M., Flanders, D.C., Economou, N.P., DeGraff, P.D., Experimental evaluation of interferometric alignment techniques for multiple mask registration. J. Vac. Sci. Technol., 19: p. 1214-1218 (1981/2).

88.   Kinoshita, H., Une, A., Iki, M., A dual grating alignment
       technique for X-ray lithography. J. Vac. Sci. Technol., B1:
       p. 1276-1279 (1983).
89.   Doemens, G., Mengel, P., Automatic mask alignment for X-
       ray microlithography. Siemens Forsch. und Entwickl.-
       Ber., Bd 13: p. 43-47 (1984).
90.   Heuberger, A., X-ray lithography. State State Tech. J., 29:
       p. 93-101 (1986).
91.   Semenzato, L., Eaton, S., Neukermans, A., Jaeger, R.,
       Monte Carlo simulation of line edge profiles and linewidth
       control in X-ray lithography. J. Vac. Sci. Technol., B3: p.
       245-252 (1985).
92.   Maydan, D., X-ray lithography for microfabrication.   J.
       Vac. Sci. Technol., 17: p. 1164-1168 (1980).
93.   David, N.E., Stover, H.L., Optical test structures for
       process control monitors, using wafer stepper metrology.
       Solid State Tech. J., 25: p. 131-141 (1982).
94.   Glendinning, W.B., Goodreau, W.M., Direct-write electron
       beam patterning reregistration and metrology. Proc. SPIE
       - Int. Soc. Opt. Eng., 480: p. 141-144 (1984).
95.   Fay, B., Alexander, D., Recent printing and registration
       results with X-ray lithography. Proc. SPIE - Int. Soc. Opt.
       Eng., 537: p. 57-68 (1985).
96.   Stemp, I.J., Nicholas, K.H., Brockman, H.E., Automatic
       testing and analysis of misregistrations found in
       semiconductor processing. IEEE Trans. Electron devices,
       26: p. 729-732 (1979).
97.   Muller, K.H., Brehm, K., Werner, K., Magnification
       corrected imaging in synchrotron radiation X-ray
       lithography. J. Vac. Sci. Technol., B6: p. 2139-2141
       (1988).
98.   Atoda, N., Kawakatsu, H., Tanino, H., Ichimura, S., Hirata,
       M., Hoh, K., Diffraction effects on pattern replication with
       synchrotron radiation. J. Vac. Sci. Technol., B1: p. 1267-
       1270 (1983).
99.   Silverman, J.P., diMilia, V., Katakoff, D., Kwietniak, K.,
       Seeger, D., Wang, L.K., Warlaumont, J.M., Wilson, A.D.,
       Crockatt, D., Devenuto, R., Hill, B., Hsia, L.C., Rippstein,
       R., Fabrication of fully scaled 0.5 µm n-type metal-oxide
       semiconductor test devices using synchrotron X-ray
       lithography: Overlay, resist processes, and device
       fabrication. J. Vac. Sci. Technol., B6: p. 2147-2152
       (1988).

100. Takahashi, N., SHI Group, Compact storage ring light source for X-ray lithography. Proc. SPIE - Int. Soc. Opt. Eng., 923: p. 47-54 (1988).

101. Aitken, J.M., 1 µm MOSFET's VLSI technology: Part VI. J. Solid Stat. Ckts., 14: p. 294 (1979).

102. Davis, R.T., Woods, M.H., Will, W.E., Measel, P.R., High-performance MOS resists radiation. Electronics, (1982).

103. Gdula, R.A., The effect of processing on radiation damage in SiO. IEEE Trans. Electron Devices, 26: p. 644-647 (1979).

104. Stover, H.L., Hause, F.L., McGreevy, D., X-ray lithography for one micron LSI. State State Tech. J., 22: p. 95-100 (1979).

105. Grove, A., Physics and Technology of Semiconductors, New York: Wiley (1967).

106. Kuhn, M., A quasi-static technique for MOS CV and surface state measurements. Solid State Elect., 13: p. 873-885 (1970).

107. Manchanda, L., - Radiation effects on MOSFET's fabricated with NMOS submicrometer technology. Electron Device Lett., 5: p. 412-414 (1984).

108. Chen, J.Y., Henderson, R.C., Patterson, D.O., Martin, R., Radiation effects of e-beam fabricated submicron NMOS transistors. Electron Device Lett., 3: p. 13-15 (1982).

109. See, for example J.D. Jackson, Classical Electrodynamics, p. 654, J. Wiley, New York (1975).

110   Sokolov, A.A., Ternov, I.M., Radiation from Relativistic Electrons, Amer. Inst. of Phys., New York, p. 82 and ff (1986).

111. One such program is TRANSMIT, available from the Center for X-ray Lithography, 3731 Schneider Dr., Stoughton, WI 53589.

112. So , D., Lai, B. and Cerrina, F., SPIE. 30, 6 (1987).

113. Krinsky, S., in handbook on Synchrotron Radiation, E. Koch, Edt., North Holland (1985).

114. Green, G.K., Brookhaven National Lab Report, BNL 50522, (1973).

115. Chapman, K., Lai, B., Cerrina, F. Nucl. Instr. and Meth. in Physics Review, A283, (1989).

116. Born, M., Wolf, E., Principles of Optics, p. 165, Pergamon Press, Oxford (1980).

117. Goldstein, S., Classical Mechanics, Addison-Wesley, (1980).

118. Availabe from CXrL, see (111) above.

119. See, for example, VLSI Processing, S.Sze Ed., McGraw Hill, (1983).
120. So, D., Lai, B., Wells, G.M., and Cerrina, F., Journal Vac. Sci. and Tech., 6, 2190-5 (1988).
121. RF Cavity.
122. More accelerators.
123. For a recent review of Japanese activity, see H. Winick, in the Proceedings of the 6th Synchrotron Radiation Instrumentation Conf., Berkeley 1989, Nucl. Instrum. and Methods, AA291, 1990.
124. See, for example, the report from the Brookhaven workshop on X-ray Lithography.
125. National Council on Radiation Protection and Measurements, Rep. 39, (1971b).
126. Aladdin safety guidelines.
127. Dushman, S., Scientific Foundations of Vacuum Technique, Wiley, New York (1967).
128. Brodsky, E., Synchrotron Radiation Center, private comm.
129. Acoustic delay line.
130. Be and other windows.
131. IBM beamlines.
132. NTT Japanese ring.
133. See, for example, Johnson, R.L., in handbook on Synchrotron Radiation, E. Koch, Edt., North Holland (1985).
134. Cole, R.K., and Cerrina, F., CXrL, unpublished.
135. Lai, B., Chapman, K., Cerrina, F., Nucl. Instrum. and Methods, A266, p. 544 (1988), SHADOW is available from CXrL, see (109).
136. Johnson, E.D., Hulbert, S.L., Garrett, R.F., Williams, G.P. and Knotek, M.L., Rev. Sci. Instru. 58, 1042 (1987).
137. See, for example, Hill, R.H., in Journ. Vac. Sci. and Techn. B7, 1387 (1989); also, Wilson, A., Solid State Tech. 29, 249 (1986).

7

Ion Lithography and Focused
Ion Beam Implantation

John Melngailis

Research Laboratory of Electronics
Massachusetts Institute of Technology
Cambridge, MA 02139

1.0 INTRODUCTION

Focused ion beams offer a new level of flexibility in microfabrication. In conventional processing resist is usually exposed in a desired pattern and developed, and the area laid bare is somehow altered; material may be removed or added or ions may be implanted. With focused ion beams the pattern definition and the surface alteration are combined in the <u>same step</u>. Beam diameters below 0.05 μm have been reported, thus, the fabrication can be carried out well into the submicron regime. In addition, focused ion beams can be used to expose resist as a prelude to conventional fabrication. The varieties of fabrication are illustrated in Figure 1.

Focused ion beam fabrication is a sequential process, point by point, and of necessity slow. The smaller the beam diameter, the more time is needed to address a given area. For example, to mill off 1 μm^3 of Si with a 100 pA, 0.1 μm diameter beam would take 16 sec., assuming a yield of 5 atoms/ion. To implant the channels of transistors, with a dose of 10^{12} ions/cm^2 and using a 600 pA 0.5 μm diameter beam, would require 3 sec/mm^2. Thus, milling is practical only over very limited areas, such as in repair or diagnostic processes, but low dose implantations can be considered over chips or even wafers since the fraction of the area to be implanted may be small.

During the past several years significant progress has been made on all fronts of focused ion beam technology and applications. Some recent reviews have been published(1-2) but they focus on only some of the aspects of the field. Perhaps the most complete review(3) is by now three years old.

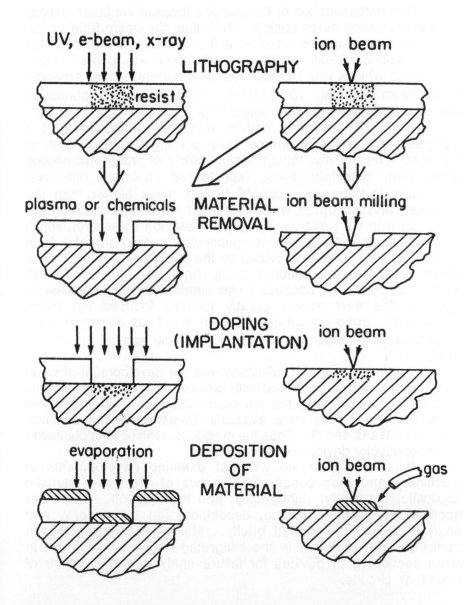

Figure 1. On the left is depicted conventional resist based fabrication. Resist is exposed and developed and the area left unprotected is eroded, implanted, or has material deposited on it. A focused ion beam can be used to expose resist to be used in further conventional processing but in addition it can remove material, implant, or deposit material locally, often with better than 0.1 μm resolution.

The demonstration of the use of a focused ion beam directly in microfabrication dates back to 1973 when the beam from an ion implanter was simply focused to a 3 μm diameter. Resist was exposed and maskless, patterned implantation was carried out(4). This source of ions was not bright and the current density achieved in the beam was only 10^{-4} A/cm^2. Thus, proposed fabrication was unfortunately quite time consuming.

An important early activity was the development of a cryogenic field emission source used to deliver hydrogen(5-8) or argon(8). This activity, though aimed mostly at proton microscopy rather than microfabrication, represented important progress: the source brightness was 10^4 to 10^5 times higher than the implanter, and ion optics were developed capable of focusing the beam down to 0.2 μm(7-9). The liquid metal ion source(10), which was developed in 1975, similarly depended on field emission from a sharp tip (in this case provided by the liquid being pulled into a sharp cusp by the electric field) and also provided high brightness. However, because it was simple and relatively easy to operate, this new source greatly spurred focused ion beam applications. Soon Ga ion beams down to 0.1 μm diameters were demonstrated with current density in the focal spot of 1.5 A/cm^2 (Ref. 11)

The next important milestone was the development of liquid metal alloy sources combined with crossed electric and magnetic field mass separation in the ion column(12-18). This meant that many species of ions were available, in particular the common dopants of GaAs and Si. Thus the maskless, resistless implantation of semiconductor devices became possible.

In this chapter we will first examine developments in machinery and then concentrate on two of the more futuristic applications, namely: lithography, and implantation. The other applications such as milling, deposition, ion microscopy, and analysis will be mentioned briefly. Milling and deposition, in particular, are important in the integrated circuit field for use in cross sectioning of devices for failure analysis and in repair of masks or circuits.

2.0 MACHINERY

The theory and design of charged particle "optics" (i.e., focusing systems) is a long established art. We will here examine only the recent advances in the external specifications of focused ion beam systems rather than the details of their internal operation.

For purposes of organizing the discussion we will divide the focused ion beam machine into three main parts: the ion source, the ion optics, and the beam writing as shown schematically in Figure 2.

2.1   Ion Source

The important property of the source of ions in focused ion beam systems is its brightness (measured in A/cm^2 steradian). Thus a large current of ions emitted into a small solid angle from a "point" source is desirable. Since charged particles repel one another, the implied confinement of such a configuration is hard to achieve. An extreme environment is needed such as is provided by a sharp tip with a high electric field on it. The two types of ion sources, the liquid metal(10-13) and the gaseous (cryogenic) field ion source(5-9), both use this configuration. Of the two, the liquid metal has so far proven to be technologically the more important one.

2.1.1   Liquid Metal Sources.   In the case of liquid metal sources, the sharp tip is largely achieved by the fact that the electric field pulls the conducting fluid into a cusp from which the ions are emitted. See Figure 3. The remarkable fact is that this structure works well, and at the low currents, desirable in focused ion beam applications, many sources can operate for long periods of time with stable extraction currents. In some cases, a servo system is installed which varies the potential on the emission tip to maintain constant current. This voltage is controlled by an auxiliary electrode so that the energy of the ions entering the rest of the ion column remains constant.

The ions emitted by the source have an energy spread in the range of 5-10 eV. (See Ref. 47 and citations therein). This appears to be unavoidable and is the most important factor in limiting beam diameter and beam current density. The mechanism for this limitation is chromatic aberration, i.e., ions of different energies are focused at different points in the column, in effect increasing the beam diameter.

The energy spread in general increases with increasing ion current from the source(19). The coulomb repulsion between ions which are at random spacings from one another has been proposed as an explanation(20). However, this model does not appear to account for the observation that the energy spread does not go to zero as the source ion current is reduced. The energy spread for a Ga source decreases as the current is lowered to 1 μA but then saturates and remains flat at a value of 4.5 eV (FWHM) even down to currents of 5 nA (Ref. 19).

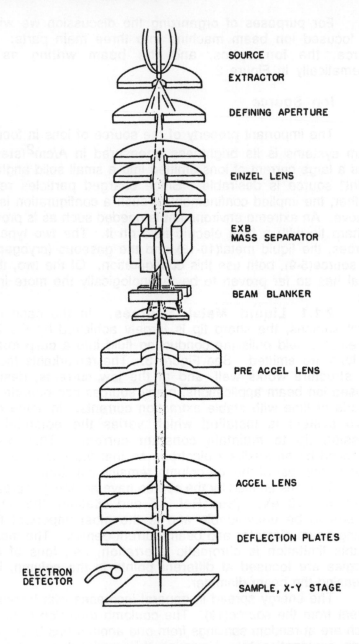

SOURCE

EXTRACTOR

DEFINING APERTURE

EINZEL LENS

EXB
MASS SEPARATOR

BEAM BLANKER

PRE ACCEL LENS

ACCEL LENS

DEFLECTION PLATES

ELECTRON
DETECTOR

SAMPLE, X-Y STAGE

Figure 2. Schematic of the focused ion beam column in use at MIT. The accel lens can be raised up to 120 kV, the pre-accel lens an additional 30 kV. The extraction voltage adds another 6-10kV. (The einzel lens does not change the energy of the beam).

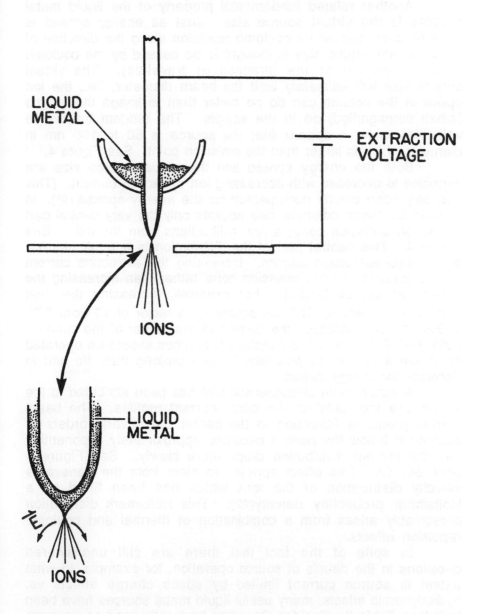

Figure 3. Schematic of the liquid metal ion source in cross section. The enlarged view of the tip shows the metal being pulled into a cusp by the electric field.

Another related fundamental property of the liquid metal sources is the virtual source size. Just as energy spread is thought to be caused by coulomb repulsion along the direction of travel, virtual source size is thought to be caused by the coulomb repulsion normal to the direction of travel(20). The virtual source size will ultimately limit the beam diameter, i.e., the ion optics of the column can do no better than to image the source (albeit demagnified) on to the sample. The random transverse velocities make it appear that the source is 50 to 100 nm in diameter which is larger than the emission point. See Figure 4.

Both the energy spread and the virtual source size are expected to decrease with decreasing ion extraction current. (This has only been clearly documented for the energy spread(19)). In focused ion beam columns, one accepts only the very central part of the ion emission cone, a few milliradians from the axis. See Figure 4. This central part of the distribution is a weak function of the total extraction current. Increasing the extraction current mainly spreads out the emission cone rather than increasing the current at the center(21). For example, increasing the total extraction current of Ga^+ ion source by a factor of 10 from 0.75 to 7.5 μA only increases the current at the center of the cone by 50% (Ref. 21). Thus liquid metal ion sources should be operated at as low a current as possible, to both prolong their life and to decrease the energy spread.

Another beam characteristic that has been attributed to the source are the "tails" in the beam current profiles. The beam current profile is Gaussian in the center but several orders of magnitude below the peak it becomes approximately exponential, i.e., the current distribution drops more slowly. See Figure 5 (Ref. 50, 52) This effect appears to stem from the transverse velocity distribution of the ions which has been fitted by a Holtsmark probability density(22). This Holtsmark distribution presumably arises from a combination of thermal and coulomb repulsion effects.

In spite of the fact that there are still unanswered questions in the details of source operation, for example, to what extent is source current limited by space charge effects vs. hydrodynamic effects, many useful liquid metal sources have been built (see Table I). Besides Ga, other frequently used sources are AuSi, AuSiBe, PdAs(23), PdAsB(23), NiB(24), and NiAs(25),, since they provide the principal dopants of GaAs and Si. Note that the lifetimes quoted in the table are those published in the original papers on the sources. By now many of these are greatly increased. For example, a commerically available Ga^+ source has operated in the author's laboratory in excess of 1000 hours.

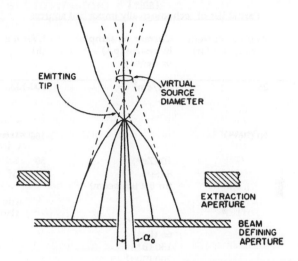

Figure 4. A further enlarged view of the tip of the cusp (Figure 3) showing the extrapolation of the perturbed ion trajectories back to the source. The waist formed by these trajectories defines the virtual source size.

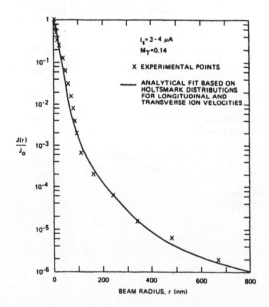

Figure 5. From Ref. 50: The current density profile of a 50 kV Ga ion beam. The beam diameter FWHM is <18 nm. J_O is the peak current density and the magnification of the system is 0.14.

Table I
Partial list of technologically important sources

Composition	Angular current density (μA/sr)	Current on sample in system (pA) or density	Lifetime (h)	References
Ga	26	5A/cm^2	200-1000	10, 11, 19, 35
Au$_{65}$Si$_{27}$Be$_8$				
Au$_{59}$Si$_{26}$Be$_{15}$	20 (Be^{++})		100	35, 36, 37, 39
AuZnSi		10 (Zn$^+$) 8 (Si$^+$) (in 0.2 μm beam)	50	40
Pd$_{50}$Ni$_{26}$ Si$_5$Be$_6$B$_{13}$		134 (Be^{++}) 66 (Be$^+$) 150 (B$^+$) 200 (Si^{++})		41
Pd$_2$As	5 As^{++}	30 pA (As^{++})	150	23
Pd$_{70}$As$_{16}$B$_{14}$	1 (B$^+$) 3 (As$^+$) 4 (As^{++})		150	23
Ni$_{45}$B$_{45}$Si$_{10}$		(25%-35% B$^+$) 2000 (at 100 μA total source)	250	42, 43
Pd$_{64}$As$_{11}$ B$_9$P$_{16}$		(18% P$^+$)	25	23
Cu$_3$P		(10% P$^+$)	20	44
Al	20-30		100	45
Au	(1 to 100 μA) total current)		50	46
Ni$_{62}$As$_{38}$		0.1-0.4A/cm^2 (As^{++})	30	25
Ni$_4$B$_6$	64 total 17.9 (B$^+$)			24

In this chapter, we have briefly touched on a few aspects of source operation that are relevant to their application. The authors file on liquid metal sources has about 100 papers in it, so that a complete review is beyond the scope of this chapter. Although at this point source operation is adequate for many focused ion beam applications, improvements which continuing research brings will be welcome; in particular, increased lifetime, reduced energy spread, increased stability, increased brightness, and new ion species.

2.1.2 Gaseous Field Ion Sources. This type of source is operated at cryogenic temperatures (liquid helium) and condenses either He (Ref. 26, 27) or H_2 (Ref. 28) on the tip. See Figure 6. Note that emission can also occur by field ionization directly from the gas. In that case cryogenic cooling is not essential. However, much higher currents and source brightnesses are observed when sources are cooled (Ref. 30). The main problems with the gaseous field ion source are the uncertainty of the emission point which leads to a non-axial emission cone(28), and the instability of the emission current due to contamination(26). Because of the light ions emitted, these sources are expected to be most suitable for lithography.

Recently progress toward a reliable field ion source has been reported. Use of a specially constructed tungsten(111) emitter tip and highly purified He gas has permitted stable operation of the source for up to 170 hrs(29). In addition, the field ion source is expected to have an energy spread of the ions of only ~1 eV. The available current density in the beam spot on the sample is, therefore, expected to be 10 to 100 times higher than in the case of liquid metal sources. These developments, together with the potential application to lithography, should motivate renewed efforts to mount these sources on ion columns.

2.2 Ion Column

The ion column mainly contains electrostatic ion lenses, but also may contain mass separators. It focuses the ions from the liquid metal source on the sample surface. In most operating regimes, the beam spot size and current density are limited by chromatic aberration and the virtual source size. Thus the total beam diameter focused on a sample is given by (See Ref. 30 and particularly Ref. 34 for more detailed discussion):

[1] $$d = (d_0^2 M^2 + d_c^2)^{1/2}$$

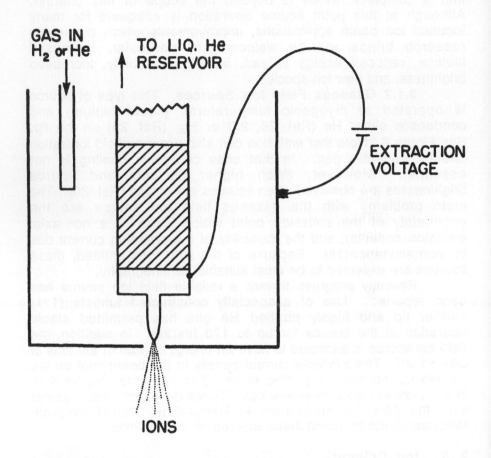

Figure 6. Schematic of gaseous field ion source. The emission tip and the entering gas are cooled to near liquid helium temperatures. The atomically sharp tip is thought to ionize gas atoms directly as well as gas condensed on the needle. Ref. 28.

where d_0 is the virtual source size, M is the magnification and d_C is the contribution due to chromatic aberration, given by:(30)

$$[2] \qquad d_C = 2C_{CO} \, M \, \alpha_0 \, \frac{\Delta E}{E}$$

where C_{CO} is the chromatic aberration coefficient referred to the source side and α_0 the half angle of the beam accepted by the beam defining aperture, see Figure 4. M is the magnification of the system (usually less than 1), E the final energy of the ions, and ΔE the energy spread.

Unless one is interested in the ultimate in minimum beam diameter, the $d_0 M$ term is small compared to d_C and the beam diameter is determined by d_C. The current density J in the focal spot is then constant since $J = 4I/\pi d_C^2$ (I is beam current), and $I \sim \alpha_0^2$ and $d_C \sim \alpha_0$, so that α_0 cancels. In practice, α_0 is a few milliradians.   For larger α_0 , spherical aberration, which is proportional to α_0^3 , may begin to play a role.

Underlying these simple expressions for the beam diameter is the complicated dependence of the chromatic aberration coefficient C_{CO}, on the layout of the electrodes and values of the electrode potentials in all of the electrostatic lenses in an ion column. Ray tracing computer programs are used to calculate the aberration coefficients(31). Generally, these programs start with some existing or envisioned electrode/potential configurations. Recently the reverse calculation has also been implemented(32), i.e., given certain constraints compute the electrode and potential configuration, which will minimize the aberration coefficients. These optimized configurations predict increases of current density of as much as a factor of 33, compared to conventional lenses with the same source parameters and working distances(32). This prediction has not been verified experimentally nor even by carrying out the ray tracing calculation using the optimized configuration.   A low aberration lens system designed in a more conventional fashion also predicts up to 20 times higher current density than hitherto reported(33). However, this is obtained in part by reducing the working distance.   (In effect, decreasing M in the above equations).

All of these calculations consider the focused ion beam going down the axis of a cylindrically symmetrical column. There are two ways that this is violated in practice. The first is the unintentional misalignment of the column. Lens element displacement, tilt, and ellipticity may occur and have been analyzed(48).   For a permitted blurring of an originally 0.1 μm beam by 0.05 μm, a few μm of misalignment, or ellipticity, and a 0.5 to 1 mrad of tilt are permitted.   Thus, as expected, lens elements must be very precisely machined and aligned.   The other non-cylindrical symmetrical effects are intentional, namely EXB mass separation and beam deflection(49).   In both cases, the energy spread again is responsible for additional beam blurring.   The mass filter is, in effect, a velocity filter and the energy spread will result in astigmatism.   In addition, beam deflection is proportional to the ratios of the applied transverse voltage to the acceleration voltage of the beam.   The energy spread translates into an uncertainty of the accelerating voltage.   A simple calculation (see Equation 3) shows that a 10 eV energy spread in a 100 kV column results in 20 nm blurring in the direction of deflection for a beam deflected by 100 μm.   This can be minimized by dynamic stigmation or by using only small deflection fields.

    Within the recognized bounds of these limitations, there has still been progress in column performance.   Beam diameters down to 0.05 μm are quoted routinely even for columns with mass separation. With Ga^+ ions, beam current densities up to 4 A/cm^2 are possible, while with alloys, the density is lower often by a factor of 10.   With special efforts, including reducing the magnification M to 0.14, features down to 15 nm width have been written in PMMA(50).

2.3   Beam Writing

    In any practical applications of focused ion beams, we must have the ability to deliver a given dose in desired patterns on the surface.   To do this, the system must be able to deflect the beam, to turn it on and off (beam blanking), and to align to existing features.   In addition, since most patterning is likely to be over entire wafers, or at least over a chip and since the maximum practical deflection of the ion beam is only a few hundred micrometers, the stage holding the sample must be moved, and the stage motion and beam deflection must be aligned to one another.

    2.3.1   Beam   Deflection.   Beam deflection is accomplished by a transverse electric field generated by suitable electrodes.   Simple parallel plates or octopoles are typically used. Although deflection appears to be straightforward, it is responsible for some of the system limitations such as writing speed and field size.   This can be illustrated by a simple example as shown in Figure 7.   If we consider a transverse electric field V_x/d_0 acting on the beam over a distance L, and an ion of mass m, charge e, and axial velocity v_z, then simple classical mechanics

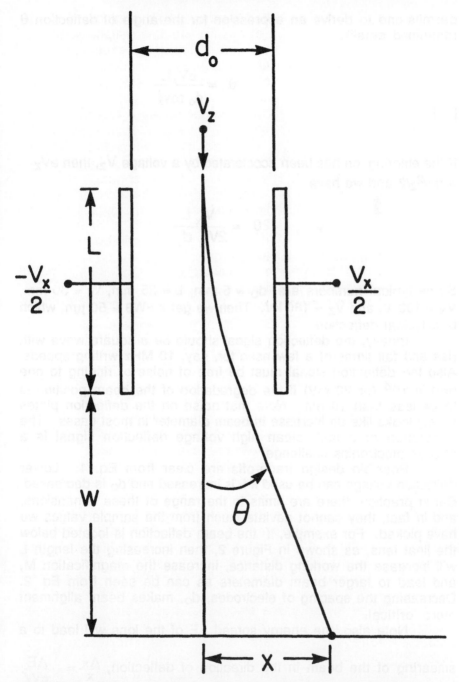

Figure 7. The parameters used to discuss beam deflection. An ion of velocity v_z is incident on the axis half way between the deflection plates and ends up being deflected through an angle θ.

permits one to derive an expression for the angle of deflection θ (assumed small):

$$\theta = \frac{eV_xL}{d_o \, mv_z^2}$$

[3]

If the entering ion has been accelerated by a voltage V_z, then $eV_z = mv^2{}_z/2$ and we have:

$$\theta = \frac{V_x L}{2V_z \, d_o}$$

[4]

Some typical numbers are: $d_0 = 5$ mm, $L = 25$ mm, $W = 30$ mm, $V_x = 100$ V, and $V_z = 150$ kV. Then we get $x \sim W\theta = 50$ μm, which is a typical deflection.

   Ideally, the deflection signal should be a square wave with rise and fall times of a few nsec for, say, 10 MHz writing speeds. Also the deflection signal must be free of noise or ringing to one part in 10^4 (or 20 mV) if the degradation of the beam position is to be less than 10 nm. Note that noise on the deflection plates simply looks like an increase in beam diameter in most cases. The generation of a fast, clean high voltage deflection signal is a serious electronics challenge.

   Possible design trade-offs are clear from Eq. 4. Lower deflection voltage can be used if L is increased and d_0 is decreased. But in practice, there are limits to the range of these dimensions, and in fact, they cannot deviate much from the sample values we have picked. For example, if the beam deflection is located below the final lens, as shown in Figure 2, then increasing the length L will increase the working distance, increase the magnification M, and lead to larger beam diameters as can be seen from Eq. 2. Decreasing the spacing of electrodes, d_0, makes beam alignment more critical.

   Note also, the energy spread ΔE of the ions will lead to a smearing of the beam in the direction of deflection, $\frac{\Delta x}{x} = -\frac{\Delta E}{eV_z}$. For ΔE=10 eV, V_z=100 kV, x=150 μm, we get $\Delta x = 15$ nm. This is not negligible in some cases.

Nevertheless, within these limitations focused ion beam writing at a few MHz has been demonstrated.

2.3.2 Beam Blanking. Beam blanking is usually done further up in the column, where the beam has lower energy. Together with deflection on the sample, it is an essential part of any writing. Blanking is done by deflecting the beam above an aperture preferably where the beam has a cross over. For the system depicted in Figure 2 the blanking plates are between the EXB mass separator and the mass separating aperture, and they deflect the beam perpendicular to the deflection direction of the EXB mass separator. Because the blanker has a finite length and displaces the beam laterally before sweeping it off the aperture, the beam has "tails" both in time and in space. (See Ref. 3 and 51) The blanking pulse needs to rise fast from zero and return cleanly and quickly to zero, but can have a noisy maximum.

2.3.3 Transit Time. The transit time of the ion in a column must also be considered. The beam blanking and beam deflection in general do not occur in the same place along the beam path, e.g., the blanking may be done early in the column and the deflection just before the beam is incident on the substrate. The distance between these two locations might be 30 cm. Sample transit times are given in Table II. This transit time has to be taken into account when the blanking and deflection are synchronized.

The transit time of the beam through the deflector and the blanker itself presents a more fundamental limit to writing time. Scaling the numbers from Table II, for example, we find that a 10 keV Au ion travels 1 cm in 0.1 μsec. Thus since the blanker and deflector are about 1 cm long in the direction of beam travel, writing speeds of 10 MHz would not be possible in this case. For lighter ions at higher energy, this limitation is not yet serious.

2.3.4 Alignment. Alignment existing features is essential for focused ion beam writing in most cases. Alignment is achieved by using the focused ion beam itself to image. In this mode, the machine is used as a scanning ion microscope which in principle operates the same as a scanning electron microscope (SEM). The beam is raster scanned over the sample and secondary electrons emitted from the point of ion incidence are collected (see Figure 8). The secondary electron signal in turn modulates the intensity of a synchronously scanned cathode ray tube. The image can be formed of both the difference in topography or the difference in materials. Figure 9 shows an example of a scanning ion microscope image as well as the same feature taken by an SEM. Note that focused ion beam imaging erodes the surface. In most cases an image can be formed at the sacrifice of a few monolayers. Use of frame storage will reduce unnecessary erosion.

Table II Time for ion to traverse 30 cm. (µsec)

Species \ Energy	10kV	50kV	300kV
H	0.22 µsec.	.097	0.040
Be	0.65	0.29	0.12
Si	1.14	0.51	0.21
Ga	1.80	0.81	0.33
Au	3.03	1.36	0.55

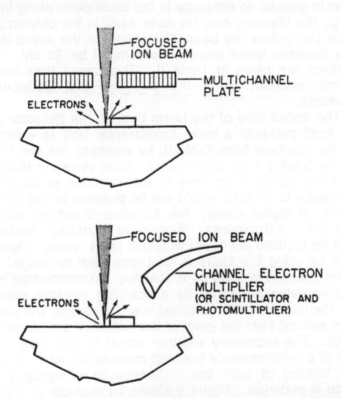

Figure 8. Schematic of the secondary electron collection with either a multichannel plate (top) or a channel electron multiplier (bottom).

Figure 9. Example of the same aluminum conductor being imaged in the scanning ion microscope mode (on left) and an SEM on the right (Ref. 86). The two concentric square vias and the shallow square were all milled with the same 68 kV Ga+ ion beam that was used for the imaging.

For accurate alignment mark observation, the nature of the secondary electron detector is important. If the detector is a channel electron multiplier (as shown in Figures 2 and 8 bottom) or a photomultiplier/scintillator detector, then the imaging is not symmetrical, and a step will look different depending on its orientation with respect to the detector. For this reason, most focused ion beam systems now use microchannel plates which are circularly symmetrical about the incident beam and therefore detect the secondary electrons symmetrically (see Figure 8 top). A typical alignment mark which might be a raised (or indented) cross on the surface would, therefore, appear as shown in Figure 10 (Refs. 53-56). As the ion beam scans over the edge of a step, the secondary electron emission increases sharply because the electrons can escape more readily from an edge than from a flat surface. The secondary electron signal for a scan across one arm of the cross has two sharp peaks, one for each edge, as shown in Fig. 10. Note that, as discussed above, for the microchannel plate the two peaks are equal while for the asymmetric detector (in this case a scintillator and photomultiplier) they are not(54). The emission of electrons has, in fact, been calculated as a function of the step wall angle and for 160 KeV Si ions incident on GaAs. An inclination of 85° was found to be optimum, i.e., on the step wall the ion beam is at a 5° grazing angle(56). Under favorable conditions, alignment to 0.1 μm was reported(54). The theoretical resolution of the alignment mark position is calculated to be 0.01 μm(56). The reason the achieved alignment is not as impressive has been attributed to vibration(56). Other factors which affect the ability to align include the edge roughness in the alignment mark and the erosion of the step during scanning.

Statistical studies of focused ion beam alignment under practical conditions have shown an overlay accuracy of σ = 0.22 to 0.26 μm (Ref. 57 and 58) and a field stitching accuracy of 0.12 μm (Ref. 57). Clearly, alignment accuracy is impaired if the alignment mark is covered by a film in a subsequent processing step. This would lead to a rounding of the profile and an uncertainty in position(58-59). In the case of a 1.7 μm high alignment mark covered by 0.5 μm of PMMA the initial alignment signal (such as shown in Figure 10) was degraded. However, repeated scanning milled off the PMMA locally and improved the alignment capability(58). In some circumstances, where a focused ion beam step is the first of a sequence, alignment marks can be milled into the surface by scanning with an appropriately heavy dose (~ 10^{18} ions/cm^2).

The ion beam profile is Gaussian over 2-3 orders of magnitude, but below that it has "tails" where the fall-off is

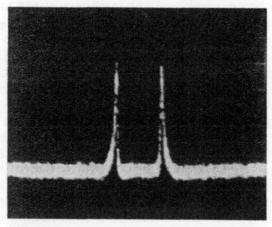

MCP

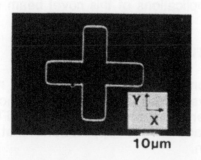

10μm

Figure 10. Top (Ref. 54). Examples of signals from an alignment mark as the focused ion beam is scanned in a line across a silicon pedestal. If a photomultiplier detector is used, the signal is seen to be assymmetrical. If a microchannel plate (MCP) is used the signal from both sides of the pedestal is the same. Bottom (Ref. 53). Image of a cross formed with a microchannel plate detector. (The signals in the upper part of the figure would be obtained by scanning the focused ion beam across one arm.)

slower (Figure 5). Clearly, alignment marks must be located at an appropriate distance from the fabricated features. Otherwise, the tails of the beam, combined with the heavy doses needed to mill alignment marks or even to view them ($\sim 10^{15}$ ions/cm^2), may produce unwanted doses in the active areas.

2.3.5 Writing Strategies. Writing strategies with a directed beam can be of two types, whether photon (laser), electron (Chapter 5), or ion, namely raster or vector. In the raster mode (rarely used in practice), the beam is scanned continuously over the surface of the sample much like the electron beam on a TV screen and the blanker is used to turn the beam on or off as needed to generate the pattern. In the vector mode, shown in Figure 11, the beam is blanked off and a deflection voltage is applied corresponding to a desired position. Then the beam is turned on and deflected an additional amount to write a desired feature. The deflection can in fact be a local raster. This method of writing has also been called "vector scan with raster fill".

The foregoing discussion of the focused ion beam machinery has examined some of the capabilities and in particular the limitations. Within the limitations impressive performance has been demonstrated. Patterned beam writing with Si, Be, As, and B ions has been carried out, aligned to existing features. We will now turn to the applications of this novel beam writing.

3.0   APPLICATIONS

Focused ion beams can be applied in microfabrication in many different ways because an energetic ion incident on a surface produces a number of effects as illustrated in Figure 12. We will briefly discuss each of the applications that exploit the ion-surface interaction.

3.1   Effects of Ion-Surface Interaction

3.1.1 Sputtering. Atoms on or near the surface can be removed leaving the surface either as neutrals or as ions. They have a distribution of energy which peaks at a few eV even though the incident ion may have many keV of energy. Sputtering, often referred to as ion milling, leads to a number of applications of focused ion beams. The local, high resolution milling capability is used in photomask repair, in circuit restructuring and repair (Refs. 60-63, see also Ref. 3 for earlier literature), in X-ray lithography mask repair, and in the trimming of optical components such as semiconductor lasers(64)(65). If the atoms that are sputtered off are mass analyzed, then we have a high

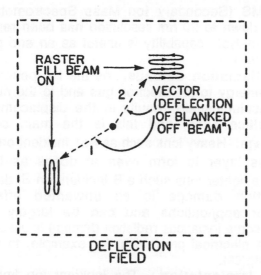

Figure 11. The beam writing process. With no deflection voltage applied the beam is in the center of the field. The beam is usually blanked off and is deflected to a desired position, blanked on and scanned to fill the feature.

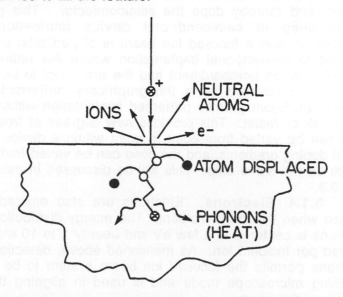

Figure 12. Schematic of an ion penetrating the surface, showing sputtering of neutral atoms, emission of electrons, lattice damage, heat generation and implantation. In addition the beam can produce chemical effects (bond breaking).

resolution SIMS (Secondary Ion Mass Spectrometry). In fact, microanalysis down to 20 nm resolution has been reported(66). In addition, the analysis capability is useful as an end point detector in ion milling(67).

3.1.2 Radiation Damage. As the ion penetrates into the solid it loses energy to the electron gas and to the nuclei. Energy loss to the nuclei usually results in the displacement of atoms from their lattice sites, and this is the main component of radiation damage. Heavy ions such as As^+ incident on Si can cause an amorphous layer to form even at doses as low as 10^{14} ions/cm^2 while lighter ions such a B incident on Si do not.

Radiation damage is an unwanted effect in most semiconductor applications and can be largely removed by annealing. In some instances radiation damage is used to advantage to change the electrical properties, for example, to render doped GaAs insulating(68).

3.1.3 Implantation. The incident ion imbeds itself in the substrate in a distribution peaked at some distance below the surface. When the material, e.g., a semiconductor, is annealed, the imbedded atom can be made to occupy a lattice site. If it has a different valence from the host lattice atoms, it will create charge carriers and thereby dope the semiconductor. This process is widely used in semiconductor device fabrication. Thus implantation with a focused ion beam is of particular interest. In contrast to conventional implantation where the entire wafer is subjected to ion bombardment and the areas not to be implanted have to be covered by a lithographically patterned film, the focused ion beam permits patterned implantation without the use of a mask or resist. This provides new degrees of freedom: the dose can be varied from point to point within a device permitting lateral doping gradients, and the dose can be varied from device to device on the same chip. This will be discussed in more detail in Sec. 3.3.

3.1.4 Electrons. Electrons are also emitted from the surface when the ion is incident. The energy distribution of these electrons is centered at a few eV and usually 1 to 10 electrons are emitted per incident ion. As mentioned above, detection of these electrons permits the focused ion beam system to be used in a scanning microscope mode and is used in aligning the sample relative to the beam.

3.1.5 Chemical Effects. Chemical effects are produced when incident ions simply break or rearrange bonds in the bombarded material.

a)   On the surface.   Incident ions can induce chemical changes in molecules absorbed on the surface including reaction with the surface. This has been used in ion induced deposition from absorbed molecules supplied by a local gas ambient(69-70). In this way deposition of metals with submicrometer patterns has been created, e.g., W from WF_6 or $W(CO)_6$ or Au from $C_7H_7O_2F_6Au$ (dimethylgold hexafluoro acetylacetonate). This capability is used to repair clear defects (absence of absorber) in photomasks. If high aspect ratio deposits of heavy metals can be produced, ion induced deposition will also find application in X-ray mask repair, and if low resistivity conducting deposits can be produced(71), it will also be used in integrated circuit restructuring and repair.

In addition, if the local gas ambient is a reactive species such as Cl_2, one can etch the surface in a patterned fashion. This is similar to ion milling, but usually many times faster since the material is removed by a chemical process in addition to being sputtered(72).

b) In the bulk. Chemical effects of incident ions in the bulk of a material can be produced by the energy lost by the ion to the electrons or to the nuclei. This is responsible for the alteration of appropriate organic (or inorganic) films which makes them soluble in a subsequent dry or wet etch. The result is that ions can be used as an exposure tool in lithography, much like electrons.

3.1.6 Heating. A large fraction of the energy lost by an incident ion is eventually converted to heat. We have analyzed this using a macroscopic heat flow model(3). Somewhat counter-intuitively, heating under a focused ion beam is, in most circumstances, negligible. On the one hand, 150 kV ions at a current density of 1 A/cm^2 results in 150 kW/cm^2. On the other hand, one is usually dealing with an area of ~ 10^{-10} cm^2. Assuming macroscopic thermal conductivity values, the details of the heat flow under a 0.1 μm diameter beam are such that the temperature rise is a few tens of degrees centigrade, even if the substrate is an insulator like SiO_2 and the beam is stationary. However, for a larger beam, say radius 0.3 μm instead of 0.05 μm, and a current density of 10 A/cm^2 the temperature rise for a stationary beam on SiO_2 calculates out to be 1500°C. This is a worst case since in most applications the beam is, in fact, scanned and the thermal conductivity of most substrates such as Si is 100 times higher.

We have briefly examined the various possible applications of focused ion beams in the framework of ion-surface interactions. In keeping with the title of this volume the rest of this discussion will concentrate on focused ion beam lithography and on focused ion beam implantation. The latter category is included because it is a direct substitute for resist based lithography and in fact fits the broader definition of the term "lithography".

3.2   Lithography

The term "lithography" as it is used in microfabrication refers to the transfer of a pattern from a mask (e.g., chrome on glass) or from a computer memory on to a radiation sensitive film (called resist) on a surface. The transfer can take place by means of UV light (photolithography), X-rays, electrons or ions. The film is then "developed" which can mean either the area exposed to the radiation is removed (positive resist) or the area not exposed is removed (negative resist). In subsequent fabrication steps the part of the substrate left uncovered is in someway altered. The resist is then removed, and a patterned surface has been produced.

While ions can be used to pattern the resist, one of the main reasons for interest in the field is that ions can pattern the surface without resist thus eliminating the application, exposure, and development steps. As discussed above, milling and induced deposition require very large doses (1 μm^3 of material can be removed or added in tens of seconds). So, these processes are suitable for only local intervention, for example, for repair. However, implantation in many cases requires doses comparable to these needed in resist exposure. Thus in the subsequent section, we will also treat implantation, and discussions of writing time will be applicable to both lithography and implantation.

Photolithography has been (and still is) the workhorse of the microelectronics industry. However, as minimum dimensions in integrated circuits are pushed below 0.5 μm, other methods of exposure will be needed. The current belief is that X-ray lithography will be used, and intense development work is being carried out in this area. Whereas in photolithography one can reduce the pattern by demagnification during projection, in X-ray lithography such a technique is not straightforward since X-ray "optical" elements (lenses or curved mirrors) are difficult to build. Nevertheless, these elements have been built and projection X-ray lithography has been demonstrated(73). Still in the foreseeable future X-ray lithography masks will likely be used at the final dimensions by direct proximity shadow printing.

At present the two candidates for generating the original pattern for X-ray masks are electron beams, and ion beams. E-beams have had a much longer history of active development and are at present favored, although, as we shall see, there is reason to believe that ions are quite competitive.

3.2.1 E-Beam vs Ion Beam. When electrons are incident on a solid, they tend to scatter their energy into a relatively large volume (of μm dimensions). See Figure 13 (from Ref. 74) This is because they are incident on a medium largely filled with particles of equal mass, so that, energy transfer in scattering is optimal. Ions, on the other hand, scatter their energy over a relatively small volume close to the point of entry, Figure 13. Exposure of resist is produced by the energy deposited in it by the incident particle.   In terms of dose/unit area, resist is, in general, about 2 orders of magnitude more sensitive to ions than to electrons. See Figure 14 (Ref. 75). In addition, ion lithography is not plagued by proximity effects. That is, in electron beam lithography fine features are exposed by using the high energy density desposited near the point of entry. The widely scattered electrons, Figure 13, produce a background which is below the threshold for exposure. However, for other than isolated single lines these backgrounds add up, and one has to correct the e-beam exposure dose for the proximity to other patterns. This effect is largely absent for ion beams(75) and is an important advantage.

Throughput. The all-important feature of any process that is considered for actual manufacturing is the number of samples fabricated per hour (throughput). Since mechanically, e-beam systems and ion beam systems are similar, the time needed for wafer loading and unloading, stage motion, and alignment should be comparable. We will consider in detail only the actual time spent under the beam.

The time t (sec) needed to expose an area, A (cm^2) with a dose φ (ions/cm^2) and a particle beam current I (ions/sec) is:

[5]
$$t = \frac{\phi A}{I} \text{ or } t = \frac{\phi A}{J(\pi/4)d^2}$$

where d is the beam diameter, and J is the particle current density (ions/cm^2 sec).

(<u>Note</u>: Since the beam has a Gaussian profile, one would normally assume that the beam "diameter" is the full width at half maximum, d$_H$. If J(o) is the maximum density of the particle current at the peak of the Gaussian curve and I the total particle current, then:

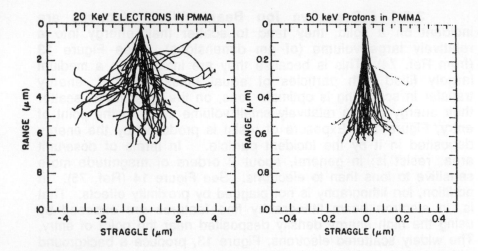

Figure 13. From Ref. 74 (Randall). A Monte Carlo calculation of the paths of ions (50 keV protons) and electrons in PMMA. Note the difference in scale and the fact that the ions scatter far less than electrons.

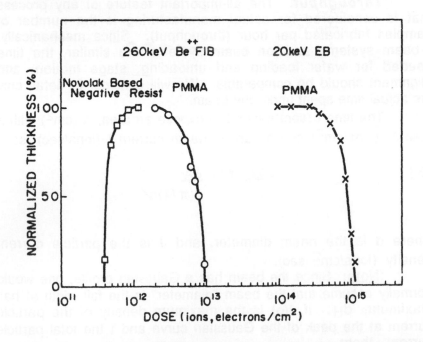

Figure 14. From Ref. 75. Exposure characteristics of PMMA and negative novolak resist. The vertical axis shows the % thickness remaining after development.

[6]
$$\frac{I}{(\pi/4) d_H^2} = 1.443 \; J(o)$$

Thus, by dividing the total current by the area enclosed by the FWHM diameter one gets a current density considerably higher even than the maximum in the Gaussian beam. Perhaps a less deceptive definition of diameter would be $d = 1.201 \; d_H$. Then $I/[(\pi/4)d^2] = J(o)$. In most papers which quote a beam diameter, the question of definition is not addressed. Given that the diameter is difficult to measure accurately, it is perhaps a fine point).

As discussed above (Sec. 2.2), the ion beam diameter is limited by chromatic aberration over much of the range of interest, and therefore, the current density in the beam is constant. In Fig. 15, we plot Eq. 5 for the time to expose 1 mm^2 vs. beam diameter for various doses. For the current density we have taken 2.4×10^{13} ions/cm^2 sec. This is deduced from the exposure of PMMA with the MIT machine, i.e., Be^{++} ions at 280 keV and 20 pA expose lines 0.05 μm wide and below. Since the beam is scanned in a single line we take the beam diameter to be 0.056 μm, i.e., somewhat scaled up from the FWHM which we expect to be below 0.05 μm.

One of the factors which limits the minimum dose which can be used to expose a feature is the finite number of ions per pixel, also called shot noise. We define the pixel to be the width of the line squared. Experimentally, features of 0.06 μm width have been exposed with various doses. At 44 ions per pixel, PMMA lines are still continuous and at 7.5 ions/pixel so are the lines for negative CMS type resist lines while at lower doses they are not (Ref. 76). Similarly, 0.015 μm wide lines have been exposed in PMMA with Ga ions and the minimum dose per pixel was found to be 25 ions(50). These features were continuous and with well defined width and straight side walls. For the purposes of this discussion, we take 32 as the minimum number of ions per pixel, and plot a shot noise limit as shown in Figure 15. In the shaded area exposed features would be discontinuous. For example, for 0.05 μm features only limited sensitivity resists can be used, namely those with minimum dose for exposure above 2×10^{12} ions/cm^2.

The comparison of the writing times for focused ion beams and electron beams can be made in a variety of ways. The simplest

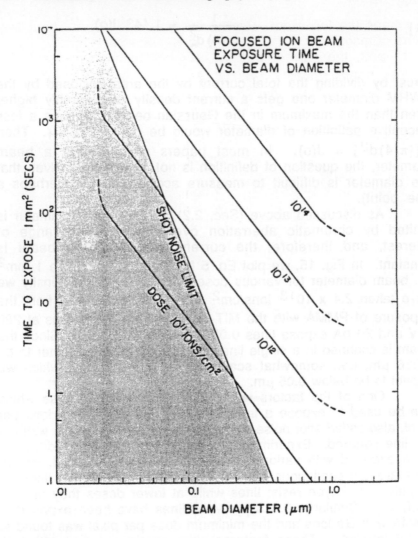

Figure 15. A plot of time to expose 1 mm² vs. beam diameter for various doses. In practice the beam diameter is varied by changing the acceptance angle. If the beam is limited by chromatic aberration, then the diameter in the focal spot is proportional to the acceptance angle. To the left and below the "shot noise limit" line, exposure of minimum width features is expected to be discontinuous due to the limited number of ions per pixel. This figure, of course, applies equally well to implantation. In that case this shot noise limit does not apply.

is perhaps the following: Resists such as PMMA are about two orders of magnitude more sensitive to ions than electrons. On the other hand, the current density of most commercial Gaussian electron beam systems (10-100 A/cm^2) is about two orders of magnitude higher than ion beams (0.4-4 A/cm^2). Thus these two effects approximately cancel, and exposure times are expected to be comparable for insensitive resists like PMMA.

A more detailed statistical analysis(77) also concludes that the pixel transfer rates of scanning ion beam and electron beam systems are of the same magnitude.

Another way to make the comparison is to look at some ultrafine features that have been written and simply divide the dose (ions or electrons/unit area) by the total particle current which was used. We have chosen as an example of ion beam lithography a recent result (78) (see Figure 16). In this case 50 nm wide features were written with a 20 pA current of 280 keV Be^{++} ions in 300 nm thick PMMA. These features are compared, see Table III, to 75 nm wide lines in 100 nm thick PMMA. The field emission electron source is on an experimental system and is perhaps the brightest available(79). The current density in the beam is about 660 A/cm^2. (Note that the sources used in the comparisons discussed earlier were of the LaB$_6$ type and the e-beam systems had current densities on the sample of 10-100 A/cm^2).

The features shown in Figure 16 were written with an ion beam whose diameter was approximately equal to the width of the line. Yet the lines had well formed vertical sidewalls and a uniform width. In the case of e-beams there is a rule of thumb that to achieve reasonably well formed lines of uniform width in resist one should expose the line with 4 to 5 adjacent passes of a beam whose diameter is 1/4 to 1/5 of the width of the line.

Accordingly, the 75 nm line in the left column of Table III was exposed with a 15 nm beam in 5 adjacent passes. Note that the bottom line in Table III shows the e-beam and ion beam exposure rates to be comparable. The ion beam, however, offers the advantages of (a) almost total absence of the proximity effect(74) and (b) relative ease of exposing thick resist with vertical sidewalls(78).

Apparently lighter ions are preferable for exposing fine features in resist(80). For example, Be ions appear to be able to expose finer features than Si ions under identical circumstances(81). We must, however, point out that 15 nm features have been exposed even with the relatively heavy Ga$^+$ ions(50). The proton or the H$_2^+$ ions have been claimed to be ideal for resist exposure(80). As discussed above in Sec. 2.1.2, the

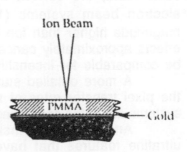

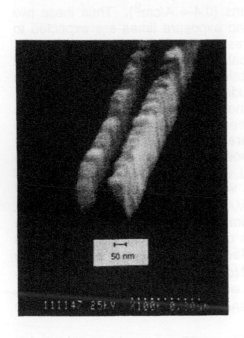

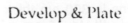

Figure 16. From Ref. 78. Focused ion beam exposure of PMMA and electroplating of features used to fabricate X-ray lithography masks. First PMMA is exposed and developed then gold is plated up from a plating base (a thin film of gold under the PMMA). The PMMA is then dissolved resulting in the features shown in the photo. In fact 150 sites were exposed with pairs of lines 40 μm long.

Table III

ELECTRON BEAM VS. FOCUSED ION BEAM
WRITING TIME FOR ULTRAFINE FEATURES

	e-beam (Ref. 79)	ion-beam (Ref. 78)
Source:	Field Emisison e-gun 25kV	Be^{++} at 280 keV
Features written:	75 nm lines in 0.1 μm thick PMMA	50 nm lines in 0.3 μm thick PMMA
Beam current:	1.17 nA	20 pA
Resist Sensitivity:	150 μC/cm^2 (9 x 10^{14} electrons/cm^2)	1 x 10^{13} ions/cm^2
Writing time:	1.3 x 10^3	1.7 x 10^3 sec/mm^2 (or 0.75 x 10^3 sec/m^2 if scaled to 75 nm features)

cryogenic source operating with gaseous H_2 is about three orders of magnitude brighter than the liquid metal ion source used in the comparisons so far. A focused ion beam system based on this source has been compared to a STEM (Scanning Transmission Electron Microscope) used as an e-beam lithography tool at 100 kV. This is a fair comparison since such a high energy electron beam has to be used to expose high aspect ratio features in thick resist similar to those exposed by ion beams (Figure 16). This STEM has a current density at the sample of $1.27 \times 10^4 A/cm^2$, but even for this very high current density, a detailed comparison[80] concludes that the writing speed of the H_2^+ source is thirty times faster than of the e-beam (STEM) for 30 nm pixels. In addition, the time of flight limitations of ion beam systems discussed in Sec. 2.3.3, which ultimately limit the writing speed, are minimized for such light ions. In fact, a 100 MHz blanking speed has been projected[80].

Thus, the main advantages of e-beam lithography are the relative ease of beam deflection and blanking, high current density, and an advanced state of development. Ion beams, even with liquid metal sources appear to offer comparable writing speeds, but most importantly, have little or no proximity effect. With gaseous field ion sources, which appear to have achieved a new breakthrough in stability[29] (see Sec. 2.1.2), a manyfold increase in writing speed is expected over that for e-beam systems .

3.2.2 Shaped Beam Systems. For many applications, the pixel-by-pixel exposure of patterns by neither electron beams nor ion beams is fast enough. Large areas with dense patterns of submicrometer features will require other tools. For example, even if we scale up the brightest e-beam system, discussed in Table III, from 0.025 µm to 0.25 µm size features, the writing time is still an unacceptable 2.3 hrs/cm^2.

Because of these considerations shaped electron beam systems have been developed (see Chapter 5) and are now being used commercially. These systems can project rectangles of any dimension from 0.15 to 2.0 µm. Although an accurate, direct comparison is not possible since the writing time is very much dependent on the size distribution of rectangles in the pattern, these shaped beam systems are usually several orders of magnitude faster than the Gaussian beam, pixel-by-pixel systems.

The shaped electron beam systems operate by passing the beam through one rectangular aperture and then focusing that aperture on to another rectangular aperture. The overlap of these two rectangles can be adjusted by deflecting the beam laterally as it travels between these two apertures.

Comparable shaped ion beam systems have not been built yet. The development most nearly related to this is the design of a system which projects a single fixed aperture(82-83). These systems differ from the usual focused ion beam system, which focuses the image of the ion source on the sample, in that the image of the aperture is focused on the sample. Clearly, the density of the current incident on the sample will be much lower, but the total current will be higher. The edge sharpness is expected to be comparable to the beam diameter in the Gaussian beam case. Clearly a shaped ion beam system would have applications not only in lithography, but perhaps even more so, in direct implantation.

3.2.3   Organic Resist.   Organic resist is almost universally used in the creation of patterns in integrated circuits. UV light, X-rays, electrons and ions can be used to expose the pattern. The exposure is then followed by development in a liquid. Because of the interest in in-situ, all vacuum processing, there is also considerable research in inorganic resist films which could be dry developed by plasma techniques. Inorganic resists will be discussed in the next section.

Incident energetic ions lose their energy both to the electrons in a solid and to the nuclei while the other exposure beams, photons and electrons, in general affect only the electrons in the solid. The rate of energy loss of ions to nuclei and electrons (also called stopping power) is well known and has been tabulated for many ion solid combinations. In the case of PMMA, the minimum dose for exposure appears to correlate more with the energy loss to electrons than with the energy loss to nuclei. This is illustrated in Figure 17 (Ref. 84). By adding the nuclear and electronic stopping powers in Figure 17, one can see that the correlation against total stopping power is better than the correlation against nuclear stopping power. However, the correlation is still much better against loss to electrons alone. In this plot the initial energy loss rate at the surface is used. In fact, exposure is produced by the energy deposited per unit volume of resist, i.e., the integral of (dE/dx) over the thickness of the resist. For thin resist and energetic ions the integral is expected to be proportional to the energy loss rate at the surface.

Regardless of what the details of the exposure mechanism are, ion lithography can expose very well defined and smooth walled features. See for example Figures 16, 18, 19. In fact the quality of these features is better than one would expect from statistical considerations. The 15 nm lines(50) shown in Figure 19, for example, have only 25 ions per pixel, (defined as the linewidth squared). If we consider the 15 nm line further divided into 5 nm sub-pixels, then there would only be 3 ions per sub-pixel! How is it that the line does not have rough edges and even

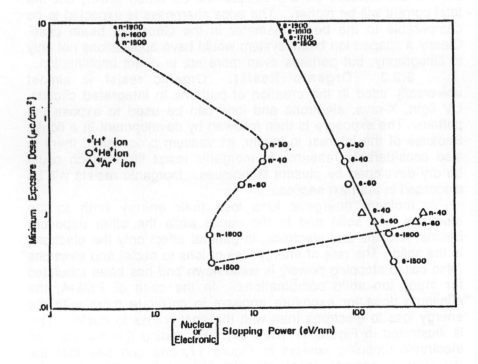

Figure 17. Minimum exposure dose of PMMA vs. stopping power (nuclear or electronic) for various ion species and ion energies. Each point is labelled with the ion energy in keV and the stopping power (n-nuclear or e-electronic) which is plotted. The correlation is seen to be much better with electronic stopping power than with nuclear. (From Ref. 84).

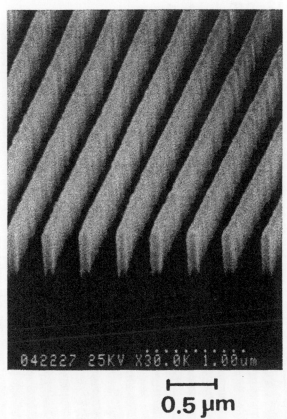

Figure 18. Negative, novolak (Shipley SAL 601) resist exposed with Be^{++} ions at 260 keV. From Ref. 75.

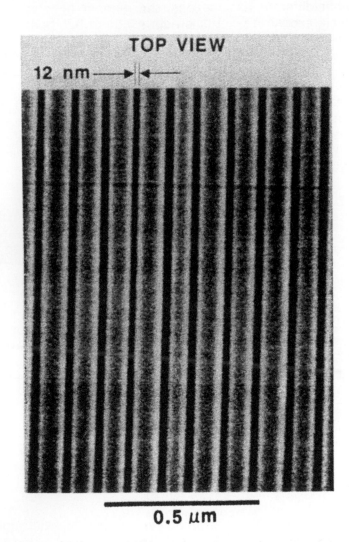

Figure 19. PMMA resist 60 nm thick exposed with 50 keV Ga+ ions. From Ref. 50 .

discontinuities?   In fact, on purely statistical grounds and somewhat different assumptions, the number of ions needed to completely expose a pixel (linewidth squared) has been calculated to be in the 600-1000 range(85).

Possibly a "smoothing" effect is operating; namely, the energy deposited in the resist redistributes itself. At these dimensions diffusion of heat (lattice vibrations), energetic electrons, or electron-hole pairs might assure that even if a sub-pixel were to receive no ions, it would still receive sufficient energy from its neighbors to be soluble in the developer.

Whatever the detailed mechanisms of exposure and development, impressive structures have been fabricated by focused ion beams whose diameter is approximately equal to the smallest feature. Example images are found in Figures 16, 18, and 19. Some of the exposure results have been listed in Table IV.

Another important issue in ion lithography is the range of the ions. One must be concerned that the resist is exposed through the entire thickness. In addition, in some cases such as direct writing on a wafer penetration of the ion into the substrate may be undesirable. For sample ranges of ions see Table V.

Anothert type of organic resist which has been used with focused ion beams is nitrocellulose(86-87). This material "self develops", that is, exposure and development occur in one step. The dose needed to expose resist is in this case proportional to the thickness since the process is sequential. For example, 2 μm thick nitrocellulose requires a dose of 6×10^{15} ions/cm^2 of 65 kV Ga$^+$ ions(86). This is more than two orders of magnitude higher than the dose needed to expose PMMA, but it is at least two orders of magnitude lower than the dose needed to mill out a comparable feature. Nitrocellulose is too insensitive to be a good resist. However, under special circumstances the absence of a wet development stage may be an advantage.

3.2.4   Inorganic Resist.   In principle lithography with inorganic resist can be more versatile than lithography with organic resist. The resist can be ultra-thin. In some cases a few monolayers are enough to inhibit plasma or wet chemical etching. The ions can act on the resist material in several different ways: introducing damage, introducing an impurity, doping the substrate to change electrochemical etching, or ion milling. In some cases the "resist" film may not be a resist in the usual sense of the word but may be the desired final film to be left patterned on the surface.

While electrons and photons also have been used to pattern inorganic resist, ions are much more effective in this process. Because of the large energy loss per unit volume accompanying ion

Table IV   Examples of organic resist exposure with ion beams

Ion	Energy (keV)	Resist	Min. Dose (ions/cm^2)	Min. Line width (μm)	Remarks	Reference
H$^+$	120	PMMA (pos.)	3x10^{12}	0.5 (1.0 μm thick)	Thru stencil mask	(88)
H$^+$	120	PGMA (neg.)	3.0x10^{11}	-	-	(92)
H$^+$	100	PMMA	2x10^{13}	0.08		(93)
He$^+$	30	PMMA	3x10^{12}		Thru stencil mask	(90)
	60	PMMA	1.8x10^{12}			
He$^+$	1500	PMMA	1.8x10^{11}			(84)
Be^{++}	260	PMMA	1x10^{13}	0.1 (0.5 μm thick)	Focused ion beam	(75)
Be^{++}	260	Novolak (Shipley SAL601) (neg).	2x10^{12}	0.1 (0.6 μm thick)	Focused ion beam	(75)
Be^{++}	130	Novolak MP 2400	2x10^{12}	-	Focused ion beam (cross links at 5x10^{12} cm^{-2})	(97)

Table IV Examples of organic resist exposure focused ion beams – cont

Ion	Energy (keV)	Resist	Min. Dose (ions/cm²)	Min. Line width (μm)	Remarks	Reference
Be⁺⁺	280	PMMA	2×10^{13}	0.05 (0.3 μm thick)	Focused ion beam	(78)
Be⁺⁺	200	PMMA	3.4×10^{7} ions/cm (line dose)	0.03 (0.4 μm depth)	Focused ion beam	(96)
Be⁺⁺	200	CMS (Chloro-methylated Polystyrene) (neg.)	3.5×10^{12} thick)	0.05 (0.3 μm	Focused ion beam	(91)
Si⁺⁺	200	CMS	1.5×10^{12}	0.1 (0.3 μm thick)	Focused ion beam	(91)
Si⁺⁺	200	PMMA	3×10^{12}	0.12 (0.3 μm thick)	Focused ion beam	(91)
Ar⁺	120	PMMA	7×10^{11}		0.3 μm thick resist	(88)

Table IV   Examples of organic resist exposure with ion beams -- cont.

Ion	Energy (keV)	Resist	Min. Dose (ions/cm^2)	Min. Line width (µm)	Remarks	Reference
Ar+	100,150	PMMA	6x10^{11}		0.21 µm thick resist	(90)
Ga+	100	PMMA	6x10^{11}	0.06 (0.12 µm thick resist)	Focused ion beam	(94)
Ga+	100	P(SiStg0-CMS10) Poly (tri-methylsilylstyrene-co-chloromethylstyrene)	1.6x10^{11}	0.11 (0.15 µm thick)	Focused ion beam	(94)
Ga+	20	PMMA		0.15 (0.06 µm thick)	Focused ion beam	(50)
Ga+	100	PMMA (dry develop O$_2$ RIE)	1.2x10^{16}	0.1 (0.5 µm thick)	Focused ion beam	(87)
Ga+	20	PMMA (dry develop)	6x10^{14}	0.5	Focused ion	(104)
Au+	100	PMMA	3x10^{12}			(91)

Table V   Ion Range in Resists

Ion	Energy (keV)	Resist	Density (g/cm³)	Range (µm)	Reference
H⁺	40	PMMA	1.19	0.52	(88)
	120	PMMA	1.19	1.12	(88)
	240	PMMA	1.19	1.85	(88)
	1500	PMMA		40	(89)
He⁺	40	PMMA	1.19	0.44	(88)
	120	PMMA	1.19	0.96	(88)
	240	PMMA	1.19	1.54	(88)
	200	PMMA		1.40	(90)
Be	100	PMMA		0.45	(91)
	200	PMMA		0.65	
				0.72	
				1.2(*)	
Si	100	PMMA		0.14	
	200	PMMA		0.25	
Ar⁺	80	PMMA	1.19	0.066	(88)
	240	PMMA	1.19	0.39	(88)
	200	PMMA		0.4	(90)
Ga	100	PMMA	1.2	0.074	(76)
Ga100		P(SiSt₉₀-CMS)	1.0	0.090	(76)
		Poly (trimethyl silylstyrene -co-chloromethylstyrene)			

*measured developed depth

penetration in a solid, and the fact that matter is also being delivered, a wide variety of inorganic films have been used to demonstrate ion beam lithography, as shown in Table VI.

A very compelling reason for interest in inorganic resists is their compatibility with all-vacuum, in-situ processing. Whereas organic resists are usually applied wet and developed by wet chemistry, inorganic resists can be applied in vacuum processes and developed by plasma etching. E-beam evaporation, RF sputtering, or molecular beam epitaxy are examples of application methods. Clearly, one can carry out a sequence of processes without removing the sample from the vacuum environment thus avoiding surface contamination. This is particularly interesting when it is combined with MBE to produce patterned buried structures by alternate growth and lithography steps(98-99) (see also Ref. 167 and citations therein).

Because of the wide variety of ion-surface interactions which are used in ion lithography, the range of sensitivities of ion resists spans many orders of magnitude, from 5×10^{10} to 10^{16} ions/cm^2. In addition, there is almost the whole periodic table of ion species which can be used. The wide range of sensitivities also implies that ultrafine features can be written by choosing less sensitive resists. For example, 5 nm lines would have 250 ions/pixel if the minimum dose for exposure were 10^{15} ions/cm^2.

Because of the well developed state of organic resist lithography, inorganic resists have not received much attention. However, if more advanced fabrication techniques are considered, such as ion lithography combined with in-situ processing, inorganic resists may prove to be useful.

3.3   Implantation

In the case of lithography, focused ion beams are in competition with established electron beam techniques, as well as developing electron beam techniques such as multilevel resists and high energy beams. For direct implantation of semiconductors this is not true. Thus this application may be of special interest. The unique capabilities that focused ion beam implantation provides can be grouped into four categories:

(1)   Novel transistors or other circuit elements can be made. Since the focused ion beam implants a wafer point by point without the use of mask or resist, the ion dose can be varied from point to point within a device to produce doping gradients. Some examples of structures built this way are bipolar transistors(110), GaAs MESFETs(112-114), and tunable Gunn

Table VI   Inorganic Film (resist) Patterning

Material	Ion (Energy keV)	Dose (ions/cm²)	"Developer" (Etch) (Polarity)	Remarks	Reference
Cr	Sb (50) (100)	3.8×10^{15}	CCl_4 Plasma (−)	0.3 to 0.4 µm wide lines 0.5 µm high	(100)
Si	In (50) (Ga, Sn 50kV also used)	5×10^{13}	H_3PO_4 at 195c	0.05 µm wide lines formed (+)	(101)
GaAs	Ga (50)	1.2×10^{14}	HCl at 50c (+)	0.03 µm wide lines formed	(101)
Al	As, Sb (50) (also N_2 & Ar in implanter)	5×10^{15}	H_3PO_4 (−)	carbon rich film formed on surface	(102)
Al/O Cermet	Ga (70) Ar (30,75)	1×10^{15}	Transene A (−)	A_2O_3 concentration on surface enhanced 0.1 µm lines exposed	(103)
Al_2O_3	Ga (50)	2.0×10^{14}	H_3PO_4 (100ºc) (+)	0.2 µm wide features	(105)

Table VI Inorganic Film (resist) Patterning -- cont.

Material	Ion (Energy keV)	Dose (ions/cm^2)	"Developer" (Etch) (Polarity)	Remarks	Reference
GaAs (n-type) InP InGaAs InGaAs P	Ga (20)	5×10^{10}	Photoelectro chemical 2M - H$_2$SO$_4$ (-)	1 μm features	(106)
InGaAs (MBE grown 30Å thick)	Ga (20)	2×10^{14}	none (+)	10 μm rectangles wet etched in underlying InP surface	(98)
GaInAs MBE grown	Ga (20)	10^{14}	in-situ 50eV Ar+ ion assisted Cl$_2$ etching (+)	5 μm wide features produced and overgrown in-situ	(98)
MoO$_3$	Ga (20-70)	1.8×10^{14}	Aqueous NaOH (-)	high constant	(107)
SiO$_2$	Ga (70) Si (140)	10^{14}	Ammonium Floride & glacial acetic acid (+)	0.5 μm wide lines etched smooth sidewalls	(108)
Spin-on glass	Ga (16)	10^{16}	CHF$_3$/O$_2$ plasma	implant reduces etch rate by 30%	(109)

diodes(111).   These and other applications will be discussed in more detail below.

(2)   Novel circuits can be made.   The focused ion beam permits the dose to be varied also from transistor to transistor in an integrated circuit(115).   This has permitted the fabrication of a 4 bit flash, digital-to-analog converter operating up to 1 GHz(116).   In addition, this same capability can be used in prototyping by making identical devices side by side varying only the implantation dose.   The conditions for best performance can then be identified, without concern for wafer-to-wafer differences.

(3)   Simpler conventional processing in cases where doses are low and over limited areas.   Channel implants in MOS transistors or GaAs MESFETs are typically in the range of 10^{11} to 10^{13} ions/cm^2, and the fraction of the area that needs to be addressed is often relatively small.   In addition, if all that is needed is a variation of the dose (or energy) from device to device, then the ion beam diameter (and the ion beam current) can be increased.   Thus, in special cases one can consider focused ion beam implantation of semiconductor devices over entire chips or even wafers, even in a production setting(117).

(4)   Patterned implantation can be carried out in-situ. The combining of several processing steps into a single vacuum, multi-compartment chamber has attracted increasing attention as a means of simplifying device fabrication.   One of the challenging aspects of this in-situ processing is the need for a "dry" resist, i.e., one that can be applied, exposed, and developed in vacuum. Focused ion beam implantation does not need a resist.   This fact has been exploited in combining focused ion beams, molecular beam epitaxy (MBE) and other processing steps to make devices in III-V compound semiconductors with buried layers doped in a desired pattern(98-99).

3.3.1 Equivalence of Broad and Focused Beam Implantation.   Since the current density in a focused ion beam is many orders of magnitude higher than the current density in a conventional implanter, one might expect differences in properties of materials implanted with the same dose by the two techniques.   Such differences have been observed both in Si and GaAs.   However, they do not appear to influence the final carrier density achieved after annealing.   Some of the observed differences between focused ion beam and conventional implantation are:

(1)   Degree of electrical activation before annealing.   Si implanted with 16 keV B$^+$ ions up to a dose of 1×10^{15} ions/cm^2 with a slow scanned focused ion beam has a sheet resistance about an order of magnitude lower than Si implanted with the same energy and dose using a conventional broad beam or a rapidly scanned focused ion beam(118). This difference, however, disappears if the sample is annealed above 800°C (see Figure 20).

(2) The critical dose for forming a continuous amorphous layer appears to be a factor of 3 to 10 times lower for focused ion beams than for broad beams(119) as seen in Figure 21. (The ion energies are 16 keV for B, 32 keV for P and 50 keV for Ga). The fact that the nature of the damage before annealing is higher in slow scan focused ion beam than in conventional beam implantation is also confirmed by transmission electron microscopy and electron diffraction(120-121).

(3) In GaAs at a current density of 7×10^{-2} A/cm^2 160 keV Be ions appear to produce less damage near the surface than conventional implants in the same (1 to 4 $\times 10^{14}$/cm^2) dose range(122). This has been studied using Raman scattering to observe the shift in the longitudinal optical phototron frequency.

(4) Si ions implanted into GaAs at 160 keV and 0.1 A/cm^2 with a focused ion beam appear to be distributed deeper in the crystal by about 0.3 µm than identical dose conventional implants at a current density of 10 µA/cm^2 (Ref. 123). This difference is observed at doses above 3×10^{13} cm^{-2}.

No convincing models have been proposed for these differences between focused beam and broad beam ion implantation. Apparently in the focused beam case the crystal has not reached equilibrium in the vicinity of an ion impact when the next impact occurs. For example, if the current density is 0.1 A/cm^2 and if the ion creates a non-equilibrium situation in a region 10 nm in diameter surrounding the point of impact, then on the average the next ion will strike this 10 nm diameter region in 2 µsec.

Understanding the details of this high current density implantation phenomenon is an interesting challenge in fundamental materials science. The practical consequences, however, can in most cases be avoided. Rapid scanning of the focused ion beam usually results in implantation which is equivalent to broad beam implantation(118). In addition, after annealing the differences in broad beam and focused ion beam implantations disappear. Numerous devices have been made where focused ion beams have been substituted for broad beams and the device characteristics have been unaffected(110)(124).

3.3.2 **Dopant Distribution in the Solid (Effects of Beam Diameter, Straggle, and Diffusion).** The key feature of focused ion beam implantation is the ability to vary the ion dose from point to point. Here, we will address the question of

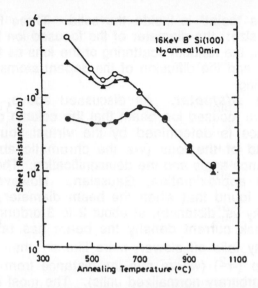

Figure 20. From Ref. 118. Annealing of 16 keV B^+ ions implanted to a dose of 1×10^{15} ions/cm^2 in Si. Open circles: conventional broad beam implant. Triangles: focused ion beam implant rapidly scanned at 9 cm/sec. Solid circles: focused ion beam implanted with a slow scan 0.04 cm/sec.

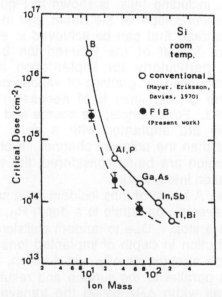

Figure 21. From Ref. 119. Critical dose for amorphization plotted vs. ion mass. Focused ion beam implantation is seen to consistently amorphize the material at a lower dose.

how large this "point" is inside the solid.   Three factors mainly determine its size: the diameter of the focused ion beam incident on the surface, the random scattering of the ions as they penetrate into the solid, and the diffusion of the dopant atoms (former ions) during annealing.

Beam Diameter. As discussed above, Sec 2.2, the diameter of the focused ion beam that the column delivers to the sample surface is determined by the virtual source size, the energy spread of the ions (via the chromatic aberration), the beam acceptance angle and the demagnification. The beam profile is, to a first approximation, Gaussian.   However, numerous authors have found that when the beam diameter is measured, (current density vs. distance), at about 2 to 3 orders of magnitude below the peak current density the beam has tails; the beam current density falls off slower, more like e^{-r} than the expected Gaussian $\exp(-r^2)$ (where r is the distance from the center of the beam in arbitrary normalized units).   The most detailed study of this phenomenon has been carried out at Hughes(22), and a theory based on the Holtzmark distribution has been proposed. The size of the tails also appears to depend on the degree of demagnification. Lower magnification, which also yields smaller beam diameters, reduces the magnitude of the tails(50). A plot of the beam profile, including tails, is shown in Figure 5.

The Gaussian shape of the beam and in particular these tails limit the "contrast" that can be achieved in exposed features. This is important for all of the focused ion beam processing techniques, but particularly for implantation and lithography. This has been calculated for gratings of various periods where the dose between the grating lines is of necessity not zero(50), as shown in Figure 22. For example, the source and drain of an MOS transistor usually are implanted with a dose three orders of magnitude higher than the adjacent channel. If channel lengths in the submicron region are being considered, the tails in the beam will need to be taken into account.

Straggle. A beam of ions incident on a surface at a given point will on the average penetrate to a depth R_p, called the range, before coming to a stop.   Due to random collisions with the lattice atoms, the distribution in depth of implanted ions has a finite half width called range straggle or ΔR_p. In addition, the ions will also scatter sideways parallel to the surface and result in a transverse distribution of half width ΔR_t, called the transverse straggle.   In the simplest description of this phenomenon both of these distributions are fitted by Gaussian distributions as illustrated in Figure 23.

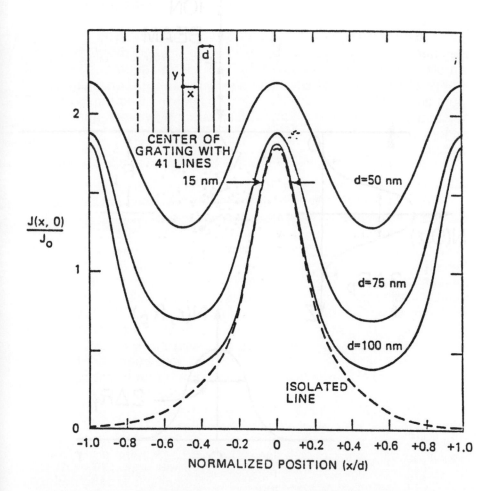

Figure 22. From Ref. 50. Normalized current density vs. nomalized position from the center of a grating line. The additon of the tails of the focused ion beam are seen to lead to a significant dose between the grating lines.

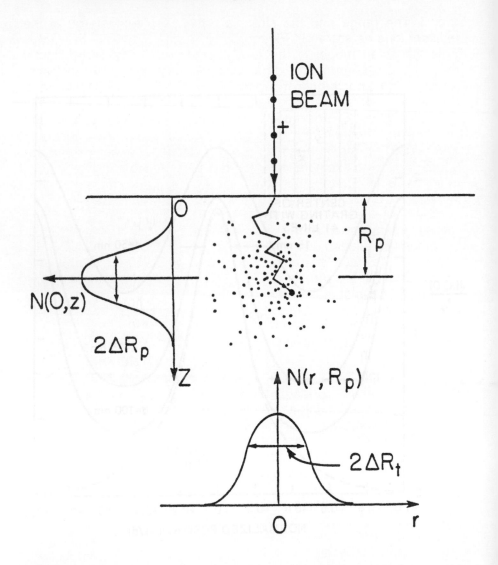

Figure 23. The distribution of a large number of ions entering the solid at the point z = 0, r = 0 in cylindrical coordinates. Due to random collisions the ions are distributed in a cloud of density N(r,z). The profiles in both width and depth are modelled as Gaussian. Values of R_p, ΔR_p and ΔR_t are plotted in Figs. 24 and 25 for various ions incident on Si and GaAs, respectively.

The range and the straggle depend on the energy of the incident ions as shown in Figure 24 for Si and Figure 25 for GaAs. This Gaussian model is the lowest order fit to the experimental data. The experimental data available is for the broad beam implantation and largely for distribution of implanted atoms as a function of depth. The Gaussian shape generally provides a good fit down to 1 or 2 orders of magnitude below the peak of the distribution. Beyond that deviations are observed. Asymmetry is introduced by the surface as well as by the details of the ion scattering. For example, ions lighter than the host lattice atoms, such as B in Si, tend to scatter backward preferentially. This skews the distribution toward the surface. While heavier ions such as As in Si tend to scatter preferentially forward, skewing the distribution away from the surface. More sophisticated models have been developed which take these effects into account.

In the case of transverse straggle little experimental data exists, and we are largely dependent on computer models. Note that newly developed techniques derived from scanning tunneling microscopy have demonstrated measurement of dopant distributions parallel to the sample surface with 200 nm resolution. A resolution of 10 nm for this technique is projected(126). This technique will prove quite useful for the verification for focused ion beam implanted lateral doping gradients. Since dopant distributions are being measured, diffusion of atoms during activation as well as the effect of straggle have been measured.

Diffusion.   Implantation that is carried out for the purpose of semiconductor doping is followed by annealing. Some diffusion of the implanted species will occur during the anneal. The diffusion in this situation is complex. The crystal is heavily damaged by the implantation and often, in fact, rendered amorphous. In addition, the implanted atoms change the electrical properties (Fermi level) of the substrate. These effects influence the diffusion coefficient. Again in the simplest model of diffusion these effects are ignored, and the diffusion coefficient is considered to be a constant dependent only on temperature. In this case, the redistribution of dopants at a given annealing temperature is governed by the diffusion equation. A Gaussian curve of the form $(Dt)^{-1/2}$ exp - $[x^2/Dt]$ is a solution of the diffusion equation. Thus the effect of diffusion is simply to broaden the Gaussian shape shown in Figure 23, and we can write an expression for the distribution of implant species as a function of position and diffusion time t as follows:

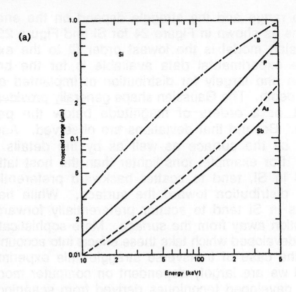

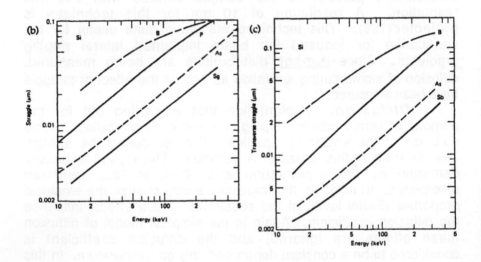

Figure 24. From Ref. 125. Plots of a) Range, R_p; b) range straggle, ΔR_t and c) transverse straggle, ΔR_t as a function of energy for various ions implanted into Si.

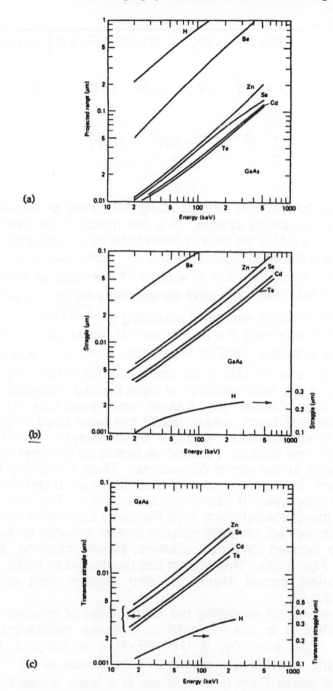

Figure 25. Same as Fig. 24 for implantation into GaAs.

[7] $N(r,z,t) = $

$$\dfrac{N}{(2\pi)^{3/2}(\Delta R_p^2 + 2Dt)^{1/2}(R^2 + \Delta R_t^2 + 2Dt)}$$

$$\times \exp\left(-1/2\,(z - R_p)^2 / \,(\Delta R_p^2 + 2Dt)\right.$$

$$\left. -\dfrac{1}{2}\, r^2 / (R^2 + \Delta R_t^2 + 2Dt)\right)$$

Here an ion beam is incident along the z axes at the origin of a cylindrical coordinate system. R is the radius of the incident Gaussian beam at 0.61 if the peak is taken to be at 1. Values of R_p, ΔR_p and ΔR_t are shown in Figures 24 and 25. Some sample values for the diffusion coefficient are D = 3×10^{-14} cm^2/sec for B in Si at T = 900°C (Ref. 166). A typical anneal might be 30 min which means that the resulting degree of spreading is $\sqrt{2Dt}$ = 0.1 μm. In rapid thermal annealing the temperature is higher but the time is shorter, for example 1200°C for 10 sec.   In that case D = 10^{-12} cm^2/sec and $\sqrt{2Dt}$ = 0.04 μm. This minimizes dopant spreading and is the main attraction of rapid thermal annealing.

3.3.3 Low Energy Ion Beams and Molecular Beam Epitaxy. Molecular beam epitaxy (MBE) has been used to grow thin films of semiconductors monolayer by monolayer. This has led to numerous new devices. Focused ion beams on the other hand permit patterning in the lateral dimensions. Thus, combining the two techniques in principle should lead to three dimensional control of the properties of semiconductor crystals. This was, in fact, done at the Optoelectronics Joint Research Laboratory where a single interconnected, ultrahigh vacuum system included an MBE chamber, two focused ion beam columns, plasma cleaning, and analysis (see Figure 26). This system has been used to make, for example, buried doped layers(99) and junction field effect transistors(127).

Another way of exploiting the combination of focused ion beams and MBE is to use implantation to cause multilayers to interdiffuse. For example, if GaAs/GaAlAs multilayers are implanted with Ga(128-129) or Si(130-132), doses of 10^{13} to 10^{15}/cm^2 will cause them to interdiffuse at a lower temperature than for unimplanted layers. This effect can be used to make confined quantum well structures. To apply this technique with focused ion beams a negative of this process would be more

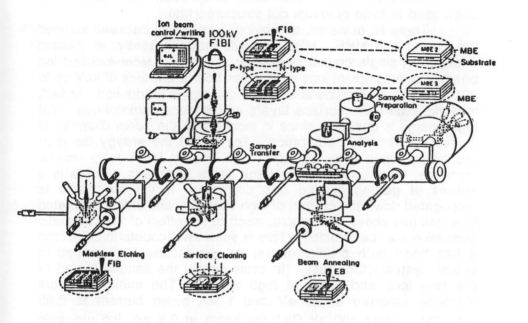

Figure 26. From Ref. 167. Schematic of an ultrahigh vacuum system constructed at the Optoelectronics Joint Research Laboratory in Kawasaki, Japan. Two focused ion beam systems, one suitable for implantation, one suitable for milling, etching or induced deposition, are connected by a vacuum transfer tube to an MBE chamber. Beam annealing, surface cleaning, and analysis chambers are also included. Multiple steps can be carried out in-situ, permitting buried patterned layers to be produced.

convenient, i.e., the implanted areas do not interdiffuse. This turns out also to be possible by initially doping the entire structure with Si and then implanting Be ions with a focused ion beam to compensate some areas(133). The compensated areas do not interdiffuse. The interdiffusion of quantum well layers has been used to build quantum dot structures(129).

There is, however, an obvious incongruity between focused ion beams and MBE. MBE deposits material gently, at thermal energies, a single crystal monolayer at a time. Most focused ion beam systems implant ions at energies from a few tens of keV up to 300 keV. Thus, considerable lattice damage is introduced. In fact, in some cases the surface layers are rendered amorphous. Yet high energy is desirable since in most cases the beam diameter is limited by chromatic aberration. The higher the energy the finer the beam focus (see Eq. 2). Low energy ion beams would, therefore, have unacceptably large beam diameters. There is a means of getting around this dilemma: If the ion beam is propagated down the column at high energies, and only decelerated in a lens just above the sample, much of the effect of the chromatic aberration can be avoided. This is somewhat counterintuitive, but it has been both predicted(134) by simulations and verified in actual systems(135- 136). (In some cases, the sample is part of the final lens and floats at high voltage). The minimum beam diameter reported at 25 eV and 1 nA beam current is 0.95 μm(135). Using 300 eV Ga$^+$ ion beam at 0.9 nA, 0.5 μm wide lines have been sputtered in 100 nm thick gold film(137). If the energy of the beam is reduced below 200 eV, gallium is deposited, as the sputtering yield falls below 1(137).

In addition, the defects produced in GaAs by Ga ions of energies between 100 eV and 150 keV has been studied(138). Trap centers were found to be produced at all energies. However, the defects produced at low energies annealed out at 600°C, while the deflects produced at high energies remained even after anneal at 850°C. Thus, low energy ion beams are more compatible with MBE and promise to be important in the creation of three dimensional heterostructures.

3.3.4 Implantation of Devices, Examples. The point of direct, maskless, resistless implantation is in fact to replace lithography and to go beyond it by making structures not hitherto possible. (We take the term lithography to imply the use of a resist medium). Since the end result, namely a patterned implant of the substrate, is in principle substituted for the conventional process which consists of combined lithography and blanket implantation, we feel that a few examples of focused ion beam implantation belong in this chapter. More extensive references to focused ion beam implantation are found in an earlier review(3).

a)   The variation of dose as a function of lateral position on the surface is a new fabrication capability which may permit new devices to be made or the performance of existing devices to be improved. An example of a new device which uses a lateral doping gradient is the tunable Gunn diode(111). The Gunn diode in this case consists of a pair of contacts on a GaAs surface with a conducting n-type channel between them. This channel is doped by focused ion beam implantation of Si where the implantation dose is varied linearly from one end to the other. The cathode has the lower doping, and when bias is applied the electric field is highest at the cathode (see Figure 27). A Gunn domain is launched from the cathode and propagates toward the anode until the electric field drops below a threshold value. (The time of this traverse is the inverse of the oscillation frequency). At that point, a new domain is launched and the existing domain is quenched. If the bias is raised, the domain will travel farther, and the oscillation frequency drops. The frequency of the device could be tuned from 6 to 23 GHz (see Figure 27). This type of device may be of interest, for example, for collision avoidance radar.

b)   GaAs MESFETs with variations of the doping profile in the channel have been fabricated by focused ion beam implantation(113-114). A step in the doping profile was created under the gate as shown in Figure 28, and the position of this step was varied from device to device to determine the optimum performance. The maximum available power and the mean transconductance achieved their highest values with the step at 2 μm from the edge of the gate nearest the drain. Although the devices fabricated with a 10 μm gate length were not of immediate practical interest, the principle of controlling and improving performance by tailored implants was demonstrated. Numerical simulations indicate that similar improvements are to be expected for 2 μm gates(113).

c)   A GaAs/GaAlAs distributed Bragg reflector laser has been fabricated by using the focused ion beam to implant a grating(139). The gratings were implanted with 100 keV Si ions to a dose of ~ 10^{14} ions/cm^2. The beam diameter (FWHM) was approximately 70 nm and the period of the gratings was 230 nm; the gratings operated at the second order. The application of focused ion beams to optoelectronics is particularly promising because more complex fabrication technology, such as MBE, is already used and often the number of devices needed may not be large so that the issue of writing time may be less critical.

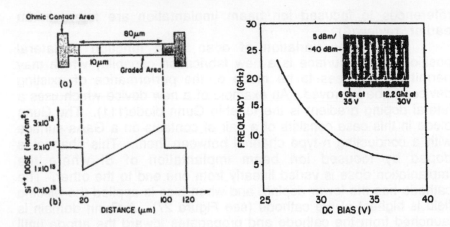

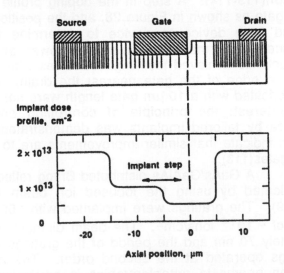

Figure 27. From Ref. 111. Tunable Gunn diode. On the left schematic of a gradient of doping implanted between two contacts on a GaAs surface. On the right the frequency of the Gunn oscillations as a function of bias voltage observed on this structure. On a more optimally designed structure output power of ~ -3 to -10 dBm has been observed instead of -40 dBm (unpublished result Lezec et al.)

Figure 28. From Ref. 113. A step in the doping profile of a GaAs MESFET implanted with a focused ion beam. The position of the step was varied from transistor to transistor.

d)   Unannealed focused ion beam implantation of 260 keV Be^{++} in a GaAlAs/GaAs heterostructure has been used to define electron conduction channels 0.4 μm wide. These multiple channels in the high mobility 2-d electron gas permitted electron focusing effects to be observed as a function of magnetic field (see K. Nakamura, et al, Appl. Phys. Lett. 56, 385 (1990)). AlGaAs/GaAs quantum wires have also been fabricated by focused ion beam implantation. Si ions at 200 keV at a dose of 2×10^{11} ions/cm^2 were used to create two parallel insulating regions with a narrow conducting channel between them. The effective width of the conducting gap as low as 30 nm was reported (140-141). These structures were used to study the phase coherence length.

The above examples fall in the category of special devices with implant doses which are varied as a function of position along the surface. Several other recent examples of devices fabricated by focused ion beam implantation are given in Ref. 2 and earlier citations are found in Ref. 3. We now turn to circuits where the focused ion beam is used to vary the dose from device to device.

e)   The focused ion beam can be used to adjust the threshold of transistors(115)(142). By implanting the channels of 13 NMOS transistors with increasing doses in 1×10^{11} ions/cm^2 increments, the threshold voltage could be varied about 19 mV per increment.

This type of threshold adjust has been exploited to fabricate 4 bit 1 GHz flash analog-to-digital converters(116). In this case 32 transistors were implanted in the channels with incrementally increasing doses of boron to achieve 16 comparators with threshold voltages spaced at approximately 94 mV; as shown in Figure 29.

f)   Finally, we turn to the direct substitution of focused ion beam implantation for conventional implantation. This has not been demonstrated so far, but it may be practical in special cases such as GaAs microwave and millimeter wave monolithic integrated circuits (MMIC's). These are high performance components where a small number of different transistors on each chip are needed to perform different functions. The implanted areas are small and often a variety of different implants are needed in various geometrics defined by separate lithography steps. Since the dose does not need to be varied from point to point within a device, the ion beam diameter and thus the beam current can be increased. Thus the implantation times are not expected to be prohibitive and one can replace large numbers of steps (resist application, exposure, development, implantation and resist removal for each dose and at a different geometry) by a single focused ion beam implantation step. The machine would

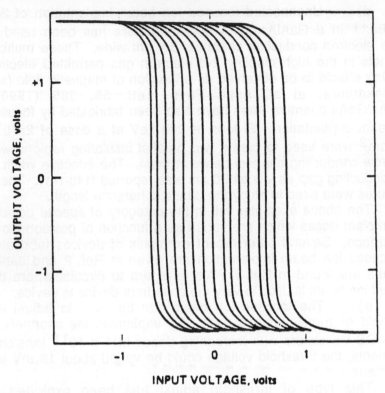

Figure 29. From Ref. 116. D.C. transfer curves for 16 focused ion beam implanted comparators.

Table VII

Projected Focused Ion Beam Implant times for Raytheon
Transmit/Receive Module (Beam Current 600 pA)

Device Implanted	Dose (ions/cm^2)	Energy (keV)	Area/chip (mm^2)	Time/chip (sec)
Low noise FET	7×10^{12}	105	0.2	3.9
Power FET	1.3×10^{12}	65	0.32	1.2
	5.6×10^{12}	160	0.32	5
RF switch	8.6×10^{11}	95	0.31	0.8
	3.0×10^{12}	30	0.31	2.9

simply be programmed to deliver the desired dose, energy, and ion species at the desired location(117).

We look at a concrete example of a transmit/receive MMIC fabricated by Raytheon(143). In all cases Si ions are implanted. Since resolution is not an issue we extrapolate from a beam of 30 pA at 2 mrad acceptance angle and 0.1 μm beam diameter to 600 pA at 9 mrad acceptance angle. The implantation steps and the times required are summarized in Table VII. The total implantation time per chip is seen to be 14 sec. This does not include the time needed to move the stage. For our machine about 0.4 sec are needed to move from field to field. Each level requires about 40 fields or 16 sec. Thus a full 3 in. wafer could be implanted in 2.5 hours. Note that 85% of that time is due to the stage motion overhead. This can be reduced greatly by writing on the fly (a technique already developed for e-beam where the stage does not stop; see Chapter 5) or by optimizing the stage motion for this application. This would mean faster stage movement (0.1 sec to move from field to field is routine for e-beam systems) and larger fields. In most current systems, the stage needs to move and settle to a high tolerance. In addition, field size is reduced to avoid beam distortion and placement inaccuracies. Since for this application high accuracy is not needed, these conditions can be greatly relaxed. For example, field sizes of 500x500 μm are reasonable, instead of the 150x150 μm. This would reduce the stage motion overhead by a factor of 40, and the total time to implant a 3 in. wafer of 100 chips would be 25 min instead of 2.5 hrs. and a total of 15 masking/lithography steps would be eliminated. Newer more aggressive designs have even more implantation steps, and moreover, to keep the steps to a limited number the designers often make compromises. With focused ion beam implantation there would be no penalty for designing devices with an even larger number of different implants. Note that to achieve the desired doses both Si^+ and Si^{++} ions can be used since the standard Au/Si liquid metal sources have both types of ions available.

Similar arguments can be made for CMOS silicon chips where both B and As would be implanted for threshold adjust of the n to p channel devices(117). However, except for special cases (such as the A/D converter discussed above) the argument here is less compelling.

We have examined two methods of patterning in microfabrication with finely focused ion beams, lithography, i.e., exposure of resist, and direct maskless/resistless implantation. In both cases, the focused ion beam addresses the sample point by point. We will now consider briefly two alternate techniques which deliver an ion dose in a prescribed pattern to the entire sample at once.

4.0 MASKED ION BEAM LITHOGRAPHY

A patterned ion dose can be delivered to a surface by using a stencil mask in close proximity to the sample. This is similar in principle to the so-called contact optical lithography, and also to X-ray lithography. Masked ion beam lithography is illustrated in Figure 30. Usually protons are used in the 100 keV range. (For a review of the subject and for citations of to earlier literature see Refs. 144 & 145). The dose required to expose most organic resists is in the range of 5×10^{11} to 3×10^{13} ions/cm^2(144,146). The source of protons is easily available. Most any implanter is capable of producing the desired exposure, over say 1 cm^2, in well under a second.

4.1 The Mask

The main challenge in masked ion beam lithography has been the mask. Most materials including the least dense are highly absorbing to protons and also have the effect of scattering an incident collimated beam. While for photons one can speak of attenuation of materials, i.e., decrease of intensity or number of photons as a function of depth in a material, for ions the situation is more complicated. A collimated beam of ions passing through a thin film will emerge both diminished in number and in energy and to some extent uncollimated (see Table VIII).

The difference in the energy loss rate, and range for protons in various materials is seen to be barely over a factor of two. For photons (X-ray or UV) the difference in attenuation can be several orders of magnitude. Thus the design of a mask for ion beam lithography, which ideally should be made of alternate areas of absorbing and transmitting material, presents a special challenge.

Several types of masks have been considered:

a) The open stencil masks(147-149) have no deleterious effects on the transmitted ions and have been used to demonstrate sub 100 nm ion lithography(149). The two main limitations are geometry and distortion. Clearly, arbitrary geometrics are not possible, i.e., the center will fall out if one tries to draw a doughnut. In addition, any geometry with large open spaces may introduce distortion; the film which is used to make the mask may have residual stress so that opening large geometrics will change

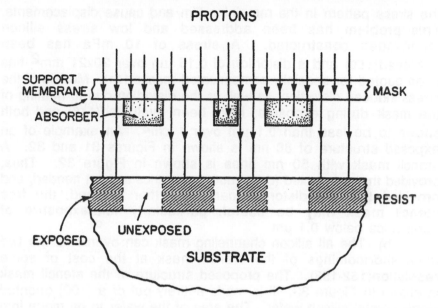

Figure 30. Schematic of masked ion beam lithography, showing the mask consisting of a support membrane, transparent to protons, and absorber regions.

Table VIII. Range and energy loss of protons in
matter for two incident energies

Material / Incident Energy	Range (µm)		Initial Energy loss rate (keV/µm)	
	(100keV)	(200keV)	(100keV)	(200keV)
Resist (KTFR)	0.88	1.41	164	206
Al	0.91	1.77	119	106
Au	0.40	0.79	203	206
Si	1.20	2.24	98.1	87.6
SiO_2	1.04	2.19	109	96.7

Reference: Bernard Smith "Ion Implantation Range Data for Silicon and Germanium Device Technologies", Research Studies Press Inc. (1977).

the stress pattern in the remaining film and cause displacements. This problem has been addressed and low stress silicon membranes constructed.   A stress of 10 mPa has been achieved(150) and a distortion of 0.15 μm over 20x21 mm^2 has been quoted for specially compensated masks(151). Models of the stress induced distortion, as well as the distortion due to heating of the mask during exposure, have been evaluated(148) and both shown to be less than 0.1 μm over 1 cm^2. An example of an exposed structure of 80 nm is shown in Figures 31 and 32. A stencil mask with 50 nm lines is shown in Figure 32.   Thus, provided that connected (doughnut) geometrics are not needed, and provided that the distortion can be further reduced, the free stencil mask may be useful, particularly for exposure of geometrics below 0.1 μm.

b)   The all silicon channeling mask can overcome the two main shortcomings of the stencil mask at the cost of some resolution(152-153). The proposed structure of the stencil mask is shown in Figure 33. It is made entirely out of a (100) oriented single crystal silicon wafer. The area of the wafer to be made into a mask is thinned from the back with a chemical etch up to a boron doped etch stop layer. The wafer is mounted to a pyrex frame, and resist is e-beam exposed on the front surface(152). The silicon is reactive ion etched through the resist mask to achieve the structure shown in Figure 33.

The challenge is to get vertical sidewalls over 2 μm deep in small features, and to stop the etch leaving a 0.5 μm membrane. The energy loss and scattering of protons along a crystallographic axis is lower than in a random direction(155). For example, a 180 keV beam of protons passing through a 0.61 μm thick membrane emerges with an energy distribution centered at 124 keV with a 8 keV halfwidth and a divergence characterized by a half angle of 0.5^o (155).

The achievable linewidth in a realistic production environment (25 μm wafer-to-mask gap) has been simulated(153). The resolution limit is predicted to be 0.5 μm for single level resist and 0.25 μm for a thick imaging layer.

Although for these masks distortion is less serious than for the stencil masks, care still must be exercised in controlling the stress in the silicon membrane(152). Other versions of the channeling mask, which make use of an absorber of another material(153) or a dechanneling layer on the upper side of the silicon membrane (156), have been considered.

4.2   Source of Ions

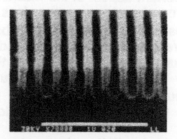

Figure 31. From Ref. (145). Lines and spaces in PMMA 80 nm wide exposed by a proton beam trough a Si_3N_4 stencil mask with a wafer-to-mask gap of 25 µm.

Figure 32. A Si stencil mask fabricated in 0.3 µm thick low stress Si membrane. Ref. 154.

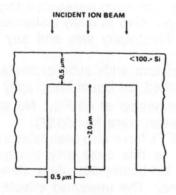

Figure 33. Silicon channeling mask structure showing the typical dimensions. From Ref. 152.

Another important aspect of the masked ion beam exposure process is the source of protons or other ions. While an implanter can be used, a specially designed vertical beam machine[144] is expected to be more convenient for mask and sample handling. If Si channeling masks are used the energy of the protons should be up to 300 keV so as to minimize the angular scattering of the beam. A beam current of about 0.5 to 1 $\mu A/cm^2$ over 2 cm^2 uniform to $\pm 5\%$ is needed(144). This would permit exposure of resist under 1 sec.

The relative simplicity of the ion source, given the extensive existing ion implanter technology, is one of the attractive features of this technique. X-ray lithography which is being developed for the same submicrometer regime generally appears to require synchrotrons costing at least 10-20 times more than a proton source if the exposure time is to be comparable. Nevertheless, at this time the semiconductor industry is investing overwhelmingly in X-ray lithography rather than masked ion beam lithography. This is largely due to the fact that the mask technology for X-rays is easier than for ions. The alignment of mask to wafer and the overlay accuracy of successively exposed layers is equally challenging for the two technologies.

4.3   Device Fabrication

In demonstrating masked ion lithography on actual devices, an obvious concern is the ions will penetrate the resist and implant and potentially damage the substrate. In some cases, it may be practical to match the resist thickness to the ion energy so that by the time the ions have penetrated through the resist they have very little energy. Also in many instances the devices can be annealed after the lithography step and any implantation effects can be eliminated.

NMOS transistors with submicrometer gates have been fabricated using masked ion beam lithography for all of the levels. The devices were annealed at 450°C. No statistically significant effects of the irradiation were found(157).

GaAs MESFET's have also been fabricated using masked ion beam lithography. In this case only the gate lithography was carried out with the ion beam(158). SiN_x stencil masks were used with 100 keV protons. The unwanted effects of proton irradiation could be avoided by using a sacrificial SiO_2 layer, the correct resist thickness, and either a 450°C 5 sec anneal or a removal of 50 nm of the surface by wet etching(158).

Although the semiconductor industry at this time appears to be largely turning to X-ray lithography for its future deep submicrometer fabrication needs, masked ion beam lithography may merit consideration in some circumstances particularly if stencil masks can be used. Features below 0.1 μm have been exposed. The problems of mask stress and heating are comparable to the X-ray case, but an intense source of radiation is available much more cheaply.

5.0   ION PROJECTION LITHOGRAPHY

So far we have examined focused ion beam lithography (and implantation) and masked ion beam lithography. In some sense ion projection lithography is a cross between them. Like the focused ion beam it uses ion optics in an ion column, and like masked ion lithography it uses a reticle, i.e., an open stencil mask to define the pattern.

As shown schematically in Figure 34, ions are incident on the reticle and some pass through the open spaces of the reticle. The image of this reticle is projected through the ion column and demagnified, e.g., a 3 cm x 3 cm reticle forms a 3 mm x 3 mm image on the sample.

Ion projection systems have been built by Ion Microfabrication Systems in Vienna, Austria. (See ref. (159) and references therein). The main motivation has been lithography, i.e., exposure of resist, although in some cases direct resistless processing has been demonstrated, e.g., exposure of SiO_2 with 80 keV He^+ ions to a dose of 6.6×10^{15} ions/cm^2 followed by wet chemical etching in 5% HF(159). Most inert ions appear to be useable in this system (H, He, N, Ne, Ar, Xe)(160). If species such as B, As, Si, or Be could be projected, then direct implantation of semiconductors would be an attractive possibility. Although projection ion lithography has a number of attractive features, such as demagnification of the mask, short exposure time and a simple source of radiation, at present it is only receiving limited attention compared to X-ray lithography.

We will discuss each of the parts of the ion projection machine and the associated critical issues:

5.1   Source of Ions

The source of ions in this case is a duoplasmatron with a small 250 μm exit aperture. Ions are extracted with about a 4° spread and appear to come from a virtual source of 50 μm

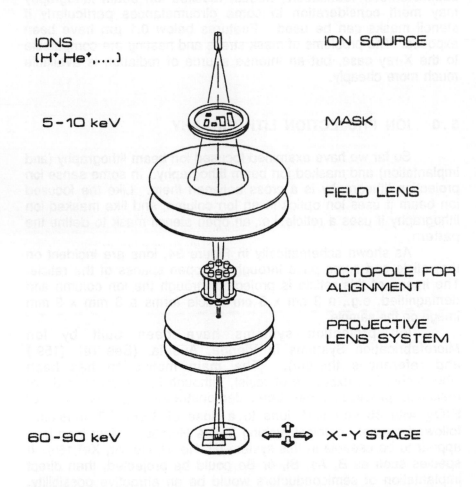

IONS
(H⁺,He⁺,....)   ION SOURCE

5-10 keV   MASK

                                 FIELD LENS

                                 OCTOPOLE FOR
                                 ALIGNMENT

                                 PROJECTIVE
                                 LENS SYSTEM

60-90 keV   ⇦⇕⇨   X-Y STAGE

Figure 34. Schematic of an ion projection lithography column. The image of the reticle is demagnified (typically x 1/5 of x 1/10) on the sample. (Ref. 159)

diameter(160). The details of the source design are not available. The ion current on the reticle is about 10 $\mu A/cm^2$(159) which at 5 kV corresponds to 50 mW/cm^2. The ion current density over the area of the reticle should be constant so that the current density imaged on the sample is also constant.

5.2   The Reticle

The reticle, because it is demagnified, is somewhat easier to fabricate than in the case of the shadow printed 1:1 mask described in Sec. 4. above. Some of the issues however are the same. Distortion during fabrication can occur if the mask material has built-in stress. Then the removal of some of that material to make open spaces will change the stress on the remaining structure. Both plated-up Ni masks(160) and etched out Si(161-162) masks have been made. The Ni masks are made by defining the areas to be open as islands of resist and plating Ni up around them. Dissolving the resist and removing the substrate leaves a free standing Ni membrane.

The Si stencil mask has the advantage of being single crystal and of being fabricated by standard integrated circuit processing technology. Here a boron doped etch stop layer is grown epitaxially. Ge is included to provide stress compensation (161-162). A silicon nitride layer is added by chemical vapor deposition. Stress in this nitride layer is compensated by implantion. Wet etching from the back is used to thin the wafer up to the B doped etch stop. The surface is then coated with resist and the desired pattern is exposed. X-ray lithography has been used(162) but e-beam, focused ion beam, or even optical lithography could be used since the reticle is at 10X or 5X of final size. The pattern is etched into the nitride and through the Si membrane by reactive ion etching using CHF_3/O_2 and SF_6/CHF_3, respectively.

The mask is subjected to an ion current density which delivers about 50 mW/cm^2 of power. Since it must remain in vacuum, heating has to be considered. A simple way of minimizing the effect of heating is to design the mask to operate at 100-150°C and to operate it at elevated temperatures. The heat loss is by radiation and is proportional to T^4. Therefore at a higher temperature, a small change in T will accommodate a relatively large change in power loss per unit area.

Another issue of concern is mask erosion by sputtering. This is minimized by the fact the mask image is demagnified on the substrate, by the fact the ions are light, and by the fact the ion dose needed on the sample is low (10^{13} ions/cm^2). Thus if we assume a sputtering yield of 0.2, typical for light ions, and a demagnification of 1/5, then, each exposure will result in a dose of 4×10^{11} ions/cm^2 on the mask, and 6×10^6 exposures would result in an erosion of 0.1 μm of the mask surface. In fact, several hundred hours of operation has been reported for a given mask[160], so that mask erosion is not a serious drawback.

5.3   Optics

The optics needed to project the image of the mask on the wafer has some special requirements. While in the focused ion beam case a beam of a few millirad acceptance angle is passed through the column and has a maximum diameter in the range of 0.1 to 0.2 mm, in the case of projection ion lithography we need to transmit an entire image in excess of 30-50 mm in diameter. The electrostatic ion lenses, which for focused ion beams are ~ 10 cm in diameter with bores of 5-10 mm, for projection need to be scaled up by more that a factor of 10. Thus externally the lenses on the existing machines are seen to be of the order of 1 m in diameter. Again the details of the design are not available. A detailed theoretical analysis of projection ion lithography has been carried out and addresses the issues of pattern distortion caused by the fact that part of the image is considerably off-axis[163]. The optics is designed so that the ions arrive at the substrate at energies in the 70-90 keV range.

5.4   Distortion

Distortion of the final pattern projected on the wafer can arise from a number of sources: distortion of the mask during fabrication, distortion during irradiation, (e.g., due to heating), and distortion during projection. Overall, this is the most serious issue. The distortion in the reticle mask can be minimized to acceptable levels as discussed above by controlling the stress during fabrication[161] and the temperature during irradiation. In addition, the absolute distortion in the reticle is demagnified on the substrate. The distortion in the ion optics has been simulated to be of the order ± 7 μm over a 12 mm radius on the sample surface[163]. The desired value is $\pm$ 0.1 μm. The source of this distortion is the off axis propagation of the ions and space charge effects. Two measures can be taken to overcome the distortion. If it is predictable and experimentally verified then the mask pattern could be predistorted to cancel out the effects of the ion optics[163]. In addition, multipoles can be included in the column to permit fine tuning of the distortion correction. So far limited experimental data is available. In one observation 5 μm of pin cushion type distortion was observed over 3.3 mm. By changing the column parameters this was changed into a comparable magnitude barrel distortion(164). Presumably an in between minimum distortion was achieved. However, since the desired distortion is below ± 0.1 μm much work remains to be done.

5.5   Alignment

Alignment of one level of exposure to existing features must be possible in any practical lithography tool. This is a challenge for any modern lithography system, such as X-ray lithography. One advantage of the present technique is that a focused ion beam can be imaged in the scanning ion microscope mode. This offers the possiblity of using a part of the projected ion beam to image alignment marks much as described above for focused ion beams.

This has recently been demonstrated(169). A shutter is used to block the pattern to be exposed but to pass the ions through special holes in the perimeter of the mask forming alignment beamlets. These beamlets are scanned across four alignment marks on the sample which are already mechanically located at ±3 μm of the desired position. Secondary electrons produced are detected by separate pairs of channeltrons at each alignment mark. The alignment is carried out <u>electrically</u> by using a special multipole at the beam crossover. Lateral positioning (x,y) as well as rotation and scale adjustment can be carried out. Alignment to better than 0.1 μm over a 2x2 mm chip was achieved at 9X demagnification between mask and wafer(169).

5.6   Results

Results reported using ion projection lithography include lines down to 0.15 μm width exposed in resist(159). Lines 0.2 μm in width have been exposed in SiO_2 and directly wet etched(159-161) (see Fig. 35). Also transistor patterns with gate width of 0.15 μm have been reported (161). He ions at 80 keV were used. The time needed to expose resist, even an insensitive resist such as PMMA, was 0.5 sec or less.

The machines that have been built so far and the results that they have produced(159), as well as the theoretical analysis (163), indicates ion projection lithography may be a promising alternative for submicrometer lithography. In fact, analysis indicates that below 0.25 μm ion projection lithography may out-perform X-ray lithography(165). The remaining issues of

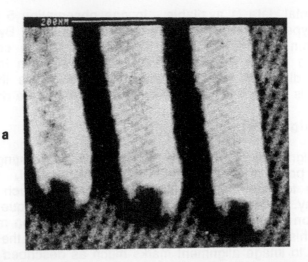

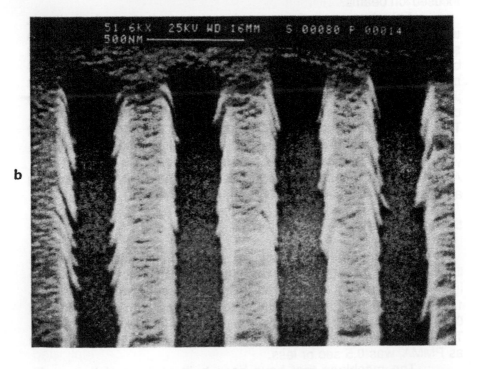

Figure 35. a) Results of exposure of resist with ion projection showing 0.13 μm lines and 0.07 μm spaces. b) SiO2 directly exposed with 80 keV He ions and directly wet etched, line width 0.15 μm. Ref. (168) and (161).

alignment, and distortion can really be addressed only by building a genuine prototype production machine.

6.0 CONCLUSION

Progress in a field of technology is often governed by the interplay of several factors: the perceived need, useful results obtained, useful results envisioned, competing technologies, and resources devoted to it. In the case of focused ion beam development for photomask repair, and integrated circuit diagnosis a clear need existed, useful results were readily demonstrated, no competing technologies were available, and the resources were expended to make commercial machines. We have only briefly mentioned this development in this chapter but have concentrated on the less clear cut issue of ion lithography and focused ion beam implantation.

Focused ion beam lithography was first demonstrated seventeen years ago, and many impressive results have been obtained since then. However, it has always had to compete with the earlier and more mature E-beam lithography. In spite of the fact that future advantages have been demonstrated and projected, the resources devoted to focused ion beam lithography development have been limited. The near term lithography needs have been satisfied by E-beams. As the demands on resolution in lithography increase we expect the interest in focused ion beams particularly with light ion species to increase.

In the case of masked ion lithography, despite impressive results there is a heavily funded competing X-ray lithography effort. One can at this point imagine special cases where the limitations of stencil masks are overshadowed by the advantages of the simpler source of radiation. In the case of projection ion lithography there appears to be no heavily developed alternative. (Projection E-beam lithography may be a possible candidate.) The technology so far has been developed exclusively by one group of workers in one small company (IMS in Vienna). However, analysis of this technique indicates possible practical applications.

Focused ion beam implantation, in contrast to lithography, can clearly perform unique functions which cannot be duplicated by conventional fabrication using broad beam implantation. The ability to vary the dose from point to point within a device and to vary the dose from device to device has permitted unique devices to be fabricated. What has limited the investment in this seemingly powerful technique is the fact that focused ion beam implantation is serial and therefore slow and that the devices fabricated, though unique and with superior performance, are not eagerly sought by the market place. This may change. In the meantime focused ion beam implantation is an exciting research tool and is a useful prototyping tool. Many implant doses, energies and unique geometries can be explored on one chip or wafer.

7.0   ACKNOWLEDGEMENTS

The author is grateful to R.L. Kubena, S. Matsui, J.N. Randall, M. Tamura, H. Loschner, and J.C. Wolfe for providing copies of figures, and to the American Institute of Physics, the IEEE, the SPIE and Elsevier Publishers for permission to reprint figures. Thanks are also due to R.L. Kubena and J.C. Corelli for a critical reading of the manuscript. Although the writing of this chapter was not directly sponsored, the author's work in the focused ion beam area is supported by DARPA, Army Research Office, SEMATECH, and the NSF.

References

1.   Harriott, L.R., Applied Surface Science, 36, 432 (1989).
2.   Namba, S., Nucl. Intr. and Methods in Physics Research, B39, 504 (1989).
3.   Melngailis, J., J. Vac. Sci. Technol., B5, 469 (1987).
4.   Seliger, R.L. and Fleming, W.P., J. Vac. Sci. Technol., 10, 1127 (1973); J. Appl. Phys. 45, 1416 (1974).
5.   Levi-Setti, R., in Scanning Electron Microscopy 1974, Part I, Proceedings of the 7th Annual Scanning Electron Microscope Symposium (IIT Research Institute, Chicago, IL, 1974), p. 125.
6.   Orloff, J.H. and Swanson, L.W., J. Vac. Sci. Technol., 12, 1209 (1975).
7.   Orloff, J.H. and Swanson, L.W., J. Vac. Sci. Technol., 15, 845 (1978).
8.   Orloff, J.H. and Swanson, L.W. J. Appl. Phys., 50, 6026 (1979).
9.   Escovitz, W.H., Fox, T.R. and Levi-Setti, R., Proc. Natl. Acad. Sci. USA, 72, 1826 (1975).
10. Clampit, R., Aitken, K.L. and Jefferies, D.K., J. Vac. Sci. Technol., 12, 1208 (1975); Clampit, R. and Jefferies, D.K., Nucl. Instrum. Methods, 149, 739 (1978); Krohn, V.E. and Ringo, G.R., Appl. Phys. Lett., 27, 479 (1975).
11. Seliger, R.L., Ward, V.W., Wang, V. and Kubena, R.L., Appl. Phys. Lett., 34, 510 (1979).
12. Wang, V., Ward, J.W. and Seliger, R.L., J. Vac. Sci. Technol., 19, 1158 (1981).
13. Shiokawa, T., Kim, P.H., Toyoda, K. and Namba, S., J. Vac. Sci. Technol., B 1, 1117 (1983).
14. Gamo, K., Inomoto, Y., Ochiai, Y. and Namba, S., Proc. of 10th Symposium on Electron and Ion Beam Science and Technology p. 422 (1982).
15. Miyauchi, E., Arimoto, H., Hashimoto, H., Furuya, T. and Utsumi, T. Japan. J. App. Phy., 22 L287 (1983).
16. Anazawa, N., Aihara, R., Ban, E. and Okunuki, M., SPIE Conf. 393, 137 (1983).
17. Ishitani, T., Umemura, K. and Tamura, H., Nuclear Instr. and Methods in Physics Research, 218, 363 (1983).
18. Cleaver, J.R.A., Heard, P.J., Ahmed, H., in Microcircuit Engineering ('83) (Academic Press, London, New York 1983) p. 135.
19. Bell, A.E., Rao, K., Schwind, G.A., and Swanson, L.W., J. Vac. Sci. Technol., B6, 927 (1988).

20.   Ward, J.W., J. Vac. Sci. Technol., B3, 207 (1985). Yan, Y.W., Groves, T.R. and Pease, R.F.W., J. Vac. Sci. Technol., B2, 1141 (1983).

21.   Marriott, P., J. de Physique Colloque C6 Suppl. au no, 11 Tome 48 p. C6-189 (Nov. 1987). (Presented at Field Emission Symp. Osaka July (1987).

22.   Ward, J.W., Kubena, R.L., and Utlaut, M.W., J. Vac. Sci. Technol., B6, 2090 (1988).

23.   Clark, W.M., Seliger, R.L., Utlaut, M.W., Bell, A.E., Swanson, L.W., Schwind, G.A. and Jergensen, J.B., J. Vac. Sci. Technol., B5, 197 (1987).

24.   Swanson, L.W., Bell, A.E., Schwind, G.A., J. Vac. Sci. Technol., B6, 491 (1988).

25.   Umemura, K., Kawanami, Y., Ishitani, T., Nucl. Instr. and Methods in Physics Res. B37/38, 208 (1989).

26.   Horiuchi, K., Itakura, T., and Ishikawa, H. J. Vac. Sci. Technol., B6, 937 (1988).

27.   Konishi, M., Takizawa, Tsumori, M. T., J. Vac. Sci. Technol., B6, 498 (1988).

28.   Lewis, G.N., Paik, H., Mioduszewski, J. and Siegel, B.M., J. Vac. Sci. Technol., B4, 116 (1986).

29.   Börret, R., Jousten, K., Böhringer, K. and Kalbitzer, S., J. Phys. D. Applied Phys., 21, 1835 (1988).

30.   Levi-Setti, R., Advances in Electronics and Electron Physics, Suppl., 13A p. 261 (Academic Press 1980).

31.   Munro, E., in "Image Processing and Computer-aided Design in Electron Optics. ed. P.W. Hawkes. (Academic Press, New York 1973) p. 284.

32.   Szilagyi, M. and Szep, J., J. Vac. Sci. Technol., B6, 953 (1988).

33.   Tsumagari, T., Ohiwa, H. and Noda, T., J. Vac. Sci. Technol., B6, 449 (1988).

34.   Muray, J. and Brodie, I., Physics of Microfabrication Ch. 2, (Plenum Press 1982).

35.   Kurihara, K., J. Vac. Sci. Technol., B3, 41 (1985).

36.   Miyauchi, E., Hashimoto, H. and Utsumi, T., Jpn. J. Appl. Phys., 22, L225 (1983).

37.   Gamo, K., Matsui, T. and Namba, S., Jpn. Appl. Phys., 22, L692 (1983).

38.   Arimoto, H., Miyauchi, E. and Hashimoto, H., Jpn. J. Appl. Phys., 24, L288 (1985).

39.   Miyauchi, E., Morita, T., Takamori, A., Arimoto, H., Bamba, Y., and Hashimoto, H., J. Vac. Sci. Technol., B4, 189 (1986).

40. Arimoto, H., Miyauchi, E., Furuya, A., Ishida, K., Takamori, T., Nakashima, H. and Hashimoto, H., J. Vac. Sci. Technol., B6, 230 (1988).

41. Arimoto, H., Takamori, A., Miyauchi, E., and Hashimoto, H., J. Vac. Sci. Technol., B3, 54 (1986).

42. Ishitani, T., Umemura, K., Kawami, Y., and Tamura, H., J. Phys. (Paris) Collog., Suppl., No. 12 45, C9 (1984).

43. Ishitani, T., Umemura, K., Hosoki, S., Takayama, S. and Tamura, H., J. Vac. Sci. Technol., A2, 1365 (1984).

44. Ishitani, T., Umemura, K. and Tamura, H., Jpn. J. Appl. Phys., 23, L330 (1984).

45. Torii, Y. and Yamada, H. Jpn. J. Appl. Phys., 22, L444 (1983).

46. Wagner, A. and Hall, T.M., J. Vac. Sci. Technol., 16, 1871 (1979).

47. Arimoto, H., Miyauchi, E., and Hashimoto, H., J. Vac. Sci. Technol., B6, 919 (1988).

48. Munro, E., J. Vac. Sci. Technol., B6, 941 (1988).

49. Thompson, W., Honjo, I. and Utlaut, M., J. Vac. Sci. Technol., B1, 1125 (1983).

50. Kubena, R.L., Stratton, F.P., Ward, J.W., Atkinson, G.M., and Joyce, R.J., J. Vac. Sci. Technol., B7, 1798 (1989).

51. Paik, H. G., Lewis, N., Kirkland, E.J. and Siegel, B.M., J. Vac. Sci. Technol., B3, 75 (1985).

52. Kubena, R.L. and Ward, J.W., (comment) Appl. Phys. Lett., 52, 2089 (1988).

53. Miyauchi, E., Morita, T., Takamori, A., Arimoto, H., Bamba, Y. and Hashimoto, H., J. Vac. Sci. Technol., B4, 189 (1986)

54. Morimoto, H., Sasaki, Y., Onoda, H. and Kato, T., Appl. Phys. Lett., 46, 898 (1985).

55. Morita, T., Arimoto, H., Miyauchi, E. and Hashimoto, H., Japan J. Appl. Phys., 26, 955 (1987).

56. Morita, T., Miyauchi, E., Arimoto, H., Takamori, A., Bamba, Y., Hashimoto, H., J. Vac. Sci. Technol., B5, 236 (1987).

57. Sawaragi, H., Kasahara, H. and Aihara, R., J. Vac. Sci. Technol., B6, 962 (1988).

58. Ochiai, Y., Kojima, Y., Matsui, S., Mochizuki, A. and Yamauchi, M., SPIE 923, 106 (1988).

59. Morita, T., Takamori, A., Arimoto, H. and Miyauchi, E., Japanese J. of Appl. Phys., 26, L234 (1987).

60. Clampitt, R., Watkins, R. and Whitaker, J., Microcircuit Engineering, 6, 605 (1987).

61.   Sudraud, P., Benassayag, G. and Bon, M., Microcircuit Engineering, 6, 583 (1987).

62.   Cleaver, J.R.A., Kirk, E.C.G., Young, R.J. and Ahmed, H. J., Vac. Sci. Technol., B6, 1026 (1988).

63.   Yamaguchi, H., Shimase, A., Haraichi, S. and Isami, M., Nucl. Instr. and Methods in Physics Research (Jan. 1988).

64.   Puretz, J., DeFreez, R.K., Elliott, R.A. and Orloff, J., Electron Lett., 22, 700 (1986).

65.   Harriott, L.R., Scotti, R.E., Cummings, K.D., and Ambrose, A.F., Appl. Phys. Lett., 48, 1704 (1986).

66.   Chabala, J.M., Levi-Setti, R. and Wang, Y.L., J. Vac. Soc. Technol., B6, 910 (1988).

67.   Harriott, L.R. and Vasile, M.J. J. Vac. Soc. Technol., B7, 181 (1989).

68.   Nakamura, K. Nozaki, T., Shiokawa, T. and Toyoda, K., Japan J. Appl. Phys., 24, L903 (1985).

69.   Gamo, K., Takakura, N., Samoto, N., Shimizu, R. and Namba, S., Japan J. Appl. Phys., 23, L293 (1984).

70.   For a recent review of ion induced depositon see Melngailis, J. and Blauner, P.G. MRS Symposium Proceedings., Vol 147, P. 127 (Materials Res. Soc. 1989).

71.   Blauner, P.G., Butt, Y. Ro, J.S., Thompson, C.V., and Melngailis, J., J. Vac. Sci. Technol., B7, 1816 (1989).

72.   Ochiai, Y., Shihoyama, K., Shiokawa, T. and Toyoda, K., Masuyama, A., Gamo, K. and Namba, S., J. Vac. Sci. Technol.,B4, 333 (1986).

73.   Berreman, D.W. et al., (to be published) Optics Letters.

74.   Randall, J.A., Reed, M.A., Matyi, R.J., Moore, T.M., Aggawal, R.J. and Wetsel, A.E., SPIE Proceedings Vol. 945 Advanced Proceedings of Semiconductor Devices II, p. 137 (1988).

75.   Matsui, S., Kojima, Y. and Ochiai, Y., Appl. Phys. Lett., 53, 868 (1988).

76.   Matsui, S., J. Mori, K., Saigo, K., Shiokawa, T., Toyoda, K. and Namba, S., J. Vac. Sci. Technol., B4, 845 (1986).

77.   Smith, H.I., J Vac. Sci. Technol., B4, 148 (1986).

78.   Chu, W., Yen, A., Ismail, K., Shepard, M.I., Lezec, H.J., Musil, C.R., Melngailis, J., Ku, Y.C., Carter, J.M. and Smith, H.I., J. Vac. Sci. Technol., B7, (1989).

79.   Gesley, M.A., Hohn, F.J., Viswanathan, R.G., and Wilson, A.D., J. Vac. Sci. Technol., B6, 2014 (1988).

80.   Siegel, B. M., J. Vac. Sci. Technol., B6, 350 (1988).

81.   Lezec, H.J. and Shepard, M.I., MIT unpublished result.

82.   Orloff, J. and Sudraud, P. Microcircuit Engineering, 3, 161 (1985).
83.   Slingerland, H.N., Bohlander, J.H., Van der Mast, K.D. and Koets, E., Microelectronic Engineering, 5, 155 (1986).
84.   Mladenov G.M. and Emmoth, B., Appl. Phys. Lett., 38, 1000 (1981).
85.   Macrander, A., Barr, D. and Wagner, A., SPIE 33, 142 (1982).
86.   Melngailis, J., Musil, C.R., Stevens, E.H., Utlaut, M., Kellogg, E.M., Post, R.T., Geis, M.W. and Mountain, R.W., J. Vac. Sci. Technol., B4, 176 (1986).
87.   Kaneko, H., Yasuoka, Y., Gamo,K. and Namba, S., J. Vac. Sci. Technol., B6, 982 (1988).
88.   Ryssel, H., Haberger, K. and Kranz, H., J. Vac. Sci. Technol., 19, 1358 (1981).
89.   Jensen, J.E., Solid State Technology, 27, 147 (June, 1984).
90.   Komuro, M., Atoda, N. and Kawakatsu, H., J. Electrochem. Soc., 126, 483 (1979).
91.   Matsui, S., Mori, K., Shiokawa, T., Toyoda, K. and Namba, S., J. Vac. Sci. Technol., B5, 853 (1987).
92.   Ryssel, H., Kranz, H., Haberger, K. and Bosch, J., Microcircuit Engineering, (1980) p. 293 ed. R.P. Kramer (Delft Univ. Press 1981).
93.   Randall, J., Flanders, D.C., Economou, N.P., Donnelly, J.P. and Bromley, E.I., J. Vac. Sci. Technol., B1, 1152 (1983).
94.   Matsui, S., Mori, K., Saigo, K., Shiokawa, T., Toyoda, K. and Namba, S., J. Vac. Sci. Technol., B4, 845 (1986).
95.   Morimoto, H., Tsukamoto, K., Shinohara, H., Inuishi, M. and Kato, T., IEEE Trans. Electr. Devices, ED34, 230 (1987).
96.   Shiokawa, T., Aoyagi, Y., Kim, P.H., Toyoda, K. and Namba, S., Japn. J. Appl. Phys., 23, L223 (1984).
97.   Kojima, Y., Ochiai, Y., Matsui, S., Japn.J. Appl. Phys., 27, L1780 (1988).
98.   Temkin, H., Harriott, L.R., Hamm, R.A., Weiner, J. and Panish, M.B., Appl. Phys. Lett., 54, 1463 (1989).
99.   Miyauchi, E. and Hashimoto, H., Nuclear Instr. and Methods in Physics Research, B21, 104 (1987).
100. Gamo, K., Moriizumi, K., Ochiai, Y., Takai, M., Namba, S., Shiokawa, T. and Minamisono, T., Japn. J. Appl. Phys., 23, L642 (1984).
101. Komuro, M., Hiroshima, H., Tanoue, H. and Kanayama, T., J. Vac. Sci. Technol., B1, 985 (1983).

102. Gamo, K., Huang, G., Moriizumo, K., Samoto, N., Shimizu, R. and Namba, S., Nucl. Instr. and Methods in Phys. Research, B7/8, 864 (1985).

103. Melngailis, J., Ehrlich, D.J., Pang, S.W. and Randall, J.N., J. Vac. Sci. Technol., B5, 379 (1987).

104. Kuwano, H., J. Appl. Phys., 55, 1149 (1984).

105. Ohta, T., Kanayama, T. and Komuro, M., Microelectronic Engineering, 6, 447 (1987); Ohta, T., Kanayama, T., Tanone, H., Komuro, M., J. Vac. Sci. Technol., B7, 89 (1989).

106. Cummings, K.D., Harriott, L.R., Chi, G.C. and Ostermayer, F.W., Appl. Phys. Lett., 48, 659 (1986).

107. Koshida, N., Ichinose, Y., Ohtaka, K., Komuro, M. and Atoda, N., Microprocess Conference, (1988).

108. Cleaver, J.R.A., Heard, P.J. and Ahmed, H., Appl. Phys. Lett., 49, 654 (1986).

109. Milgram, A. and Puretz, J., J. Vac. Sci. Technol., B3, 879 (1985).

110. Reuss, R.H., Morgan, D.,Goldenetz, A., Clark, W.M. Rensch, D.B. and Utlaut, M., J. Vac. Sci. Technol., B4, 290 (1986).

111. Lezec, H.J., Ismail, K., Mahoney, L.J., Shepard, M.I., Antoniadis, D.A., Melngailis, J., IEEE Electron Dev. Lett., 9, 476 (1988).

112. Evason, A.F., Cleaver, J.R.A. and Ahmed, H., IEEE Electron Dev. Lett., 9, 281 (1988)

113. Evason, A.F., Cleaver, J.R.A. and Ahmed, H., J. Vac. Sci. Technol., B6, 1832 (1988).

114. Rensch, D.B., Matthews, D.S., Utlaut, M.W., Courtney, M.D., Clark, W.M., IEEE Trans Electr. Devices, ED34, 2232 (1987).

115. Lee, J.Y. and Kubena, R.L., Appl. Phys. Lett., 48, 668 (1986).

116. Walden, R.H., Schmitz, A.E., Larson, L.E., Kramer, A.R. and Pasiecznik, J., Proc. IEEE 1988 Custom. Integrated Circuits Conf. p. 18.7.I (May 1988) IEEE 88 CH 2584-1.

117. Melngailis, J., Proceedings 8th University-Government-Industry Microelectronics Symp., p. 70. (IEEE 1989).

118. Tamura, M., Shukuri, S., Tachi, S., Ishitani, T. and Tamura, H., Japn. J. Appl. Phys., 22, L 698 (1983).

119. Tamura, M., Shukuri, S., Wada, Y., Madokoro, Y. and Ishitani, T., Proceedings of 12th Internat. Conf. on Appl. of Ion Beams in Materials Science Tokyo, P. 17 (Sept 1987).

120. Tamura, M., Shukumi, S. and Madokoro, Y., J. Vac. Sci. Technol., B6, 996 (1988).
121. Madokoro, Y., Shukumi, S., Umemura, K. and Tamura, M., Nucl. Instr. and Methods in Physics Research, B39, 511 (1989).
122. Bamba, Y., Miyauchi, E., Arimoto, H., Takamori, A. and Hashimoto, H., Japn. J. Appl. Phys., 23, L515 (1984).
123. Bamba, Y., Miyauchi, E., Makajima, M., Arimoto, H., Takamori, A. and Hashimoto, H., Japn. J. Appl. Phys., 24, L6, (1985).
124. Kubena, R.L., Lee, J.Y., Jullens, R.A., Brault, R.G., Middleton, P.L. and Stevens, E.H., IEEE Trans on Electr. Devices, ED31, 1186 (1984).
125. Ghandi, S.K., VLSI Fabrication Principles, (Wiley, New York 1985).
126. Williams, C.C., Slinkmann, J., Hough, W.P. and Wideramasinghe, H.K., Appl. Phys. Lett., 55, 1662 (1989).
127. Miyauchi, E. and Hashimoto, H., Nucl. Instr. and Methods in Physics Research, B7/8, 851 (1985).
128. Hirayama, Y., Suzuki, Y., Tarucha, S. and Okamoto, H., Jpn. J. Appl. Phys., 24, L516 (1985).
129. Petroff, P.M., Cibert, J., Dolan, G.J., Pearton, S., Gossard, A.C. and English, J.H., in Proceedings of the 1986 Materials Research Society Symposium C, (North-Holland, New York, to be published), Vol. 76, Paper No. C1.2.
130. Hirayama, Y., Suzuki, Y. and Okamoto, H., Jpn. J. Appl. Phys., 24, 1498 (1985).
131. Ishida, K., Matsui, K., Fukunaga, T., Takamori, T. and Nakashima, H., Jpn. J. Appl. Phys., 25, L690 (1986).
132. Venkatesan, T., Schwartz, S.A., Hwang, D.M., Bhat, R., Koza, M., Yoon, H.W. and Mei, P., Appl. Phys. Lett., 49, 701 (1986).
133. Ishida, K., Takamori, T., Matsui, K., Fukunaga, T., Morita, T., Miyauchi, E., Hashimoto, H. and Nakashima, H., Jpn. J. Appl. Phys., 25, L783, (1986).
134. Narum, D.H. and Pease, R.F.W., J. Vac. Sci. Technol., B4, 154 (1986).
135. Narum, D.H. and Pease, R.F.W., J. Vac. Sci. Technol., B6, 966 (1988).
136. Kasahara, H., Sawaragi, H., Aihara, R., Gamo, K., Namba, S. and Hassel-Shearer, M., J. Vac. Sci. Technol., B6, 974 (1988).

137.   Narum, D.H. and Pease, R.F.W., J. Vac. Sci. Technol., B6, 2115 (1988).

138.   Gamo, K., Miyake, H., Yuba, Y. and Namba, S., J. Vac. Sci. Technol., B6, 2124 (1988). See also Nakamura, K., Nozaki, T., Shiokawa, T. and Toyoda, K., J. Vac. Sci. Technol., B5, 203 (1987).

139.   Wu, M.C., Boenki, M.M., Wang, S., Clark, W.M., Stevens, E.H. and Utlaut, M.W., Appl. Phys. Lett., 53, 265 (1988).

140.   Hiramoto, T., Hirakawa, K., Iye, Y. and Ikoma, T., Appl. Phys. Lett., 54, 2103 (1989).

141.   Hiramoto, T., Hirakawa, K. and Ikoma, T., J. Vac. Sci. Technol., B6, 1014 (1988).

142.   Lee, J.Y., Clark, W.M. and Utlaut, M.W., Solid State Electronics, 31, 155 (1988).

143.   Kazior, T.E. and Mozzi, R.L., Raytheon Research Laboratory, (private communication).

144.   Bartelt, J.L., Solid State Technology, Vol. 29, p. 215(May 1986)

145.   Randall, J.N., J. Vac. Sci. Technol., A4 777 (1986).

146.   Braunt, R.G., Miller, L.J., Polymer Engrg. and Sci., 20 (16), p. 1064 (1980).

147.   Randall, J. N., Bromley, E.I. and Economou, N.P., J. Vac. Sci. Technol., B4, 10 (1980).

148.   Randall, J.N. and Sivasankav, R., J. Vac. Sci. Technol., B5, 223 (1987).

149.   Fong, F.O., Stumbo, D.P., Sen, S., Damm, G., Engler, D.W., Wolfe, J.C., Randall, J.N. and Mauger, P. and Shimkunas, A., Microcircuit Engineering, 11, 449-452 (1990).

150.   Sen, S., Fong, F.O. Wolfe, J.C., J. Yen, Junling, Mauger, P., Shiumkunas, A.R., Loschner, H. and Randall, J.N., J. Vac. Sci. Technol., B7, 1802 (1989).

151.   Buchmann, L.M., Csepregi, L., Heuberger, A., Muller, K.P., Chalupka, A., Hammel, E., Loschner, H. and Stengl, G., J. Vac. Sci. Technol., B6, 2080 (1988).

152.   Atkinson, G.M., Bartelt,J.L. and Middleton, P.L., J. Vac. Sci. Technol., B5, 219 (1987).

153.   Atkinson, G.M., Bartelt, J.L., Neureuther, A.R. and Chang, N.W., J. Vac. Sci. Technol., B5, 232 (1987).

154.   Stumbo, D.P., Damm, D.A., Engler, D.W., Fong, F.O., Sen, S., Wolfe, J.C., Randall, J.N., Mauger, P., Shimkunas, A. and Loschner H. (to be published).

155.   Parma, E.J., Hart, R.R. and Bartelt, J.L., J. Vac. Sci. Technol., B5, 228 (1987).

156. Csepregi, L., Iberl, F. and Eichinger, P., Appl. Phys. Lett., 37, 630 (1980).

157. Bartelt, J.L., Slayman, C.W., Wood, J.E., Chen, J.Y., McKenna, C.M., Minning, C.P., Coakley, J.F., Holman, R.E., and Perrygo, C.M., J. Vac. Sci. Technol., 19, 1166 (1981).

158. Pang, S.W., Lyszczarz, T.M., Chen, C.L., Donnelly, J.P. and Randall, J.N., J. Vac. Sci. Technol., B5, 215 (1987).

159. Stengl, G., Loschner, H. and Wolf, P., Nucl. Instr. and Methods in Physics Research, B19/20, 987 (1987).

160. Stengl, G., Loschner, H., Maurer, W. and Wolf, P., J. Vac. Sci. Techonl., B4, 194 (1986).

161. Buchmann, L.M., Csepregi, L., Heuberger, A., Muller, K.P., Chalupka, A., Hammel, E., Loschner, H. and Stengl, G., J. Vac. Sci. Technol., B6, 2080 (1988).

162. Heuberger, A., Buchmann, L.M., Csepregi, L. and Muller, K.P., Microcircuit Engineering, 6, 333 (1987).

163. Miller, P.A., J. Vac. Sci. Technol., B7, 1053 (1989).

164. Stengl, G., Loschner, H., Hammel, E., Wolf, E.D. and Muray, J.J., Emerging Technologies for In-Situ Processing, Ehrlich D.J. and Nguyen V.T., eds. p. 113 (M. Nijhoff 1988).

165. King, M.C. private communcation.

166. Sze, S.M. (ed.) VLSI Technology, p.193 (McGraw Hill 1983).

167. Hayashi, I., in Emerging Technologies for In-Situ Processing, Ehrlich, D.J. and Nguyen V.T., eds. p. 16 (M. Nijhoff 1988).

168. Stengl, G., Loschner, H., and Muray, J.J., Extended Abstracts 18th International Conference on Solid State Devices and Materlals (Tokyo 1986) pp. 29-32.

169. Fallmann, W., Paschke, F., Stangl, G., Buchmann, L.M., Heuberger, A., Chalupka, A., Fegerl, J., Fischer, R., Loschner, H., Malek, L., Novak, R., Stengl, G., Traher, C., and Wolf, P. Submitted to AEC (Archive for Electronics and Communication FRG) (Jan. 1990).

INDEX

Abbe's theory - 259
Aberrations, lens - 242
Absolute accuracy - 368, 371
Absorbance of resist - 47
Absorber defect - 457
Absorber multilayers - 469
Absorption - 47
Absorption spectra - 48
Acceptance angle - 615, 624
Accuracy goals - 432
Acid-catalyzed resists - 67, 140
Actinic wavelength - 273
Active data path - 410
Actual flux - 511
A/D converters - 613
Additive absorber - 463
Adhesion and resist adhesion
  contact angle - 93
  dewetting - 87
  ESCA - 93
  hexamethyldisilazane (HMDS) - 80
  priming mechanisms - 87, 92, 93
  promoters - 80
  resist lifting - 87
  rework wafers - 95
  spin coating problems - 93
  substrate nonwetting - 87
Address grid - 376, 387
Airy disk - 245
Aligner - 478
Alignment
  capacities - 572
  data - 340
  errors - 306, 333
  grid error - 338
  mark - 572, 574, 625
  mark stability - 503
  system - 334, 569, 579, 625
  target - 338

Amorphous layer - 600
Amplitude transmittance - 255
Anamorphic errors - 348
Angle distribution - 515
Angular distribution - 513
Angular divergence - 266
Angular resolution - 248
Angular scattering - 620
Annealed defects - 610
Annealing - 576, 599, 605
Annealing device damage - 500
Antireflection coating - 350
Antireflective layers (ARLs) - 331
Aperture - 410
Application Specific Integrated Circuits
  (ASIC) - 35
Area coverage distribution - 396
Aspect ratio - 339
Assemby lens - 289
Assymetric signal - 339
Assymetrical aberration - 262
Assymmetry - 285, 605
Astigmatism - 288, 313, 566
Attributes control chart - 304
Autofocus systems - 316
Automation - 370, 372
Average current - 526
Axial images - 284
Axial velocity - 566
Axially symmetric - 282
Azide
  photochemistry - 54, 55

Backscattered electrons - 186, 416
Backscattering, electron - 390, 414
Bandwidth - 323
Barrel distortion -348, 625
Be foil - 534
Be window - 536

638

Printed and bound by CPI Group (UK) Ltd, Croydon, CR0 4YY

03/10/2024

01040433-0008